T0156059

Elements of Algebraic Topology

James R. Munkres
Massachusetts Institute of Technology
Cambridge, Massachusetts

The Advanced Book Program
CRC Press
Taylor & Francis Group
Boca Raton London New York
CRC Press is an imprint of the
Taylor & Francis Group, an informa business
A CHAPMAN & HALL BOOK

First published 1984 by Westview Press

Published 2018 by CRC Press
Taylor & Francis Group
6000 Broken Sound Parkway NW, Suite 300
Boca Raton, FL 33487-2742

CRC Press is an imprint of the Taylor & Francis Group, an informa business

Copyright © 1984 Taylor & Francis Group LLC

No claim to original U.S. Government works

This book contains information obtained from authentic and highly regarded sources. Reasonable efforts have been made to publish reliable data and information, but the author and publisher cannot assume responsibility for the validity of all materials or the consequences of their use. The authors and publishers have attempted to trace the copyright holders of all material reproduced in this publication and apologize to copyright holders if permission to publish in this form has not been obtained. If any copyright material has not been acknowledged please write and let us know so we may rectify in any future reprint.

Except as permitted under U.S. Copyright Law, no part of this book may be reprinted, reproduced, transmitted, or utilized in any form by any electronic, mechanical, or other means, now known or hereafter invented, including photocopying, microfilming, and recording, or in any information storage or retrieval system, without written permission from the publishers.

For permission to photocopy or use material electronically from this work, please access www. copyright.com (http://www.copyright.com/) or contact the Copyright Clearance Center, Inc. (CCC), 222 Rosewood Drive, Danvers, MA 01923, 978-750-8400. CCC is a not-for-profit organization that provides licenses and registration for a variety of users. For organizations that have been granted a photocopy license by the CCC, a separate system of payment has been arranged.

Trademark Notice: Product or corporate names may be trademarks or registered trademarks, and are used only for identification and explanation without intent to infringe.

Visit the Taylor & Francis Web site at
http://www.taylorandfrancis.com

and the CRC Press Web site at
http://www.crcpress.com

Library of Congress Cataloging in Publication Data

Munkres, James R., 1930–
Elements of algebraic topology.

Bibliography: p.
Includes index.
1. Algebraic topology. I. Title.
QA612.M86 1984 514'.2 84-6250

ISBN 13: 978-0-2016-2728-2 (pbk)
ISBN 13: 978-0-2010-4586-4 (hbk)

Contents[†]

[†]The sections marked with an asterisk can be omitted or postponed until needed. Chapter 2 can be omitted.

Preface

This book is intended as a text for a first-year graduate course in algebraic topology; it presents the basic material of homology and cohomology theory. For students who will go on in topology, differential geometry, Lie groups, or homological algebra, the subject is a prerequisite for later work. For other students, it should be part of their general background, along with algebra and real and complex analysis.

Geometric motivation and applications are stressed throughout. The abstract aspects of the subject are introduced gradually, after the groundwork has been laid with specific examples.

The book begins with a treatment of the simplicial homology groups, the most concrete of the homology theories. After a proof of their topological invariance and verification of the Eilenberg-Steenrod axioms, the singular homology groups are introduced as their natural generalization. CW complexes appear as a useful computational tool. This basic "core" material is rounded out with a treatment of cohomology groups and the cohomology ring.

There are two additional chapters. The first deals with homological algebra, including the universal coefficient theorems and the Künneth theorem. The second deals with manifolds—specifically, the duality theorems associated with the names of Poincaré, Lefschetz, Alexander, and Pontryagin. Čech cohomology is introduced to study the last of these.

The book does not treat homotopy theory; to do so would have made it unwieldy. There is a thorough and readable elementary treatment of the fundamental group in Massey's book [Ma]; for general homotopy theory, the reader may consult the excellent treatise by Whitehead, for which the present text is useful preparation [Wh].

Prerequisites

We assume the student has some background in both general topology and algebra. In topology, we assume familiarity with continuous functions and compactness and connectedness in general topological spaces, along with the separation axioms up through the Tietze extension theorem for normal spaces. Students without this background should be prepared to do some independent study; any standard book in topology will suffice ([D], [W], [Mu], [K], for example). Even with this background, the student might not know enough about quotient spaces for our purposes; therefore, we review this topic when the need arises (§20 and §37).

As far as algebra is concerned, a course dealing with groups, factor groups, and homomorphisms, along with basic facts about rings, fields, and vector spaces, will suffice. No particularly deep theorems will be needed. We review the basic results as needed, dealing with direct sums and direct products in §5 and proving the fundamental theorem of finitely generated abelian groups in §11.

How the book is organized

Everyone who teaches a course in algebraic topology has a different opinion regarding the appropriate choice of topics. I have attempted to organize the book as flexibly as possible, to enable the instructor to follow his or her own preferences in this matter. The first six chapters cover the basic "core" material mentioned earlier. Certain sections marked with asterisks are not part of the basic core and can thus be omitted or postponed without loss of continuity. The last two chapters, on homological algebra and duality, respectively, are independent of one another; either or both may be covered.

The instructor who wishes to do so can abbreviate the treatment of simplicial homology by omitting Chapter 2. With this approach the topological invariance of the simplicial homology groups is proved, not directly via simplicial approximations as in Chapter 2, but as a consequence of the isomorphism between simplicial and singular theory (§34).

When the book is used for a two-semester course, one can reasonably expect to cover it in its entirety. This is the plan I usually follow when I teach the first-year graduate course at MIT; this allows enough time to treat the exercises thoroughly. The exercises themselves vary from routine to challenging. The more difficult ones are marked with asterisks, but none is unreasonably hard.

If the book is to be used for a one-semester course, some choices will have to be made about what material to cover. One possible syllabus consists of the first four chapters in their entirety. Another consists of the first five chapters with most or all asterisked sections omitted.

A third possible syllabus, which omits Chapter 2, consists of the following:

Chapter 1
Chapter 3 (omit §27)
Chapter 4 (insert §15 before §31 and §20 before §37)
Chapters 5 and 6

If time allows, the instructor can include material from Chapter 7 or the first four sections of Chapter 8. (The later sections of Chapter 8 depend on omitted material.)

Acknowledgments

Anyone who teaches algebraic topology has had many occasions to refer to the classic books by Hilton and Wylie [H-W] and by Spanier [S]. I am no exception; certainly the reader will recognize their influence throughout the present text. I learned about CW complexes from George Whitehead; the treatment of duality in manifolds is based on lectures by Norman Steenrod. From my students at MIT, I learned what I know about motivation of definitions, order of topics, pace of presentation, and suitability of exercises.

To Miss Viola Wiley go my thanks for typing the original set of lecture notes on which the book is based.

Finally, I recall my debt to my parents, who always encouraged me to follow my own path, though it led far from where it began. To them, with love and remembrance, this book is dedicated.

J.R.M.

1

Homology Groups of a Simplicial Complex

A fundamental problem of topology is that of determining, for two spaces, whether or not they are homeomorphic. To show two spaces *are* homeomorphic, one needs to construct a continuous bijective map, with continuous inverse, mapping one space to the other. To show two spaces are not homeomorphic involves showing that such a map does not exist. To do that is often harder. The usual way of proceeding is to find some topological property (i.e., some property invariant under homeomorphisms) that is satisfied by one space but not the other. For example, the closed unit disc in $\mathbf{R}^2$ cannot be homeomorphic with the plane $\mathbf{R}^2$, because the closed disc is compact and the plane is not. Nor can the real line $\mathbf{R}$ be homeomorphic with $\mathbf{R}^2$, because deleting a point from $\mathbf{R}$ leaves a disconnected space remaining, while deleting a point from $\mathbf{R}^2$ does not.

Such elementary properties do not carry one very far in tackling homeomorphism problems. Classifying all compact surfaces up to homeomorphism, for instance, demands more sophisticated topological invariants than these. So does the problem of showing that, in general, $\mathbf{R}^n$ and $\mathbf{R}^m$ are not homeomorphic if $n \neq m$.

Algebraic topology originated in the attempts by such mathematicians as Poincaré and Betti to construct such topological invariants. Poincaré introduced a certain group, called the *fundamental group* of a topological space; it is by its definition a topological invariant. One can show fairly readily that a number of familiar spaces, such as the sphere, torus, and Klein bottle, have fundamental groups that are different, so these spaces cannot be homeomorphic ([Mu], Chapter 8). In fact, one can classify all compact surfaces using the fundamental group ([Ma], Chapter 4).

Betti, on the other hand, associated with each space a certain sequence of abelian groups called its *homology groups*. In this case, it was not at all obvious that homeomorphic spaces had isomorphic homology groups, although it was eventually proved true. These groups can also be used to tackle homeomorphism problems; one advantage they possess is that they are often easier to compute than the fundamental group.

We shall begin our discussion of algebraic topology by studying the homology groups. Later on, we shall deal with other topological invariants, such as the cohomology groups and the cohomology ring.

There are several different ways of defining homology groups, all of which lead to the same results for spaces that are sufficiently "nice." The two we shall consider in detail are the *simplicial* and the *singular* groups. We begin with the simplicial homology groups, which came first historically. Both conceptually and computationally, they are concrete and down-to-earth. They are defined, however, only for particularly "nice" spaces (polyhedra), and it is hard work to prove their topological invariance. After that, we shall treat the singular homology groups, which were introduced as a generalization of the simplicial groups. They are defined for arbitrary spaces, and it is immediate from their definition that they are topological invariants. Furthermore, they are much more convenient for theoretical purposes than the simplicial groups. They are not as readily computable as the simplicial groups, but they agree with the simplicial homology groups when both are defined.

A third way of defining homology groups for arbitrary spaces is due to E. Čech. The Čech homology theory is still not completely satisfactory, but Čech cohomology theory is both important and useful. It will appear near the end of this book.

§1. SIMPLICES

Before defining simplicial homology groups, we must discuss the class of spaces for which they are defined, which is the class of all polyhedra. A polyhedron is a space that can be built from such "building blocks" as line segments, triangles, tetrahedra, and their higher dimensional analogues, by "gluing them together" along their faces. In this section, we shall discuss these basic building blocks; in the next, we shall use them to construct polyhedra.

First we need to study a bit of the analytic geometry of euclidean space.

Given a set $\{a_0, \ldots, a_n\}$ of points of $\mathbf{R}^N$, this set is said to be **geometrically independent** if for any (real) scalars t_i, the equations

$$\sum_{i=0}^{n} t_i = 0 \quad \text{and} \quad \sum_{i=0}^{n} t_i a_i = \mathbf{0}$$

imply that $t_0 = t_1 = \cdots = t_n = 0$.

It is clear that a one-point set is always geometrically independent. Simple algebra shows that in general $\{a_0, \ldots, a_n\}$ is geometrically independent if and only if the vectors

$$a_1 - a_0, \ldots, a_n - a_0$$

are linearly independent in the sense of ordinary linear algebra. Thus two distinct points in $\mathbf{R}^N$ form a geometrically independent set, as do three non-collinear points, four non-coplanar points, and so on.

Given a geometrically independent set of points $\{a_0, \ldots, a_n\}$, we define the **n-plane** P spanned by these points to consist of all points x of $\mathbf{R}^N$ such that

$$x = \sum_{i=0}^{n} t_i a_i,$$

for some scalars t_i with $\Sigma t_i = 1$. Since the a_i are geometrically independent, the t_i are uniquely determined by x. Note that each point a_i belongs to the plane P.

The plane P can also be described as the set of all points x such that

$$x = a_0 + \sum_{i=1}^{n} t_i(a_i - a_0)$$

for some scalars $t_1, \ldots, t_n$; in this form we speak of P as the "plane through a_0 parallel to the vectors $a_i - a_0$."

It is elementary to check that if $\{a_0, \ldots, a_n\}$ is geometrically independent, and if w lies outside the plane that these points span, then $\{w, a_0, \ldots, a_n\}$ is geometrically independent.

An **affine transformation** T of $\mathbf{R}^N$ is a map that is a composition of translations (i.e., maps of the form $T(x) = x + p$ for fixed p), and non-singular linear transformations. If T is an affine transformation, it is immediate from the definitions that T preserves geometrically independent sets, and that T carries the plane P spanned by $a_0, \ldots, a_n$ onto the plane spanned by $Ta_0, \ldots, Ta_n$.

Now the translation $T(x) = x - a_0$ carries P onto the vector subspace of $\mathbf{R}^N$ having $a_1 - a_0, \ldots, a_n - a_0$ as a basis; if we follow T by a linear transformation of $\mathbf{R}^N$ carrying $a_1 - a_0, \ldots, a_n - a_0$ to the first n unit basis vectors $\epsilon_1, \ldots, \epsilon_n$ in $\mathbf{R}^N$, we obtain an affine transformation S of $\mathbf{R}^N$ such that $S(a_0) = \mathbf{0}$ and $S(a_i) = \epsilon_i$ for $i > 0$. The map S carries P onto the plane $\mathbf{R}^n \times \mathbf{0}$ of the first n coordinates in $\mathbf{R}^N$; it is thus clear why we call P a "plane of dimension n" in $\mathbf{R}^N$.

Definition. Let $\{a_0, \ldots, a_n\}$ be a geometrically independent set in $\mathbf{R}^N$. We define the **n-simplex** σ spanned by $a_0, \ldots, a_n$ to be the set of all points x of $\mathbf{R}^N$ such that

$$x = \sum_{i=0}^{n} t_i a_i, \quad \text{where} \quad \sum_{i=0}^{n} t_i = 1$$

and $t_i \geq 0$ *for all* i. The numbers t_i are uniquely determined by x; they are called the **barycentric coordinates** of the point x of σ with respect to $a_0, \ldots, a_n$.

Example 1. In low dimensions, one can picture a simplex easily. A 0-simplex is a point, of course. The 1-simplex spanned by a_0 and a_1 consists of all points of the form

$$x = ta_0 + (1 - t)a_1$$

with $0 \leq t \leq 1$; this just the line segment joining a_0 and a_1. Similarly, the 2-simplex σ spanned by a_0, a_1, a_2 equals the triangle having these three points as vertices. This can be seen most easily as follows: Assume $x \neq a_0$. Then

$$x = \sum_{i=0}^{2} t_i a_i = t_0 a_0 + (1 - t_0)[(t_1/\lambda)a_1 + (t_2/\lambda)a_2]$$

where $\lambda = 1 - t_0$. The expression in brackets represents a point p of the line segment joining a_1 and a_2, since $(t_1 + t_2)/\lambda = 1$ and $t_i/\lambda \geq 0$ for $i = 1,2$.

Thus x is a point of the line segment joining a_0 and p. See Figure 1.1. Conversely, any point of such a line segment is in σ, as you can check. It follows that σ equals the union of all line segments joining a_0 to points of a_1a_2; that is, σ is a triangle.

A similar proof shows that a 3-simplex is a tetrahedron.

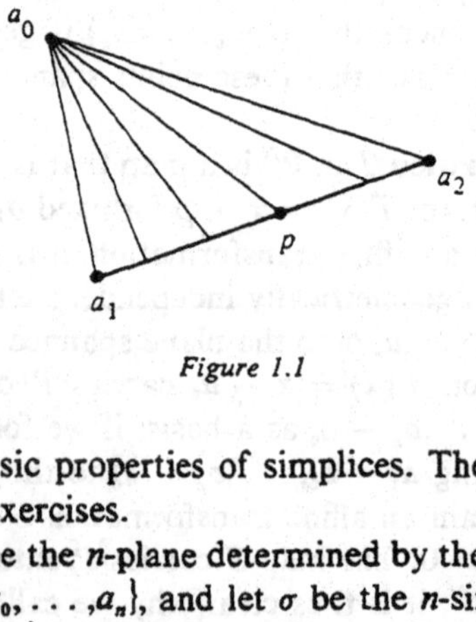

Figure 1.1

Let us list some basic properties of simplices. The proofs are elementary and are largely left as exercises.

Throughout, let P be the n-plane determined by the points of the geometrically independent set $\{a_0, \ldots, a_n\}$, and let σ be the n-simplex spanned by these points. If $x \in \sigma$, let $\{t_i(x)\}$ be the barycentric coordinates of x; they are determined uniquely by the conditions

$$x = \sum_{i=0}^{n} t_i a_i \quad \text{and} \quad \sum_{i=0}^{n} t_i = 1.$$

The following properties hold:

(1) *The barycentric coordinates* $t_i(x)$ *of* x *with respect to* $a_0, \ldots, a_n$ *are continuous functions of* x.

(2) *σ equals the union of all line segments joining a_0 to points of the simplex s spanned by $a_1, \ldots, a_n$. Two such line segments intersect only in the point a_0.*

Recall now that a subset A of $\mathbf{R}^N$ is said to be **convex** if for each pair x,y of points of A, the line segment joining them lies in A.

(3) *σ is a compact, convex set in $\mathbf{R}^N$, which equals the intersection of all convex sets in $\mathbf{R}^N$ containing $a_0, \ldots, a_n$.*

(4) *Given a simplex σ, there is one and only one geometrically independent set of points spanning σ.*

The points $a_0, \ldots, a_n$ that span σ are called the **vertices** of σ; the number n is called the **dimension** of σ. Any simplex spanned by a subset of $\{a_0, \ldots, a_n\}$ is called a **face** of σ. In particular, the face of σ spanned by $a_1, \ldots, a_n$ is called the face **opposite** a_0. The faces of σ different from σ itself are called the **proper faces** of σ; their union is called the **boundary** of σ and denoted $\operatorname{Bd} \sigma$. The **interior** of σ is defined by the equation $\operatorname{Int} \sigma = \sigma - \operatorname{Bd} \sigma$; the set $\operatorname{Int} \sigma$ is sometimes called an **open simplex.**

Since $\operatorname{Bd} \sigma$ consists of all points x of σ such that at least one of the barycentric coordinates $t_i(x)$ is zero, $\operatorname{Int} \sigma$ consists of those points of σ for which $t_i(x) > 0$ for all i. It follows that, given $x \in \sigma$, there is exactly one face s of σ such that $x \in \operatorname{Int} s$, for s must be the face of σ spanned by those a_i for which $t_i(x)$ is positive.

(5) $\operatorname{Int} \sigma$ *is convex and is open in the plane P; its closure is σ. Furthermore,* $\operatorname{Int} \sigma$ *equals the union of all open line segments joining a_0 to points of* $\operatorname{Int} s$, *where s is the face of σ opposite a_0.*

Let us recall here some standard notation. If x is in $\mathbf{R}^n$ and $x = (x_1, \ldots, x_n)$, then the **norm** of x is defined by the equation

$$\|x\| = \left[\sum_{i=1}^{n} (x_i)^2 \right]^{\frac{1}{2}}.$$

The **unit n-ball** B^n is the set of all points x of $\mathbf{R}^n$ for which $\|x\| \leq 1$, and the **unit sphere** S^{n-1} is the set of points for which $\|x\| = 1$. The **upper hemisphere** E^{n-1}_{+} of S^{n-1} consists of all points x of S^{n-1} for which $x_n \geq 0$; while the **lower hemisphere** E^{n-1}_{-} consists of those points for which $x_n \leq 0$.

With these definitions, B^0 is a one-point space, B^1 equals the interval $[-1,1]$, and S^0 is the two-point space $\{-1,1\}$. The 2-ball B^2 is the unit disc in $\mathbf{R}^2$ centered at the origin; and S^1 is the unit circle.

(6) *There is a homeomorphism of σ with the unit ball B^n that carries* $\operatorname{Bd} \sigma$ *onto the unit sphere S^{n-1}.*

We leave properties (1)–(5) as exercises, and prove (6). In fact, we shall prove a stronger result, which will be useful to us later.

Recall that if $w \in \mathbf{R}^n$, a **ray** $\mathcal{R}$ emanating from w is the set of all points of the form $w + tp$, where p is a fixed point of $\mathbf{R}^n - \mathbf{0}$ and t ranges over the nonnegative reals.

Lemma 1.1. *Let U be a bounded, convex, open set in $\mathbf{R}^n$; let $w \in U$.*

(a) *Each ray emanating from w intersects* $\text{Bd } U = \bar{U} - U$ *in precisely one point.*

(b) *There is a homeomorphism of $\bar{U}$ with B^n carrying* Bd U *onto S^{n-1}.*

Proof. (a) Given a ray $\mathcal{R}$ emanating from w, its intersection with U is convex, bounded, and open in $\mathcal{R}$. Hence it consists of all points of the form $w + tp$, where t ranges over a half-open interval $[0,a)$. Then $\mathcal{R}$ intersects $\bar{U} - U$ in the point $x = w + ap$.

Suppose $\mathcal{R}$ intersects $\bar{U} - U$ in another point, say y. Then x lies between w and y on the ray $\mathcal{R}$. Indeed, since $y = w + bp$ for some $b > a$, we have

$$x = (1 - t)w + ty,$$

where $t = a/b$. We rewrite this equation in the form

$$w = (x - ty)/(1 - t).$$

Then we choose a sequence y_n of points of U converging to y, and we define

$$w_n = (x - ty_n)/(1 - t).$$

See Figure 1.2. The sequence w_n converges to w, so that $w_n \in U$ for some n. But then since $x = tw_n + (1 - t)y_n$, the point x belongs to U, because U is convex. This fact contradicts our choice of x.

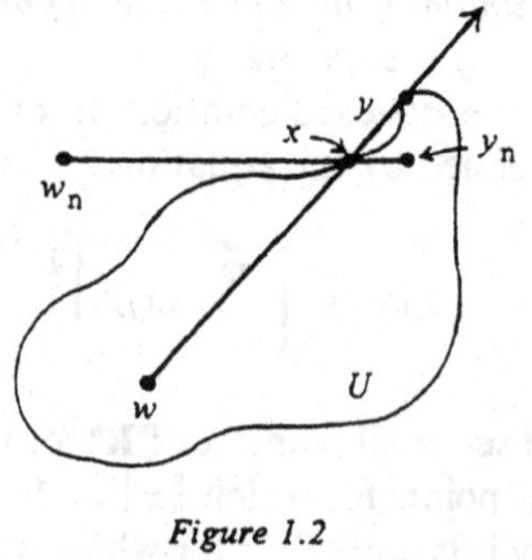

Figure 1.2

(b) Assume $w = \mathbf{0}$ for convenience. The equation $f(x) = x/\|x\|$ defines a continuous map f of $\mathbf{R}^n - \mathbf{0}$ onto S^{n-1}. By (a), f restricts to a bijection of Bd U with S^{n-1}. Since Bd U is compact, this restriction is a homeomorphism; let $g : S^{n-1} \to \text{Bd } U$ be its inverse. Extend g to a bijection $G : B^n \to \bar{U}$ by letting G map the line segment joining $\mathbf{0}$ to the point u of S^{n-1} linearly onto the line segment joining $\mathbf{0}$ to $g(u)$. Formally, we define

$$G(x) = \begin{cases} \|g(x/\|x\|)\|x & \text{if } x \neq \mathbf{0}, \\ \mathbf{0} & \text{if } x = \mathbf{0}. \end{cases}$$

Continuity of G for $x \neq \mathbf{0}$ is immediate. Continuity at $\mathbf{0}$ is easy: If M is a bound for $\|g(x)\|$, then whenever $\|x - \mathbf{0}\| < \delta$, we have $\|G(x) - G(\mathbf{0})\| < M\delta$. □

EXERCISES

1. Verify properties (1)–(3) of simplices. [*Hint:* If T is the affine transformation carrying a_0 to $\mathbf{0}$ and a_i to ϵ_i, then T carries the point

$$x = \sum_{i=0}^{n} t_i a_i$$

to the point $(t_1, \ldots, t_n, 0, \ldots, 0)$.]

2. Verify property (4) as follows:
   (a) Show that if $x \in \sigma$ and $x \neq a_0, \ldots, a_n$, then x lies in some open line segment contained in σ. (Assume $n > 0$.)
   (b) Show that a_0 lies in no open line segment contained in σ, by showing that if $a_0 = tx + (1 - t)y$, where $x,y \in \sigma$ and $0 < t < 1$, then $x = y = a_0$.
3. Verify property (5).
4. Generalize property (2) as follows: Let σ be spanned by $a_0, \ldots, a_n$. Let s be the face of σ spanned by $a_0, \ldots, a_p$ (where $p < n$); and let t be the face spanned by $a_{p+1}, \ldots, a_n$. Then t is called the face of σ ***opposite*** s.
   (a) Show that σ is the union of all line segments joining points of s to points of t, and two of these line segments intersect in at most a common end point.
   (b) Show that Int σ is the union of all open line segments joining points of Int s to points of Int t.
5. Let U be a bounded open set in $\mathbf{R}^n$. Suppose U is **star-convex** relative to the origin; this means that for each x in U, the line segment from $\mathbf{0}$ to x lies in U.
   (a) Show that a ray from $\mathbf{0}$ may intersect Bd U in more than one point.
   *(b) Show by example that $\bar{U}$ need not be homeomorphic to B^n.

§2. SIMPLICIAL COMPLEXES AND SIMPLICIAL MAPS

Complexes in $\mathbf{R}^N$

Definition. A **simplicial complex** K in $\mathbf{R}^N$ is a collection of simplices in $\mathbf{R}^N$ such that:

(1) Every face of a simplex of K is in K.

(2) The intersection of any two simplexes of K is a face of each of them.

Example 1. The collection K_1 pictured in Figure 2.1, consisting of a 2-simplex and its faces, is a simplicial complex. The collection K_2, consisting of two 2-simplices with an edge in common, along with their faces, is a simplicial complex; while the collection K_3 is not. What about K_4?

The following lemma is sometimes useful in verifying that a collection of simplices is a simplicial complex:

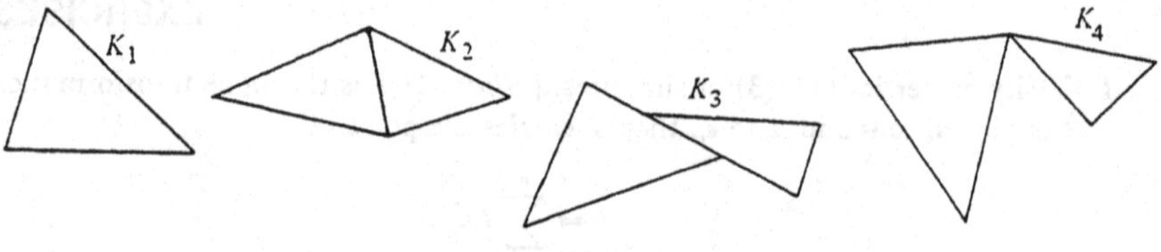

Figure 2.1

Lemma 2.1. *A collection K of simplices is a simplicial complex if and only if the following hold:*

(1) *Every face of a simplex of K is in K.*

(2′) *Every pair of distinct simplices of K have disjoint interiors.*

Proof. First, assume K is a simplicial complex. Given two simplices σ and τ of K, we show that if their interiors have a point x in common, then $\sigma = \tau$. Let $s = \sigma \cap \tau$. If s were a proper face of σ, then x would belong to Bd σ, which it does not. Therefore, $s = \sigma$. A similar argument shows that $s = \tau$.

To prove the converse, assume (1) and (2′). We show that if the set $\sigma \cap \tau$ is nonempty, then it equals the face σ' of σ that is spanned by those vertices $b_0, \ldots, b_m$ of σ that lie in τ. First, σ' is contained in $\sigma \cap \tau$ because $\sigma \cap \tau$ is convex and contains $b_0, \ldots, b_m$. To prove the reverse inclusion, suppose $x \in \sigma \cap \tau$. Then $x \in \text{Int}\, s \cap \text{Int}\, t$, for some face s of σ and some face t of τ. It follows from (2′) that $s = t$; hence the vertices of s lie in τ, so that by definition they are elements of the set $\{b_0, \ldots, b_m\}$. Then s is a face of σ', so that $x \in \sigma'$, as desired. □

It follows from this lemma that if σ is a simplex, then the collection consisting of σ and its proper faces is a simplicial complex: Condition (1) is immediate; and condition (2′) holds because for each point $x \in \sigma$, there is exactly one face s of σ such that $x \in \text{Int}\, s$.

Definition. If L is a subcollection of K that contains all faces of its elements, then L is a simplicial complex in its own right; it is called a **subcomplex** of K. One subcomplex of K is the collection of all simplices of K of dimension at most p; it is called the ***p*-skeleton** of K and is denoted $K^{(p)}$. The points of the collection $K^{(0)}$ are called the **vertices** of K.

Definition. Let $|K|$ be the subset of $\mathbf{R}^N$ that is the union of the simplices of K. Giving each simplex its natural topology as a subspace of $\mathbf{R}^N$, we then topologize $|K|$ by declaring a subset A of $|K|$ to be closed in $|K|$ if and only if $A \cap \sigma$ is closed in σ, for each σ in K. It is easy to see that this defines a topology on $|K|$, for this collection of sets is closed under finite unions and arbitrary intersections. The space $|K|$ is called the **underlying space** of K, or the **polytope** of K.

A space that is the polytope of a simplicial complex will be called a **poly-**

hedron. (We note that some topologists reserve this term for the polytope of a *finite* simplicial complex.)

In general, the topology of $|K|$ is finer (larger) than the topology $|K|$ inherits as a subspace of $\mathbf{R}^N$: If A is closed in $|K|$ in the subspace topology, then $A = B \cap |K|$ for some closed set B in $\mathbf{R}^N$. Then $B \cap \sigma$ is closed in σ for each σ, so $B \cap |K| = A$ is closed in the topology of $|K|$, by definition.

The two topologies are different in general. (See Examples 2 and 3.) However, if K is finite, they are the same. For suppose K is finite and A is closed in $|K|$. Then $A \cap \sigma$ is closed in σ and hence closed in $\mathbf{R}^N$. Because A is the union of finitely many sets $A \cap \sigma$, the set A also is closed in $\mathbf{R}^N$.

Example 2. Let K be the collection of all 1-simplices in $\mathbf{R}$ of the form $[m, m+1]$, where m is an integer different from 0, along with all simplices of the form $[1/(n+1), 1/n]$ for n a positive integer, along with all faces of these simplices. Then K is a complex whose underlying space equals $\mathbf{R}$ as a *set* but not as a topological space. For instance, the set of points of the form $1/n$ is closed in $|K|$ but not in $\mathbf{R}$.

Example 3. Let K be the collection of 1-simplices $\sigma_1, \sigma_2, \ldots$ and their vertices, where σ_i is the 1-simplex in $\mathbf{R}^2$ having vertices $\mathbf{0}$ and $(1,1/i)$. See Figure 2.2. Then K is a simplicial complex. The intersection of $|K|$ with the open parabolic arc $\{(x,x^2) \mid x > 0\}$ is closed in $|K|$, because its intersection with each simplex σ_i is a single point. It is not closed in the topology $|K|$ derives from $\mathbf{R}^2$, however, because in that topology it has the origin as a limit point.

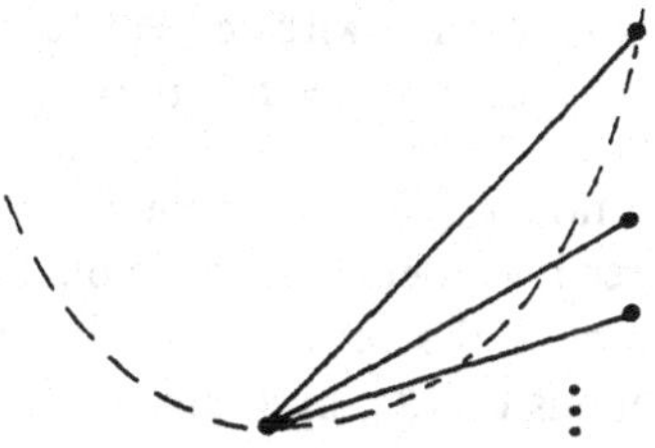

Figure 2.2

We prove some elementary topological properties of polyhedra.

Lemma 2.2. *If L is a subcomplex of K, then $|L|$ is a closed subspace of $|K|$. In particular, if $\sigma \in K$, then σ is a closed subspace of $|K|$.*

Proof. Suppose A is closed in $|L|$. If σ is a simplex of K, then $\sigma \cap |L|$ is the union of those faces s_i of σ that belong to L. Since A is closed in $|L|$, the set $A \cap s_i$ is closed in s_i and hence closed in σ. Since $A \cap \sigma$ is the finite union of the sets $A \cap s_i$, it is closed in σ. We conclude that A is closed in $|K|$.

Conversely, if B is closed in $|K|$, then $B \cap \sigma$ is closed in σ for each $\sigma \in K$, and in particular for each $\sigma \in L$. Hence $B \cap |L|$ is closed in $|L|$. □

Lemma 2.3. *A map $f\colon |K| \to X$ is continuous if and only if $f|\sigma$ is continuous for each $\sigma \in K$.*

Proof. If f is continuous, so is $f|\sigma$ since σ is a subspace of K. Conversely, suppose each map $f|\sigma$ is continuous. If C is a closed set of X, then $f^{-1}(C) \cap \sigma = (f|\sigma)^{-1}(C)$, which is closed in σ by continuity of $f|\sigma$. Thus $f^{-1}(C)$ is closed in $|K|$ by definition. □

Definition. If X is a space and if $\mathcal{C}$ is a collection of subspaces of X whose union is X, the topology of X is said to be **coherent** with the collection $\mathcal{C}$, provided a set A is closed in X if and only if $A \cap C$ is closed in C for each $C \in \mathcal{C}$. It is equivalent to require that U be open in X if and only if $U \cap C$ is open in C for each C.

In particular, the topology of $|K|$ is coherent with the collection of subspaces σ, for $\sigma \in K$.

The analogue of Lemma 2.3 holds for coherent topologies in general; a map $f: X \to Y$ is continuous if and only if $f|C$ is continuous for each $C \in \mathcal{C}$.

Definition. If x is a point of the polyhedron $|K|$, then x is interior to precisely one simplex of K, whose vertices are (say) $a_0, \ldots, a_n$. Then

$$x = \sum_{i=0}^{n} t_i a_i,$$

where $t_i > 0$ for each i and $\Sigma t_i = 1$. If v is an arbitrary vertex of K, we define the **barycentric coordinate** $t_v(x)$ of x with respect to v by setting $t_v(x) = 0$ if v is not one of the vertices a_i, and $t_v(x) = t_i$ if $v = a_i$.

For fixed v, the function $t_v(x)$ is continuous when restricted to a fixed simplex σ of K, since either it is identically zero on σ or equals the barycentric coordinate of x with respect to the vertex v of σ in the sense formerly defined. Therefore, $t_v(x)$ is continuous on $|K|$, by Lemma 2.3.

Lemma 2.4. *$|K|$ is Hausdorff.*

Proof. Given $x_0 \neq x_1$, there is at least one vertex v such that $t_v(x_0) \neq t_v(x_1)$. Choose r between these two numbers; then the sets $\{x \mid t_v(x) < r\}$ and $\{x \mid t_v(x) > r\}$ are the required disjoint open sets. □

Lemma 2.5. *If K is finite, then $|K|$ is compact. Conversely, if a subset A of $|K|$ is compact, then $A \subset |K_0|$ for some finite subcomplex K_0 of K.*

Proof. If K is finite, then $|K|$ is a finite union of compact subspaces σ, and hence is compact. Now suppose A is compact and A does not lie in the polytope of any finite subcomplex of K. Choose a point $x_s \in A \cap \operatorname{Int} s$ whenever this set is nonempty. Then the set $B = \{x_s\}$ is infinite. Furthermore, every subset of B is closed, since its intersection with any simplex σ is finite. Being closed and discrete, B has no limit point, contrary to the fact that every infinite subset of a compact space has a limit point. □

Three particular subspaces of $|K|$ are often useful when studying local properties of $|K|$. We mention them here:

Definition. If v is a vertex of K, the **star** of v in K, denoted by St v, or sometimes by St(v,K), is the union of the interiors of those simplices of K that have v as a vertex. Its closure, denoted $\overline{\text{St}}\, v$, is called the **closed star** of v in K. It is the union of all simplices of K having v as a vertex, and is the polytope of a subcomplex of K. The set $\overline{\text{St}}\, v - \text{St}\, v$ is called the **link** of v in K and is denoted Lk v. See Figure 2.3.

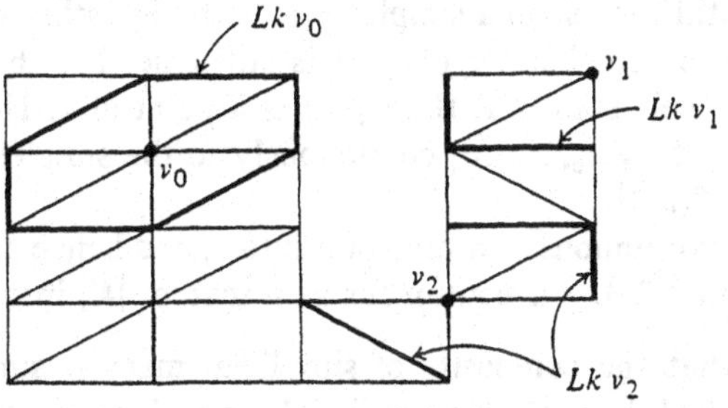

Figure 2.3

The set St v is open in $|K|$, since it consists of all points x of $|K|$ such that $t_v(x) > 0$. Its complement is the union of all simplices of K that do not have v as a vertex; it is thus the polytope of a subcomplex of K. The set Lk v is also the polytope of a subcomplex of K; it is the intersection of $\overline{\text{St}}\, v$ and the complement of St v. The sets St v and $\overline{\text{St}}\, v$ are easily seen to be path connected; the set Lk v need not be connected, however.

Definition. A simplicial complex K is said to be **locally finite** if each vertex of K belongs only to finitely many simplices of K. Said differently, a complex K is locally finite if and only if each closed star $\overline{\text{St}}\, v$ is the polytope of a finite subcomplex of K.

Lemma 2.6. *The complex K is locally finite if and only if the space $|K|$ is locally compact.*

Proof. Suppose K is locally finite. Given $x \in |K|$, it lies in St v for some vertex v of K. Since $\overline{\text{St}}\, v$ is a compact set, $|K|$ is locally compact. We leave the converse as an exercise. □

Simplicial maps

Now we introduce the notion of a "simplicial map" of one complex into another.

Lemma 2.7. *Let K and L be complexes, and let $f : K^{(0)} \to L^{(0)}$ be a map. Suppose that whenever the vertices $v_0, \ldots, v_n$ of K span a simplex of K, the points $f(v_0), \ldots, f(v_n)$ are vertices of a simplex of L. Then f can be extended to a continuous map $g : |K| \to |L|$ such that*

$$x = \sum_{i=0}^{n} t_i v_i \quad \Longrightarrow \quad g(x) = \sum_{i=0}^{n} t_i f(v_i).$$

We call g the (linear) **simplicial map** induced by the vertex map f.

Proof. Note that although the vertices $f(v_0), \ldots, f(v_n)$ of L are not necessarily distinct, still they span a simplex τ of L, by hypothesis. When we "collect terms" in the expression for $g(x)$, it is still true that the coefficients are non-negative and their sum is 1; thus $g(x)$ is a point of τ. Hence g maps the n-simplex σ spanned by $v_0, \ldots, v_n$ continuously to the simplex τ whose vertex set is $\{f(v_0), \ldots, f(v_n)\}$.

The map g is continuous as a map of σ into τ, and hence as a map of σ into $|L|$. Then by Lemma 2.3, g is continuous as a map of $|K|$ into $|L|$. □

We remark that the composite of simplicial maps is simplicial: Suppose $g : |K| \to |L|$ and $h : |L| \to |M|$ are simplicial maps. By definition, if $x = \Sigma\, t_i v_i$ (where the v_i are distinct vertices of $\sigma \in K$), then $g(x) = \Sigma t_i g(v_i)$.

Now this same formula would hold even if the v_i were not distinct, so long as $\{v_0, \ldots, v_n\}$ is the vertex set of a simplex of K. For example, suppose

$$x = \sum_{i=0}^{n} t_i v_i,$$

where $t_i \geq 0$ for all i and $\Sigma t_i = 1$; and suppose that $v_0 = v_1$ and the vertices $v_1, \ldots, v_n$ are distinct. Write

$$x = (t_0 + t_1)v_0 + t_2 v_2 + \cdots + t_n v_n;$$

then by definition

$$g(x) = (t_0 + t_1)g(v_0) + t_2 g(v_2) + \cdots + t_n g(v_n)$$
$$= \sum_{i=0}^{n} t_i g(v_i).$$

Applying this remark to the present case, we note that even though the vertices $g(v_0), \ldots, g(v_n)$ of L are not necessarily distinct, the following formula holds:

$$h(g(x)) = h(\Sigma t_i g(v_i)) = \Sigma t_i h(g(v_i)).$$

Therefore $h \circ g$ is a simplicial map, as claimed.

Lemma 2.8. *Suppose $f : K^{(0)} \to L^{(0)}$ is a bijective correspondence such that the vertices $v_0, \ldots, v_n$ of K span a simplex of K if and only if*

$f(v_0), \ldots, f(v_n)$ span a simplex of L. Then the induced simplicial map $g: |K| \to |L|$ is a homeomorphism.

The map g is called a **simplicial homeomorphism,** or an **isomorphism,** of K with L.

Proof. Each simplex σ of K is mapped by g onto a simplex τ of L of the same dimension as σ. We need only show that the linear map $h: \tau \to \sigma$ induced by the vertex correspondence f^{-1} is the inverse of the map $g: \sigma \to \tau$. And for that we note that if $x = \Sigma t_i v_i$, then $g(x) = \Sigma t_i f(v_i)$ by definition; whence

$$\begin{aligned} h(g(x)) &= h(\Sigma t_i f(v_i)) = \Sigma t_i f^{-1}(f(v_i)) \\ &= \Sigma t_i v_i = x. \quad \square \end{aligned}$$

Corollary 2.9. *Let Δ^N denote the complex consisting of an N-simplex and its faces. If K is a finite complex, then K is isomorphic to a subcomplex of Δ^N for some N.*

Proof. Let $v_0, \ldots, v_N$ be the vertices of K. Choose $a_0, \ldots, a_N$ to be geometrically independent points in $\mathbf{R}^N$, and let Δ^N consist of the N-simplex they span, along with its faces. The vertex map $f(v_i) = a_i$ induces an isomorphism of K with a subcomplex of Δ^N. $\square$

General simplicial complexes

Our insistence that a simplicial complex K must lie in $\mathbf{R}^N$ for some N puts a limitation on the cardinality of K and on the dimension of the simplices of K. We now remove these restrictions.

Let J be an arbitrary index set, and let $\mathbf{R}^J$ denote the J-fold product of $\mathbf{R}$ with itself. An element of $\mathbf{R}^J$ is a function from J to $\mathbf{R}$, ordinarily denoted in "tuple notation" by $(x_\alpha)_{\alpha \in J}$. The product $\mathbf{R}^J$ is of course a vector space with the usual component-wise addition and multiplication by scalars.

Let $\mathbf{E}^J$ denote the subset of $\mathbf{R}^J$ consisting of all points $(x_\alpha)_{\alpha \in J}$ such that $x_\alpha = 0$ for all but finitely many values of α. Then $\mathbf{E}^J$ is also a vector space under component-wise addition and multiplication of scalars. If ϵ_α is the map of J into $\mathbf{R}$ whose value is 1 on the index α and 0 on all other elements of J, then the set $\{\epsilon_\alpha \mid \alpha \in J\}$ is a basis for $\mathbf{E}^J$. (It is not, of course, a basis for $\mathbf{R}^J$.)

We call $\mathbf{E}^J$ **generalized euclidean space** and topologize it by the metric

$$|x - y| = \max \{|x_\alpha - y_\alpha|\}_{\alpha \in J}.$$

Everything we have done for complexes in $\mathbf{R}^N$ generalizes to complexes in $\mathbf{E}^J$. The space $\mathbf{E}^J$ is the union of certain of its finite-dimensional subspaces—namely, the subspaces spanned by finite subsets of the basis $\{\epsilon_\alpha \mid \alpha \in J\}$. Each such subspace is just a copy of $\mathbf{R}^N$ for some N. Any finite set $\{a_0, \ldots, a_n\}$ of points of $\mathbf{E}^J$ lies in such a subspace; if they are independent, the simplex they span lies in the same subspace. Furthermore, the metric for $\mathbf{E}^J$ gives the usual topology of each such subspace. Therefore, any finite collection of simplices in E^J lies in a copy of R^N for some N. All that we are really doing is to allow ourselves to deal with complexes for which the total collection of simplices cannot be fitted entirely into any one R^N. Conceptually it may seem more complicated to work in E^J than in R^N, but in practice it causes no difficulty at all.

We leave it to you to check that our results hold for complexes in E^J. We shall use these results freely henceforth.

Let us make one further comment. If K is a complex in R^N, then each simplex of K has dimension at most N. On the other hand, if K is a complex in E^J, there need be no upper bound on the dimensions of the simplices of K. We define the **dimension** of K, denoted $\dim K$, to be the largest dimension of a simplex of K, if such exists; otherwise, we say that K has **infinite dimension.**

EXERCISES

1. Let K be a simplicial complex; let $\sigma \in K$. When is $\text{Int}\,\sigma$ open in $|K|$? When is σ open in $|K|$?
2. Show that in general, $\text{St}\, v$ and $\overline{\text{St}}\, v$ are path connected.
3. (a) Show directly that the polyhedron of Example 3 is not locally compact.
   (b) Show that, in general, if a complex K is not locally finite, then the space $|K|$ is not locally compact.
4. Show that the polyhedron of Example 3 is not metrizable. [*Hint:* Show that first countability axiom fails.]
5. If $g : |K| \to |L|$ is a simplicial map carrying the vertices of σ onto the vertices of τ, show that g maps some face of σ homeomorphically onto τ.
6. Check that the proofs of Lemmas 2.1–2.8 apply without change to complexes in E^J, if one simply replaces R^N by E^J throughout.
7. Let K be a complex. Show that $|K|$ is metrizable if and only if K is locally finite. [*Hint:* The function
$$d(x,y) = \text{lub}\, |t_v(x) - t_v(y)|$$
is a metric for the topology of each finite subcomplex of K.]
8. Let K be a complex. Show that $|K|$ is normal. [*Hint:* If A is closed in $|K|$ and if $f : A \to [0,1]$ is continuous, extend f step-by-step to $A \cup |K^{(p)}|$, using the Tietze theorem.]
9. Let K be a complex in R^N. Show that $|K|$ is a subspace of R^N if and only if each point x of $|K|$ lies in an open set of R^N that intersects only finitely many simplices of K. Generalize to E^J.
10. Show that the collection of all simplices in R of the form $[1/(n+1), 1/n]$ for n a positive integer, along with their vertices, is a complex whose polytope is the subspace $(0,1]$ of R.

§3. ABSTRACT SIMPLICIAL COMPLEXES

In practice, specifying a polyhedron X by giving a collection of simplices whose union is X is not a very convenient way of dealing with specific polyhedra. One quickly gets immersed in details of analytic geometry; it is messy to specify all the simplices and to make sure they intersect only when they should. It is much easier to specify X by means of an "abstract simplicial complex," a notion we now introduce.

Definition. An **abstract simplicial complex** is a collection $\mathcal{S}$ of finite nonempty sets, such that if A is an element of $\mathcal{S}$, so is every nonempty subset of A.

The element A of $\mathcal{S}$ is called a **simplex** of $\mathcal{S}$; its **dimension** is one less than the number of its elements. Each nonempty subset of A is called a **face** of A. The **dimension** of $\mathcal{S}$ is the largest dimension of one of its simplices, or is infinite if there is no such largest dimension. The **vertex set** V of $\mathcal{S}$ is the union of the one-point elements of $\mathcal{S}$; we shall make no distinction between the vertex $v \in V$ and the 0-simplex $\{v\} \in \mathcal{S}$. A subcollection of $\mathcal{S}$ that is itself a complex is called a **subcomplex** of $\mathcal{S}$.

Two abstract complexes $\mathcal{S}$ and $\mathcal{T}$ are said to be **isomorphic** if there is a bijective correspondence f mapping the vertex set of $\mathcal{S}$ to the vertex set of $\mathcal{T}$ such that $\{a_0, \ldots, a_n\} \in \mathcal{S}$ if and only if $\{f(a_0), \ldots, f(a_n)\} \in \mathcal{T}$.

Definition. If K is a simplicial complex, let V be the vertex set of K. Let $\mathcal{K}$ be the collection of all subsets $\{a_0, \ldots, a_n\}$ of V such that the vertices $a_0, \ldots, a_n$ span a simplex of K. The collection $\mathcal{K}$ is called the **vertex scheme** of K.

The collection $\mathcal{K}$ is a particular example of an abstract simplicial complex. It is in fact the crucial example, as the following theorem shows:

Theorem 3.1. (a) *Every abstract complex $\mathcal{S}$ is isomorphic to the vertex scheme of some simplicial complex K.*

(b) *Two simplicial complexes are linearly isomorphic if and only if their vertex schemes are isomorphic as abstract simplicial complexes.*

Proof. Part (b) follows at once from Lemma 2.8. To prove (a), we proceed as follows: Given an index set J, let Δ^J be the collection of all simplices in $\mathbf{E}^J$ spanned by finite subsets of the standard basis $\{\epsilon_\alpha\}$ for $\mathbf{E}^J$. It is easy to see that Δ^J is a simplicial complex; if σ and τ are two simplices of Δ^J, then their combined vertex set is geometrically independent and spans a simplex of Δ^J. We shall call Δ^J an "infinite-dimensional simplex."

Now let $\mathcal{S}$ be an abstract complex with vertex set V. Choose an index set J large enough that there is an injective function $f: V \rightarrow \{\epsilon_\alpha\}_{\alpha \in J}$. (Let $J = V$ if you wish.) We specify a subcomplex K of Δ^J by the condition that for each (abstract) simplex $\{a_0, \ldots, a_n\} \in \mathscr{S}$, the (geometric) simplex spanned by $f(a_0), \ldots, f(a_n)$ is to be in K. It is immediate that K is a simplicial complex and $\mathscr{S}$ is isomorphic to the vertex scheme of K; f is the required correspondence between their vertex sets. □

Definition. If the abstract simplicial complex $\mathscr{S}$ is isomorphic with the vertex scheme of the simplicial complex K, we call K a **geometric realization** of $\mathscr{S}$. It is uniquely determined up to a linear isomorphism.

Let us illustrate how abstract complexes can be used to specify particular simplicial complexes.

Example 1. Suppose we wish to indicate a simplicial complex K whose underlying space is homeomorphic to the cylinder $S^1 \times I$. (Here I denotes the closed unit interval $[0,1]$.) One way of doing so is to draw the picture in Figure 3.1, which specifies K as a collection consisting of six 2-simplices and their faces. Another way of picturing this same complex K is to draw the diagram in Figure 3.2. This diagram consists of two things: first, a complex L whose underlying space is a rectangle, and second, a particular labelling of the vertices of L (some vertices being given the same label). We shall consider this diagram to be a short-hand way of denoting the abstract complex $\mathscr{S}$ whose vertex set consists of the letters a, b, c, d, e, and f, and whose simplices are the sets $\{a, f, d\}$, $\{a, b, d\}$, $\{b, c, d\}$, $\{c, d, e\}$, $\{a, c, e\}$, $\{a, e, f\}$, along with their nonempty subsets. Of course, this abstract complex is isomorphic to the vertex scheme of the complex K pictured earlier, so it specifies precisely the same complex (up to linear isomorphism). That is, the complex K of Figure 3.1 is a geometric realization of $\mathscr{S}$.

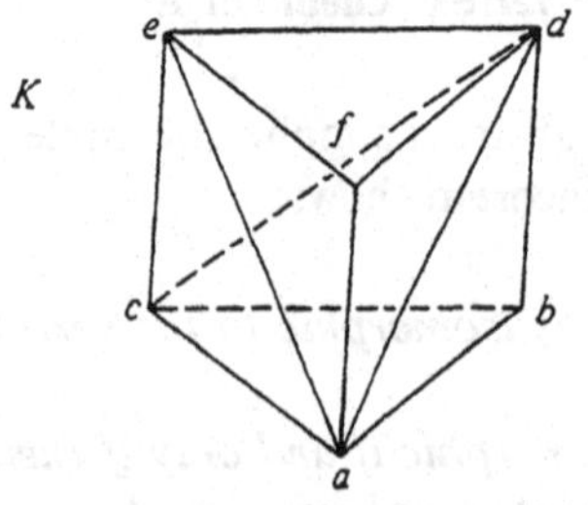

Figure 3.1

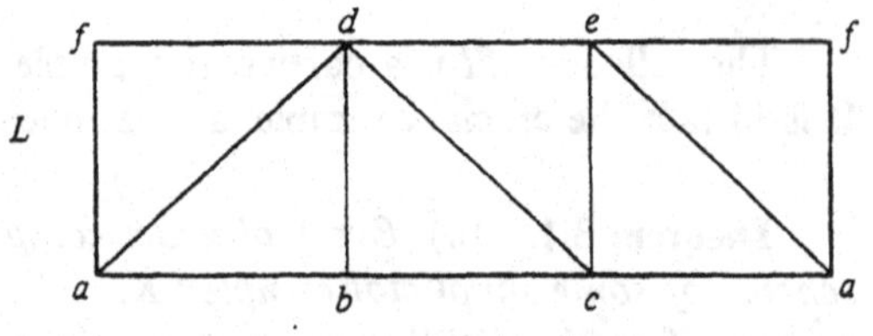

Figure 3.2

Let $f : L^{(0)} \to K^{(0)}$ be the map that assigns to each vertex of L the correspondingly labelled vertex of K. Then f extends to a simplicial map $g : |L| \to |K|$. Because the spaces are compact Hausdorff, g is a quotient map, or "pasting map." It identifies the right edge of $|L|$ linearly with the left edge of $|L|$. And of course this is the usual way one forms a cylinder from a rectangular piece of paper—one bends it around and pastes the right edge to the left edge!

Example 2. Now suppose we *begin* with a complex L and a labelling of its vertices. Consider for instance the same complex L with a different labelling of the vertices, as in Figure 3.3. Just as before, this diagram indicates a certain abstract complex $\mathscr{S}$,

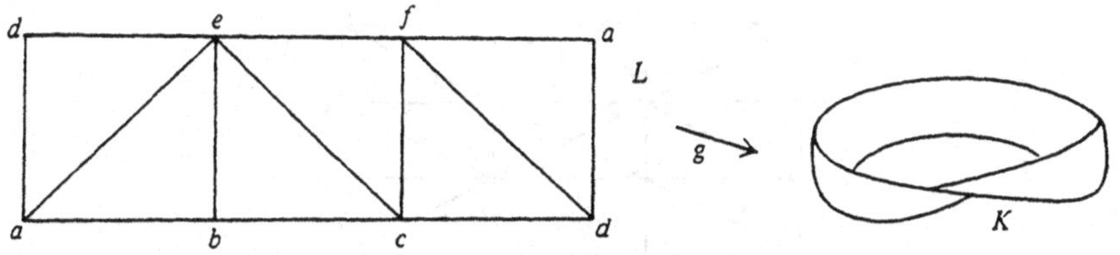

Figure 3.3

whose simplices one can list. Let K be a geometric realization of $\mathcal{S}$. As before, the vertices of K correspond to the letters $a, \ldots, f$; we consider the linear simplicial map $g: |L| \to |K|$ that assigns to each vertex of L the correspondingly labelled vertex of K. Again, g is a quotient map; in this case it identifies the left edge of $|L|$ linearly with the right edge of $|L|$, *but with a twist*. The space $|K|$ is the one we call the **Möbius band**; it can be pictured in $\mathbf{R}^3$ as the familiar space indicated in Figure 3.3.

Example 3. The **torus** is often defined as the quotient space obtained from a rectangle by making the identifications pictured in Figure 3.4. If we wish to construct a complex whose underlying space is homeomorphic to the torus, we can thus obtain it by using the diagram in Figure 3.5. You can check that the resulting quotient map of L onto the geometric realization of this diagram carries out precisely the identifications needed to form the torus.

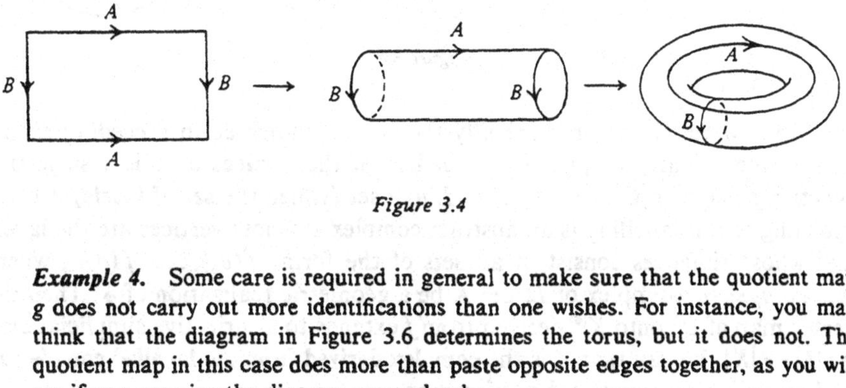

Figure 3.4

Example 4. Some care is required in general to make sure that the quotient map g does not carry out more identifications than one wishes. For instance, you may think that the diagram in Figure 3.6 determines the torus, but it does not. The quotient map in this case does more than paste opposite edges together, as you will see if you examine the diagram more closely.

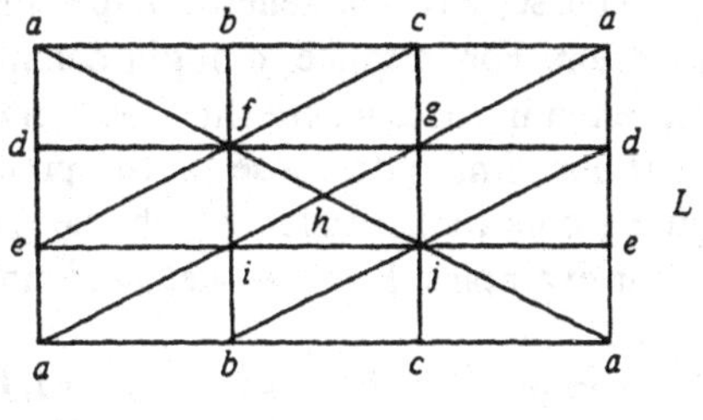

Figure 3.5

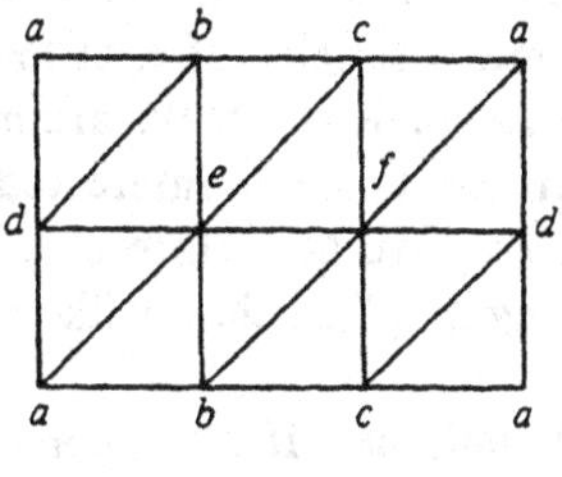

Figure 3.6

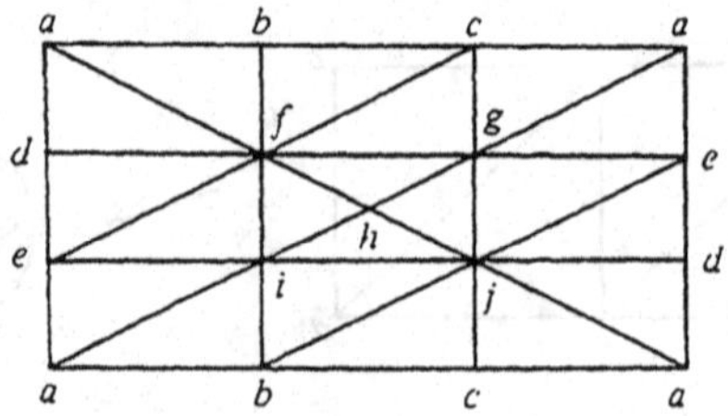

Figure 3.7

Example 5. The diagram in Figure 3.7 indicates an abstract complex whose underlying space is called the **Klein bottle.** It is the quotient space obtained from the rectangle by pasting the edges together as indicated in Figure 3.8. The resulting space cannot be imbedded in $\mathbf{R}^3$, but one can picture it in $\mathbf{R}^3$ by letting it "pass through itself."

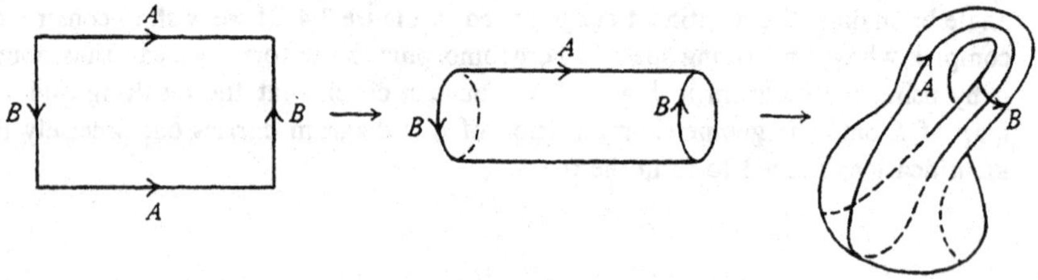

Figure 3.8

We now describe more carefully the process indicated in preceding examples: Given a finite complex L, a **labelling** of the vertices of L is a surjective function f mapping the vertex set of L to a set (called the **set of labels**). Corresponding to this labelling is an abstract complex $\mathcal{S}$ whose vertices are the labels and whose simplices consist of all sets of the form $\{f(v_0), \ldots, f(v_n)\}$, where $v_0, \ldots, v_n$ span a simplex of L. Let K be a geometric realization of $\mathcal{S}$. Then the vertex map of $L^{(0)}$ onto $K^{(0)}$ derived from f extends to a surjective simplicial map $g: |L| \to |K|$. We say that K is the **complex derived from the labelled complex** L, and we call g the **associated pasting map.**

Because $|L|$ is compact and $|K|$ is Hausdorff, g maps closed sets to closed sets; thus g is a closed quotient map. Of course, g can in general map a simplex of L onto a simplex of K of lower dimension. For instance, g might collapse all of L to a single point. We are more interested in cases where this does not occur. We are particularly interested in situations similar to those of the preceding examples. We now state a lemma giving conditions under which the general "pasting map" g behaves like those in our examples. First, we need a definition.

Definition. If L is a complex, a subcomplex L_0 of L is said to be a **full subcomplex** of L provided each simplex of L whose vertices belong to L_0 belongs to L_0 itself.

For example, the boundary of the rectangle L pictured in Figure 3.5 is a full subcomplex of L, but the boundary of the rectangle pictured in Figure 3.6 is not.

Lemma 3.2. *Let L be a complex; let f be a labelling of its vertices; let $g: |L| \to |K|$ be the associated pasting map. Let L_0 be a full subcomplex of L. Suppose that whenever v and w are vertices of L having the same label:*

(1) *v and w belong to L_0.*

(2) *$\overline{\text{St}}\, v$ and $\overline{\text{St}}\, w$ are disjoint.*

Then $\dim g(\sigma) = \dim \sigma$ for all $\sigma \in L$. Furthermore, if $g(\sigma_1) = g(\sigma_2)$, then σ_1 and σ_2 must be disjoint simplices belonging to L_0.

The proof is easy and is left as an exercise. In the usual applications of this lemma, $|L|$ is a polyhedral region in the plane or $\mathbf{R}^N$, and $|L_0|$ is the boundary of the region.

EXERCISES

1. The **projective plane** P^2 is defined as the space obtained from the 2-sphere S^2 by identifying x with $-x$ for each $x \in S^2$.
   (a) Show P^2 is homeomorphic with the space obtained from B^2 by identifying x with $-x$ for each $x \in S^1$.
   (b) Show that the labelled complex L of Figure 3.9 determines a complex K whose space is homeomorphic to P^2.

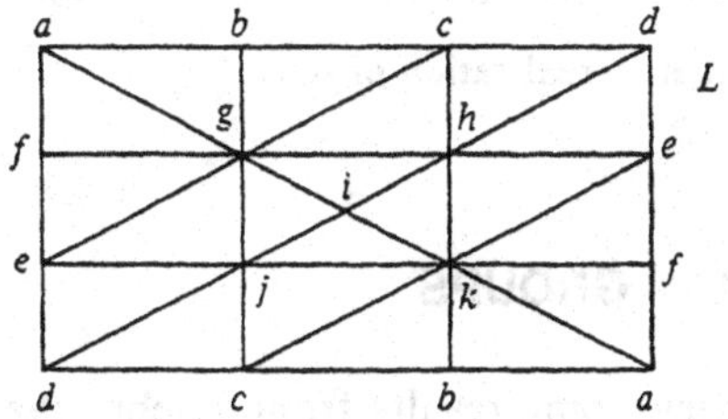

Figure 3.9

   (c) Describe the space determined by the labelled complex of Figure 3.10.

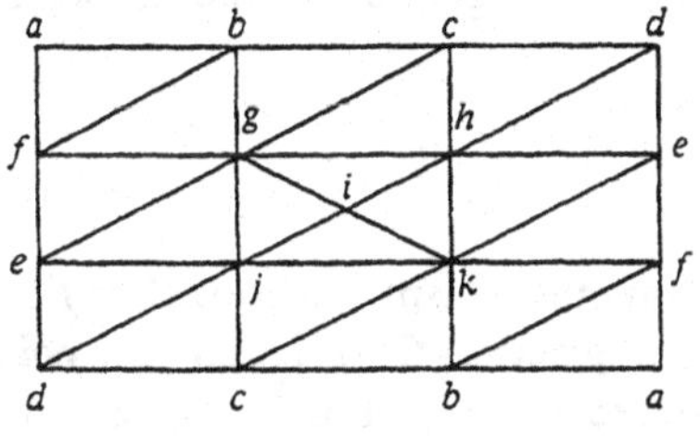

Figure 3.10

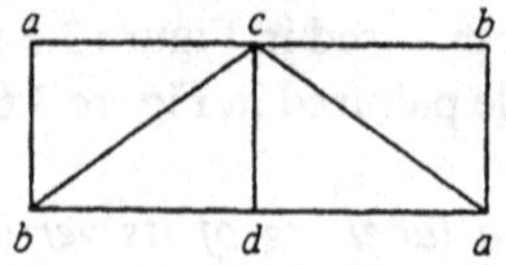

Figure 3.11

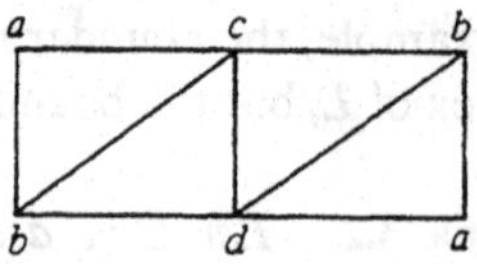

Figure 3.12

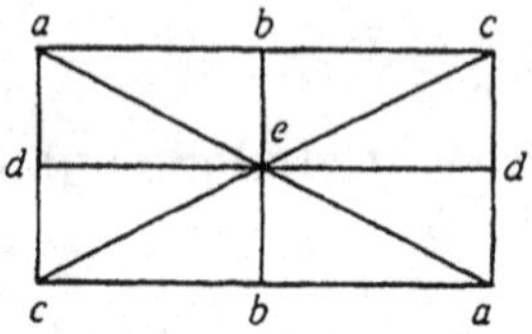

Figure 3.13

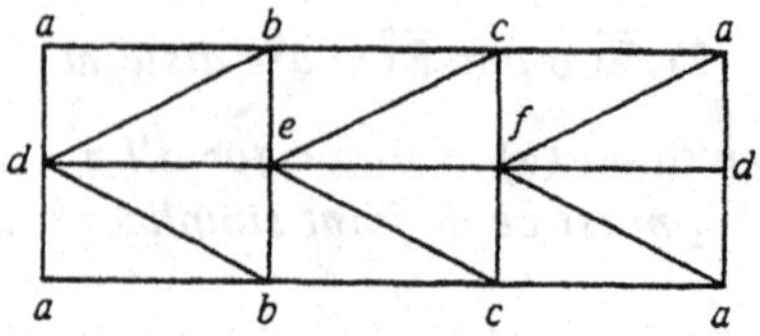

Figure 3.14

2. Describe the spaces determined by the labelled complexes in Figures 3.11–3.14.
3. Prove Lemma 3.2.
4. Let S be a set with a partial order relation $\leq$. A standard technique in combinatorics is to associate with S the abstract complex $\mathcal{S}$ whose vertices are the elements of S and whose simplices are the finite simply-ordered subsets of S. Suppose one is given the partial order on $\{a_1, \ldots, a_8\}$ generated by the following relations:

$$a_1 \leq a_3 \leq a_7 \leq a_8; \qquad a_1 \leq a_5 \leq a_7;$$

$$a_2 \leq a_6 \leq a_8; \qquad a_2 \leq a_5.$$

Describe a geometric realization of $\mathcal{S}$.

§4. REVIEW OF ABELIAN GROUPS

In this section, we review some results from algebra that we shall be using—specifically, facts about abelian groups.

We write abelian groups additively. Then 0 denotes the neutral element, and $-g$ denotes the additive inverse of g. If n is a positive integer, then ng denotes the n-fold sum $g + \cdots + g$, and $(-n)g$ denotes $n(-g)$.

We denote the group of integers by **Z**, the rationals by **Q**, and the complex numbers by **C**.

Homomorphisms

If $f : G \to H$ is a homomorphism, the **kernel** of f is the subgroup $f^{-1}(0)$ of G, the **image** of f is the subgroup $f(G)$ of H, and the **cokernel** of f is the quotient group $H/f(G)$. We denote these groups by $\ker f$ and $\operatorname{im} f$ and $\operatorname{cok} f$, respectively. The map f is a monomorphism if and only if the kernel of f vanishes (i.e., equals the trivial group). And f is an epimorphism if and only if the cokernel of f vanishes; in this case, f induces an isomorphism $G/\ker f \simeq H$.

Free abelian groups

An abelian group G is **free** if it has a **basis**—that is, if there is a family $\{g_\alpha\}_{\alpha \in J}$ of elements of G such that each $g \in G$ can be written uniquely as a finite sum

$$g = \Sigma n_\alpha g_\alpha,$$

with n_α an integer. Uniqueness implies that each element g_α has infinite order; that is, g_α generates an infinite cyclic subgroup of G.

More generally, if each $g \in G$ can be written as a finite sum $g = \Sigma n_\alpha g_\alpha$, but not necessarily uniquely, then we say that the family $\{g_\alpha\}$ **generates** G. In particular, if the set $\{g_\alpha\}$ is finite, we say that G is **finitely generated.**

If G is free and has a basis consisting of n elements, say $g_1, \ldots, g_n$, then it is easy to see that every basis for G consists of precisely n elements. For the group $G/2G$ consists of all cosets of the form

$$(\Sigma \epsilon_i g_i) + 2G,$$

where $\epsilon_i = 0$ or 1; this fact implies that the group $G/2G$ consists of precisely 2^n elements. The number of elements in a basis for G is called the **rank** of G.

It is true more generally that if G has an infinite basis, any two bases for G have the same cardinality. We shall not use this fact.

A crucial property of free abelian groups is the following: If G has a basis $\{g_\alpha\}$, then any function f from the set $\{g_\alpha\}$ to an abelian group H extends uniquely to a homomorphism of G into H.

One specific way of constructing free abelian groups is the following: Given a set S, we define the **free abelian group G generated by** S to be the set of all functions $\phi : S \to \mathbf{Z}$ such that $\phi(x) \neq 0$ for only finitely many values of x; we add two such functions by adding their values. Given $x \in S$, there is a characteristic function ϕ_x for x, defined by setting

$$\phi_x(y) = \begin{cases} 0 & \text{if} \quad y \neq x, \\ 1 & \text{if} \quad y = x. \end{cases}$$

The functions $\{\phi_x \mid x \in S\}$ form a basis for G, for each function $\phi \in G$ can be written uniquely as a finite sum

$$\phi = \Sigma n_x \phi_x,$$

where $n_x = \phi(x)$ and the summation extends over all x for which $\phi(x) \neq 0$. We often abuse notation and identify the element $x \in S$ with its characteristic function ϕ_x. With this notation, the general element of G can be written uniquely as a finite "formal linear combination"

$$\phi = \Sigma n_\alpha x_\alpha$$

of the elements of the set S.

If G is an abelian group, an element g of G has **finite order** if $ng = 0$ for some positive integer n. The set of all elements of finite order in G is a subgroup T of G, called the **torsion subgroup.** If T vanishes, we say G is **torsion-free.** A free abelian group is necessarily torsion-free, but not conversely.

If T consists of only finitely many elements, then the number of elements in T is called the **order** of T. If T has finite order, then each element of T has finite order; but not conversely.

Internal direct sums

Suppose G is an abelian group, and suppose $\{G_\alpha\}_{\alpha \in J}$ is a collection of subgroups of G, indexed bijectively by some index set J. Suppose that each g in G can be written uniquely as a finite sum $g = \Sigma g_\alpha$, where $g_\alpha \in G_\alpha$ for each α. Then G is said to be the **internal direct sum** of the groups G_α, and we write

$$G = \oplus_{\alpha \in J} G_\alpha.$$

If the collection $\{G_\alpha\}$ is finite, say $\{G_\alpha\} = \{G_1, \ldots, G_n\}$, we also write this direct sum in the form $G = G_1 \oplus \cdots \oplus G_n$.

If each g in G can be written as a finite sum $g = \Sigma g_\alpha$, but not necessarily uniquely, we say simply that G is the **sum** of the groups $\{G_\alpha\}$, and we write $G = \Sigma G_\alpha$, or, in the finite case, $G = G_1 + \cdots + G_n$. In this situation, we also say that the groups $\{G_\alpha\}$ **generate** G.

If $G = \Sigma G_\alpha$, then this sum is direct if and only if the equation $0 = \Sigma g_\alpha$ implies that $g_\alpha = 0$ for each α. This in turn occurs if and only if for each fixed index α_0, one has

$$G_{\alpha_0} \cap \left(\sum_{\alpha \neq \alpha_0} G_\alpha \right) = \{0\}.$$

In particular, if $G = G_1 + G_2$, then this sum is direct if and only if $G_1 \cap G_2 = \{0\}$.

The resemblance to free abelian groups is strong. Indeed, if G is free with basis $\{g_\alpha\}$, then G is the direct sum of the subgroups $\{G_\alpha\}$, where G_α is the infinite cyclic group generated by g_α. Conversely, if G is a direct sum of infinite cyclic subgroups, then G is a free abelian group.

If G is the direct sum of subgroups $\{G_\alpha\}$, and if for each α, one has a homomorphism f_α of G_α into the abelian group H, the homomorphisms $\{f_\alpha\}$ extend uniquely to a homomorphism of G into H.

Here is a useful criterion for showing G is a direct sum:

Lemma 4.1. *Let G be an abelian group. If G is the direct sum of the subgroups $\{G_\alpha\}$, then there are homomorphisms*

$$j_\beta : G_\beta \to G \qquad \text{and} \qquad \pi_\beta : G \to G_\beta$$

such that $\pi_\beta \circ j_\alpha$ is the zero homomorphism if $\alpha \neq \beta$, and the identity homomorphism if $\alpha = \beta$.

Conversely, suppose $\{G_\alpha\}$ is a family of abelian groups, and there are homomorphisms j_β and π_β as above. Then j_β is a monomorphism. Furthermore, if the groups $j_\alpha(G_\alpha)$ generate G, then G is their direct sum.

Proof. Suppose $G = \bigoplus G_\alpha$. We define j_β to be the inclusion homomorphism. To define π_β, write $g = \Sigma g_\alpha$, where $g_\alpha \in G_\alpha$ for each α; and let $\pi_\beta(g) = g_\beta$. Uniqueness of the representation of g shows π_β is a well-defined homomorphism.

Consider the converse. Because $\pi_\alpha \circ j_\alpha$ is the identity, j_α is injective (and π_α is surjective). If the groups $j_\alpha(G_\alpha)$ generate G, every element of G can be written as a finite sum $\Sigma\, j_\alpha(g_\alpha)$, by hypothesis. To show this representation is unique, suppose

$$\Sigma\, j_\alpha(g_\alpha) = \Sigma\, j_\alpha(g'_\alpha).$$

Applying π_β, we see that $g_\beta = g'_\beta$. □

Direct products and external direct sums

Let $\{G_\alpha\}_{\alpha \in J}$ be an indexed family of abelian groups. Their **direct product** $\Pi_{\alpha \in J} G_\alpha$ is the group whose underlying set is the cartesian product of the sets G_α, and whose group operation is component-wise addition. Their **external direct sum** G is the subgroup of the direct product consisting of all tuples $(g_\alpha)_{\alpha \in J}$ such that $g_\alpha = 0_\alpha$ for all but finitely many values of α. (Here 0_α is the zero element of G_α.) The group G is sometimes also called the "weak direct product" of the groups G_α.

The relation between internal and external direct sums is described as follows: Suppose G is the external direct sum of the groups $\{G_\alpha\}$. Then for each β, we define $\pi_\beta : G \to G_\beta$ to be projection onto the βth factor. And we define $j_\beta : G_\beta \to G$ by letting it carry the element $g \in G_\beta$ to the tuple $(g_\alpha)_{\alpha \in J}$, where $g_\alpha = 0_\alpha$ for all α different from β, and $g_\beta = g$. Then $\pi_\beta \circ j_\alpha = 0$ for $\alpha \neq \beta$, and $\pi_\alpha \circ j_\alpha$ is the identity. It follows that G equals the *internal* direct sum of the groups $G'_\alpha = j_\alpha(G_\alpha)$, where G'_α is isomorphic to G_α.

Thus the notions of internal and external direct sums are closely related. The difference is mainly one of notation. For this reason, we customarily use the notations

$$G = G_1 \oplus \cdots \oplus G_n \qquad \text{and} \qquad G = \bigoplus G_\alpha$$

to denote either internal or external direct sums, relying on the context to make clear which is meant (if indeed, it is important). With this notation, one can for instance express the fact that G is free abelian of rank 3 merely by writing $G \simeq \mathbf{Z} \oplus \mathbf{Z} \oplus \mathbf{Z}$.

If G_1 is a subgroup of G, we say that G_1 is a **direct summand** in G if there is a subgroup G_2 of G such that $G = G_1 \oplus G_2$. In this case, if H_i is a subgroup of G_i, for $i = 1,2$, then the sum $H_1 + H_2$ is direct, and furthermore,

$$\frac{G}{H_1 \oplus H_2} \simeq \frac{G_1}{H_1} \oplus \frac{G_2}{H_2}.$$

In particular, if $G = G_1 \oplus G_2$, then $G/G_1 \simeq G_2$.

Of course, one can have $G/G_1 \simeq G_2$ without its following that $G = G_1 \oplus G_2$; that is, G_1 may be a subgroup of G without being a direct summand in G. For instance, the subgroup $n\mathbf{Z}$ of the integers is not a direct summand in $\mathbf{Z}$, for that would mean that

$$\mathbf{Z} \simeq n\mathbf{Z} \oplus G_2$$

for some subgroup G_2 of $\mathbf{Z}$. But then G_2 is isomorphic to $\mathbf{Z}/n\mathbf{Z}$, which is a group of finite order, while no subgroup of $\mathbf{Z}$ has finite order.

Incidentally, we shall denote the group $\mathbf{Z}/n\mathbf{Z}$ of integers modulo n simply by $\mathbf{Z}/n$, in accordance with current usage.

The fundamental theorem of finitely generated abelian groups

There are actually two theorems that are important to us. The first is a theorem about subgroups of free abelian groups. We state it here, and give a proof in §11:

Theorem 4.2. *Let F be a free abelian group. If R is a subgroup of F, then R is also a free abelian group. If F has rank n, then R has rank $r \leq n$; furthermore, there is a basis $e_1, \ldots, e_n$ for F and integers $t_1, \ldots, t_k$ with $t_i > 1$ such that*

(1) *$t_1e_1, \ldots, t_ke_k, e_{k+1}, \ldots, e_r$ is a basis for R.*

(2) *$t_1 \mid t_2 \mid \cdots \mid t_k$, that is, t_i divides t_{i+1} for all i.*

The integers $t_1, \ldots, t_k$ are uniquely determined by F and R, although the basis $e_1, \ldots, e_n$ is not.

An immediate corollary of this theorem is the following:

Theorem 4.3 (The fundamental theorem of finitely generated abelian groups). *Let G be a finitely generated abelian group. Let T be its torsion subgroup.*

(a) *There is a free abelian subgroup H of G having finite rank β such that $G = H \oplus T$.*

(b) *There are finite cyclic groups $T_1, \ldots, T_k$, where T_i has order $t_i > 1$, such that $t_1 \mid t_2 \mid \cdots \mid t_k$ and*

$$T = T_1 \oplus \cdots \oplus T_k.$$

(c) *The numbers β and $t_1, \ldots, t_k$ are uniquely determined by G.*

The number β is called the **betti number** of G; the numbers $t_1, \ldots, t_k$ are called the **torsion coefficients** of G. Note that β is the rank of the free abelian group $G/T \simeq H$. The rank of the subgroup H and the orders of the subgroups T_i are uniquely determined, but the subgroups themselves are not.

Proof. Let S be a finite set of generators $\{g_i\}$ for G; let F be the free abelian group on the set S. The map carrying each g_i to itself extends to a homomorphism carrying F *onto* G. Let R be the kernel of this homomorphism. Then $F/R \simeq G$. Choose bases for F and R as in Theorem 4.2. Then

$$F = F_1 \oplus \cdots \oplus F_n$$

where F_i is infinite cyclic with generator e_i; and

$$R = t_1F_1 \oplus \cdots \oplus t_kF_k \oplus F_{k+1} \oplus \cdots \oplus F_r.$$

We compute the quotient group as follows:

$$F/R \simeq (F_1/t_1F_1 \oplus \cdots \oplus F_k/t_kF_k) \oplus (F_{r+1} \oplus \cdots \oplus F_n).$$

Thus there is an isomorphism

$$f: G \to (\mathbf{Z}/t_1 \oplus \cdots \oplus \mathbf{Z}/t_k) \oplus (\mathbf{Z} \oplus \cdots \oplus \mathbf{Z}).$$

The torsion subgroup T of G must be mapped to the subgroup $\mathbf{Z}/t_1 \oplus \cdots \oplus \mathbf{Z}/t_k$ by f, since any isomorphism preserves torsion subgroups. Parts (a) and (b) of the theorem follow. Part (c) is left to the exercises. □

This theorem shows that any finitely generated abelian group G can be written as a finite direct sum of cyclic groups; that is,

$$G \simeq (\mathbf{Z} \oplus \cdots \oplus \mathbf{Z}) \oplus \mathbf{Z}/t_1 \oplus \cdots \oplus \mathbf{Z}/t_k.$$

with $t_i > 1$ and $t_1 \mid t_2 \mid \cdots \mid t_k$. This representation is in some sense a "canonical form" for G. There is another such canonical form, derived as follows:

Recall first the fact that if m and n are relatively prime positive integers, then

$$\mathbf{Z}/m \oplus \mathbf{Z}/n \simeq \mathbf{Z}/mn.$$

It follows that any finite cyclic group can be written as a direct sum of cyclic groups whose orders are powers of primes. Theorem 4.3 then implies that for any finitely generated group G,

$$G \simeq (\mathbf{Z} \oplus \cdots \oplus \mathbf{Z}) \oplus (\mathbf{Z}/a_1 \oplus \cdots \oplus \mathbf{Z}/a_s)$$

where each a_i is a power of a prime. This is another canonical form for G, since the numbers a_i are uniquely determined by G (up to a rearrangement), as we shall see.

EXERCISES

1. Show that if G is a finitely generated abelian group, every subgroup of G is finitely generated. (This result does not hold for non-abelian groups.)
2. (a) Show that if G is free, then G is torsion-free.
   (b) Show that if G is finitely generated and torsion-free, then G is free.

(c) Show that the additive group of rationals **Q** is torsion-free but not free. [*Hint:* If $\{g_\alpha\}$ is a basis for **Q**, let β be fixed and express $g_\beta/2$ in terms of this basis.]

3. (a) Show that if m and n are relatively prime, then $\mathbf{Z}/m \oplus \mathbf{Z}/n$ is cyclic of order mn.
(b) If $G \simeq \mathbf{Z}/18 \oplus \mathbf{Z}/36$, express G as a direct sum of cyclic groups of prime power order.
(c) If $G \simeq \mathbf{Z}/2 \oplus \mathbf{Z}/4 \oplus \mathbf{Z}/3 \oplus \mathbf{Z}/3 \oplus \mathbf{Z}/9$, find the torsion coefficients of G.
(d) If $G \simeq \mathbf{Z}/15 \oplus \mathbf{Z}/20 \oplus \mathbf{Z}/18$, find the invariant factors and the torsion coefficients of G.

4. (a) Let p be prime; let $b_1, \ldots, b_k$ be non-negative integers. Show that if

$$G \simeq (\mathbf{Z}/p)^{b_1} \oplus (\mathbf{Z}/p^2)^{b_2} \oplus \cdots \oplus (\mathbf{Z}/p^k)^{b_k},$$

then the integers b_i are uniquely determined by G. [*Hint:* Consider the kernel of the homomorphism $f_i : G \to G$ that is multiplication by p^i. Show that f_1 and f_2 determine b_1. Proceed similarly.]
(b) Let $p_1, \ldots, p_N$ be a sequence of distinct primes. Generalize (a) to a finite direct sum of terms of the form $(\mathbf{Z}/(p_i)^k)^{b_{ik}}$, where $b_{ik} \geq 0$.
(c) Verify (c) of Theorem 4.3. That is, show that the betti number, invariant factors, and torsion coefficients of a finitely generated abelian group G are uniquely determined by G.
(d) Show that the numbers t_i appearing in the conclusion of Theorem 4.2 are uniquely determined by F and R.

§5. HOMOLOGY GROUPS

Now we are ready to define the homology groups. First we must discuss the notion of "orientation."

Definition. Let σ be a simplex (either geometric or abstract). Define two orderings of its vertex set to be equivalent if they differ from one another by an even permutation. If $\dim \sigma > 0$, the orderings of the vertices of σ then fall into two equivalence classes. Each of these classes is called an **orientation** of σ. (If σ is a 0-simplex, then there is only one class and hence only one orientation of σ.) An **oriented simplex** is a simplex σ together with an orientation of σ.

If the points $v_0, \ldots, v_p$ are independent, we shall use the symbol

$$v_0 \ldots v_p$$

to denote the simplex they span, and we shall use the symbol

$$[v_0, \ldots, v_p]$$

to denote the oriented simplex consisting of the simplex $v_0 \ldots v_p$ and the equivalence class of the particular ordering $(v_0, \ldots, v_p)$.

Occasionally, when the context makes the meaning clear, we may use a single letter such as σ to denote either a simplex or an oriented simplex.

Example 1. We often picture an orientation of a 1-simplex by drawing an arrow on it. The oriented simplex $[v_0, v_1]$ is pictured in Figure 5.1; one draws an arrow pointing from v_0 to v_1. An orientation of a 2-simplex is pictured by a circular arrow. The oriented simplex $[v_0, v_1, v_2]$ is indicated in the figure by drawing an arrow in the direction from v_0 to v_1 to v_2. You can check that $[v_1, v_2, v_0]$ and $[v_2, v_0, v_1]$ are indicated by the same clockwise arrow. An arrow in the counterclockwise direction would indicate the oppositely oriented simplex.

Similarly, the oriented simplex $[v_0, v_1, v_2, v_3]$ is pictured by drawing a spiral arrow, as in the figure. The arrow in this picture is called a "right-hand screw"; if one curls the fingers of the right hand in the direction from v_0 to v_1 to v_2, the thumb points toward v_3. You can check that $[v_0, v_2, v_3, v_1]$, and each of the other ten orderings equivalent to these two, also give rise to right-hand screws. A "left-hand screw" is used to picture the opposite orientation.

These examples illustrate that our definition of orientation agrees with the intuitive geometric notions derived from vector calculus.

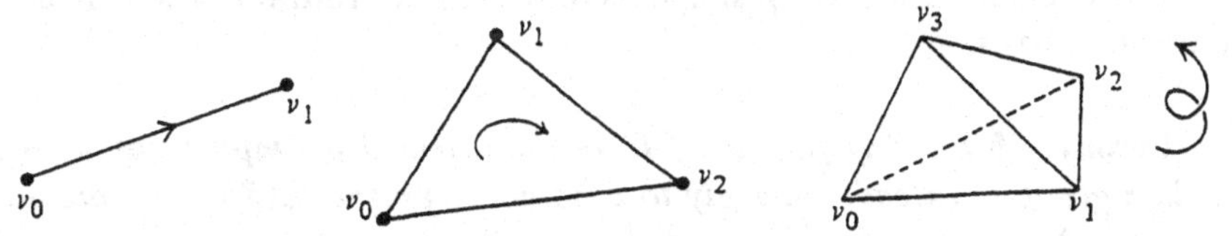

Figure 5.1

Definition. Let K be a simplicial complex. A ***p*-chain** on K is a function c from the set of oriented p-simplices of K to the integers, such that:

(1) $c(\sigma) = -c(\sigma')$ if σ and σ' are opposite orientations of the same simplex.

(2) $c(\sigma) = 0$ for all but finitely many oriented p-simplices σ.

We add p-chains by adding their values; the resulting group is denoted $C_p(K)$ and is called the **group of (oriented) *p*-chains** of K. If $p < 0$ or $p > \dim K$, we let $C_p(K)$ denote the trivial group.

If σ is an oriented simplex, the **elementary chain** c corresponding to σ is the function defined as follows:

$$
\begin{aligned}
c(\sigma) &= 1, \\
c(\sigma') &= -1 \qquad \text{if } \sigma' \text{ is the opposite orientation of } \sigma, \\
c(\tau) &= 0 \qquad \text{for all other oriented simplices } \tau.
\end{aligned}
$$

By abuse of notation, we often use the symbol σ to denote not only a simplex, or an oriented simplex, but also to denote the elementary p-chain c corresponding to the oriented simplex σ. With this convention, if σ and σ' are opposite orientations of the same simplex, then we can write $\sigma' = -\sigma$, because this equation holds when σ and σ' are interpreted as elementary chains.

Lemma 5.1. *$C_p(K)$ is free abelian; a basis for $C_p(K)$ can be obtained by orienting each p-simplex and using the corresponding elementary chains as a basis.*

Proof. The proof is straightforward. Once all the p-simplices of K are oriented (arbitrarily), each p-chain can be written uniquely as a finite linear combination

$$c = \Sigma n_i \sigma_i$$

of the corresponding elementary chains σ_i. The chain c assigns the value n_i to the oriented p-simplex σ_i, the value $-n_i$ to the opposite orientation of σ_i, and the value 0 to all oriented p-simplices not appearing in the summation. □

The group $C_0(K)$ differs from the others, since it has a natural basis (since a 0-simplex has only one orientation). The group $C_p(K)$ has no "natural" basis if $p > 0$; one must orient the p-simplices of K in some arbitrary fashion in order to obtain a basis.

Corollary 5.2. *Any function f from the oriented p-simplices of K to an abelian group G extends uniquely to a homomorphism $C_p(K) \to G$, provided that $f(-\sigma) = -f(\sigma)$ for all oriented p-simplices σ.* □

Definition. We now define a homomorphism

$$\partial_p : C_p(K) \to C_{p-1}(K)$$

called the **boundary operator.** If $\sigma = [v_0, \ldots, v_p]$ is an oriented simplex with $p > 0$, we define

$$(*) \qquad \partial_p \sigma = \partial_p [v_0, \ldots, v_p] = \sum_{i=0}^{p} (-1)^i [v_0, \ldots, \hat{v}_i, \ldots, v_p],$$

where the symbol $\hat{v}_i$ means that the vertex v_i is to be deleted from the array. Since $C_p(K)$ is the trivial group for $p < 0$, the operator ∂_p is the trivial homomorphism for $p \leq 0$.

We must check that ∂_p is well-defined and that $\partial_p(-\sigma) = -\partial_p \sigma$. For this purpose, it suffices to show that the right-hand side of (*) changes sign if we exchange two adjacent vertices in the array $[v_0, \ldots, v_p]$. So let us compare the expressions for

$$\partial_p [v_0, \ldots, v_j, v_{j+1}, \ldots, v_p] \quad \text{and} \quad \partial_p [v_0, \ldots, v_{j+1}, v_j, \ldots, v_p].$$

For $i \neq j, j+1$, the ith terms in these two expressions differ precisely by a sign; the terms are identical except that v_j and v_{j+1} have been interchanged.

What about the ith terms for $i = j$ and $i = j + 1$? In the first expression, one has

$$(-1)^j[\ldots,v_{j-1},\hat{v}_j,v_{j+1},v_{j+2},\ldots] + (-1)^{j+1}[\ldots,v_{j-1},v_j,\hat{v}_{j+1},v_{j+2},\ldots].$$

In the second expression, one has

$$(-1)^j[\ldots,v_{j-1},\hat{v}_{j+1},v_j,v_{j+2},\ldots] + (-1)^{j+1}[\ldots,v_{j-1},v_{j+1},\hat{v}_j,v_{j+2},\ldots].$$

Comparing, one sees these two expressions differ by a sign.

Example 2. For a 1-simplex, we compute $\partial_1[v_0,v_1] = v_1 - v_0$. For a 2-simplex, one has

$$\partial_2[v_0,v_1,v_2] = [v_1,v_2] - [v_0,v_2] + [v_0,v_1].$$

And for a 3-simplex one has the formula

$$\partial_3[v_0,v_1,v_2,v_3] = [v_1,v_2,v_3] - [v_0,v_2,v_3] + [v_0,v_1,v_3] - [v_0,v_1,v_2].$$

The geometric content of these formulas is pictured in Figure 5.2. If you remember the versions of Green's, Stokes', and Gauss' theorems you studied in calculus, these pictures should look rather familiar.

Example 3. Consider the 1-chain $\partial_2[v_0,v_1,v_2]$ pictured in Figure 5.2. If you apply the operator ∂_1 to this 1-chain, you get zero; everything cancels out because each vertex appears as the initial point of one edge and as the end point of another edge. You can check that a similar cancellation occurs when you compute $\partial_2\partial_3[v_0,v_1,v_2,v_3]$.

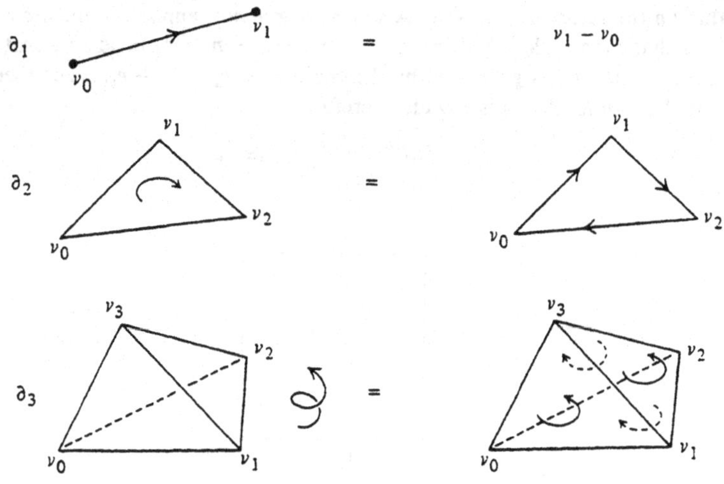

Figure 5.2

The computations discussed in Example 3 illustrate a general fact:

Lemma 5.3. $\partial_{p-1} \circ \partial_p = 0.$

Proof. The proof is straightforward. We compute

$$\partial_{p-1}\partial_p[v_0,\dots,v_p] = \sum_{i=0}^{p} (-1)^i \partial_{p-1}[v_0,\dots,\hat{v}_i,\dots,v_p]$$

$$= \sum_{j<i} (-1)^i(-1)^j[\dots,\hat{v}_j,\dots,\hat{v}_i,\dots]$$

$$+ \sum_{j>i} (-1)^i(-1)^{j-1}[\dots,\hat{v}_i,\dots,\hat{v}_j,\dots].$$

The terms of these two summations cancel in pairs. □

Definition. The kernel of $\partial_p : C_p(K) \to C_{p-1}(K)$ is called the group of ***p*-cycles** and denoted $Z_p(K)$ (for the German word "Zyklus"). The image of $\partial_{p+1} : C_{p+1}(K) \to C_p(K)$ is called the group of ***p*-boundaries** and is denoted $B_p(K)$. By the preceding lemma, each boundary of a $p + 1$ chain is automatically a p-cycle. That is, $B_p(K) \subset Z_p(K)$. We define

$$H_p(K) = Z_p(K)/B_p(K),$$

and call it the pth **homology group** of K.

Let us compute a few examples.

Example 4. Consider the complex K of Figure 5.3, whose underlying space is the boundary of a square with edges e_1, e_2, e_3, e_4. The group $C_1(K)$ is free abelian of rank 4; the general 1-chain c is of the form $\Sigma n_i e_i$. Computing $\partial_1 c$, we see that its value on the vertex v is $n_1 - n_2$. A similar argument, applied to the other vertices, shows that c is a cycle if and only if $n_1 = n_2 = n_3 = n_4$. One concludes that $Z_1(K)$ is infinite cyclic, and is generated by the chain $e_1 + e_2 + e_3 + e_4$. Since there are no 2-simplices in K, $B_1(K)$ is trivial. Therefore,

$$H_1(K) = Z_1(K) \simeq \mathbf{Z}.$$

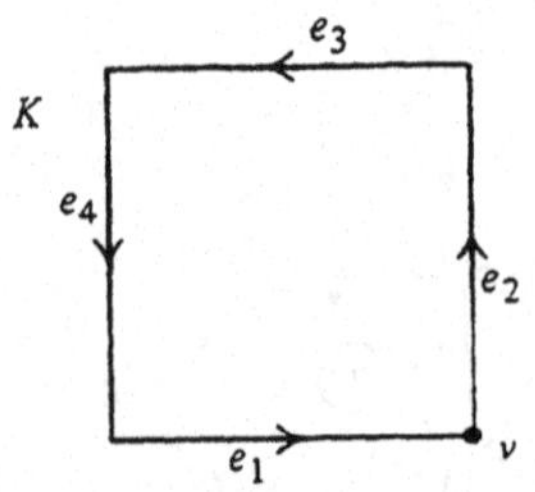

Figure 5.3

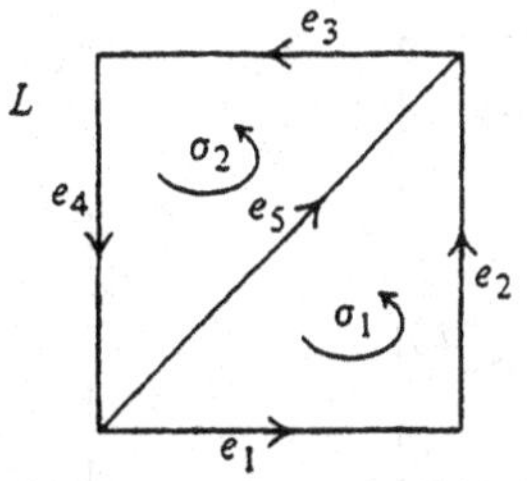

Figure 5.4

Example 5. Consider the complex L of Figure 5.4, whose underlying space is a square. The general 1-chain is of the form $\Sigma n_i e_i$. One reasons as before to conclude that this 1-chain is a cycle if and only if $n_1 = n_2$, $n_3 = n_4$, and $n_5 = n_3 - n_2$. One can assign values to n_2 and n_3 arbitrarily; then the others are determined. Therefore, $Z_1(L)$ is free abelian of rank 2; a basis consists of the chain $e_1 + e_2 - e_5$ obtained by taking $n_2 = 1$ and $n_3 = 0$, and the chain $e_3 + e_4 + e_5$ obtained by taking $n_2 = 0$ and $n_3 = 1$. The first equals $\partial_2\sigma_1$ and the second equals $\partial_2\sigma_2$. Therefore,

$$H_1(L) = Z_1(L)/B_1(L) = 0.$$

Likewise, $H_2(L) = 0$; the general 2-chain $m_1\sigma_1 + m_2\sigma_2$ is a cycle if and only if $m_1 = m_2 = 0$.

These examples may begin to give you a feeling for what the homology groups mean geometrically. Only by computing many more examples can one begin to "see" what a homology class is. Our hope is that after you get a feeling for what homology means geometrically, you will begin to believe what is at the moment far from clear—that the homology groups of a complex K actually depend only on the underlying space $|K|$.

Now let us consider another example. It involves a complex having more simplices than those in the preceding examples. In general, as the number of simplices increases, calculating the group of cycles Z_p and the group of boundaries B_p becomes more tedious. We can short-cut some of these calculations by avoiding calculating these groups and proceeding directly to a calculation of the homology groups H_p.

We deal here only with the groups $H_p(K)$ for $p > 0$, postponing discussion of the group $H_0(K)$ to §7.

We need some terminology. We shall say that a chain c is **carried by** a subcomplex L of K if c has value 0 on every simplex that is not in L. And we say that two p-chains c and c' are **homologous** if $c - c' = \partial_{p+1}d$ for some $p + 1$ chain d. In particular, if $c = \partial_{p+1}d$, we say that c is **homologous to zero,** or simply that c **bounds.**

Example 6. Consider the complex M indicated in Figure 5.5, whose underlying space is a square. Instead of computing the group of 1-cycles specifically, we reason

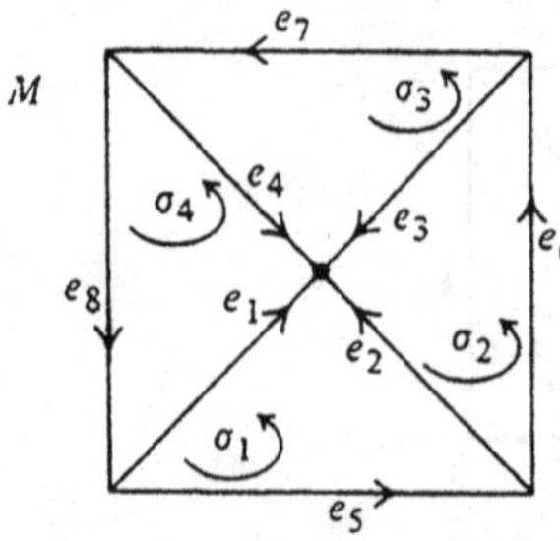

Figure 5.5

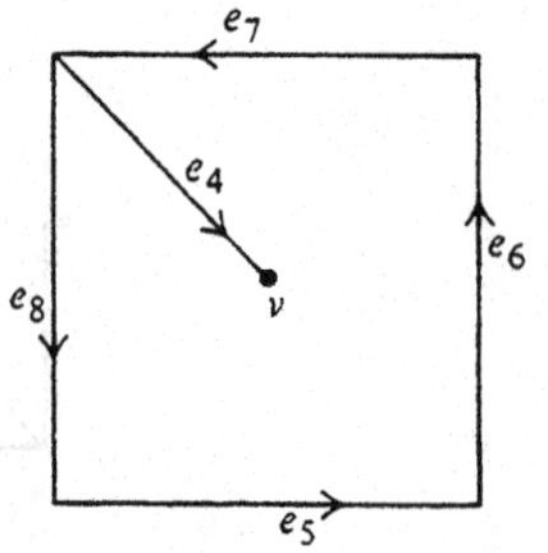

Figure 5.6

as follows: Given a 1-chain c, let a be the value of c on e_1. Then by direct computation, the chain

$$c_1 = c + \partial_2(a\sigma_1)$$

has value 0 on the oriented simplex e_1. Intuitively speaking, by modifying c by a boundary, we have "pushed it off e_1." We then "push c_1 off e_2" in a similar manner, as follows: Let b be the value of c_1 on e_2; then the chain

$$c_2 = c_1 + \partial_2(b\sigma_2)$$

has value 0 on the oriented simplex e_2. It also has value 0 on e_1, since e_1 does not appear in the expression for $\partial_2\sigma_2$. Now letting d denote the value of c_2 on e_3, one sees that

$$c_3 = c_2 + \partial_2(d\sigma_3)$$

has value 0 on e_3 and on e_2 and on e_1. We have thus "pushed c off" all of e_1, e_2, and e_3. Said differently, we have proved the following result:

Given a 1-chain c, it is homologous to a chain c_3 that is carried by the subcomplex of M pictured in Figure 5.6.

Now if c happens to be a *cycle*, then c_3 is also a cycle; it follows that the value of c_3 on the simplex e_4 must be 0. (Otherwise, ∂c_3 would have non-zero value on the center vertex v.) Thus every 1-cycle of M is homologous to a 1-cycle carried by the boundary of the square. By the same argument as used before, such a cycle must be some multiple of the cycle $e_5 + e_6 + e_7 + e_8$. And this cycle bounds; indeed, it clearly equals $\partial(\sigma_1 + \sigma_2 + \sigma_3 + \sigma_4)$. Thus $H_1(M) = 0$, as expected.

The fact that $H_2(M) = 0$ is easy; one sees readily (as before) that $\Sigma m_i\sigma_i$ is a cycle if and only if $m_i = 0$ for all i.

Note that the homology groups of M are the same as the homology groups of the complex L of Example 5. This fact lends some plausibility to our remark (yet to be proved) that the homology groups of a complex depend only on its underlying space.

EXERCISES

1. Let $\mathcal{S}$ be the abstract complex consisting of the 1-simplices $\{v_0,v_1\}$, $\{v_1,v_2\}$, $\ldots$, $\{v_{n-1},v_n\}$, $\{v_n,v_0\}$ and their vertices. If K is a geometric realization of $\mathcal{S}$, compute $H_1(K)$.

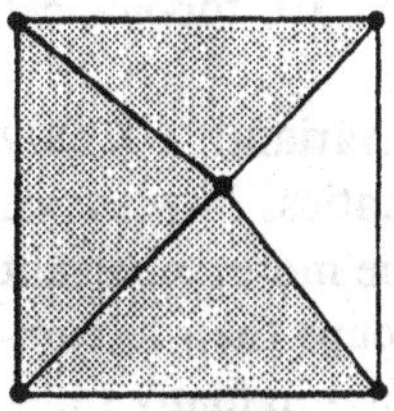

Figure 5.7

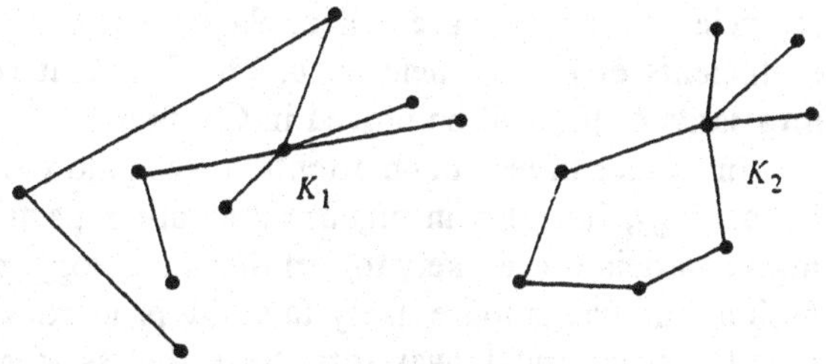

Figure 5.8

2. Consider the complex M pictured in Figure 5.7; it is the union of three triangles and a line segment. Compute the homology groups $H_1(M)$ and $H_2(M)$.
3. A 1-dimensional complex is called a **tree** if its 1-dimensional homology vanishes. Is either of the complexes pictured in Figure 5.8 a tree?
4. Let K be the complex consisting of the proper faces of a 3-simplex. Compute $H_1(K)$ and $H_2(K)$.
5. For what values of i is it true that

$$H_i(K^{(p)}) \simeq H_i(K)?$$

6. An "infinite p-chain" on K is a function c from the oriented p-simplices of K to the integers such that $c(\sigma) = -c(\sigma')$ if σ and σ' are opposite orientations of the same simplex. We do not require that $c(\sigma) = 0$ for all but finitely many oriented simplices. Let $C_p^\infty(K)$ denote the group of infinite p-chains. It is abelian, but it will not in general be free.
(a) Show that if K is locally finite, then one can define a boundary operator

$$\partial_p^\infty : C_p^\infty(K) \to C_{p-1}^\infty(K)$$

by the formula used earlier, and Lemma 5.3 holds. The resulting groups

$$H_p^\infty(K) = \ker \partial_p^\infty / \operatorname{im} \partial_{p+1}^\infty$$

are called the **homology groups based on infinite chains.**
(b) Let K be the complex whose space is $\mathbf{R}$ and whose vertices are the integers. Show that

$$H_1(K) = 0 \qquad \text{and} \qquad H_1^\infty(K) \simeq \mathbf{Z}.$$

7. Let $\mathcal{S}$ be the abstract complex whose simplices consist of the sets $\{im, m\}$, $\{im, -m\}$, and $\{m, -m\}$ for all positive integers m, along with their faces. If K is a geometric realization of $\mathcal{S}$, compute $H_1(K)$ and $H_1^\infty(K)$.

§6. HOMOLOGY GROUPS OF SURFACES

If K is a finite complex, then the chain group $C_p(K)$ has finite rank, so the cycle group $Z_p(K)$ also has finite rank. Then $H_p(K)$ is finitely generated, so that the fundamental theorem of abelian groups applies. The betti number and torsion coefficients of $H_p(K)$ are called, classically, the betti number of K and torsion coefficients of K in dimension p. The fact that these numbers are topological invariants of $|K|$ will be proved in Chapter 2.

In former times, much attention was paid to *numerical* invariants, not only in topology, but also in algebra and other parts of mathematics. Nowadays, mathematicians are likely to feel that homology *groups* are the more important notion, and one is more likely to study properties of these groups than to compute their numerical invariants. Nevertheless, it is still important in many situations to compute the homology groups specifically—that is, to find the betti numbers and torsion coefficients for a given space.

One of the greatest virtues of the *simplicial* homology groups is that it is in fact possible to do precisely this. In §11 we shall prove a theorem to the effect that, for a finite complex K, the homology groups are effectively computable. This means that there is a specific algorithm for finding the betti numbers and torsion coefficients of K.

In the present section, we shall compute the betti numbers and torsion coefficients of the compact surfaces. The techniques we shall use may seem a bit awkward and ad hoc in nature. But, in fact, they are effective on a large class of spaces. In a later section, when we study CW complexes, we shall return to these techniques and show they are part of a systematic approach to computing homology groups.

Convention. For convenience in notation, we shall henceforth delete the dimensional subscript p on the boundary operator ∂_p, and rely on the context to make clear which of these operators is intended.

We shall compute the homology of the torus, the Klein bottle, and several other spaces that can be constructed from a rectangle L by identifying its edges appropriately. Thus we begin by proving certain facts about L itself.

Lemma 6.1. *Let L be the complex of Figure* 6.1, *whose underlying space is a rectangle. Let* Bd L *denote the complex whose space is the boundary of the rectangle. Orient each 2-simplex σ_i of L by a counterclockwise arrow. Orient the* 1-*simplices arbitrarily. Then:*

(1) *Every* 1-*cycle of L is homologous to a* 1-*cycle carried by* Bd L.

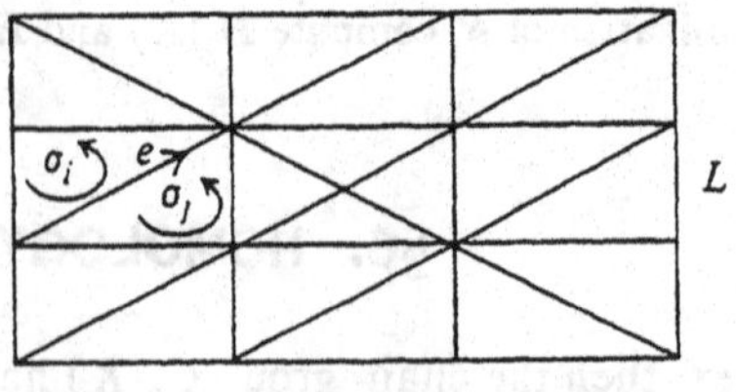

Figure 6.1

(2) *If d is a 2-chain of L and if ∂d is carried by* Bd L, *then d is a multiple of the chain $\Sigma\sigma_i$.*

Proof. The proof of (2) is easy. If σ_i and σ_j have an edge e in common, then ∂d must have value 0 on e. It follows that d must have the same value on σ_i as it does on σ_j. Continuing this process, we see that d has the same value on every oriented 2-simplex σ_i.

To prove (1), we proceed as in Example 6 of the preceding section. Given a 1-chain c of L, one "pushes it off" the 1-simplices, one at a time. First, one shows that c is homologous to a 1-chain c_1 carried by the subcomplex pictured in Figure 6.2. Then one shows that c_1 is in turn homologous to a 1-chain c_2 carried by the subcomplex of Figure 6.3. Finally, one notes that in the case where the original chain c is a cycle, then the chain c_2 is also a cycle. It follows that c_2 must be carried by Bd L, for otherwise c_2 would have a non-zero coefficient on one or more of the vertices $v_1, \ldots, v_5$. □

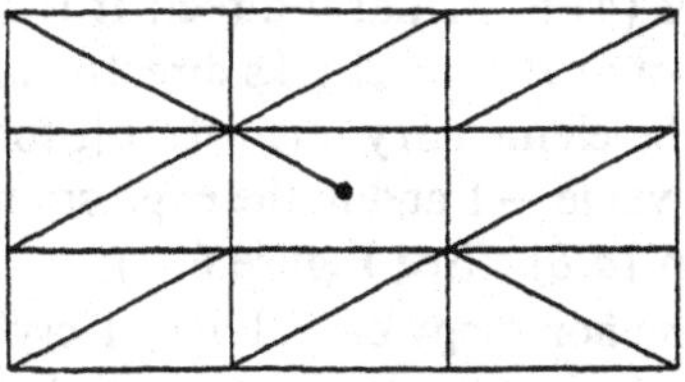

Figure 6.2

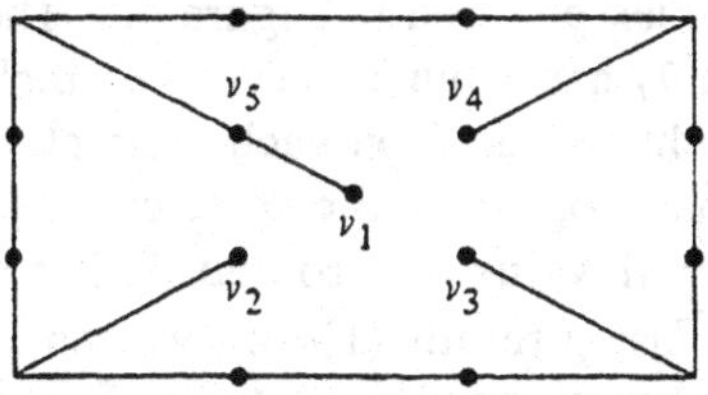

Figure 6.3

Theorem 6.2. *Let T denote the complex represented by the labelled rectangle L of Figure 6.4; its underlying space is the torus. Then:*

$$H_1(T) \simeq \mathbf{Z} \oplus \mathbf{Z} \qquad \textit{and} \qquad H_2(T) \simeq \mathbf{Z}.$$

Orient each 2-simplex of L counterclockwise; use the induced orientation of the 2-simplices of T; let γ denote their sum. Let

$$w_1 = [a,b] + [b,c] + [c,a],$$
$$z_1 = [a,d] + [d,e] + [e,a].$$

Then γ generates $H_2(T)$ and w_1 and z_1 represent a basis for $H_1(T)$.

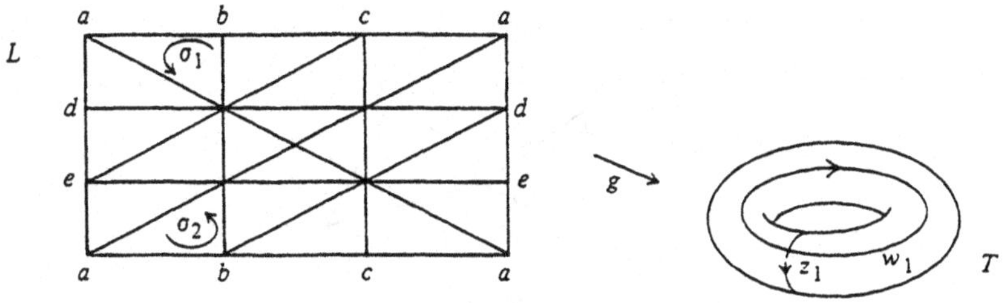

Figure 6.4

Proof. Let $g: |L| \to |T|$ be the pasting map; let $A = g(|\text{Bd } L|)$. Then A is homeomorphic to a space that is the union of two circles with a point in common. (Such a space is called a **wedge of two circles.**) Orient the 1-simplices of T arbitrarily. Because g makes identifications only among simplices of Bd L, the arguments we gave earlier in proving Lemma 6.1 apply verbatim to prove the following:

(1) *Every 1-cycle of T is homologous to a 1-cycle carried by A.*

(2) *If d is a 2-chain of T and if ∂d is carried by A, then d is a multiple of γ.*

However, in the complex T, two further results hold:

(3) *If c is a 1-cycle of T carried by A, then c is of the form $nw_1 + mz_1$.*

(4) $\partial\gamma = 0$.

The proof of (3) is easy, given the fact that A is just the 1-dimensional complex pictured in Figure 6.5. The proof of (4) is similarly direct: It is clear that $\partial\gamma$ has value 0 on every 1-simplex of T not in A. One checks directly that it also has value 0 on each 1-simplex in A. The elementary chain $[a,b]$, for instance, appears in the expression for $\partial\sigma_1$ with value -1 and in the expression for $\partial\sigma_2$ with value $+1$, so that $\partial\gamma$ has value 0 on $[a,b]$. (See Figure 6.4.)

Using results (1)–(4), we can compute the homology of T. Every 1-cycle of T is homologous to a 1-cycle of the form $c = nw_1 + mz_1$, by (1) and (3). Such a cycle bounds only if it is trivial: For if $c = \partial d$ for some d, then (2) applies to show that $d = p\gamma$ for some p; since $\partial\gamma = 0$ by (4), we have $c = \partial d = 0$. We conclude that

$$H_1(T) \simeq \mathbf{Z} \oplus \mathbf{Z};$$

and the (cosets of the) 1-cycles w_1 and z_1 form a basis for the 1-dimensional homology.

To compute $H_2(T)$, note that by (2) any 2-cycle d of T must be of the form $p\gamma$ for some p. Each such 2-chain is in fact a cycle, by (4), and there are no 3-chains for it to bound. We conclude that

$$H_2(T) \simeq \mathbf{Z},$$

and this group has as generator the 2-cycle γ. □

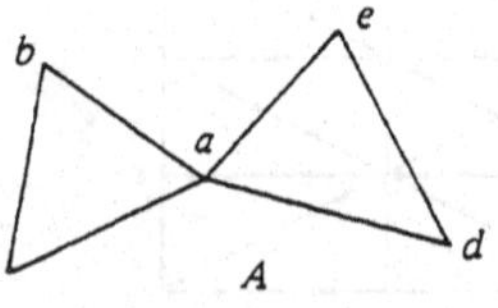

Figure 6.5

Theorem 6.3. *Let S denote the complex represented by the labelled rectangle of Figure 6.6; its underlying space is the Klein bottle. Then*

$$H_1(S) \simeq \mathbf{Z} \oplus \mathbf{Z}/2 \qquad \textit{and} \qquad H_2(S) = 0.$$

The torsion element of $H_1(S)$ is represented by the chain z_1, and a generator for the group $H_1(S)$ modulo torsion is represented by w_1, where

$$w_1 = [a,b] + [b,c] + [c,a],$$
$$z_1 = [a,d] + [d,e] + [e,a].$$

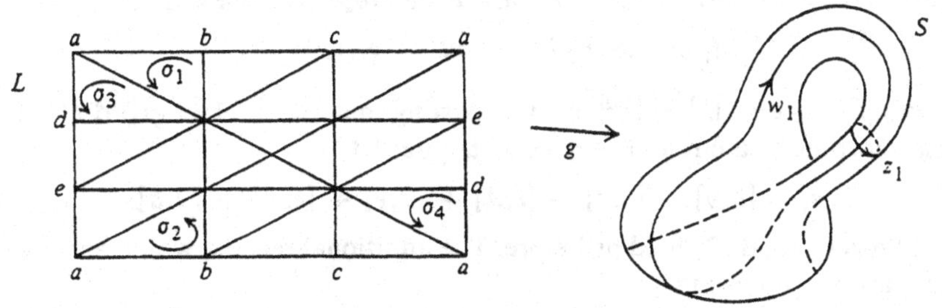

Figure 6.6

Proof. Let $g : |L| \to |S|$ be the pasting map. Let $A = g(|\text{Bd } L|)$; as before, it is the wedge of two circles. Orient the 2-simplices of S as before; let γ be their sum. Orient the 1-simplices of S arbitrarily. Note that (1) and (2) of the preceding proof hold; neither involve the particular identifications on the boundary. Because A is the wedge of two circles, (3) holds as well. The final condition is different, however; one has $\partial\gamma = 2z_1$.

This equation follows by direct computation. For example, $[a,b]$ appears in $\partial\sigma_1$ with coefficient -1 and in $\partial\sigma_2$ with coefficient $+1$, while $[a,d]$ appears in both $\partial\sigma_3$ and $\partial\sigma_4$ with coefficient $+1$.

Putting these facts together, we compute the homology of S: Every 1-cycle of S is homologous to a cycle of the form $c = nw_1 + mz_1$, by (1) and (3). If $c = \partial d$ for some d, then $d = p\gamma$ by (2); whence $\partial d = 2pz_1$. Thus $nw_1 + mz_1$ bounds if and only if m is even and n is zero. We conclude that

$$H_1(S) \simeq \mathbf{Z} \oplus \mathbf{Z}/2.$$

The cycle z_1 represents the torsion element, and w_1 represents a generator of the infinite cyclic group $H_1(S)/T_1(S)$.

To compute $H_2(S)$, note that any 2-cycle d of S must be of the form $p\gamma$ by (2); since $p\gamma$ is not a cycle, by (4), we have

$$H_2(S) = 0. \quad \square$$

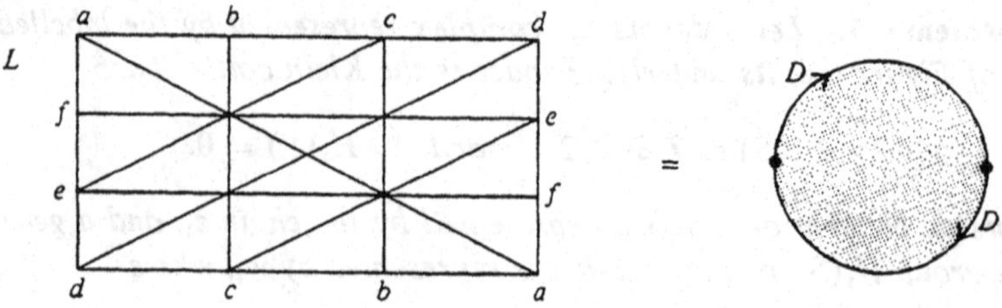

Figure 6.7

Theorem 6.4. *Let P^2 be the complex indicated by the labelled rectangle of Figure 6.7; its underlying space is called the* **projective plane.** *Then*

$$H_1(P^2) \simeq \mathbf{Z}/2 \quad \textit{and} \quad H_2(P^2) = 0.$$

Proof. Let $g: |L| \to |P^2|$ be the pasting map. Let $A = g(|\text{Bd } L|)$; it is homeomorphic to a circle. Let γ be as before; let

$$z_1 = [a,b] + [b,c] + [c,d] + [d,e] + [e,f] + [f,a].$$

Conditions (1) and (2) hold as before. The additional results, which are easy to verify, are the following:

(3) *Every 1-cycle carried by A is a multiple of z_1.*

(4) $\partial\gamma = -2z_1$.

From these facts, we conclude that

$$H_1(P^2) \simeq \mathbf{Z}/2 \quad \text{and} \quad H_2(P^2) = 0.$$

The non-zero element of H_1 is represented by the cycle z_1. □

Definition. The **connected sum** of P^2 with itself is defined as the space obtained from two copies of the projective plane by deleting a small open disc from each, and pasting together the pieces that remain, along their free edges. We denote this space by $P^2 \# P^2$.

The space $P^2 \# P^2$ can be represented as a quotient space of a rectangle, obtained by pasting its edges together in the manner indicated in Figure 6.8. (Note that if you cut the rectangle along the dotted line C, you have two projective planes with an open disc removed from each, as indicated in the figure.)

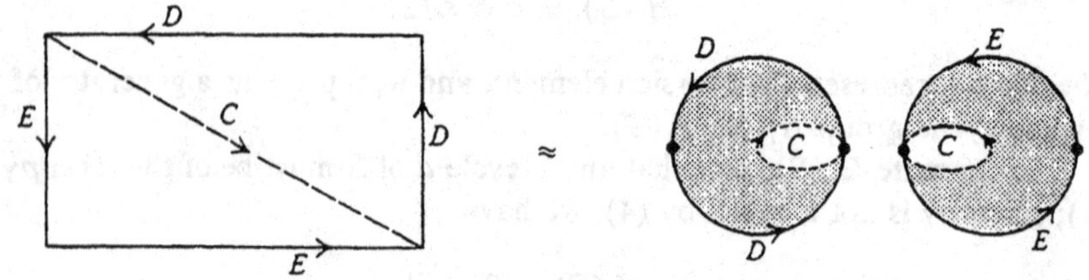

Figure 6.8

Theorem 6.5. *Let $P^2 \# P^2$ be the connected sum of two projective planes. Then*

$$H_1(P^2 \# P^2) \simeq \mathbf{Z} \oplus \mathbf{Z}/2 \quad \textit{and} \quad H_2(P^2 \# P^2) = 0.$$

Proof. We represent $P^2 \# P^2$ by the same rectangle L as before, with an appropriate vertex-labelling. In this case, the complex $A = g(|\text{Bd } L|)$ is again the wedge of two circles. Let w_1 be the 1-cycle "running across the top edge" and let z_1 be the 1-cycle "running down the left edge" in Figure 6.8, in the directions indicated. Conditions (1) and (2) of Theorem 6.2 hold. Conditions (3) and (4) are the following:

(3) *Every* 1-*cycle carried by* A *is of the form* $nw_1 + mz_1$.

(4) $\partial\gamma = 2w_1 + 2z_1$.

It is then clear that $H_2(P^2 \# P^2) = 0$. But how can one compute H_1? Some diddling is needed. We want to compute the quotient of the group G having w_1 and z_1 as basis, by the subgroup H generated by $2(w_1 + z_1)$. For this purpose, we need to choose bases for the two groups that are "compatible," as in our basic theorem on subgroups of free abelian groups (Theorem 4.2). In this case, it is easy to see what is needed: we need to choose $w_1 + z_1$ to be one of the basis elements for G. Can we do this? Of course; $\{w_1, w_1 + z_1\}$ will serve as a basis for G just as well as $\{w_1, z_1\}$ does. (One can express each set in terms of the other: $w_1 = w_1$ and $z_1 = (-(w_1) + (w_1 + z_1))$.) If we use this basis for G, computation is easy:

$$H_1(P^2 \# P^2) \simeq \mathbf{Z} \oplus \mathbf{Z}/2,$$

the torsion element is represented by $w_1 + z_1$, and w_1 represents a generator of H_1/T_1.

(We remark that w_1 is not the only cycle representing a generator of H_1/T_1. The cycle z_1 does just as well; so does the cycle $2w_1 + 3z_1$, as well as many others. For $\{z_1, w_1 + z_1\}$ and $\{2w_1 + 3z_1, w_1 + z_1\}$ are other bases for G, as you can check.) □

The astute reader might notice that the answers here are the same as for the Klein bottle. This is no accident, for the two spaces are in fact homeomorphic. Figure 6.9 indicates the proof.

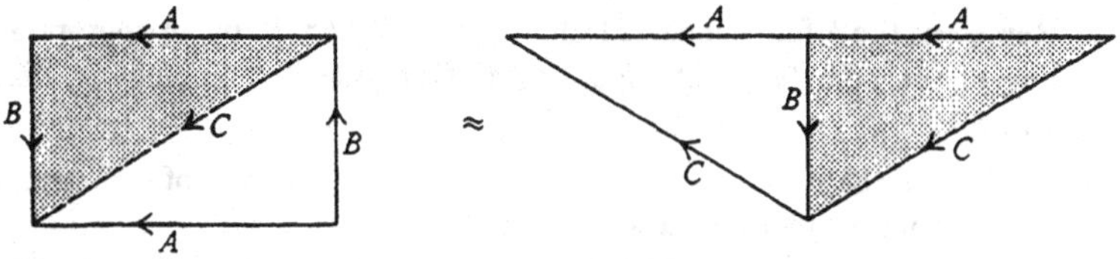

Figure 6.9

By now we have worked enough examples that you should be able to compute the homology groups of the general compact surface. We leave this computation to the exercises.

EXERCISES

1. Let w_1 and z_1 be the cycles on the Klein bottle pictured in Figure 6.6. Show that $w_1 + z_1$ represents a generator of the infinite cyclic group $H_1(S)/T_1(S)$.

2. The connected sum $T \# T$ of two tori is obtained by deleting an open disc from each of two disjoint tori and gluing together the pieces that remain, along their boundaries. It can be represented as a quotient space of an octagonal region in the plane by making identifications on the boundary as indicated in Figure 6.10. (Splitting this octagon along the dotted line gives two tori with open discs deleted.)

   (a) Construct a complex K whose underlying space is $T \# T$ by an appropriate vertex-labelling of the complex L pictured in Figure 6.11.

   (b) Compute the homology of $T \# T$ in dimensions 1 and 2 by following the pattern of Theorem 6.2. Specifically, let A be the image of Bd L under the quotient map; then A is a wedge of four circles. Orient each 2-simplex of L counterclockwise; let γ be the sum of the correspondingly oriented simplices of K. Show first that every 1-cycle of K is homologous to one carried by A. Then show that every 2-chain of K whose boundary is carried by A is a multiple of γ. Complete the computation by analyzing the 1-cycles carried by A, and by computing $\partial\gamma$.

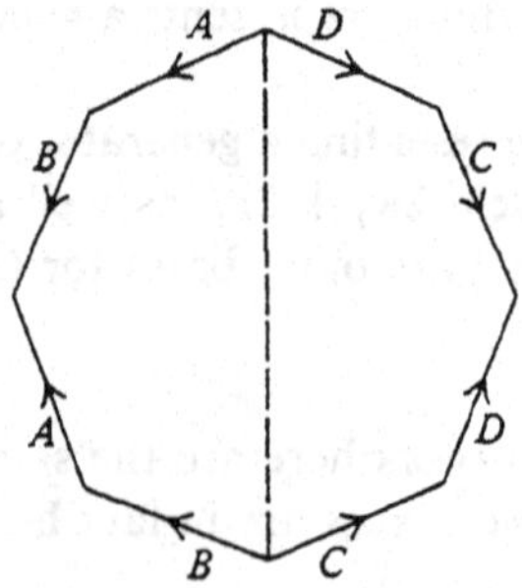

Figure 6.10

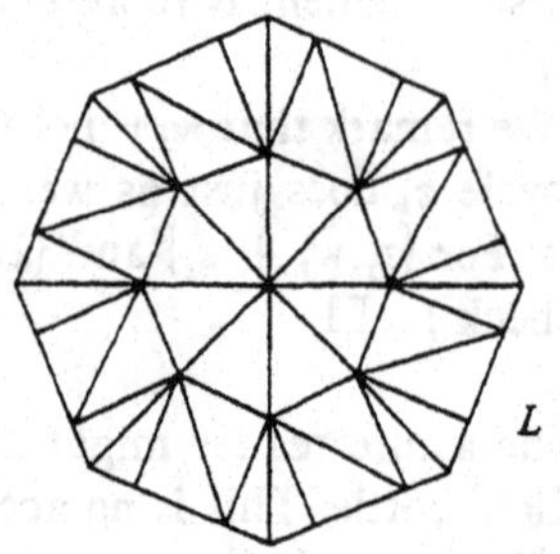

Figure 6.11

3. Represent the 4-fold connected sum $P^2 \# P^2 \# P^2 \# P^2$ by an appropriate labelling of the complex L of Figure 6.11. Compute its homology in dimensions 1 and 2.

4. (a) Define the n-fold connected sum $X_n = T \# \cdots \# T$ of tori, and compute its homology in dimensions 1 and 2.

   (b) Define the n-fold connected sum $Y_n = P^2 \# \cdots \# P^2$ of projective planes, and compute its homology in dimensions 1 and 2.

[It is a standard theorem that every compact surface is homeomorphic to one of the spaces in the following list:

$$S^2; \quad X_1, X_2, \ldots; \quad Y_1, Y_2, \ldots.$$

(See [Ma].) Once we have proved that the homology groups are topological invariants, it then follows from the computations of this exercise that no two of these surfaces are homeomorphic.]

5. Compute the homology of $T \# P^2$. To which of the surfaces listed in Exercise 4 must it be homeomorphic? Can you construct the homeomorphism?
6. (a) Compute the homology in dimensions 1 and 2 of the quotient space indicated in Figure 6.12. We call this space the "5-fold dunce cap."
   (b) Define analogously the "k-fold dunce cap" and compute its homology.
7. Compute the homology of the space indicated in Figure 6.13.
8. Given finitely generated abelian groups G_1 and G_2, with G_2 free, show there is a finite 2-dimensional complex K such that $|K|$ is connected, $H_1(K) \simeq G_1$, and $H_2(K) \simeq G_2$.

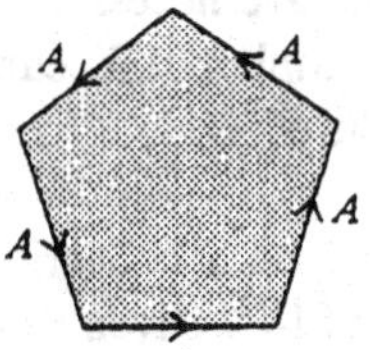

Figure 6.12

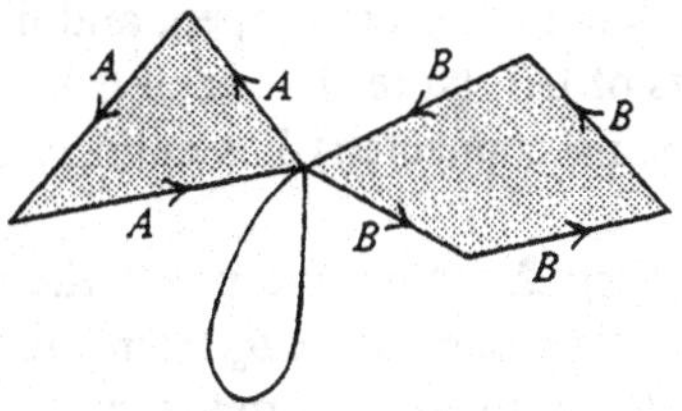

Figure 6.13

§7. ZERO-DIMENSIONAL HOMOLOGY

We have not yet computed any zero-dimensional homology group. In this section, we shall show that this group has a simple topological interpretation that makes its computation trivial.

We prove the following theorem:

Theorem 7.1. *Let K be a complex. Then the group $H_0(K)$ is free abelian. If $\{v_\alpha\}$ is a collection consisting of one vertex from each component of $|K|$, then the homology classes of the chains v_α form a basis for $H_0(K)$.*

Proof. Step 1. If v and w are vertices of K, let us define $v \sim w$ if there is a sequence

$$a_0, \ldots, a_n$$

of vertices of K such that $v = a_0$ and $w = a_n$, and $a_i a_{i+1}$ is a 1-simplex of K for each i. This relation is clearly an equivalence relation. Given v, define

$$C_v = \bigcup \{\text{St } w \mid w \sim v\}.$$

We show that the sets C_v are the components of $|K|$.

Note first that C_v is open because it is a union of open sets. Furthermore, $C_v = C_{v'}$ if $v \sim v'$.

Second, we show C_v is connected, in fact, path connected. Given v, let $w \sim v$ and let x be a point of St w. Choose a sequence $a_0, \ldots, a_n$ of vertices of K, as before. Then the broken line path with successive vertices $a_0, \ldots, a_n, x$ lies in C_v: For $a_i \sim v$ by definition, so that St $a_i \subset C_v$, and in particular, the line segment $a_i a_{i+1}$ lies in C_v. Similarly, the line segment $a_n x$ lies in St a_n, which is contained in C_v. Hence C_v is path connected.

Third, we show that distinct sets C_v and $C_{v'}$ are disjoint. Suppose x is a point of $C_v \cap C_{v'}$. Then $x \in$ St w for some w equivalent to v, and $x \in$ St w' for some w' equivalent to v'. Since x has positive barycentric coordinates with respect to both w and w', some simplex of K has w and w' as vertices. Then ww' must be a 1-simplex of K, so $w \sim w'$. It follows that $v \sim v'$, so that the two sets C_v and $C_{v'}$ are the same.

Being connected, open, and disjoint, the sets C_v are necessarily the components of $|K|$. Note that each is the space of a subcomplex of K; each simplex of K (being connected) lies entirely in one component of $|K|$.

Step 2. Now we prove the theorem. Let $\{v_\alpha\}$ be a collection of vertices containing one vertex v_α from each component C_α of $|K|$. Given a vertex w of K, it belongs to some component of $|K|$, say C_α. By hypothesis, $w \sim v_\alpha$, so there is a sequence $a_0, \ldots, a_n$ of vertices of K, as before, leading from v_α to w. The 1-chain

$$[a_0, a_1] + [a_1, a_2] + \cdots + [a_{n-1}, a_n]$$

has as its boundary the 0-chain $a_n - a_0 = w - v_\alpha$. Thus the 0-chain w is homologous to the 0-chain v_α. We conclude that every chain in K is homologous to a linear combination of the elementary 0-chains v_α.

We now show that no non-trivial chain of the form $c = \Sigma\, n_\alpha v_\alpha$ bounds. Suppose $c = \partial d$ for some 1-chain d. Since each 1-simplex of K lies in a unique component of $|K|$, we can write $d = \Sigma\, d_\alpha$, where d_α consists of those terms of d that are carried by C_α. Since $\partial d = \Sigma\, \partial d_\alpha$ and ∂d_α is carried by C_α, we conclude that $\partial d_\alpha = n_\alpha v_\alpha$. It follows that $n_\alpha = 0$ for each α. For let $\epsilon : C_0(K) \to \mathbf{Z}$ be the homomorphism defined by setting $\epsilon(v) = 1$ for each vertex v of K. Then $\epsilon(\partial[v,w]) = \epsilon(w - v) = 1 - 1 = 0$ for any elementary 1-chain $[v,w]$. As a result, $\epsilon(\partial d) = 0$ for every 1-chain d. In particular, $0 = \epsilon(\partial d_\alpha) = \epsilon(n_\alpha v_\alpha) = n_\alpha$. □

For some purposes, it is convenient to consider another version of 0-dimensional homology. We consider that situation now.

Definition. Let $\epsilon : C_0(K) \to \mathbf{Z}$ be the surjective homomorphism defined by $\epsilon(v) = 1$ for each vertex v of K. Then if c is a 0-chain, $\epsilon(c)$ equals the sum of the values of c on the vertices of K. The map ϵ is called an **augmentation map** for $C_0(K)$. We have just noted that $\epsilon(\partial d) = 0$ if d is a 1-chain. We define the **reduced homology group** of K in dimension 0, denoted $\tilde{H}_0(K)$, by the equation

$$\tilde{H}_0(K) = \ker \epsilon / \operatorname{im} \partial_1$$

(If $p > 0$, we let $\tilde{H}_p(K)$ denote the usual group $H_p(K)$.)

The relation between reduced and ordinary homology is as follows:

Theorem 7.2. *The group $\tilde{H}_0(K)$ is free abelian, and*

$$\tilde{H}_0(K) \oplus \mathbf{Z} \simeq H_0(K).$$

Thus $\tilde{H}_0(K)$ vanishes if $|K|$ is connected. If $|K|$ is not connected, let $\{v_\alpha\}$ consist of one vertex from each component of $|K|$; let α_0 be a fixed index. Then the homology classes of the chains $v_\alpha - v_{\alpha_0}$, for $\alpha \neq \alpha_0$, form a basis for $\tilde{H}_0(K)$.

Proof. Given a 0-chain c, it is homologous to a 0-chain of the form $c' = \Sigma\, n_\alpha v_\alpha$; and the chain c' bounds only if $n_\alpha = 0$ for all α. Now if $c \in \ker \epsilon$, then $\epsilon(c) = \epsilon(c') = \epsilon(\Sigma n_\alpha v_\alpha) = \Sigma n_\alpha = 0$. If $|K|$ has only one component, this means that $c' = 0$. If $|K|$ has more than one component, it implies that c' is a linear combination of the 0-chains $v_\alpha - v_{\alpha_0}$. □

EXERCISE

1. (a) Let G be an abelian group and let $\phi : G \to \mathbf{Z}$ be an epimorphism. Show that G has an infinite cyclic subgroup H such that

$$G = (\ker \phi) \oplus H.$$

[*Hint:* Define a homomorphism $\psi : \mathbf{Z} \to G$ such that $\phi \circ \psi$ is the identity; let $H = \operatorname{im} \psi$.]

(b) Show that if $\phi : C_0(K) \to \mathbf{Z}$ is any epimorphism such that $\phi \circ \partial_1 = 0$, then

$$H_0(K) \simeq (\ker \phi)/(\operatorname{im} \partial_1) \oplus \mathbf{Z}.$$

§8. THE HOMOLOGY OF A CONE

Now we compute the homology of the n-simplex and of its boundary. A convenient way of doing this is to introduce the notion of a *cone.*

Definition. Suppose that K is a complex in $\mathbf{E}^J$, and w is a point of $\mathbf{E}^J$ such that each ray emanating from w intersects $|K|$ in at most one point. We define the **cone on K with vertex** w to be the collection of all simplices of the form $wa_0 \dots a_p$, where $a_0 \dots a_p$ is a simplex of K, along with all faces of such simplices. We denote this collection $w * K$.

We show that $w * K$ is a well-defined complex, and it contains K as a subcomplex; K is often called the **base** of the cone.

First we show that the set $\{w, a_0, \dots, a_p\}$ is geometrically independent: If w were in the plane P determined by $a_0, \dots, a_p$, we could consider the line segment joining w to an interior point x of $\sigma = a_0 \dots a_p$. The set Int σ, being open in P, would contain an interval of points on this line segment. But the ray from w through x intersects $|K|$ in only one point, by hypothesis.

We now show that $w * K$ is a complex. The simplices of $w * K$ are of three types: simplices $a_0 \dots a_p$ of K, simplices of the form $wa_0 \dots a_p$, and the 0-simplex w. A pair of simplices of the first type have disjoint interiors because K is a complex. The open simplex Int$(wa_0 \dots a_p)$ is the union of all open line segments joining w to points of Int$(a_0 \dots a_p)$; no two such open simplices can intersect because no ray from w contains more than one point of $|K|$. For the same reason, simplices of the first and second types have disjoint interiors.

Example 1. If K_σ is the complex consisting of the simplex $\sigma = v_0 \dots v_n$ and its faces, then $K_\sigma = v_0 * K_s$, where s is the face of σ opposite v_0. Thus every simplex of positive dimension is a cone.

Example 2. If K is the complex in R^2 consisting of the intervals $[n, n+1] \times 0$ on the x-axis and their vertices, and if w is a point on the y-axis different from the origin, then $w * K$ is the complex illustrated in Figure 8.1. Although $|K|$ is a subspace of $\mathbf{R}^2$, $|w * K|$ is not a subspace of $\mathbf{R}^2$. (See Exercise 9 of §2.)

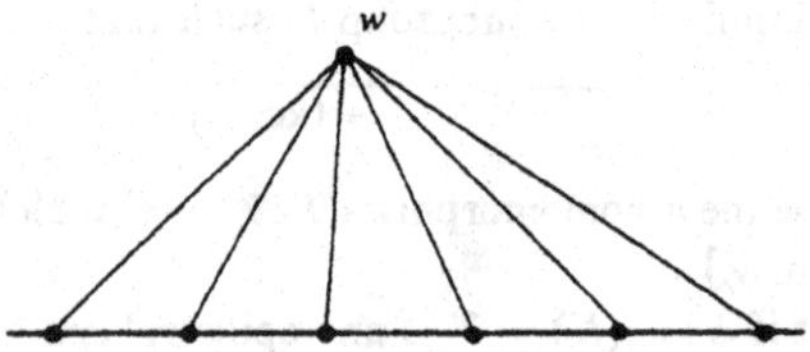

Figure 8.1

One particularly useful consequence of the cone construction is the following:

Lemma 8.1. *Let U be a bounded convex open set in $\mathbf{R}^n$; let $w \in U$. If K is a finite complex such that $|K| = \bar{U} - U$, then $w * K$ is a finite complex such that $|w * K| = \bar{U}$.*

Proof. It follows at once from Lemma 1.1 that each ray emanating from w intersects $|K|$ in precisely one point, and that $\bar{U}$ is the union of all line segments joining w to points of $|K|$. □

Given K, we note that any two cones $w * K$ and $z * K$ over K are isomorphic. The vertex map that carries each vertex of K to itself and carries w to z induces an isomorphism of $w * K$ with $z * K$.

Note also that for a complex K in $\mathbf{R}^N$, there may be no point w in $\mathbf{R}^N$ such that the cone complex $w * K$ can be formed. However, we can always consider K as a complex in $\mathbf{R}^N \times 0 \subset \mathbf{R}^{N+1}$; then the point $w = (0,\ldots,0,1) \in \mathbf{R}^{N+1}$ will do. A similar remark applies to a complex in $\mathbf{E}^J$.

Now we compute the homology of a cone, and show that it vanishes in positive dimensions. For this purpose, we shall introduce a certain bracket operation that will also be useful later.

Definition. Let $w * K$ be a cone. If $\sigma = [a_0,\ldots,a_p]$ is an oriented simplex of K, let $[w,\sigma]$ denote the oriented simplex $[w,a_0,\ldots,a_p]$ of $w * K$. This operation is well-defined; exchanging two vertices in the array $[a_0,\ldots,a_p]$ results in exchanging two vertices in the array $[w,a_0,\ldots,a_p]$. More generally, if

$$c_p = \Sigma n_i \sigma_i$$

is a p-chain of K, we define

$$[w,c_p] = \Sigma n_i [w,\sigma_i].$$

This **bracket operation** is a homomorphism carrying $C_p(K)$ into $C_{p+1}(w * K)$.

We compute readily from the boundary formula:

$$\partial [w,\sigma] = \begin{cases} \sigma - w & \text{if } \dim \sigma = 0, \\ \sigma - [w,\partial\sigma] & \text{if } \dim \sigma > 0. \end{cases}$$

This leads to the following more general formulas:

$$(*) \qquad \begin{aligned} \partial [w,c_0] &= c_0 - \epsilon(c_0)w, \\ \partial [w,c_p] &= c_p - [w,\partial c_p] \quad \text{if} \quad p > 0. \end{aligned}$$

Theorem 8.2. *If $w * K$ is a cone, then for all p,*

$$\tilde{H}_p(w * K) = 0.$$

In general, a complex whose reduced homology vanishes in all dimensions is said to be **acyclic.**

Proof. The reduced homology of $w * K$ vanishes in dimension 0, because $|w * K|$ is connected. Consider the case $p > 0$. Let z_p be a p-cycle of $w * K$; we show that z_p bounds. Let us write

$$z_p = c_p + [w,d_{p-1}],$$

where c_p consists of those terms of z_p that are carried by K, and d_{p-1} is a chain of K. We show that

$$z_p - \partial [w,c_p] = 0;$$

then our result is proved. By direct computation,

$$\begin{aligned} z_p - \partial[w,c_p] &= c_p + [w,d_{p-1}] - c_p + [w,\partial c_p] \\ &= [w,e_{p-1}], \end{aligned}$$

where $e_{p-1} = d_{p-1} + \partial c_p$ is a chain of K. Now since z_p is a cycle,

$$0 = \begin{cases} e_{p-1} - \epsilon(e_{p-1})w & \text{if } p = 1, \\ e_{p-1} - [w,\partial e_{p-1}] & \text{if } p > 1. \end{cases}$$

Now the portion of this chain carried by K is e_{p-1}; therefore, $e_{p-1} = 0$. We conclude that

$$z_p - \partial[w,c_p] = [w,e_{p-1}] = 0,$$

as desired. □

Theorem 8.3. *Let σ be an n-simplex. The complex K_σ consisting of σ and its faces is acyclic. If $n > 0$, let Σ^{n-1} denote the complex whose polytope is* Bd σ. *Orient σ. Then $\tilde{H}_{n-1}(\Sigma^{n-1})$ is infinite cyclic and is generated by the chain $\partial\sigma$; furthermore, $\tilde{H}_i(\Sigma^{n-1}) = 0$ for $i \neq n-1$.*

Proof. Because K_σ is a cone, it is acyclic. Let us compare the chain groups of K_σ and Σ^{n-1}; they are equal except in dimension n:

$$\begin{array}{ccccccc} C_n(K_\sigma) & \xrightarrow{\partial_n} & C_{n-1}(K_\sigma) & \xrightarrow{\partial_{n-1}} \cdots \to & C_0(K_\sigma) & \xrightarrow{\epsilon} & \mathbf{Z}. \\ & & \| & & \| & & \| \\ 0 & \xrightarrow{\partial'_n} & C_{n-1}(\Sigma^{n-1}) & \xrightarrow{\partial'_{n-1}} \cdots \to & C_0(\Sigma^{n-1}) & \to & \mathbf{Z}. \end{array}$$

It follows at once that $\tilde{H}_i(\Sigma^{n-1}) = \tilde{H}_i(K_\sigma) = 0$ for $i \neq n-1$. Let us compute the homology in dimension $n-1$. First take the case $n > 1$. One has

$$\begin{aligned} H_{n-1}(\Sigma^{n-1}) &= Z_{n-1}(\Sigma^{n-1}), \text{ because there are no } n-1 \text{ boundaries,} \\ &= \ker \partial_{n-1} \\ &= \operatorname{im} \partial_n, \text{ because } H_{n-1}(K_\sigma) = 0. \end{aligned}$$

Now $C_n(K_\sigma)$ is infinite cyclic and is generated by σ. Hence im ∂_n is cyclic and is generated by $\partial_n\sigma$; it is infinite because $C_{n-1}(K_\sigma)$ has no torsion.

The argument for $n = 1$ is similar, except that ∂_{n-1} is replaced by ϵ throughout. □

EXERCISE

1. Let K be a complex; let $w_0 * K$ and $w_1 * K$ be two cones on K whose polytopes intersect only in $|K|$.
   (a) Show that $(w_0 * K) \cup (w_1 * K)$ is a complex; it is called a **suspension** of K and denoted $S(K)$.

(b) Using the bracket notation, define $\phi : C_p(K) \to C_{p+1}(S(K))$ by the equation

$$\phi(c_p) = [w_0, c_p] - [w_1, c_p].$$

Show that ϕ induces a homomorphism

$$\phi_* : \tilde{H}_p(K) \to \tilde{H}_{p+1}(S(K)).$$

(c) Show ϕ_* is an isomorphism when K consists of the proper faces of a 2-simplex.

We will see later that ϕ_* is an isomorphism in general (see §25).

§9. RELATIVE HOMOLOGY

Suppose K_0 is a subcomplex of the complex K. In many of the applications of topology, it is convenient to consider what are called the *relative* homology groups of K modulo K_0. We introduce them briefly here and compute some examples, postponing a more complete discussion to Chapter 3.

If K_0 is a subcomplex of the complex K, then the chain group $C_p(K_0)$ can be considered to be a subgroup of the chain group $C_p(K)$ in the natural way. Formally, if c_p is a chain on K_0 (that is, a function on the oriented simplices of K_0), one extends it to a chain on K by letting its value be zero on each oriented p-simplex of K not in K_0. When we write c_p as a linear combination of oriented p-simplices of K_0, we need merely to "consider" these simplices as belonging to K in order to "consider" c_p as a chain of K.

Definition. If K_0 is a subcomplex of K, the quotient group $C_p(K)/C_p(K_0)$ is called the group of **relative chains of K modulo K_0**, and is denoted by $C_p(K,K_0)$.

Note that the group $C_p(K,K_0)$ is free abelian. Indeed, if we orient the p-simplices of K so as to obtain a basis for $C_p(K)$, the subcollection consisting of the oriented p-simplices of K_0 is a basis for $C_p(K_0)$. Then the quotient $C_p(K)/C_p(K_0)$ is free, for it has as basis all cosets of the form

$$\{\sigma_i\} = \sigma_i + C_p(K_0),$$

where σ_i is a p-simplex of K that is *not* in K_0.

The boundary operator $\partial : C_p(K_0) \to C_{p-1}(K_0)$ is just the restriction of the boundary operator on $C_p(K)$. We use the same symbol to denote both these homomorphisms, when no confusion will result. This homomorphism induces a homomorphism

$$C_p(K,K_0) \to C_{p-1}(K,K_0)$$

of the relative chain groups, which we also denote by ∂. As before, it satisfies the equation $\partial \circ \partial = 0$. We let

$$Z_p(K,K_0) = \ker \partial : C_p(K,K_0) \to C_{p-1}(K,K_0),$$
$$B_p(K,K_0) = \operatorname{im} \partial : C_{p+1}(K,K_0) \to C_p(K,K_0),$$
$$H_p(K,K_0) = Z_p(K,K_0)/B_p(K,K_0).$$

These groups are called, respectively, the group of **relative p-cycles**, the group of **relative p-boundaries**, and the **relative homology group** in dimension p, of K modulo K_0.

Note that a relative p-chain, which is a coset $c_p + C_p(K_0)$, is a relative cycle if and only if ∂c_p is carried by K_0. Furthermore, it is a relative boundary if and only if there is a $p + 1$ chain d_{p+1} of K such that $c_p - \partial d_{p+1}$ is carried by K_0.

Example 1. Let K consist of an n-simplex and its faces; let K_0 be the set of proper faces of K. Then the group $C_p(K,K_0)$ vanishes except when $p = n$, in which case it is infinite cyclic. It follows that

$$H_i(K,K_0) = 0 \quad \text{for} \quad i \neq n,$$
$$H_n(K,K_0) \simeq \mathbf{Z}.$$

Example 2. Let K be a complex and let K_0 consist of a single vertex v of K. Using the results of §7, one sees readily that $H_0(K,v)$ is free abelian; one obtains a basis for $H_0(K,v)$ by choosing one vertex from each component of $|K|$ other than the component containing v. Then $H_0(K,v) \simeq \tilde{H}_0(K)$.

It is not hard to show that $H_p(K,v) \simeq H_p(K)$ for $p > 0$; see the exercises.

Example 3. Let K be the complex indicated in Figure 9.1, whose underlying space is a square. Let K_0 be the subcomplex whose space is the boundary of the square. It is easy to see that the 2-chain $\Sigma m_i\sigma_i$ represents a relative cycle of K modulo K_0 if and only if $m_1 = m_2 = m_3 = m_4$. Since there are no boundaries in dimension 2,

$$H_2(K,K_0) \simeq \mathbf{Z}$$

and the chain $\gamma = \Sigma\sigma_i$ represents a generator.

We showed in Example 6 of §5 that any 1-chain c of K is homologous to a 1-chain c_3 carried by $K_0 \cup e_4$. Now if c represents a relative 1-cycle (so that ∂c is

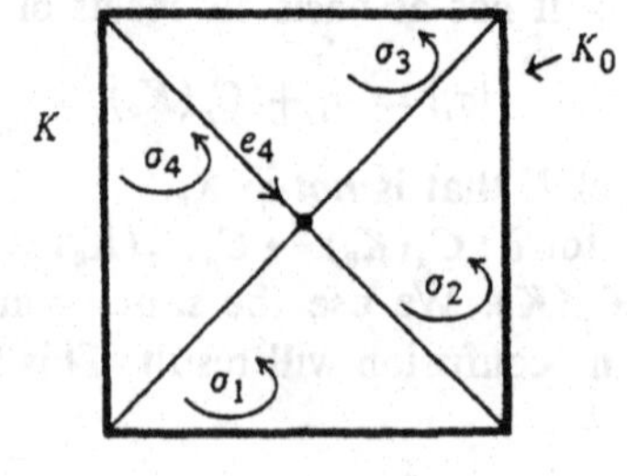

Figure 9.1

carried by K_0), $\partial c_3 = \partial c$ is also carried by K_0. It follows that the value of c_3 on e_4 must be zero, whence c_3 is actually carried by K_0. We conclude that

$$H_1(K,K_0) = 0.$$

Comparing this computation with that of Example 1 lends some plausibility to the statement (yet to be proved) that the relative homology groups are topological invariants.

Example 4. Let K be the complex indicated in Figure 9.2. Its underlying space is called an **annulus**. Let K_0 denote the 1-dimensional complex whose space is the union of the inner and outer edges of K. We compute the homology of K modulo K_0.

First, $H_0(K,K_0) = 0$ because the relative chain group itself vanishes in dimension 0. To compute H_1 and H_2, one first verifies three facts:

(i) If c is a 1-chain of K, then c is homologous to a 1-chain of K that is carried by the subcomplex M pictured in Figure 9.3.

(ii) Orient each 2-simplex of K counterclockwise. If d is any 2-chain of K such that ∂d is carried by M, then $d = m\gamma$, where γ is the sum of all the oriented 2-simplices of K.

(iii) $\partial\gamma$ is carried by K_0.

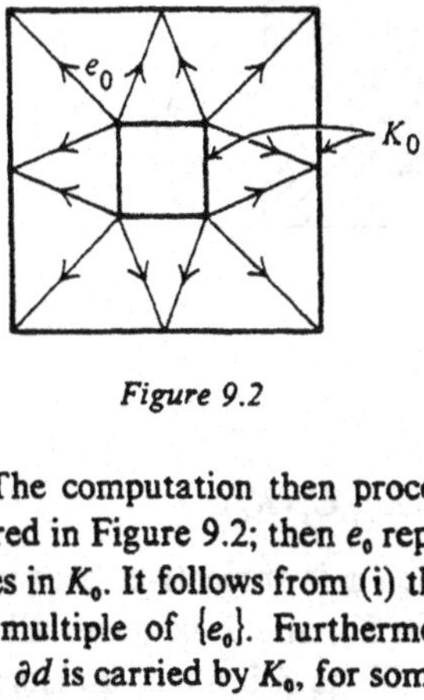

Figure 9.2

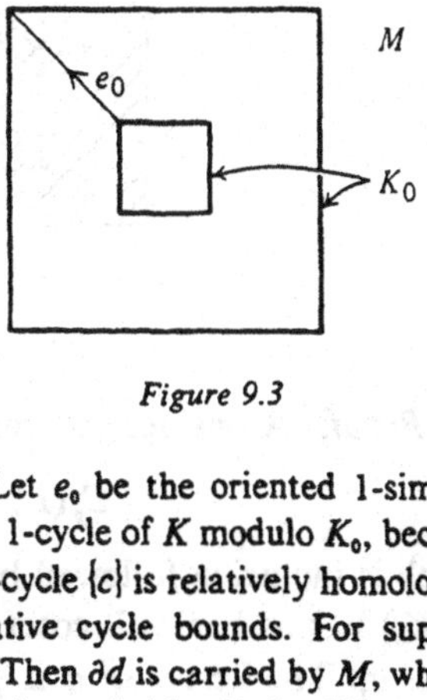

Figure 9.3

The computation then proceeds as follows: Let e_0 be the oriented 1-simplex pictured in Figure 9.2; then e_0 represents a relative 1-cycle of K modulo K_0, because ∂e_0 lies in K_0. It follows from (i) that *any* relative 1-cycle $\{c\}$ is relatively homologous to a multiple of $\{e_0\}$. Furthermore, no such relative cycle bounds. For suppose $ne_0 - \partial d$ is carried by K_0, for some 2-chain d of K. Then ∂d is carried by M, whence by (ii), $d = m\gamma$ for some integer m. But $\partial d = m\partial\gamma$ is carried by K_0, by (iii), so that $n = 0$. We conclude that

$$H_1(K,K_0) \simeq \mathbf{Z},$$

and the relative cycle e_0 represents a generator.

A similar argument, using (ii) and (iii), shows that

$$H_2(K,K_0) \simeq \mathbf{Z},$$

and the relative cycle γ represents a generator.

Students often visualize the relative homology group $H_p(K,K_0)$ as representing the homology of the quotient space $X = |K|/|K_0|$ obtained by collapsing

$|K_0|$ to a point p, modulo that point. Assuming X is homeomorphic to a polyhedron (so its simplicial homology is defined) this is in fact correct, but the proof is not easy. (See Lemma 70.1 and the exercises of §39.)

Roughly speaking, the relative homology group $H_p(K,K_0)$ depends only on the part of K lying outside or on the boundary of K_0; it "ignores" the part of K lying inside K_0. We express this fact formally in the following theorem:

Theorem 9.1 (Excision theorem). *Let K be a complex; let K_0 be a subcomplex. Let U be an open set contained in $|K_0|$, such that $|K| - U$ is the polytope of a subcomplex L of K. Let L_0 be the subcomplex of K whose polytope is $|K_0| - U$. Then inclusion induces an isomorphism*

$$H_p(L,L_0) \simeq H_p(K,K_0).$$

We think of $(|L|,|L_0|)$ as having been formed by "excising away" the open set U from $|K|$ and $|K_0|$. See Figure 9.4.

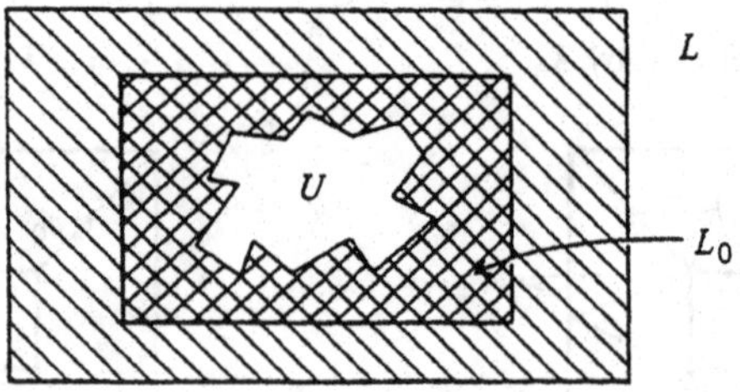

Figure 9.4

Proof. Consider the composite map ϕ,

$$C_p(L) \to C_p(K) \to C_p(K)/C_p(K_0),$$

which is inclusion followed by projection. Then ϕ is surjective, because $C_p(K)/C_p(K_0)$ has as basis all cosets $\{\sigma_i\}$ for σ_i not in K_0, and L contains all such simplices σ_i. The kernel of ϕ is precisely $C_p(L_0)$. Thus ϕ induces an isomorphism

$$C_p(L)/C_p(L_0) \simeq C_p(K)/C_p(K_0),$$

for all p. Since the boundary operator is preserved under this isomorphism, it follows that $H_p(L,L_0) \simeq H_p(K,K_0)$. □

This elementary fact has some useful consequences we shall consider later.

EXERCISES

1. Let K be the complex pictured in Figure 9.2; let K_1 be its "outer edge." Compute $H_i(K)$ and $H_i(K,K_1)$.

2. Let K be a complex whose underlying space is the Möbius band; let K_0 be its "edge." Compute $H_i(K)$ and $H_i(K,K_0)$. [*Hint:* See Example 2 of §3. Here the "edge" consists of the line segments ab, bc, cd, de, ef, and fa.]
3. Show that if K is a complex and v is a vertex of K, then $H_i(K,v) \simeq \tilde{H}_i(K)$ for all i. [*Hint:* Care is needed when $i = 1$.]
4. Describe $H_0(K,K_0)$ in general.
5. Let $|K|$ be the torus, represented by a labelled rectangle in the usual way; let K_0 be the subcomplex represented by the top edge of the rectangle. Compute $H_i(K,K_0)$. (See Figure 6.4; $|K_0|$ is the union of the line segments ab, bc, and ca.)
6. Let K be a 2-dimensional complex; let σ be a 2-simplex of K; let K_0 be the subcomplex whose space is $|K| - \operatorname{Int}\sigma$. Compute $H_i(K,K_0)$.

*§10. HOMOLOGY WITH ARBITRARY COEFFICIENTS

There is one further version of homology that we shall mention here, although we shall not study it in detail until Chapter 6. It arises when one introduces an arbitrary abelian group as "coefficient group."

Let G be an abelian group. Let K be a simplicial complex. A **p-chain of K with coefficients in G** is a function c_p from the oriented p-simplices of K to G that vanishes on all but finitely many p-simplices, such that

$$c_p(\sigma') = -c_p(\sigma)$$

if σ' and σ are opposite orientations of the same simplex. Two chains are added by adding their values. The resulting group is denoted $C_p(K;G)$.

If σ is an oriented simplex and if $g \in G$, we use $g\sigma$ to denote the elementary chain whose value is g on σ, $-g$ on the opposite orientation of σ, and 0 on all other oriented simplices. In this notation, $g(-\sigma) = (-g)\sigma$, where $-\sigma$ as usual denotes σ with the opposite orientation. If one orients all the p-simplices of K, then each chain c_p can be written uniquely as a finite sum

$$c_p = \Sigma g_i \sigma_i$$

of elementary chains. Thus $C_p(K;G)$ is the direct sum of subgroups isomorphic to G, one for each p-simplex of K.

The boundary operator $\partial : C_p(K;G) \rightarrow C_{p-1}(K;G)$ is defined easily by the formula

$$\partial(g\sigma) = g(\partial\sigma)$$

where $\partial\sigma$ is the ordinary boundary, defined earlier. As before, $\partial \circ \partial = 0$; and we define $Z_p(K;G)$ to be the kernel of the homomorphism

$$\partial : C_p(K;G) \rightarrow C_{p-1}(K;G),$$

$B_{p-1}(K;G)$ to be its image, and

$$H_p(K;G) = Z_p(K;G)/B_p(K;G).$$

These groups are called, respectively, the **cycles, boundaries,** and **homology of** K **with coefficients in** G.

Of course, one can also study *relative* homology with coefficients in G. The details are clear. The groups in question are denoted by $H_p(K,K_0;G)$.

We are not going to do much concerning homology with general coefficients for some time to come. But you should be aware of its existence at an early stage, for it is often useful.

Example 1. One group that is frequently used as a coefficient group is the group $\mathbf{Z}/2$ of integers mod 2. Let us calculate the homology of the torus and the Klein bottle using these coefficients. The argument given in §6 goes through essentially unchanged to show that

$$H_1(T;\mathbf{Z}/2) \simeq \mathbf{Z}/2 \oplus \mathbf{Z}/2,$$
$$H_2(T;\mathbf{Z}/2) \simeq \mathbf{Z}/2.$$

For the Klein bottle S, the argument goes through but with some changes. One has the results

$$H_1(S;\mathbf{Z}/2) \simeq \mathbf{Z}/2 \oplus \mathbf{Z}/2,$$
$$H_2(S;\mathbf{Z}/2) \simeq \mathbf{Z}/2.$$

For with $\mathbf{Z}/2$ coefficients, the basic 2-chain γ, which is the sum of the 2-simplices of S, has boundary *zero*. (In the group $\mathbf{Z}/2$, one has $2 = 1 + 1 = 0$.)

Note that homology with $\mathbf{Z}/2$ coefficients is inadequate to distinguish between T and S.

Example 2. Let us compute the homology of the torus and Klein bottle using the rational numbers $\mathbf{Q}$ as coefficients. The same arguments as before apply, but the end results are different. For the torus, one has

$$H_1(T;\mathbf{Q}) \simeq \mathbf{Q} \oplus \mathbf{Q}, \qquad H_2(T;\mathbf{Q}) \simeq \mathbf{Q}.$$

For the Klein bottle, one has

$$H_1(S;\mathbf{Q}) \simeq \mathbf{Q}, \qquad H_2(S;\mathbf{Q}) = 0.$$

For with $\mathbf{Q}$ coefficients, the cycle z_1 bounds the chain $\frac{1}{2}\gamma$.

EXERCISES

1. Compute the homology of P^2 with $\mathbf{Z}/2$ and $\mathbf{Q}$ coefficients.

2. Show that homology with $\mathbf{Q}$ coefficients suffices to distinguish among S^2 and the connected sums $P^2 \# \ldots \# P^2$ and $T \# \ldots \# T$.

3. Let S be the Klein bottle; compute the homology of S with $\mathbf{Z}/3$ and $\mathbf{Z}/4$ coefficients.

4. Compute the homology of the k-fold dunce cap with $\mathbf{Z}/n$ coefficients and with $\mathbf{Q}$ coefficients. (See Exercise 6 of §6.)

5. Let (K,K_0) be the pair consisting of the Möbius band and its edge. Compute $H_i(K,K_0;\mathbf{Z}/2)$ and $H_i(K,K_0;\mathbf{Q})$. (See Exercise 2 of §9.)

*§11. THE COMPUTABILITY OF HOMOLOGY GROUPS

We have computed the homology groups of some familiar spaces, such as the sphere and the torus and the Klein bottle. Now we ask the question whether one can in fact compute homology groups in general. For finite complexes, the answer is affirmative. In this section, we present an explicit algorithm for carrying out the computation.

First, we prove a basic theorem giving a "normal form" for homomorphisms of finitely generated free abelian groups. The proof is constructive in nature. One corollary is the theorem about subgroups of free abelian groups that we stated earlier as Theorem 4.2. A second corollary is a theorem concerning standard bases for free chain complexes. And a third corollary gives our desired algorithm for computing the homology groups of a finite complex.

First, we need two lemmas with which you might already be familiar.

Lemma 11.1. *Let A be a free abelian group of rank n. If B is a subgroup of A, then B is free abelian of rank $r \leq n$.*

Proof. We may without loss of generality assume that B is a subgroup of the n-fold direct product $\mathbf{Z}^n = \mathbf{Z} \times \cdots \times \mathbf{Z}$. We construct a basis for B as follows:

Let $\pi_i : \mathbf{Z}^n \to \mathbf{Z}$ denote projection on the ith coordinate. For each $m \leq n$, let B_m be the subgroup of B defined by the equation

$$B_m = B \cap (\mathbf{Z}^m \times \mathbf{0}).$$

That is, B_m consists of all $\mathbf{x} \in B$ such that $\pi_i(\mathbf{x}) = 0$ for $i > m$. In particular, $B_n = B$. Now the homomorphism

$$\pi_m : B_m \to \mathbf{Z}$$

carries B_m onto a subgroup of $\mathbf{Z}$. If this subgroup is trivial, let $\mathbf{x}_m = \mathbf{0}$; otherwise, choose $\mathbf{x}_m \in B_m$ so that its image $\pi_m(\mathbf{x}_m)$ generates this subgroup. We assert that the non-zero elements of the set $\{\mathbf{x}_1, \ldots, \mathbf{x}_n\}$ form a basis for B.

First, we show that for each m, the elements $\mathbf{x}_1, \ldots, \mathbf{x}_m$ generate B_m. (Then, in particular, the elements $\mathbf{x}_1, \ldots, \mathbf{x}_n$ generate B.) It is trivial that $\mathbf{x}_1$ generates B_1; indeed if d is the integer $\pi_1(\mathbf{x}_1)$, then

$$\mathbf{x}_1 = (d, 0, \ldots, 0)$$

and B_1 consists of all multiples of this element.

Assume that $\mathbf{x}_1, \ldots, \mathbf{x}_{m-1}$ generate B_{m-1}; let $\mathbf{x} \in B_m$. Now $\pi_m(\mathbf{x}) = k\pi_m(\mathbf{x}_m)$ for some integer k. It follows that

$$\pi_m(\mathbf{x} - k\mathbf{x}_m) = 0,$$

so that $\mathbf{x} - k\mathbf{x}_m$ belongs to B_{m-1}. Then

$$\mathbf{x} - k\mathbf{x}_m = k_1\mathbf{x}_1 + \cdots + k_{m-1}\mathbf{x}_{m-1}$$

by the induction hypothesis. Hence $\mathbf{x}_1, \ldots, \mathbf{x}_m$ generate B_m.

Second, we show that for each m, the non-zero elements in the set $\{\mathbf{x}_1, \ldots, \mathbf{x}_m\}$ are independent. The result is trivial when $m = 1$. Suppose it true for $m - 1$. Then we show that if

$$\lambda_1 \mathbf{x}_1 + \cdots + \lambda_m \mathbf{x}_m = \mathbf{0},$$

then it follows that for each i, $\lambda_i = 0$ whenever $\mathbf{x}_i \neq \mathbf{0}$; independence follows.

Applying the map π_m, we derive the equation

$$\lambda_m \pi_m(\mathbf{x}_m) = 0.$$

From this equation, it follows that either $\lambda_m = 0$ or $\mathbf{x}_m = \mathbf{0}$. For if $\lambda_m \neq 0$, then $\pi_m(\mathbf{x}_m) = 0$, whence the subgroup $\pi_m(B_m)$ is trivial and $\mathbf{x}_m = \mathbf{0}$ by definition. We conclude two things:

$$\lambda_m = 0 \quad \text{if} \quad \mathbf{x}_m \neq \mathbf{0},$$
$$\lambda_1 \mathbf{x}_1 + \cdots + \lambda_{m-1} \mathbf{x}_{m-1} = \mathbf{0}.$$

The induction hypothesis now applies to show that for $i < m$,

$$\lambda_i = 0 \quad \text{whenever} \quad \mathbf{x}_i \neq \mathbf{0}. \quad \square$$

For later use, we generalize this result to arbitrary free abelian groups:

Lemma 11.2. *If A is a free abelian group, any subgroup B of A is free.*

Proof. The proof given for the finite case generalizes, provided we assume that the basis for A is indexed by a well-ordered set J having a largest element. (And the well-ordering theorem, which is equivalent to the axiom of choice, tells us this assumption is justified.)

We begin by assuming A equals a direct sum of copies of $\mathbf{Z}$; that is, A equals the subgroup of the cartesian product $\mathbf{Z}^J$ consisting of all tuples $(n_\alpha)_{\alpha \in J}$ such that $n_\alpha = 0$ for all but finitely many α. Then we proceed as before.

Let B be a subgroup of A. Let B_β consist of those elements $\mathbf{x}$ of B such that $\pi_\alpha(\mathbf{x}) = 0$ for $\alpha > \beta$. Consider the subgroup $\pi_\beta(B_\beta)$ of $\mathbf{Z}$; if it is trivial define $\mathbf{x}_\beta = \mathbf{0}$, otherwise choose $\mathbf{x}_\beta \in B_\beta$ so $\pi_\beta(\mathbf{x}_\beta)$ generates the subgroup.

We show first that the set $\{\mathbf{x}_\alpha \mid \alpha \leq \beta\}$ generates B_β. This fact is trivial if β is the smallest element of J. We prove it in general by transfinite induction. Given $\mathbf{x} \in B_\beta$, we have

$$\pi_\beta(\mathbf{x}) = k\pi_\beta(\mathbf{x}_\beta)$$

for some integer k. Hence $\pi_\beta(\mathbf{x} - k\mathbf{x}_\beta) = 0$. Consider the set of those indices α for which $\pi_\alpha(\mathbf{x} - k\mathbf{x}_\beta) \neq 0$. (If there are none, $\mathbf{x} = k\mathbf{x}_\beta$ and we are through.) All of these indices are less than β, because $\mathbf{x}$ and $\mathbf{x}_\beta$ belong to B_β. Furthermore, this set of indices is *finite,* so it has a largest element γ, which is less than β. But this means that $\mathbf{x} - k\mathbf{x}_\beta$ belongs to B_γ, whence by the induction hypothesis, $\mathbf{x} - k\mathbf{x}_\beta$ can be written as a linear combination of elements $\mathbf{x}_\alpha$ with each $\alpha \leq \gamma$.

Second, we show that the non-zero elements in the set $\{\mathbf{x}_\alpha \mid \alpha \leq \beta\}$ are independent. Again, this fact is trivial if β is the smallest element of J. In general, suppose

$$\lambda_{\alpha_1}\mathbf{x}_{\alpha_1} + \cdots + \lambda_{\alpha_k}\mathbf{x}_{\alpha_k} + \lambda_\beta \mathbf{x}_\beta = \mathbf{0},$$

where $\alpha_i < \beta$. Applying π_β, we see that

$$\lambda_\beta \pi_\beta(\mathbf{x}_\beta) = 0.$$

As before, it follows that either $\lambda_\beta = 0$ or $\mathbf{x}_\beta = \mathbf{0}$. We conclude that

$$\lambda_\beta = 0 \quad \text{if} \quad \mathbf{x}_\beta \neq \mathbf{0},$$

and

$$\lambda_{\alpha_1}\mathbf{x}_{\alpha_1} + \cdots + \lambda_{\alpha_k}\mathbf{x}_{\alpha_k} = \mathbf{0}.$$

The induction hypothesis now implies that $\lambda_{\alpha_i} = 0$ whenever $\mathbf{x}_{\alpha_i} \neq \mathbf{0}$. □

We now prove our basic theorem. First we need a definition.

Definition. Let G and G' be free abelian groups with bases $a_1, \ldots, a_n$ and $a_1', \ldots, a_m'$, respectively. If $f: G \to G'$ is a homomorphism, then

$$f(a_j) = \sum_{i=1}^{m} \lambda_{ij} a_i'$$

for unique integers λ_{ij}. The matrix (λ_{ij}) is called the **matrix of** f relative to the given bases for G and G'.

Theorem 11.3. *Let G and G' be free abelian groups of ranks n and m, respectively; let $f: G \to G'$ be a homomorphism. Then there are bases for G and G' such that, relative to these bases, the matrix of f has the form*

$$B = \left[\begin{array}{ccc|c} b_1 & & 0 & \\ & \ddots & & 0 \\ 0 & & b_l & \\ \hline & 0 & & 0 \end{array}\right]$$

where $b_i \geq 1$ and $b_1 \mid b_2 \mid \cdots \mid b_l$.

This matrix is in fact uniquely determined by f (although the bases involved are not). It is called a **normal form** for the matrix of f.

Proof. We begin by choosing bases in G and G' arbitrarily. Let A be the matrix of f relative to these bases. We shall give shortly a procedure for modifying these bases so as to bring the matrix into the normal form described. It is called "the reduction algorithm." The theorem follows. □

Consider the following "elementary row operations" on an integer matrix A:

(1) Exchange row i and row k.

(2) Multiply row i by -1.

(3) Replace row i by (row i) + q(row k), where q is an integer and $k \neq i$.

Each of these operations corresponds to a change of basis in G'. The first corresponds to an exchange of a_i' and a_k'. The second corresponds to replacing a_i' by $-a_i'$. And the third corresponds to replacing a_k' by $a_k' - qa_i'$, as you can readily check.

There are three similar "column operations" on A that correspond to changes of basis in G.

We now show how to apply these six operations to an arbitrary matrix A so as to reduce it to our desired normal form. We assume A is not the zero matrix, since in that case the result is trivial.

Before we begin, we note the following fact: If c is an integer that divides each entry of the matrix A, and if B is obtained from A by applying any one of these elementary operations, then c also divides each entry of B.

The reduction algorithm

Given a matrix $A = (a_{ij})$ of integers, not all zero, let $\alpha(A)$ denote the smallest non-zero element of the set of numbers $|a_{ij}|$. We call a_{ij} a **minimal entry** of A if $|a_{ij}| = \alpha(A)$.

The reduction procedure consists of two steps. The first brings the matrix to a form where $\alpha(A)$ is as small as possible. The second reduces the dimensions of the matrix involved.

Step 1. We seek to modify the matrix by elementary operations so as to *decrease* the value of the function α. We prove the following:

If the number $\alpha(A)$ fails to divide some entry of A, then it is possible to decrease the value of α by applying elementary operations to A; and conversely.

The converse is easy. If the number $\alpha(A)$ divides each entry of A, then it will divide each entry of any matrix B obtained by applying elementary operations to A. In this situation, it is not possible to reduce the value of α by applying elementary operations.

To prove the result itself, we suppose a_{ij} is a minimal entry of A that fails to divide some entry of A. If the entry a_{ij} fails to divide some entry a_{kj} in its *column,* then we perform a division, writing

$$\frac{a_{kj}}{a_{ij}} = q + \frac{r}{a_{ij}},$$

where $0 < |r| < |a_{ij}|$. Signs do not matter here; q and r may be either positive or negative. We then replace (row k) of A by (row k) $-$ q(row i). The result is to replace the entry a_{kj} in the kth row and jth column of A by $a_{kj} - qa_{ij} = r$. The value of α for this new matrix is at most $|r|$, which is less than $\alpha(A)$.

A similar argument applies if a_{ij} fails to divide some entry in its *row.*

Finally, suppose a_{ij} divides each entry in its row and each entry in its column, but fails to divide the entry a_{st}, where $s \neq i$ and $t \neq j$. Consider the following four entries of A:

$$\begin{matrix} a_{ij} & \cdots & a_{it} \\ \vdots & & \vdots \\ a_{sj} & \cdots & a_{st} \end{matrix}$$

Because a_{ij} divides a_{sj}, we can by elementary operations bring the matrix to the form where the entries in these four places are as follows:

$$\begin{matrix} a_{ij} & \cdots & a_{it} \\ \vdots & & \vdots \\ 0 & \cdots & a_{st} + la_{it} \end{matrix}$$

If we then replace (row i) of this matrix by (row i) + (row s), we are back in the previous situation, where a_{ij} fails to divide some entry in its row.

Step 2. At the beginning of this step, we have a matrix A whose minimal entry divides every entry of A.

Apply elementary operations to bring a minimal entry of A to the upper left corner of the matrix and to make it positive. Because it divides all entries in its row and column, we can apply elementary operations to make all the other entries in its row and column into zeros. Note that at the end of this process, the entry in the upper left corner divides all entries of the matrix.

One now begins Step 1 again, applying it to the smaller matrix obtained by ignoring the first row and first column of our matrix.

Step 3. The algorithm terminates either when the smaller matrix is the zero matrix or when it disappears. At this point our matrix is in normal form. The only question is whether the diagonal entries $b_1, \ldots, b_l$ successively divide one another. But this is immediate. We just noted that at the end of the first application of Step 2, the entry b_1 in the upper left corner divides all entries of the matrix. This fact remains true as we continue to apply elementary operations. In particular, when the algorithm terminates, b_1 must divide each of $b_2, \ldots, b_l$.

A similar argument shows b_2 divides each of $b_3, \ldots, b_l$. And so on.

It now follows immediately from Exercise 4 of §4 that the numbers $b_1, \ldots, b_l$ are uniquely determined by the homomorphism f. For the number l of non-zero entries in the matrix is just the rank of the free abelian group $f(G) \subset G'$. And those numbers b_i that are greater than 1 are just the torsion coefficients $t_1, \ldots, t_k$ of the quotient group $G'/f(G)$.

Applications of the reduction algorithm

Now we prove the basic theorem concerning subgroups of free abelian groups, which we stated in §4.

Proof of Theorem 4.2. Given a free abelian group F of rank n, we know from Lemma 11.1 that any subgroup R is free of rank $r \leq n$. Consider the *inclusion* homomorphism $j : R \to F$, and choose bases $a_1, \ldots, a_r$ for R and $e_1, \ldots, e_n$ for F relative to which the matrix of j is in the normal form of the preceding theorem. Because j is a monomorphism, this normal form has no zero columns. Thus $j(a_i) = b_i e_i$ for $i = 1, \ldots, r$, where $b_i \geq 1$ and $b_1 \mid b_2 \mid \cdots \mid b_r$. Since $j(a_i) = a_i$, it follows that $b_1 e_1, \ldots, b_r e_r$ is a basis for R. □

Now we prove the "standard basis theorem" for free chain complexes.

Definition. A **chain complex** $\mathcal{C}$ is a sequence

$$\cdots \longrightarrow C_{p+1} \xrightarrow{\partial_{p+1}} C_p \xrightarrow{\partial_p} C_{p-1} \longrightarrow \cdots$$

of abelian groups C_i and homomorphisms ∂_i, indexed with the integers, such that $\partial_p \circ \partial_{p+1} = 0$ for all p. The pth **homology group** of $\mathcal{C}$ is defined by the equation

$$H_p(\mathcal{C}) = \ker \partial_p / \operatorname{im} \partial_{p+1}.$$

If $H_p(\mathcal{C})$ is finitely generated, its betti number and torsion coefficients are called the betti number and torsion coefficients of $\mathcal{C}$ in dimension p.

Theorem 11.4 (Standard bases for free chain complexes). *Let $\{C_p, \partial_p\}$ be a chain complex; suppose each group C_p is free of finite rank. Then for each p there are subgroups U_p, V_p, W_p of C_p such that*

$$C_p = U_p \oplus V_p \oplus W_p,$$

where $\partial_p(U_p) \subset W_{p-1}$ and $\partial_p(V_p) = 0$ and $\partial_p(W_p) = 0$. Furthermore, there are bases for U_p and W_{p-1} relative to which $\partial_p : U_p \to W_{p-1}$ has a matrix of the form

$$B = \begin{bmatrix} b_1 & & 0 \\ & \ddots & \\ 0 & & b_l \end{bmatrix},$$

where $b_i \geq 1$ and $b_1 \mid b_2 \mid \cdots \mid b_l$.

Proof. Step 1. Let

$$Z_p = \ker \partial_p \quad \text{and} \quad B_p = \operatorname{im} \partial_{p+1}.$$

Let W_p consist of all elements c_p of C_p such that some non-zero multiple of c_p

belongs to B_p. It is a subgroup of C_p, and is called the group of **weak boundaries.** Clearly

$$B_p \subset W_p \subset Z_p \subset C_p.$$

(The second inclusion uses the fact that C_p is torsion-free, so that the equation $mc_p = \partial_{p+1} d_{p+1}$ implies that $\partial_p c_p = 0$.) We show that W_p is a direct summand in Z_p.

Consider the natural projection

$$Z_p \to H_p(\mathcal{C}) \to H_p(\mathcal{C})/T_p(\mathcal{C}),$$

where $T_p(\mathcal{C})$ is the torsion subgroup of $H_p(\mathcal{C})$. The kernel of this projection is W_p; therefore, $Z_p/W_p \simeq H_p/T_p$. The latter group is finitely generated and torsion-free, so it is free. If $c_1 + W_p, \ldots, c_k + W_p$ is a basis for Z_p/W_p, and $d_1, \ldots, d_l$ is a basis for W_p, then it is straightforward to check that $c_1, \ldots, c_k, d_1, \ldots, d_l$ is a basis for Z_p. Then $Z_p = V_p \oplus W_p$, where V_p is the group with basis $c_1, \ldots, c_k$.

Step 2. Suppose we choose bases $e_1, \ldots, e_n$ for C_p, and $e_1', \ldots, e_m'$ for C_{p-1}, relative to which the matrix of $\partial_p : C_p \to C_{p-1}$ has the normal form

$$\begin{array}{c} \\ e_1' \\ \vdots \\ e_l' \\ e_{l+1}' \\ \vdots \\ e_m' \end{array}
\begin{array}{c} \begin{array}{ccc|ccc} e_1 & \cdots & e_l & e_{l+1} & \cdots & e_n \end{array} \\
\left[\begin{array}{ccc|c} b_1 & & 0 & \\ & \ddots & & 0 \\ 0 & & b_l & \\ \hline & & & \\ & 0 & & 0 \\ & & & \end{array}\right] \end{array}$$

where $b_i \geq 1$ and $b_1 \mid b_2 \mid \cdots \mid b_l$. Then the following hold:

(1) $e_{l+1}, \ldots, e_n$ is a basis for Z_p.

(2) $e_1', \ldots, e_l'$ is a basis for W_{p-1}.

(3) $b_1 e_1', \ldots, b_l e_l'$ is a basis for B_{p-1}.

We prove these results as follows: Let c_p be the general p-chain. We compute its boundary; if

$$c_p = \sum_{i=1}^{n} a_i e_i, \qquad \text{then} \qquad \partial_p c_p = \sum_{i=1}^{l} a_i b_i e_i'.$$

To prove (1), we note that since $b_i \neq 0$, the p-chain c_p is a cycle if and only if $a_i = 0$ for $i = 1, \ldots, l$. To prove (3), we note that any $p-1$ boundary $\partial_p c_p$ lies in the group generated by $b_1 e_1', \ldots, b_l e_l'$; since $b_i \neq 0$, these elements are independent. Finally, we prove (2). Note first that each of $e'_1,\ldots,e'_l$ belongs to W_{p-1}, since $b_i e'_i = \partial e_i$. Conversely, let

$$c_{p-1} = \sum_{i=1}^{m} d_i e'_i$$

be a $p-1$ chain and suppose $c_{p-1} \in W_{p-1}$. Then c_{p-1} satisfies an equation of the form

$$\lambda c_{p-1} = \partial_p c_p = \sum_{i=1}^{l} a_i b_i e'_i$$

for some $\lambda \neq 0$. Equating coefficients, we see that $\lambda d_i = 0$ for $i > l$, whence $d_i = 0$ for $i > l$. Thus $e'_1,\ldots,e'_l$ is a basis for W_{p-1}.

Step 3. We prove the theorem. Choose bases for C_p and C_{p-1} as in Step 2. Define U_p to be the group spanned by $e_1,\ldots,e_l$; then

$$C_p = U_p \oplus Z_p.$$

Using Step 1, choose V_p so that $Z_p = V_p \oplus W_p$. Then we have a decomposition of C_p such that $\partial_p(V_p) = 0$ and $\partial_p(W_p) = 0$. The existence of the desired bases for U_p and W_{p-1} follows from Step 2. □

Note that W_p and $Z_p = V_p \oplus W_p$ are uniquely determined subgroups of C_p. The subgroups U_p and V_p are not uniquely determined, however.

Theorem 11.5. *The homology groups of a finite complex K are effectively computable.*

Proof. By the preceding theorem, there is a decomposition

$$C_p(K) = U_p \oplus V_p \oplus W_p$$

where $Z_p = V_p \oplus W_p$ is the group of p-cycles and W_p is the group of weak p-boundaries. Now

$$H_p(K) = Z_p/B_p \simeq V_p \oplus (W_p/B_p) \simeq (Z_p/W_p) \oplus (W_p/B_p).$$

The group Z_p/W_p is free and the group W_p/B_p is a torsion group; computing $H_p(K)$ thus reduces to computing these two groups.

Let us choose bases for the chain groups $C_p(K)$ by orienting the simplices of K, once and for all. Then consider the matrix of the boundary homomorphism $\partial_p : C_p(K) \to C_{p-1}(K)$ relative to this choice of bases; the entries of this matrix will in fact have values in the set $\{0,1,-1\}$. Using the reduction algorithm described earlier, we reduce this matrix to normal form. Examining Step 2 of the preceding proof, we conclude from the results proved there the following facts about this normal form:

(1) The rank of Z_p equals the number of zero columns.

(2) The rank of W_{p-1} equals the number of non-zero rows.

(3) There is an isomorphism

$$W_{p-1}/B_{p-1} \simeq \mathbf{Z}/b_1 \oplus \mathbf{Z}/b_2 \oplus \cdots \oplus \mathbf{Z}/b_l.$$

Thus the normal form for the matrix of $\partial_p : C_p \to C_{p-1}$ gives us the torsion coefficients of K in dimension $p - 1$; they are the entries of the matrix that are greater than 1. This normal form also gives us the rank of Z_p. On the other hand, the normal form for $\partial_{p+1} : C_{p+1} \to C_p$ gives us the rank of W_p. The difference of these numbers is the rank of Z_p/W_p—that is, the betti number of K in dimension p. □

EXERCISES

1. Show that the reduction algorithm is not needed if one wishes merely to compute the betti numbers of a finite complex K; instead all that is needed is an algorithm for determining the rank of a matrix. Specifically, show that if A_p is the matrix of $\partial_p : C_p(K) \to C_{p-1}(K)$ relative to some choice of basis, then

$$\beta_p(K) = \operatorname{rank} C_p(K) - \operatorname{rank} A_p - \operatorname{rank} A_{p+1}.$$

2. Compute the homology groups of the quotient space indicated in Figure 11.1. [*Hint:* First check whether all the vertices are identified.]

3. Reduce to normal form the matrix

$$\begin{bmatrix} 2 & 6 & 4 \\ 4 & -7 & 4 \\ 4 & 8 & 4 \end{bmatrix}.$$

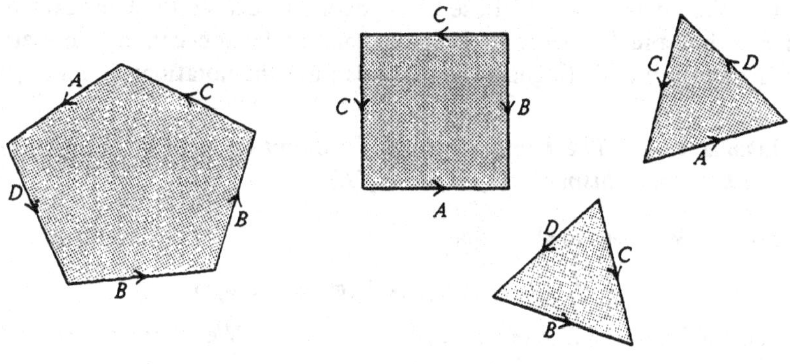

Figure 11.1

§12. HOMOMORPHISMS INDUCED BY SIMPLICIAL MAPS

If f is a simplicial map of $|K|$ into $|L|$, then f maps each p-simplex σ_i of K onto a simplex τ_i of L of the same or lower dimension. We shall define a homomorphism of p-chains that carries a formal sum $\Sigma m_i\sigma_i$ of oriented p-simplices of K onto the formal sum $\Sigma m_i\tau_i$ of their images. (We delete from the latter sum those simplices τ_i whose dimension is less than p.) This map in turn induces a homomorphism of homology groups, as we shall see.

As a general notation, we shall use the phrase

"$f : K \to L$ is a simplicial map"

to mean that f is a continuous map of $|K|$ into $|L|$ that maps each simplex of K linearly onto a simplex of L. Thus f maps each vertex of K to a vertex of L, and it equals the simplicial map induced by this vertex map, as defined in §2.

Definition. Let $f : K \to L$ be a simplicial map. If $v_0 \dots v_p$ is a simplex of K, then the points $f(v_0), \dots, f(v_p)$ span a simplex of L. We define a homomorphism $f_\# : C_p(K) \to C_p(L)$ by defining it on oriented simplices as follows:

$$f_\#([v_0, \dots, v_p]) = \begin{cases} [f(v_0), \dots, f(v_p)] & \text{if } f(v_0), \dots, f(v_p) \text{ are distinct,} \\ 0 & \text{otherwise.} \end{cases}$$

This map is clearly well-defined; exchanging two vertices in the expression $[v_0, \dots, v_p]$ changes the sign of the right side of the equation. The family of homomorphisms $\{f_\#\}$, one in each dimension, is called the **chain map induced by the simplicial map** f.

Properly speaking, one should use dimensional subscripts to distinguish these homomorphisms, denoting the map in dimension p by $(f_\#)_p : C_p(K) \to C_p(L)$. Normally, however, we shall omit the subscript, relying on the context to make the situation clear, just as we do with the boundary operator ∂.

In a similar vein, we shall use the symbol ∂ to denote the boundary operators in both K and L, in order to keep the notation from becoming cumbersome. If it is necessary to distinguish them, we can use the notations ∂_K and ∂_L.

Lemma 12.1. *The homomorphism $f_\#$ commutes with ∂; therefore $f_\#$ induces a homomorphism $f_* : H_p(K) \to H_p(L)$.*

Proof. We need to show that

$$(*) \qquad \partial f_\#([v_0, \dots, v_p]) = f_\#(\partial [v_0, \dots, v_p]).$$

Let τ be the simplex of L spanned by $f(v_0), \dots, f(v_p)$. We consider various cases.

Case 1. $\dim \tau = p$. In this case, the vertices $f(v_0), \dots, f(v_p)$ are distinct and the result follows at once from the definitions of $f_\#$ and ∂.

Case 2. $\dim \tau \le p - 2$. In this case, the left side of (*) vanishes because $f(v_0), \dots, f(v_p)$ are not distinct, and the right side vanishes because for each i, at least two of the points $f(v_0), \dots, f(v_{i-1}), f(v_{i+1}), \dots, f(v_p)$ are the same.

Case 3. $\dim \tau = p - 1$. In this case, we may assume the vertices so ordered that $f(v_0) = f(v_1)$, and $f(v_1), \dots, f(v_p)$ are distinct. Then the left side of (*) vanishes. The right side has only two non-zero terms; it equals

$$[f(v_1), f(v_2), \dots, f(v_p)] - [f(v_0), f(v_2), \dots, f(v_p)].$$

Since $f(v_0) = f(v_1)$, these terms cancel each other, as desired.

The homomorphism $f_\#$ carries cycles to cycles, since the equation $\partial c_p = 0$ implies that $\partial f_\#(c_p) = f_\#(\partial c_p) = 0$. And $f_\#$ carries boundaries to boundaries, since the equation $c_p = \partial d_{p+1}$ implies that $f_\#(c_p) = f_\#(\partial d_{p+1}) = \partial f_\#(d_{p+1})$. Thus $f_\#$ induces a homomorphism $f_* : H_p(K) \to H_p(L)$ of homology groups. □

Theorem 12.2. (a) *Let $i : K \to K$ be the identity simplicial map. Then $i_* : H_p(K) \to H_p(K)$ is the identity homomorphism.*

(b) *Let $f : K \to L$ and $g : L \to M$ be simplicial maps. Then $(g \circ f)_* = g_* \circ f_*$; that is, the following diagram commutes:*

$$\begin{array}{ccc} H_p(K) & \xrightarrow{(g \circ f)_*} & H_p(M) \\ & {\scriptstyle f_*} \searrow \quad \nearrow {\scriptstyle g_*} & \\ & H_p(L) & \end{array}$$

Proof. It is immediate from the definition that $i_\#$ is the identity and $(g \circ f)_\# = g_\# \circ f_\#$, as you can check. The theorem follows. □

This theorem expresses what are called the "functorial properties" of the induced homomorphism. This phrase will be defined formally later when we discuss *categories* and *functors*. For the present, we point out simply that the operator H_p assigns to each simplicial complex an abelian group, and the operator $*$ assigns to each simplicial map of one complex into another, a homomorphism of the corresponding abelian groups. Because (a) and (b) hold, we say that $(H_p, *)$ is a "functor" from the "category" of simplicial complexes and simplicial maps to the "category" of abelian groups and homomorphisms.

Lemma 12.3. *The chain map $f_\#$ preserves the augmentation map ϵ; therefore, it induces a homomorphism f_* of reduced homology groups.*

Proof. Let $f : K \to L$ be simplicial. Then $\epsilon f_\#(v) = 1$ and $\epsilon(v) = 1$ for each vertex v of K. Thus $\epsilon \circ f_\# = \epsilon$. This equation implies that $f_\#$ carries the kernel of $\epsilon_K : C_0(K) \to \mathbf{Z}$ into the kernel of $\epsilon_L : C_0(L) \to \mathbf{Z}$, and thus induces a homomorphism $f_* : \tilde{H}_0(K) \to \tilde{H}_0(L)$. □

Example 1. Consider the complexes K and T indicated in Figure 12.1. Their underlying spaces are the circle and torus, respectively. Now $H_1(K) \simeq \mathbf{Z}$; let us

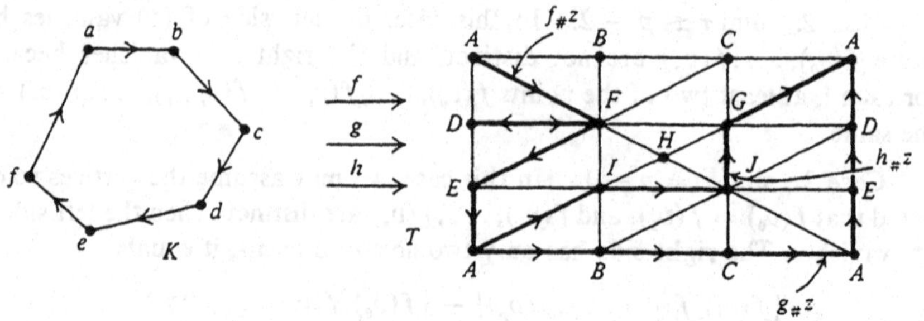

Figure 12.1

use the cycle z indicated in the figure as a generator. Similarly, $H_1(T) \simeq \mathbf{Z} \oplus \mathbf{Z}$; let us use the cosets of the cycles $w_1 = [A,B] + [B,C] + [C,A]$ and $z_1 = [A,D] + [D,E] + [E,A]$ as a basis.

Now consider the simplicial maps:

$f: a \to A$	$g: a \to A$	$h: a \to A$
$b \to F$	$b \to B$	$b \to I$
$c \to D$	$c \to C$	$c \to J$
$d \to D$	$d \to A$	$d \to G$
$e \to F$	$e \to E$	$e \to G$
$f \to E$	$f \to D$	$f \to A$

You can check that $f_\#(z)$ is homologous to z_1, that $g_\#(z)$ equals $w_1 - z_1$, and that $h_\#(z)$ is homologous to $w_1 - z_1$. Thus g_* and h_* are equal as homomorphisms of 1-dimensional homology. They are also equal on 0-dimensional homology.

In general, a given homomorphism can be induced by quite different simplicial maps, as the preceding example shows. This fact leads us to consider the general question: Under what conditions do two simplicial maps induce the same homomorphism of homology groups? Answering this question involves an important technique, one we shall use many times. So we begin by explaining the underlying motivation.

Given simplicial maps $f,g: K \to L$, we wish to find conditions under which $f_\#(z)$ and $g_\#(z)$ are homologous for each cycle $z \in Z_p(K)$. Said differently, we want to find conditions under which there is a function (commonly called D) that assigns to each p-cycle z of K, a $p + 1$ chain Dz of L, such that

$$\partial Dz = g_\#(z) - f_\#(z).$$

Let us consider an example in which this is possible. Suppose K is the boundary of a triangle, and L consists of the sides of a triangular prism, as pictured in Figure 12.2. Suppose f and g are the simplicial maps that carry K onto the two ends of the prism, respectively. If z is the 1-cycle generating $H_1(K)$, as indicated in the figure, it is quite easy to find a 2-chain Dz whose boundary is

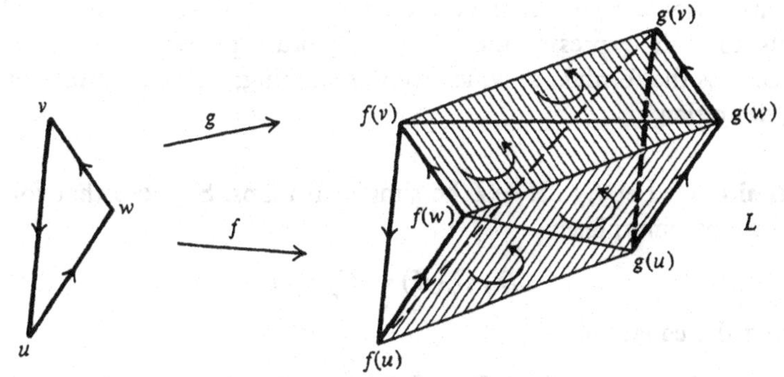

Figure 12.2

$g_\#(z) - f_\#(z)$; we let Dz equal the sum of all the 2-simplices of L, oriented appropriately. You can check that this 2-chain does the job.

In general, it is awkward to define D just for cycles, since that requires us to compute the group of cycles. What proves more satisfactory is to define D as a homomorphism on the entire chain group, because then one can define D one simplex at a time. How might that procedure work in the present example? It is fairly clear how to proceed. Given a vertex v of K, we define Dv to be the edge of L that leads from $f(v)$ to $g(v)$, as indicated in Figure 12.3. And given the oriented 1-simplex σ of K, we define $D\sigma$ to be the sum of the two oriented simplices that are heavily shaded in the figure, which form one side of the prism. We proceed similarly for the other simplices of K. Since z is the sum of the oriented edges of K, the chain Dz will be the sum of the oriented 2-simplices of L, just as before, and $\partial Dz = g_\#(z) - f_\#(z)$, as desired.

One can ask what sort of formula holds for ∂Dc when c is an arbitrary chain. The answer is clear from the definition. Given the oriented simplex σ, we compute $\partial D\sigma = g_\#(\sigma) - Dv - f_\#(\sigma) + Dw$; see Figure 12.3 for verification. That is,

(*) $$\partial(D\sigma) = g_\#(\sigma) - f_\#(\sigma) - D(\partial\sigma).$$

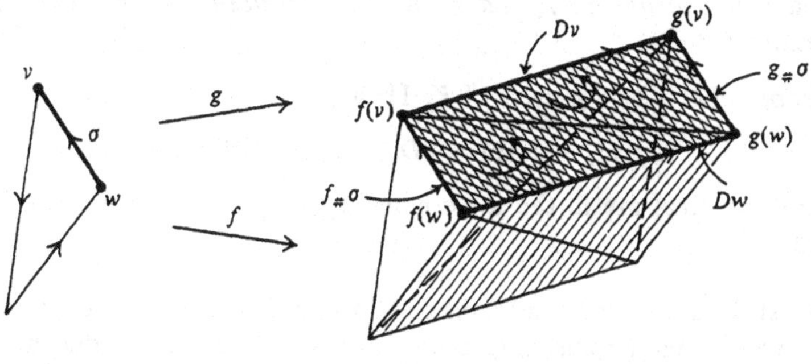

Figure 12.3

This same formula holds for the other simplices of K, as you can check.

This formula expresses the crucial algebraic property of the homomorphism D. It is the formula at which we were aiming; we make it part of the following definition.

Definition. Let $f,g : K \to L$ be simplicial maps. Suppose that for each p, one has a homomorphism

$$D : C_p(K) \to C_{p+1}(L)$$

satisfying the equation

$$\partial D + D\partial = g_\# - f_\#.$$

Then D is said to be a **chain homotopy** between $f_\#$ and $g_\#$.

We have omitted dimensional subscripts in this formula; the following diagram may make the maps involved clearer:

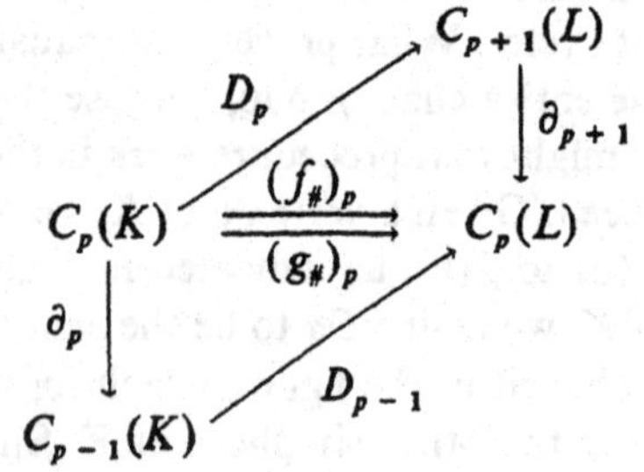

With subscripts, the formula becomes

$$\partial_{p+1}D_p + D_{p-1}\partial_p = (g_\#)_p - (f_\#)_p.$$

This formula is more precise, but messier. We shall customarily omit dimensional subscripts.

The importance of chain homotopies comes from the following theorem:

Theorem 12.4. *If there is a chain homotopy between $f_\#$ and $g_\#$, then the induced homomorphisms f_* and g_*, for both reduced and ordinary homology, are equal.*

Proof. Let z be a p-cycle of K. Then

$$g_\#(z) - f_\#(z) = \partial Dz + D\partial z = \partial Dz + 0,$$

so $g_\#(z)$ and $f_\#(z)$ are in the same homology class. Thus $g_*(\{z\}) = f_*(\{z\})$, as desired. □

We still want to find conditions on two simplicial maps f and g under which the induced homomorphisms f_* and g_* are equal. We have reduced this problem to the problem of finding conditions under which one can construct a chain homotopy between $f_\#$ and $g_\#$. Here is one set of conditions under which this is possible:

Definition. Given two simplicial maps $f, g: K \to L$, these maps are said to be **contiguous** if for each simplex $v_0 \dots v_p$ of K, the points

$$f(v_0), \dots, f(v_p), g(v_0), \dots, g(v_p)$$

span a simplex τ of L. (The simplex τ may be of any dimension from 0 to $2p + 1$, depending on how many of these points are distinct.)

Roughly speaking, this condition says that f and g are "fairly close"; one can move the simplex $f(\sigma)$ to the simplex $g(\sigma)$ across some possibly larger simplex τ of which both are faces.

Theorem 12.5. *If $f, g: K \to L$ are contiguous simplicial maps, then there is a chain homotopy between $f_\#$ and $g_\#$.*

Proof. The argument we give here is a standard one; you should master it. For each simplex $\sigma = v_0 \dots v_p$ of K, let $L(\sigma)$ denote the subcomplex of L consisting of the simplex whose vertex set is $\{f(v_0), \dots, f(v_p), g(v_0), \dots, g(v_p)\}$, and its faces. We note the following facts:

(1) $L(\sigma)$ is nonempty, and $\tilde{H}_i(L(\sigma)) = 0$ for all i.

(2) If s is a face of σ, then $L(s) \subset L(\sigma)$.

(3) For each oriented simplex σ, the chains $f_\#(\sigma)$ and $g_\#(\sigma)$ are carried by $L(\sigma)$.

Using these facts, we shall construct the required chain homotopy $D: C_p(K) \to C_{p+1}(L)$, by induction on p. For each σ, the chain $D\sigma$ will be carried by $L(\sigma)$.

Let $p = 0$; let v be a vertex of K. Because $f_\#$ and $g_\#$ preserve augmentation, $\epsilon(g_\#(v) - f_\#(v)) = 1 - 1 = 0$. Thus $g_\#(v) - f_\#(v)$ represents an element of the reduced homology group $\tilde{H}_0(L(v))$. Because this group vanishes, we can choose a 1-chain Dv of L carried by the subcomplex $L(v)$ such that

$$\partial(Dv) = g_\#(v) - f_\#(v).$$

Then $\partial Dv + D\partial v = \partial Dv + 0 = g_\#(v) - f_\#(v)$, as desired. Define D in this way for each vertex of K.

Now suppose D is defined in dimensions less than p, such that for each oriented simplex s of dimension less than p, the chain Ds is carried by $L(s)$, and such that

$$\partial Ds + D\partial s = g_\#(s) - f_\#(s).$$

Let σ be an oriented simplex of dimension p. We wish to define $D\sigma$ so that $\partial(D\sigma)$ equals the chain

$$c = g_\#(\sigma) - f_\#(\sigma) - D\partial\sigma.$$

Note that c is a well-defined chain; $D\partial\sigma$ is defined because $\partial\sigma$ has dimension $p-1$. Furthermore, c is a cycle, for we compute

$$\begin{aligned}\partial c &= \partial g_{\#}(\sigma) - \partial f_{\#}(\sigma) - \partial D(\partial\sigma)\\ &= \partial g_{\#}(\sigma) - \partial f_{\#}(\sigma) - [g_{\#}(\partial\sigma) - f_{\#}(\partial\sigma) - D\partial(\partial\sigma)],\end{aligned}$$

applying the induction hypothesis to the $p-1$ chain $\partial\sigma$. Using the fact that $\partial \circ \partial = 0$, we see that $\partial c = 0$.

Finally, we note that c is carried by $L(\sigma)$: Both $g_{\#}(\sigma)$ and $f_{\#}(\sigma)$ are carried by $L(\sigma)$, by (3). To show $D\partial\sigma$ is carried by $L(\sigma)$, note that the chain $\partial\sigma$ is a sum of oriented faces of σ. For each such face s, the chain Ds is carried by $L(s)$, and $L(s) \subset L(\sigma)$ by (2). Thus $D\partial\sigma$ is carried by $L(\sigma)$.

Since c is a p-cycle carried by $L(\sigma)$, and since $H_p(L(\sigma)) = 0$, we can choose a $p+1$ chain $D\sigma$ carried by $L(\sigma)$ such that

$$\partial D\sigma = c = g_{\#}(\sigma) - f_{\#}(\sigma) - D\partial\sigma.$$

We then define $D(-\sigma) = -D(\sigma)$. We repeat this process for each p-simplex σ of K; then we have the required chain homotopy D in dimension p. The theorem follows. □

Some students are bothered by the fact that constructing the chain homotopy D involves arbitrary choices. They would be happier if there were a definite formula for D. Unfortunately, there is no such neat formula, because basically there are many possible chain homotopies between $f_{\#}$ and $g_{\#}$, and there is no reason for preferring one over the other.

To illustrate this fact, let us consider the preceding proof in a particular case. At the first step of the proof, when v is a vertex, the chain Dv is in fact uniquely determined. Since Dv is to be a chain carried by $L(v)$, then necessarily

$$\begin{aligned} Dv &= 0 && \text{if } f(v) = g(v),\\ Dv &= [f(v), g(v)] && \text{if } f(v) \neq g(v).\end{aligned}$$

In the first case, $L(v)$ is a vertex; and in the second case, it consists of a 1-simplex and its faces.

But at the very next step of the proof, choices can arise. Let $\sigma = vw$ be a 1-simplex, and suppose the points $f(v), f(w), g(v), g(w)$ are all distinct, so they span a 3-simplex τ, as indicated in Figure 12.4. The chains Dv and Dw are indicated in the figure. How shall we define $D\sigma$? There are two obvious choices for a 2-chain $D\sigma$ whose boundary is the 1-chain $g_{\#}(\sigma) - f_{\#}(\sigma) - D\partial\sigma$ pictured in Figure 12.5. One choice would consist of the front two faces of τ, oriented as indicated; the other choice would consist of the back two faces, appropriately oriented. There is no reason to prefer one choice to the other; we must simply choose one arbitrarily.

Application to relative homology

Let K_0 be a subcomplex of K, and let L_0 be a subcomplex of L. Let $f: K \to L$ be a simplicial map that carries each simplex of K_0 into a simplex of

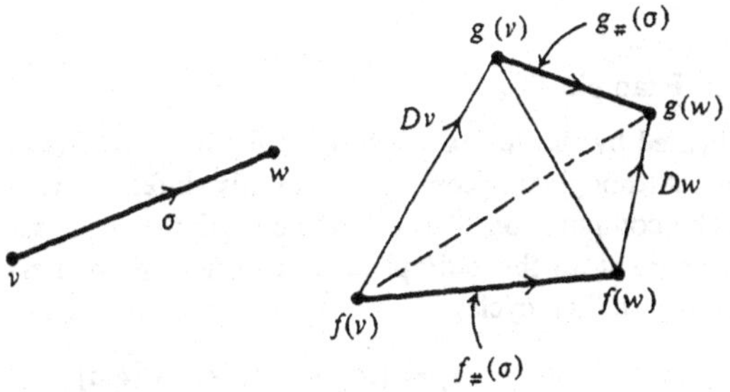

Figure 12.4

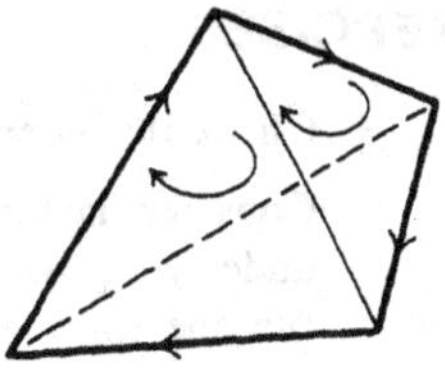

Figure 12.5

L_0. We often express this by the phrase

$$\text{“}f : (K,K_0) \to (L,L_0) \text{ is a simplicial map.”}$$

In this case, it is immediate that $f_\#$ maps $C_p(K_0)$ into $C_p(L_0)$, so that one has an induced map (also denoted $f_\#$)

$$f_\# : C_p(K,K_0) \to C_p(L,L_0).$$

This map commutes with ∂ and thus induces a homomorphism

$$f_* : H_p(K,K_0) \to H_p(L,L_0).$$

The functorial properties stated in Theorem 12.2 carry over immediately to relative homology.

Definition. Let $f,g : (K,K_0) \to (L,L_0)$ be two simplicial maps. We say f and g are **contiguous as maps of pairs** if for each simplex $\sigma = v_0 \dots v_p$ of K, the points

$$f(v_0), \dots, f(v_p), g(v_0), \dots, g(v_p)$$

span a simplex of L, and if $\sigma \in K_0$, they span a simplex of L_0.

Theorem 12.6. *Let $f, g : (K,K_0) \to (L,L_0)$ be contiguous as maps of pairs. Then there is for all p a homomorphism*

$$D : C_p(K,K_0) \to C_{p+1}(L,L_0)$$

such that $\partial D + D\partial = g_\# - f_\#$. It follows that f_* and g_* are equal as maps of relative homology groups.

Proof. The chain homotopy D constructed in the preceding proof automatically maps $C_p(K_0)$ into $C_{p+1}(L_0)$. For if $\sigma \in K_0$, the complex $L(\sigma)$ is by definition a subcomplex of L_0. Given σ, the chain $D\sigma$ is carried by $L(\sigma)$; therefore, D maps $C_p(K_0)$ into $C_{p+1}(L_0)$. Then D induces the required homomorphism of the relative chain groups. □

EXERCISES

1. Check the assertions made in Example 1.

2. Consider the complex K indicated by the vertex labelling in Figure 12.6, whose underlying space is the torus. Although this complex is slightly different from the one considered earlier, the computation of the homology of the torus remains the same. Let γ be the sum of all the 2-simplices of K, oriented counterclockwise; then γ generates $H_2(K)$. The cycles

$$w_1 = [a,b] + [b,c] + [c,a] \qquad \text{and} \qquad z_1 = [a,d] + [d,e] + [e,a]$$

generate $H_1(K)$.

(a) Define simplicial maps f, g, h, k of K with itself such that

$$\begin{array}{llll} f: a \to a & g: a \to a & h: a \to d & k: a \to p \\ b \to d & b \to c & b \to p & b \to q \\ c \to e & c \to b & c \to q & c \to d \\ d \to c & d \to d & d \to d & d \to u \\ e \to b & e \to e & e \to d & e \to b \end{array}$$

(b) Compute the values of f_*, g_*, h_*, and k_* on the homology classes $\{w_1\}$, $\{z_1\}$, and $\{\gamma\}$.

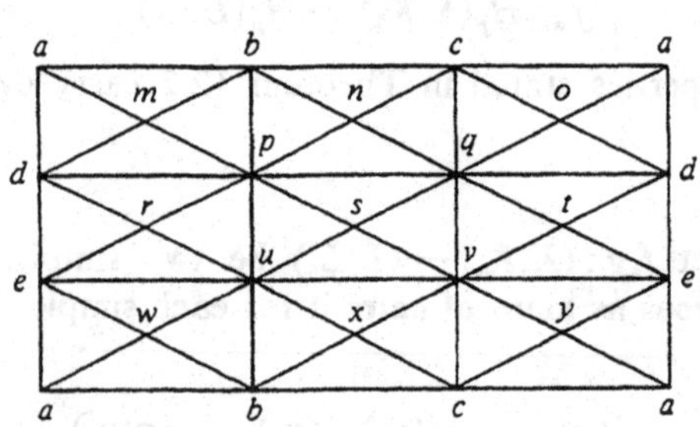

Figure 12.6

3. Use the complex in Figure 12.6, but with the letters d and e on the right-hand edge reversed, to represent the Klein bottle S. Let w_1 and z_1 be as before.
(a) Define a simplicial map $f: S \to S$ such that $f_*(\{w_1\}) = \{z_1\}$.
(b) Show there is no simplicial map $g: S \to S$ such that $g_*(\{z_1\}) = \{w_1\}$.

4. Let K be the torus, as in Exercise 2; let γ be the 2-cycle indicated there. Let L be the complex consisting of the proper faces of the 3-simplex having vertices A, B, C, D; and let γ' be the 2-cycle $\partial[A,B,C,D]$ of L.
(a) Show that any map of the vertices of K to the vertices of L induces a simplicial map of K to L.
(b) Let f be the simplicial map carrying m and r to A, p to B, b and u to C, and all other vertices to D. Compute the induced homomorphism $f_*: H_2(K) \to H_2(L)$, in terms of γ and γ'.

(c) Let g be the simplicial map agreeing with f on the vertices of K except that $g(r) = C$. Compute g_*.
(d) Let h agree with g on the vertices of K except that $h(u) = A$. Compute h_*.

5. Let $f,g : (K,K_0) \to (L,L_0)$ be simplicial maps. Show that if f and g are contiguous as maps of K into L, and if L_0 is a full subcomplex of L, then f and g are contiguous as maps of pairs.

§13. CHAIN COMPLEXES AND ACYCLIC CARRIERS

Many of the definitions and constructions we have made within the context of simplicial complexes also occur in more general situations. We digress at this point to discuss this more general context. We shall use these results many times later on. Proofs are left as exercises.

First we define the algebraic analogues to our chain groups. (We mentioned them earlier in §11.)

Definition. A **chain complex** $\mathcal{C}$ is a family $\{C_p, \partial_p\}$ of abelian groups C_p and homomorphisms

$$\partial_p : C_p \to C_{p-1},$$

indexed with the integers, such that $\partial_p \circ \partial_{p+1} = 0$ for all p.

If $C_p = 0$ for $p < 0$, then $\mathcal{C}$ is said to be a **non-negative** chain complex. If C_p is a free abelian group for each p, then $\mathcal{C}$ is called a **free** chain complex. The group

$$H_p(\mathcal{C}) = \ker \partial_p / \operatorname{im} \partial_{p+1}$$

is called the pth **homology group** of the chain complex $\mathcal{C}$.

If $\mathcal{C}$ is a non-negative chain complex, an **augmentation** for $\mathcal{C}$ is an epimorphism $\epsilon : C_0 \to \mathbf{Z}$ such that $\epsilon \circ \partial_1 = 0$. The **augmented chain complex** $\{\mathcal{C}, \epsilon\}$ is the chain complex obtained from $\mathcal{C}$ by adjoining the group $\mathbf{Z}$ in dimension -1 and using ϵ as the boundary operator in dimension 0. The homology groups of the augmented chain complex are called the **reduced homology groups** of the original chain complex $\mathcal{C}$, relative to the augmentation ϵ. They are denoted either $H_i(\{\mathcal{C},\epsilon\})$ or $\tilde{H}_i(\mathcal{C})$. It follows readily that $\tilde{H}_p(\mathcal{C}) = H_p(\mathcal{C})$ for $p \neq 0$, and

$$H_0(\mathcal{C}) \simeq \tilde{H}_0(\mathcal{C}) \oplus \mathbf{Z}.$$

(See the exercises.)

An arbitrary chain complex $\mathcal{D}$ is said to be **acyclic** if $H_i(\mathcal{D}) = 0$ for all i. In particular, the augumented chain complex $\{\mathcal{C},\epsilon\}$ is acyclic if $H_i(\{\mathcal{C},\epsilon\}) = \tilde{H}_i(\mathcal{C}) = 0$ for all i, or equivalently if $H_0(\mathcal{C}) \simeq \mathbf{Z}$ and $H_i(\mathcal{C}) = 0$ for $i \neq 0$.

Definition. Let $\mathcal{C} = \{C_p, \partial_p\}$ and $\mathcal{C}' = \{C'_p, \partial'_p\}$ be chain complexes. A **chain map** $\phi : \mathcal{C} \to \mathcal{C}'$ is a family of homomorphisms

$$\phi_p : C_p \to C'_p$$

such that $\partial'_p \circ \phi_p = \phi_{p-1} \circ \partial_p$ for all p.

A chain map $\phi : \mathcal{C} \to \mathcal{C}'$ induces a homomorphism

$$(\phi_*)_p : H_p(\mathcal{C}) \to H_p(\mathcal{C}').$$

Furthermore, the following hold:

(1) The identity map i of $\mathcal{C}$ is a chain map, and $(i_*)_p$ is the identity map of $H_p(\mathcal{C})$.

(2) If $\phi : \mathcal{C} \to \mathcal{C}'$ and $\psi : \mathcal{C}' \to \mathcal{C}''$ are chain maps, then $\psi \circ \phi$ is a chain map, and $(\psi \circ \phi)_* = \psi_* \circ \phi_*$.

If $\{\mathcal{C}, \epsilon\}$ and $\{\mathcal{C}', \epsilon'\}$ are augmented chain complexes, the chain map $\phi : \mathcal{C} \to \mathcal{C}'$ is said to be **augmentation-preserving** if $\epsilon' \circ \phi_0 = \epsilon$. If we extend ϕ to the (-1)-dimensional groups by letting it equal the identity map of $\mathbf{Z}$, then ϕ is called a **chain map of augmented chain complexes.** It follows that an augmentation-preserving chain map ϕ induces a homomorphism $\phi_* : H_p(\mathcal{C}) \to H_p(\mathcal{C}')$ of reduced homology groups.

Example 1. The chain complex

$$\mathcal{C}(K, K_0) = \{C_p(K, K_0), \partial_p\}$$

defined in §9 is called the **oriented chain complex of the simplicial pair** (K, K_0). It is both free and non-negative, but, in general, it does not have an augmentation. (For example, in Example 4 of §9 the entire group $C_0(K, K_0)$ vanishes, so there can be no surjective map $C_0(K, K_0) \to \mathbf{Z}$.) If K_0 is empty, then the complex $\mathcal{C}(K, \varnothing) = \mathcal{C}(K)$ has a standard augmentation, defined by $\epsilon(v) = 1$ for each vertex v of K, as we have seen.

Example 2. If $f : K \to L$ is a simplicial map of simplicial complexes K and L, then $f_\#$ is an augmentation-preserving chain map of $\mathcal{C}(K)$ into $\mathcal{C}(L)$. However, in general there exist augmentation-preserving chain maps $\phi : \mathcal{C}(K) \to \mathcal{C}(L)$ that are not induced by simplicial maps.

Definition. If $\phi, \psi : \mathcal{C} \to \mathcal{C}'$ are chain maps, then a **chain homotopy** of ϕ to ψ is a family of homomorphisms

$$D_p : C_p \to C'_{p+1}$$

such that

$$\partial'_{p+1} D_p + D_{p-1} \partial_p = \psi_p - \phi_p$$

for all p.

Henceforth we shall normally omit subscripts on boundary operators and chain maps and chain homotopies. The preceding formula, for instance, then assumes the more familiar form $\partial' D + D\partial = \psi - \phi$.

Definition. A chain map $\phi : \mathcal{C} \to \mathcal{C}'$ is called a **chain equivalence** if there is a chain map $\phi' : \mathcal{C}' \to \mathcal{C}$ such that $\phi' \circ \phi$ and $\phi \circ \phi'$ are chain homotopic to the identity maps of $\mathcal{C}$ and $\mathcal{C}'$, respectively. We call ϕ' a **chain-homotopy inverse** to ϕ.

We list several properties of chain homotopies; proofs are left to the exercises.

(1) *Chain homotopy is an equivalence relation on the set of chain maps from $\mathcal{C}$ to $\mathcal{C}'$.*

(2) *Composition of chain maps induces a well-defined composition operation on chain-homotopy classes.*

(3) *If ϕ and ψ are chain homotopic, then they induce the same homomorphism in homology.*

(4) *If ϕ is a chain equivalence, with chain-homotopy inverse ϕ', then ϕ_* and $(\phi')_*$ are homology isomorphisms that are inverse to each other.*

(5) *If $\phi : \mathcal{C} \to \mathcal{C}'$ and $\psi : \mathcal{C}' \to \mathcal{C}''$ are chain equivalences, then $\psi \circ \phi$ is a chain equivalence.*

Now we investigate what all this means in the special case of augmented chain complexes.

Lemma 13.1. *Let $\mathcal{C}$ and $\mathcal{C}'$ be non-negative chain complexes. Let $\phi, \psi : \mathcal{C} \to \mathcal{C}'$ be chain maps; let D be a chain homotopy between them. Suppose $\mathcal{C}$ and $\mathcal{C}'$ are augmented by ϵ and ϵ', respectively. If ϕ preserves augmentation, so does ψ. If we extend ϕ and ψ to be the identity in dimension -1, and extend D to be zero in dimension -1, then D is a chain homotopy between the extended chain maps.*

Proof. If $c_0 \in C_0$, we have

$$\partial D c_0 = \phi(c_0) - \psi(c_0)$$

because $\partial c_0 = 0$. Then

$$0 = \epsilon'(\partial D c_0) = \epsilon'\phi(c_0) - \epsilon'\psi(c_0) = \epsilon(c_0) - \epsilon'\psi(c_0),$$

as desired. Furthermore, we have in dimension 0 the equation

$$D\epsilon(c_0) + \partial D(c_0) = \phi(c_0) - \psi(c_0),$$

because $D\epsilon(c_0) = 0$, and we have in dimension -1 the equation

$$\epsilon'(D(1)) = \phi(1) - \psi(1)$$

because both sides vanish. Thus D is a chain homotopy between the extended chain maps. $\square$

Lemma 13.2. *Let $\mathcal{C}$ and $\mathcal{C}'$ be non-negative chain complexes. Let $\phi : \mathcal{C} \to \mathcal{C}'$ be a chain equivalence with chain-homotopy inverse ϕ'. Suppose $\mathcal{C}$ and $\mathcal{C}'$ are augmented by ϵ, ϵ' respectively. If ϕ preserves augmentation, so does ϕ'. Furthermore, ϕ and ϕ' are chain-homotopy inverses as maps of augmented complexes. Therefore, they induce inverse isomorphisms in reduced homology.*

Proof. If D' is a chain homotopy between $\phi \circ \phi'$ and the identity, then in dimension 0,

$$\partial' D' c_0' = \phi\phi'(c_0') - c_0',$$

so

$$0 = \epsilon'(\partial' D' c_0') = \epsilon'\phi\phi'(c_0') - \epsilon'(c_0') = \epsilon\phi'(c_0') - \epsilon'(c_0').$$

Thus ϕ' preserves augmentation. The remainder of the statement follows from Lemma 13.1. □

These definitions set up the general algebraic framework into which the oriented simplicial chain groups fit as a special case. Now we seek to put into this general context the method by which in the preceding section we constructed a chain homotopy. As motivation, we first consider the case where the maps ϕ and ψ are general chain maps, but the chain groups are the familiar oriented simplicial chain groups. Later, we consider general chain complexes and general chain maps.

Definition. Let K and L be simplicial complexes. An **acyclic carrier** from K to L is a function Φ that assigns to each simplex σ of K, a subcomplex $\Phi(\sigma)$ of L such that:

(1) $\Phi(\sigma)$ is nonempty and acyclic.

(2) If s is a face of σ, then $\Phi(s) \subset \Phi(\sigma)$.

If $f : C_p(K) \to C_q(L)$ is a homomorphism, we say that f is **carried** by Φ if for each oriented p-simplex σ of K, the chain $f(\sigma)$ is carried by the subcomplex $\Phi(\sigma)$ of L.

Theorem 13.3 (Acyclic carrier theorem, geometric version). *Let Φ be an acyclic carrier from K to L.*

(a) *If ϕ and ψ are two augmentation-preserving chain maps from $\mathcal{C}(K)$ to $\mathcal{C}(L)$ that are carried by Φ, there exists a chain homotopy D of ϕ to ψ that is also carried by Φ.*

(b) *There exists an augmentation-preserving chain map from $\mathcal{C}(K)$ to $\mathcal{C}(L)$ that is carried by Φ.*

Proof. Part (a) of the theorem is proved by copying the proof of Theorem 12.5; one simply replaces the subcomplex $L(\sigma)$ by the subcomplex $\Phi(\sigma)$ throughout. To prove (b), we proceed as follows: For each vertex v of K, define $\phi(v)$ to be a 0-chain c of $\Phi(v)$ such that $\epsilon(c) = 1$. This we can do because $\Phi(v)$

is nonempty, so $\epsilon : \Phi(v) \to \mathbf{Z}$ is surjective. (In fact, we can simply choose $\phi(v)$ to be a vertex of $\Phi(v)$.) Then ϕ preserves augmentation, and

$$\partial\phi(v) = 0 = \phi(\partial v).$$

Let $\sigma = [v,w]$ be an oriented 1-simplex of K. The chain $c = \phi(\partial\sigma)$ is well-defined because $\partial\sigma$ is a 0-chain and ϕ has been defined in dimension 0. Furthermore, c is carried by $\Phi(\sigma)$; for $\phi(\partial\sigma)$ is carried by $\Phi(v)$ and $\Phi(w)$, both of which are contained in $\Phi(\sigma)$ by (2). Finally,

$$\epsilon(c) = \epsilon\phi(\partial\sigma) = \epsilon(\partial\sigma) = 0,$$

because ϕ preserves augmentation. Hence c represents an element of $\tilde{H}_0(\Phi(\sigma))$. Because $\Phi(\sigma)$ is acyclic (by (1)), we can choose a 1-chain carried by $\Phi(\sigma)$ whose boundary is c. We denote this 1-chain by $\phi(\sigma)$; then

$$\partial\phi(\sigma) = c = \phi(\partial\sigma).$$

For the induction step, let $p > 1$. Assume that if $\dim s < p$, then $\phi(s)$ is defined and $\partial\phi(s) = \phi\partial(s)$. Let σ be an oriented simplex of dimension p; we seek to define $\phi(\sigma)$. The chain $c = \phi(\partial\sigma)$ is a well-defined $p-1$ chain. It is carried by $\Phi(\sigma)$, since $\Phi(\partial\sigma)$ is carried by the union of the complexes $\Phi(s_i)$, where s_i ranges over the $p-1$ faces of σ, and each of these complexes is contained in $\Phi(\sigma)$. Furthermore, c is a cycle since

$$\partial c = \partial\phi(\partial\sigma) = \phi\partial(\partial\sigma) = 0.$$

Here we apply the induction hypothesis to the $p-1$ chain $\partial\sigma$. Since $\Phi(\sigma)$ is acyclic, we can choose $\phi(\sigma)$ to be a p-chain carried by $\Phi(\sigma)$ such that $\partial\phi(\sigma) = c$. Then $\partial\phi(\sigma) = \phi\partial(\sigma)$, as desired.

The theorem follows by induction. □

With this proof as motivation, we now formulate an even more general version of this theorem. In this form, all geometry disappears and only algebra remains!

Definition. Let $\mathcal{C} = \{C_p, \partial_p\}$ be a chain complex. A **subchain complex** $\mathcal{D}$ of $\mathcal{C}$ is a chain complex whose pth chain group is a subgroup of C_p, and whose boundary operator in each dimension p is the restriction of ∂_p.

Definition. Let $\{\mathcal{C},\epsilon\} = \{C_p,\partial_p,\epsilon\}$ be an augmented chain complex. Suppose $\mathcal{C}$ is free; let $\{\sigma_p^\alpha\}$ be a basis for C_p, as α ranges over some index set J_p. Let $\{\mathcal{C}',\epsilon'\} = \{C'_p,\partial'_p,\epsilon'\}$ be an arbitrary augmented chain complex. An **acyclic carrier** from $\mathcal{C}$ to $\mathcal{C}'$, relative to the given bases, is a function Φ that assigns to each basis element σ_p^α, a subchain complex $\Phi(\sigma_p^\alpha)$ of $\mathcal{C}'$, satisfying the following conditions:

(1) The chain complex $\Phi(\sigma_p^\alpha)$ is augmented by ϵ' and is acyclic.

(2) If σ_{p-1}^β appears in the expression for $\partial_p\sigma_p^\alpha$ in terms of the preferred basis for C_{p-1}, then $\Phi(\sigma_{p-1}^\beta)$ is a subchain complex of $\Phi(\sigma_p^\alpha)$.

A homomorphism $f : C_p \to C'_q$ is said to be **carried** by Φ if $f(\sigma_p^\alpha)$ belongs to the q-dimensional group of the subchain complex $\Phi(\sigma_p^\alpha)$ of $\mathcal{C}'$, for each α.

Theorem 13.4 (Acyclic carrier theorem, algebraic version). *Let $\mathcal{C}$ and $\mathcal{C}'$ be augmented chain complexes; let $\mathcal{C}$ be free. Let Φ be an acyclic carrier from $\mathcal{C}$ to $\mathcal{C}'$, relative to some set of preferred bases for $\mathcal{C}$. Then there is an augmentation-preserving chain map $\phi : \mathcal{C} \to \mathcal{C}'$ carried by Φ, and any two such are chain homotopic; the chain homotopy is also carried by Φ.*

Proof. The proof of this theorem is just a jazzed-up version of the preceding proof. The requirement that the restriction of ϵ' give an augmentation for $\Phi(\sigma_p^\alpha)$ means that ϵ' must map the 0-dimensional group of this chain complex *onto* $\mathbf{Z}$; this corresponds to the requirement in the earlier version that $\Phi(\sigma)$ be nonempty for all σ. □

Application: Ordered simplicial homology

The algebraic version of the acyclic carrier theorem may seem unnecessarily abstract to you. But it is indeed useful. Let us give one application now; it is a theorem that we will use later when we study singular homology. This theorem involves a new way of defining the simplicial homology groups, using *ordered* simplices rather than *oriented* simplices.

Let K be a simplicial complex. An **ordered p-simplex** of K is a $p + 1$ tuple $(v_0, \ldots, v_p)$ of vertices of K, where v_i are vertices of a simplex of K *but need not be distinct.* (For example, if vw is a 1-simplex of K, then (v,w,w,v) is an ordered 3-simplex of K.)

Let $C'_p(K)$ be the free abelian group generated by the ordered p-simplices of K; it is called the group of **ordered p-chains** of K. As usual, we shall identify the ordered p-simplex $(v_0, \ldots, v_p)$ with the elementary p-chain whose value is 1 on this ordered simplex and 0 on all other ordered simplices. Then every element of $C'_p(K)$ can be written uniquely as a finite linear combination, with integral coefficients, of ordered p-simplices. We define $\partial'_p : C'_p(K) \to C'_{p-1}(K)$ by the formula

$$\partial'_p(v_0, \ldots, v_p) = \sum_{i=0}^{p} (-1)^i (v_0, \ldots, \hat{v}_i, \ldots, v_p).$$

Then ∂'_p is a well-defined homomorphism; and one checks as before that $\partial'_p \circ \partial'_{p+1} = 0$.

The chain complex $\mathcal{C}'(K) = \{C'_p(K), \partial'_p\}$ is called the **ordered chain complex** of K. It is augmented by defining $\epsilon'(v) = 1$ for every vertex v of K. Although much too huge to be useful for computational purposes, this chain complex is sometimes quite convenient for theoretical purposes. Its homology is, surprisingly enough, isomorphic to the simplicial homology of K, in a natural way. We outline a proof of this fact, leaving the details to you.

Lemma 13.5. *If $w * K$ is a cone over the complex K, then $w * K$ is acyclic in ordered homology.*

Proof. Define

$$D : C'_p(w * K) \to C'_{p+1}(w * K)$$

for $p \geq 0$ by the equation

$$D((v_0, \ldots, v_p)) = (w, v_0, \ldots, v_p).$$

Note that it is irrelevant here whether any of the v_i are equal to w. Let $c_p \in C'_p(w * K)$. We compute

$$\partial'_1 D c_0 = c_0 - \epsilon'(c_0) w,$$
$$\partial'_{p+1} D c_p = c_p - D\partial'_p c_p \qquad \text{if} \qquad p > 0.$$

The lemma follows: If c_p is a *cycle* and $p > 0$, then $c_p = \partial'_{p+1} D c_p$; if c_0 is a 0-chain lying in ker ϵ', then $c_0 = \partial'_1 D c_0$. □

Theorem 13.6. *Choose a partial ordering of the vertices of K that induces a linear ordering on the vertices of each simplex of K. Define $\phi : C_p(K) \to C'_p(K)$ by letting*

$$\phi([v_0, \ldots, v_p]) = (v_0, \ldots, v_p)$$

if $v_0 < v_1 < \cdots < v_p$ in the given ordering. Define $\psi : C'_p(K) \to C_p(K)$ by the equation

$$\psi((w_0, \ldots, w_p)) = \begin{cases} [w_0, \ldots, w_p] & \text{if the } w_i \text{ are distinct,} \\ 0 & \text{otherwise.} \end{cases}$$

Then ϕ and ψ are augmentation-preserving chain maps that are chain-homotopy inverses.

If K_0 is a subcomplex of K, then ϕ and ψ induce chain maps of the relative chain complexes that are chain-homotopy inverses.

The proof is an application of the acyclic carrier theorem; it is left as an exercise.

It follows from this theorem that oriented and ordered homology are isomorphic in a rather "natural" way. To explain what we mean by "naturality," we need to consider how a simplicial map acts in ordered homology.

Definition. Let $f : K \to L$ be a simplicial map. Define $f'_\# : \mathcal{C}'(K) \to \mathcal{C}'(L)$ by the rule

$$f'_\#((v_0, \ldots, v_p)) = (f(v_0), \ldots, f(v_p)).$$

It is easy to check that $f'_\#$ is a chain map, easier in fact than in the oriented case, for we need not worry whether or not vertices are distinct. Clearly $f'_\#$ preserves augmentation. If f maps K_0 to L_0, then $f'_\#$ induces a chain map

$$f'_\#: \mathcal{C}'(K,K_0) \to \mathcal{C}'(L,L_0),$$

and a corresponding homomorphism in homology.

Theorem 13.7. *Let $f:(K,K_0) \to (L,L_0)$ be a simplicial map. Let ϕ and ψ be as in the preceding theorem. Then the following diagram commutes:*

$$\begin{array}{ccc} H_i(K,K_0) & \xrightarrow{f_*} & H_i(L,L_0) \\ \uparrow \psi_* & & \uparrow \psi_* \\ H_i(\mathcal{C}'(K,K_0)) & \xrightarrow{f'_*} & H_i(\mathcal{C}'(L,L_0)). \end{array}$$

Similarly, $\phi_* \circ f_* = f'_* \circ \phi_*$.

Proof. One checks directly from the definition that $f_\# \circ \psi = \psi \circ f'_\#$. Thus the diagram already commutes on the chain level, so it commutes on the homology level as well. It is not true that $\phi \circ f_\# = f'_\# \circ \phi$, since ϕ depends on a particular ordering of vertices. However, because ϕ_* is the inverse of ψ_*, it is true that $\phi_* \circ f_* = f'_* \circ \phi_*$. □

EXERCISES

1. If $\{\mathcal{C},\epsilon\}$ is an augmented chain complex, show that $\tilde{H}_{-1}(\mathcal{C}) = 0$ and
$$\tilde{H}_0(\mathcal{C}) \oplus \mathbf{Z} \simeq H_0(\mathcal{C}).$$
(See the exercises of §7.) Conclude that $\{\mathcal{C},\epsilon\}$ is acyclic if and only if $H_p(\mathcal{C})$ is infinite cyclic for $p = 0$ and vanishes for $p \neq 0$.
2. Check properties (1)–(5) of chain homotopies. Only (2) and (5) require care.
3. Prove (a) of Theorem 13.3.
4. Consider Example 1 of §12. Show that although the maps g and h are not contiguous, there is nevertheless a chain homotopy between $g_\#$ and $h_\#$, as follows:
   (a) Define an acyclic carrier Φ from K to L carrying both $g_\#$ and $h_\#$.
   (b) Define a specific chain homotopy between $g_\#$ and $h_\#$ that is carried by Φ.
   (c) Define a chain map $\phi: \mathcal{C}(K) \to \mathcal{C}(L)$ carried by Φ that is not induced by a simplicial map; define a chain homotopy between ϕ and $h_\#$.
5. Check the details of the proof of Theorem 13.4.
6. Prove Theorem 13.6 as follows:
   (a) Show ϕ and ψ are augmentation-preserving chain maps, and show that $\psi \circ \phi$ equals the identity map of $\mathcal{C}(K)$.
   (b) Define an acyclic carrier from $\mathcal{C}'(K)$ to $\mathcal{C}'(K)$ that carries both $\phi \circ \psi$ and the identity map.

2

Topological Invariance of the Homology Groups

In the preceding chapter, we defined a function assigning to each simplicial complex K a sequence of abelian groups called its homology groups. We now prove that these groups depend only on the underlying topological space of K.

The way to approach this problem is to study continuous maps of one polyhedron to another, and what such maps do to the homology groups. We show that a continuous map $h : |K| \rightarrow |L|$ of polyhedra induces, in a rather natural way, a homomorphism $h_* : H_p(K) \rightarrow H_p(L)$ of the homology groups of the corresponding simplicial complexes. Constructing this induced homomorphism will prove to be a reasonably arduous task.

It turns out that when h is a homeomorphism of topological spaces, then h_* is an isomorphism of groups. The topological invariance of the simplicial homology groups follows.

It is not hard to see intuitively why there should be such an induced homomorphism. If one thinks of a homology class as a geometric object, it seems fairly reasonable that its image under h should be a well-defined homology class. A closed loop on the torus T, for instance, is mapped by $h : T \rightarrow X$ into a closed loop in X. But to make this idea algebraically precise requires some effort.

We already know that a *simplicial* map $f : |K| \rightarrow |L|$ induces a homomorphism f_* of homology groups. This chapter is devoted to showing that an arbitrary continuous map h can be approximated (in a suitable sense) by a simplicial map f, and that the resulting induced homomorphism depends only on the map h, not on the particular approximation chosen.

§14. SIMPLICIAL APPROXIMATIONS

In this section, we study what it means for an arbitrary continuous map to be "approximated" by a simplicial map.

Definition. Let $h : |K| \to |L|$ be a continuous map. We say that h satisfies the **star condition** with respect to K and L if for each vertex v of K, there is a vertex w of L such that

$$h(\operatorname{St} v) \subset \operatorname{St} w.$$

Lemma 14.1. *Let $h : |K| \to |L|$ satisfy the star condition with respect to K and L. Choose $f : K^{(0)} \to L^{(0)}$ so that for each vertex v of K,*

$$h(\operatorname{St} v) \subset \operatorname{St} f(v).$$

(a) *Given $\sigma \in K$. Choose $x \in \operatorname{Int} \sigma$; and choose τ so $h(x) \in \operatorname{Int} \tau$. Then f maps each vertex of σ to a vertex of τ.*

(b) *f may be extended to a simplicial map of K into L, which we also denote by f.*

(c) *If $g : K \to L$ is another simplicial map such that $h(\operatorname{St} v) \subset \operatorname{St} g(v)$ for each vertex v of K, then f and g are contiguous.*

Proof. (a) Let $\sigma = v_0 \dots v_p$. Then $x \in \operatorname{St} v_i$ for each i, so

$$h(x) \in h(\operatorname{St} v_i) \subset \operatorname{St} f(v_i).$$

This means that $h(x)$ has a positive barycentric coordinate with respect to each of the vertices $f(v_i)$, for $i = 0, \dots, p$. These vertices must thus form a subset of the vertex set of τ.

(b) Because f carries the vertices of σ to vertices of a simplex of L, it may be extended to a simplicial map $f : K \to L$.

(c) Let σ, x, and τ be as before. Since

$$h(x) \in h(\operatorname{St} v_i) \subset \operatorname{St} g(v_i)$$

for $i = 0, \dots, p$, the vertices $g(v_i)$ must also be vertices of τ. Hence $f(v_0), \dots, f(v_p), g(v_0), \dots, g(v_p)$ span a face of τ, so f and g are contiguous. □

Definition. Let $h : |K| \to |L|$ be a continuous map. If $f : K \to L$ is a simplicial map such that

$$h(\operatorname{St} v) \subset \operatorname{St} f(v)$$

for each vertex v of K, then f is called a **simplicial approximation** to h.

We think of the simplicial approximation f as being "close" to h in some sense. One way to make this precise is to note that given $x \in |K|$, there is a simplex of L that contains both $h(x)$ and $f(x)$:

Lemma 14.2. *Let $f: K \to L$ be a simplicial approximation to $h: |K| \to |L|$. Given $x \in |K|$, there is a simplex τ of L such that $h(x) \in \operatorname{Int} \tau$ and $f(x) \in \tau$.*

Proof. This follows immediately from (a) of the preceding lemma. □

Theorem 14.3. *Let $h: |K| \to |L|$ and $k: |L| \to |M|$ have simplicial approximations $f: K \to L$ and $g: L \to M$, respectively. Then $g \circ f$ is a simplicial approximation to $k \circ h$.*

Proof. We know $g \circ f$ is a simplicial map. If v is a vertex of K, then

$$h(\operatorname{St} v) \subset \operatorname{St} f(v)$$

because f is a simplicial approximation to h. It follows that

$$k(h(\operatorname{St} v)) \subset k(\operatorname{St} f(v)) \subset \operatorname{St} g(f(v))$$

because g is a simplicial approximation to k. □

Example 1. Let K and L be the complexes pictured in Figure 14.1, whose underlying spaces are homeomorphic to the circle and to the annulus, respectively. Let K' be the complex obtained from K by inserting extra vertices, as pictured. Let h be the indicated continuous map, where we denote $h(a)$ by A, and similarly for the other vertices.

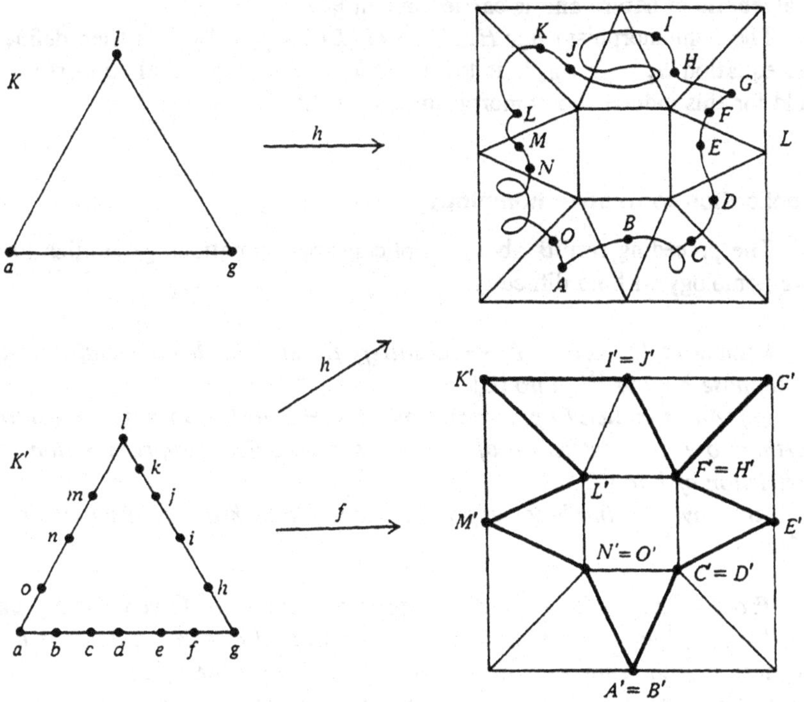

Figure 14.1

Now h does not satisfy the star condition relative to K and L, but it does satisfy the star condition relative to K' and L. Hence h has a simplicial approximation $f: K' \to L$. One such is pictured; we denote $f(a)$ by A', and similarly for the other vertices.

If $h: |K| \to |L|$ satisfies the star condition (relative to K and L), there is a well-defined homomorphism

$$h_*: H_p(K) \to H_p(L)$$

obtained by setting $h_* = f_*$, where f is any simplicial approximation to h. It is easy to see that the "functorial properties" are satisfied.

However, in general an arbitrary continuous map $h: |K| \to |L|$ will not satisfy the star condition relative to K and L, so we cannot obtain an induced homomorphism h_* in this way. How shall we proceed? There are two ideas involved:

First, one shows that given $h: |K| \to |L|$, it is possible to "subdivide" K, forming a new complex K' with the same underlying space as K, such that h does satisfy the star condition relative to K' and L. (This is what we did in Example 1 preceding.) This step is geometric in nature, and is carried out in §15 and §16.

Second, one shows that the identity map $i: |K'| \to |K|$ has a simplicial approximation $g: K' \to K$, which induces a homology *isomorphism* g_*. This step is algebraic in nature and is carried out in §17.

The homomorphism $h_*: H_p(K) \to H_p(L)$ induced by h is then defined by the equation $h_* = f_* \circ g_*^{-1}$. It turns out that the "functorial properties" also hold for this induced homomorphism, as we shall see in §18.

Application to relative homology

The preceding results about simplicial approximations generalize to relative homology with no difficulty:

Lemma 14.4. *Let $h: |K| \to |L|$ satisfy the star condition relative to K and L; suppose h maps $|K_0|$ into $|L_0|$.*

(a) *Any simplicial approximation $f: K \to L$ to h also maps $|K_0|$ into $|L_0|$; furthermore, the restriction of f to K_0 is a simplicial approximation to the restriction of h to $|K_0|$.*

(b) *Any two simplicial approximations f, g to h are contiguous as maps of pairs.*

Proof. Let f, g be simplicial approximations to h. Given $\sigma \in K_0$, choose $x \in \text{Int}\,\sigma$, and let τ be the simplex of L such that $h(x) \in \text{Int}\,\tau$. Because h maps $|K_0|$ into $|L_0|$, the simplex τ must belong to L_0. Since both f and g map σ onto faces of τ, they map K_0 into L_0, and are contiguous as maps of pairs.

We show $f|K_0$ is a simplicial approximation to the restriction of h to $|K_0|$. Let v be a vertex of K_0; then $\mathrm{St}(v,K_0) = \mathrm{St}(v,K) \cap |K_0|$. We conclude that

$$\begin{aligned} h(\mathrm{St}(v,K_0)) &\subset h(\mathrm{St}(v,K)) \cap h(|K_0|) \\ &\subset \mathrm{St}(f(v),L) \cap |L_0| = \mathrm{St}(f(v),L_0), \end{aligned}$$

as desired. □

EXERCISES

1. Consider the map $h : |K| \to |L|$ of Example 1. Determine how many different simplicial approximations $f : K' \to L$ to h there are. Let f and g be two such; choose a cycle z generating $H_1(K')$, and find a chain d of L such that $\partial d = f_{\#}(z) - g_{\#}(z)$.
2. A **homotopy** between two maps $f, h : X \to Y$ is a continuous map $F : X \times I \to Y$, where $I = [0,1]$, such that $F(x,0) = f(x)$ and $F(x,1) = h(x)$ for all x in X.
   (a) Show that any two maps $f,h : X \to \mathbf{R}^N$ are homotopic; the formula

   (*) $$F(x,t) = (1-t)f(x) + th(x)$$

   is called the **straight-line homotopy** between them.
   (b) Let K and L be finite complexes in $\mathbf{R}^N$; let $f : K \to L$ be a simplicial approximation to $h : |K| \to |L|$. Show that (*) defines a homotopy between f and h.
   *(c) Discuss (b) in the case where K and L are not finite. (We will return to this case later.)

§15. BARYCENTRIC SUBDIVISION

In this section, we show that a finite complex may be "subdivided" into simplices that are as small as desired. This geometric result will be used in the present chapter in our study of simplicial homology, and again in Chapter 4 when we deal with singular homology.

Definition. Let K be a geometric complex in $\mathbf{E}^J$. A complex K' is said to be a **subdivision** of K if:

(1) Each simplex of K' is contained in a simplex of K.

(2) Each simplex of K equals the union of *finitely* many simplices of K'.

These conditions imply that the union of the simplices of K' equals the union of the simplices of K—that is, that $|K'|$ and $|K|$ are equal as sets. The finiteness part of condition (2) guarantees that $|K'|$ and $|K|$ are equal as topological spaces, as you can readily check.

Note that if K'' is a subdivision of K', and if K' is a subdivision of K, then K'' is a subdivision of K.

Also note that if K' is a subdivision of K, and if K_0 is a subcomplex of K, then the collection of all simplices of K' that lie in $|K_0|$ is automatically a subdivision of K_0. We call it the subdivision of K_0 **induced** by K'.

For later use, we note the following.

Lemma 15.1. *Let K' be a subdivision of K. Then for each vertex w of K', there is a vertex v of K such that*

$$\mathrm{St}(w,K') \subset \mathrm{St}(v,K).$$

Indeed, if σ is the simplex of K such that $w \in \mathrm{Int}\,\sigma$, then this inclusion holds precisely when v is a vertex of σ.

Proof. If this inclusion holds, then since w belongs to $\mathrm{St}(w,K')$, w must lie in some open simplex of K that has v as a vertex.

Conversely, suppose $w \in \mathrm{Int}\,\sigma$ and v is a vertex of σ. It suffices to show that

$$|K| - \mathrm{St}(v,K) \subset |K| - \mathrm{St}(w,K').$$

The set on the left side of this inclusion is the union of all simplices of K that do not have v as a vertex. Hence it is also a union of simplices τ of K'. No such simplex τ can have w as a vertex, because $w \in \mathrm{Int}\,\sigma \subset \mathrm{St}(v,K)$. Thus any such simplex lies in $|K| - \mathrm{St}(w,K')$. □

Example 1. Let K consist of the 1-simplex $[0,1]$ and its vertices. Let L consist of the 1-simplices $[1/(n+1),1/n]$ and their vertices, for n a positive integer, along with the vertex 0. Then $|L| = |K|$ as sets but not as topological spaces; L satisfies all the conditions for a subdivision except the finiteness part of (2).

Example 2. Let Σ be the complex consisting of a 2-simplex σ and its faces. The subdivision K of Bd σ indicated in Figure 15.1 can be extended to a subdivision Σ' of Σ by forming the cone $w * K$, where w is an interior point of σ; the subdivision Σ' is said to be obtained by "starring K from w." This method of subdividing complexes will prove very useful.

Now we describe the "starring" method for subdividing complexes in general. We shall need the following lemma, whose proof is straightforward.

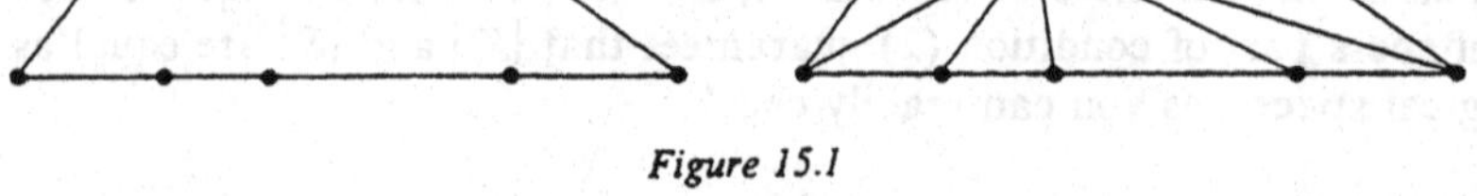

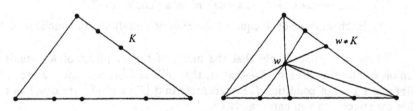

Figure 15.1

Lemma 15.2. *If K is a complex, then the intersection of any collection of subcomplexes of K is a subcomplex of K. Conversely, if $\{K_\alpha\}$ is a collection of complexes in $\mathbf{E}^J$, and if the intersection of every pair $|K_\alpha| \cap |K_\beta|$ is the polytope of a complex that is a subcomplex of both K_α and K_β, then the union $\bigcup K_\alpha$ is a complex.* □

Our method for constructing subdivisions is a step-by-step one. We describe one step in the process now.

Definition. Let K be a complex; suppose that L_p is a subdivision of the p-skeleton of K. Let σ be a $p + 1$ simplex of K. The set Bd σ is the polytope of a subcomplex of the p-skeleton of K, and hence of a subcomplex of L_p; we denote the latter subcomplex by L_σ. If w_σ is an interior point of σ, then the cone $w_\sigma * L_\sigma$ is a complex whose underlying space is σ. We define L_{p+1} to be the union of L_p and the complexes $w_\sigma * L_\sigma$, as σ ranges over all $p + 1$ simplices of K. We show L_{p+1} is a complex; it is said to be the **subdivision of $K^{(p+1)}$ obtained by starring L_p from the points w_σ**.

To verify that L_{p+1} is a complex, we note that

$$|w_\sigma * L_\sigma| \cap |L_p| = \text{Bd}\,\sigma,$$

which is the polytope of the subcomplex L_σ of both $w_\sigma * L_\sigma$ and L_p. Similarly, if τ is another $p + 1$ simplex of K, then the spaces $|w_\sigma * L_\sigma|$ and $|w_\tau * L_\tau|$ intersect in the simplex $\sigma \cap \tau$ of K, which is the polytope of a subcomplex of L_p and hence of both L_σ and L_τ. It follows from Lemma 15.2 that L_{p+1} is a complex.

Now the complex L_{p+1} depends on the choice of the points w_σ. Often it is convenient to choose a "canonical" interior point of σ to use for starring purposes. The usual such point is the following:

Definition. If $\sigma = v_0 \ldots v_p$, the **barycenter** of σ is defined to be the point

$$\hat{\sigma} = \sum_{i=0}^{p} \frac{1}{p+1} v_i.$$

It is the point of Int σ all of whose barycentric coordinates with respect to the vertices of σ are equal.

If σ is a 1-simplex, then $\hat{\sigma}$ is its midpoint. If σ is a 0-simplex, then $\hat{\sigma} = \sigma$. In general, $\hat{\sigma}$ equals the centroid of σ, but that fact is not important for us.

Now we describe our general method of constructing subdivisions.

Definition. Let K be a complex. We define a sequence of subdivisions of the skeletons of K as follows: Let $L_0 = K^{(0)}$, the 0-skeleton of K. In general, if L_p is a subdivision of the p-skeleton of K, let L_{p+1} be the subdivision of the $p + 1$ skeleton obtained by starring L_p from the barycenters of the $p + 1$ simplices of

K. By Lemma 15.2, the union of the complexes L_p is a subdivision of K. It is called the **first barycentric subdivision** of K, and denoted sd K.

Having formed a complex sd K, we can now construct *its* first barycentric subdivision sd(sd K), which we denote by sd^2K. This complex is called the **second barycentric subdivision** of K. Similarly one defines sd^nK in general.

On some occasions it is convenient to have a specific description of the simplices of the first barycentric subdivision. We give such a description now. Let us use the notation $\sigma_1 \succ \sigma_2$ to mean "σ_2 is a proper face of σ_1."

Lemma 15.3. *The complex* sd K *equals the collection of all simplices of the form*

$$\hat{\sigma}_1\hat{\sigma}_2 \ldots \hat{\sigma}_n,$$

where $\sigma_1 \succ \sigma_2 \succ \ldots \succ \sigma_n$.

Proof. We prove this fact by induction. It is immediate that the simplices of sd K lying in the subdivision of $K^{(0)}$ are of this form. (Each such simplex is a vertex of K, and $\hat{v} = v$ for a vertex.)

Suppose now that each simplex of sd K lying in $|K^{(p)}|$ is of this form. Let τ be a simplex of sd K lying in $|K^{(p+1)}|$ and not in $|K^{(p)}|$. Then τ belongs to one of the complexes $\hat{\sigma} * L_\sigma$, where σ is a $p+1$ simplex of K and L_σ is the first barycentric subdivision of the complex consisting of the proper faces of σ. By the induction hypothesis, each simplex of L_σ is of the form $\hat{\sigma}_1\hat{\sigma}_2 \ldots \hat{\sigma}_n$, where $\sigma_1 \succ \sigma_2 \succ \ldots \succ \sigma_n$ and σ_1 is a proper face of σ. Then τ must be of the form

$$\hat{\sigma}\hat{\sigma}_1\hat{\sigma}_2 \ldots \hat{\sigma}_n,$$

which is of the desired form. □

Example 3. Consider the complex K indicated in Figure 15.2. Its first and second barycentric subdivisions are pictured. Note that each of the simplices of sd K is of the form described in Lemma 15.3. Note also how rapidly the simplices of sd^nK decrease in size as n increases. This is a general fact, which we shall prove now.

Theorem 15.4. *Given a finite complex K, given a metric for $|K|$, and given $\epsilon > 0$, there is an N such that each simplex of* sd^NK *has diameter less than* ϵ.

Proof. Because K is finite, $|K|$ is a subspace of the euclidean space $\mathbf{E}^J$ in which it lies. Because $|K|$ is compact, it is irrelevant which metric we use for $|K|$. (For if d_1 and d_2 are two metrics for $|K|$, then the identity map of $(|K|,d_1)$ to $(|K|,d_2)$ is uniformly continuous. Thus given $\epsilon > 0$, there is a $\delta > 0$ such that any set with d_1-diameter less than δ has d_2-diameter less than ϵ.) Therefore we may as well use the metric of $\mathbf{E}^J$, which is

$$|x - y| = \max |x_\alpha - y_\alpha|.$$

Step 1. We show that if $\sigma = v_0 \ldots v_p$ is a simplex, then the diameter of σ equals the number $l = \max |v_i - v_j|$, which is the maximum distance between

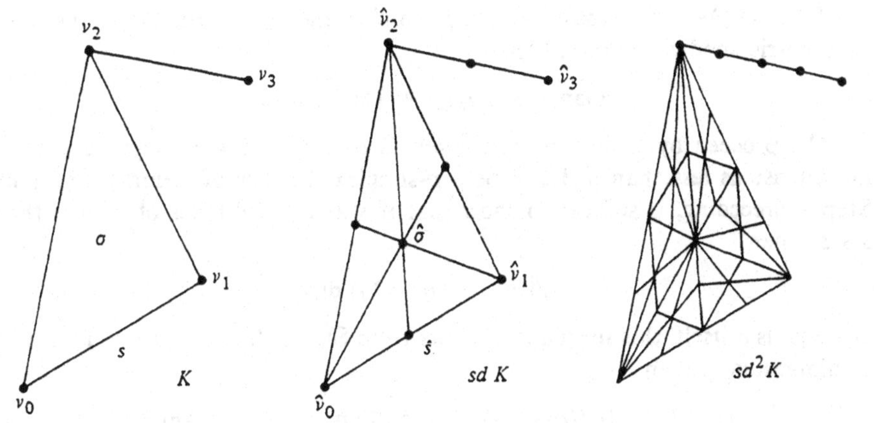

Figure 15.2

the vertices of σ. Because $v_i, v_j \in \sigma$, we know that diam $\sigma \geq l$. We wish to prove the reverse inequality.

We first show that $|x - v_i| \leq l$ for every $x \in \sigma$. Consider the closed neighborhood of v_i of radius l in $\mathbf{E}^J$, defined by the equation

$$C(v_i;l) = \{x; |x - v_i| \leq l\}.$$

You can check that this set is convex. Therefore, since it contains all the vertices of σ, it must contain σ. Then $|x - v_i| \leq l$ for $x \in \sigma$.

Now we show that $|x - z| \leq l$ for all $x,z \in \sigma$, so that diam $\sigma \leq l$, as desired. Given x, consider the closed neighborhood $C(x;l)$. This set contains all the vertices of σ, by the result of the preceding paragraph. Being convex, it contains σ. Thus $|x - z| \leq l$ for $x,z \in \sigma$.

Step 2. We show that if σ has dimension p, then for every $z \in \sigma$,

$$|\hat{\sigma} - z| \leq \frac{p}{p+1} \operatorname{diam} \sigma.$$

For this purpose, we compute

$$\begin{aligned} |v_0 - \hat{\sigma}| &= \left| v_0 - \sum_{i=0}^{p} (1/(p+1)) v_i \right| \\ &\leq \sum_{i=1}^{p} |(1/(p+1))\,(v_0 - v_i)| \\ &\leq (p/(p+1)) \max |v_0 - v_i| \leq (p/(p+1)) \operatorname{diam} \sigma. \end{aligned}$$

A similar computation holds for $|v_j - \hat{\sigma}|$. Therefore the closed neighborhood of $\hat{\sigma}$ of radius $(p/(p+1))$ diam σ contains all vertices of σ. Being convex, it contains σ.

Step 3. We show that if σ is a p-simplex and τ is a simplex in the first barycentric subdivision of σ, then

$$\operatorname{diam} \tau \leq (p/(p+1)) \operatorname{diam} \sigma.$$

We proceed by induction. The result is trivial for $p = 0$; suppose it true in dimensions less than p. Let σ be a p-simplex. In view of Lemma 15.3 and Step 1 preceding, it suffices to show that if s and s' are faces of σ such that $s \succ s'$, then

$$|\hat{s} - \hat{s}'| \leq (p/(p+1)) \operatorname{diam} \sigma.$$

If s equals σ itself, this inequality follows from Step 2. If s is a proper face of σ of dimension q, then

$$|\hat{s} - \hat{s}'| \leq (q/(q+1)) \operatorname{diam} s \leq (p/(p+1)) \operatorname{diam} \sigma.$$

The first inequality follows by the induction hypothesis, and the second from the fact that $f(x) = x/(x+1)$ is increasing for $x > 0$.

Step 4. Let K have dimension n; let d be the maximum diameter of a simplex of K. The maximum diameter of a simplex in the Nth barycentric subdivision of K is $(n/(n+1))^N d$; if N is sufficiently large, this number is less than ϵ. □

EXERCISES

1. Let K be a complex; let $x_0 \in |K|$.
   (a) Show there is a subdivision of K whose vertex set contains x_0.
   *(b) Show there is a subdivision of K whose vertex set consists of x_0 and the vertices of K.

2. If $\mathcal{A}$ and $\mathcal{B}$ are collections of sets, we say that $\mathcal{B}$ **refines** $\mathcal{A}$ if for each $B \in \mathcal{B}$, there is an $A \in \mathcal{A}$ such that $B \subset A$.

   A space X is said to have **finite covering dimension** if there is an integer m satisfying the following condition: For every open covering $\mathcal{A}$ of X, there is an open covering $\mathcal{B}$ of X that refines $\mathcal{A}$, such that no point of X lies in more than $m + 1$ elements of $\mathcal{B}$.

   The **covering dimension** of such a space X is the smallest integer m for which this condition holds.
   (a) Show that a discrete set has covering dimension 0.
   (b) Show that $[0,1]$ has covering dimension 1.
   (c) Show that if X has covering dimension m, then any closed subspace A of X has covering dimension at most m.
   (d) Show that if K is a finite complex of dimension m, then the covering dimension of $|K|$ exists and is at most m. (We will see later that it is precisely m.)

§16. THE SIMPLICIAL APPROXIMATION THEOREM

We now show that if $h : |K| \to |L|$ is a continuous map, then there is a subdivision K' of K such that h has a simplicial approximation $f : K' \to L$. The proof when K is finite follows easily from the results of the preceding section; barycentric subdivision will suffice. The general case requires a slightly more sophisticated technique of subdivision, called generalized barycentric subdivision, which we shall describe shortly.

Theorem 16.1 (The finite simplicial approximation theorem). *Let K and L be complexes; let K be finite. Given a continuous map $h : |K| \to |L|$, there is an N such that h has a simplicial approximation $f : \text{sd}^N K \to L$.*

Proof. Cover $|K|$ by the open sets $h^{-1}(\text{St } w)$, as w ranges over the vertices of L. Now given this open covering $\mathcal{A}$ of the compact metric space K, there is a number λ such that any set of diameter less than λ lies in one of the elements of $\mathcal{A}$; such a number is called a **Lebesgue number** for $\mathcal{A}$. If there were no such λ, one could choose a sequence C_n of sets, where C_n has diameter less than $1/n$ but does not lie in any element of $\mathcal{A}$. Choose $x_n \in C_n$; by compactness, some subsequence x_{n_i} converges, say to x. Now $x \in A$ for some $A \in \mathcal{A}$. Because A is open, it contains C_{n_i} for i sufficiently large, contrary to construction.

Choose N so that each simplex in $\text{sd}^N K$ has diameter less than $\lambda/2$. Then each star of a vertex in $\text{sd}^N K$ has diameter less than λ, so it lies in one of the sets $h^{-1}(\text{St } w)$. Then $h : |K| \to |L|$ satisfies the star condition relative to $\text{sd}^N K$ and L, and the desired simplicial approximation exists. □

As a preliminary step toward our generalized version of barycentric subdivision, we show how to subdivide a complex K in a way that leaves a given subcomplex K_0 unchanged.

Definition. Let K be a complex; let K_0 be a subcomplex. We define a sequence of subdivisions of the skeletons of K as follows: Let $J_0 = K^{(0)}$. In general, suppose J_p is a subdivision of the p-skeleton of K, and each simplex of K_0 of dimension at most p belongs to J_p. Define J_{p+1} to be the union of the complex J_p, all $p+1$ simplices σ belonging to K_0, and the cones $\hat{\sigma} * J_\sigma$, as σ ranges over all $p+1$ simplices of K *not* in K_0. (Here J_σ is the subcomplex of J_p whose polytope is $\text{Bd } \sigma$.) The union of the complexes J_p is a subdivision of K, denoted $\text{sd}(K/K_0)$ and called the **first barycentric subdivision of K, holding K_0 fixed.**

As with barycentric subdivisions, this process can now be repeated. The complex $\text{sd}(\text{sd}(K/K_0)/K_0)$ will be called the **second barycentric subdivision of K holding K_0 fixed,** and denoted by $\text{sd}^2(K/K_0)$. And so on.

Example 1. Figure 16.1 illustrates the first barycentric subdivision sd K of a complex K, and the first barycentric subdivision $\text{sd}(K/K_0)$ holding a complex K_0 fixed.

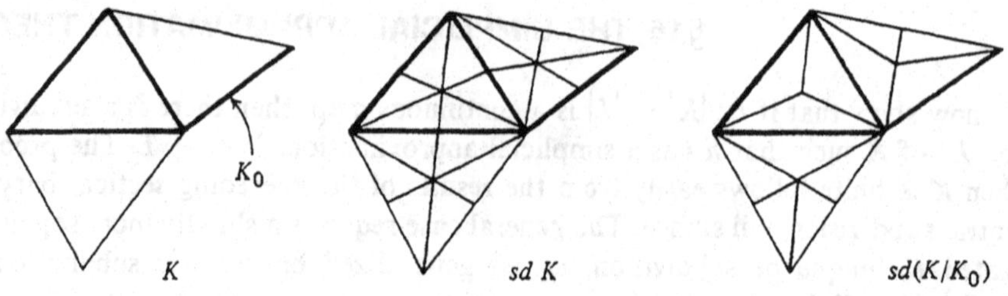

Figure 16.1

To prove the general simplicial approximation theorem, iterated barycentric subdivision will not suffice, because the Lebesgue number argument used in the proof of Theorem 16.1 requires the space $|K|$ to be compact. For a general complex K, the number λ that measures how finely a simplex must be subdivided may vary from one simplex to another. Thus we must generalize our notion of barycentric subdivision to allow for this possibility:

Definition. Let K be a complex. Let N be a function assigning to each positive-dimensional simplex σ of K, a non-negative integer $N(\sigma)$. We construct a subdivision of K as follows: Let $L_0 = K^{(0)}$. In general, suppose L_p is a subdivision of the p-skeleton of K. For each $p + 1$ simplex σ of K, let L_σ be the subcomplex of L_p whose polytope is Bd σ. Form the cone $\hat{\sigma} * L_\sigma$; then subdivide this cone barycentrically $N(\sigma)$ times, holding L_σ fixed. Define L_{p+1} to be the union of L_p and the complexes

$$\text{sd}^{N(\sigma)}(\hat{\sigma} * L_\sigma/L_\sigma),$$

as σ ranges over all $p + 1$ simplices of K. Then L_{p+1} is a subdivision of the $p + 1$ skeleton of K. The union of the complexes L_p is a subdivision of K. It is called the **generalized barycentric subdivision** of K corresponding to the function $N(\sigma)$.

The remainder of this section is devoted to showing that this generalized barycentric subdivision is adequate to prove the general simplicial approximation theorem. The techniques involved in the proof will not be used later in the book, so the reader may skip the details and simply take the theorem on faith if desired.

Keep Figure 16.1 in mind as we compare the complex sd(K/K_0) with the complex sd K in general:

Lemma 16.2. *Let K_0 be a subcomplex of K.*

(a) *If τ is a simplex of* sd(K/K_0), *then τ is of the form*

$$\tau = \hat{\sigma}_1 \ldots \hat{\sigma}_q v_0 \ldots v_p,$$

where $s = v_0 \ldots v_p$ is a simplex of K_0, and $\sigma_1, \ldots, \sigma_q$ are simplices of K not in K_0, and $\sigma_1 > \cdots > \sigma_q > s$.

(b) *Either $v_0 \dots v_p$ or $\hat{\sigma}_1 \dots \hat{\sigma}_q$ may be missing from this expression. The simplex τ is disjoint from $|K_0|$ if and only if $v_0 \dots v_p$ is missing; in this case, τ is a simplex of* sd K.

Proof. (a) The result is true if τ is in J_0. In general, let τ be a simplex of J_{p+1} not in J_p. Then either τ belongs to K_0, in which case τ is of the form $v_0 \dots v_r$, or τ belongs to one of the cones $\hat{\sigma} * J_\sigma$. Now each simplex of J_σ has the form $\hat{\sigma}_1 \dots \hat{\sigma}_q v_0 \dots v_r$, by the induction hypothesis, where $\sigma \succ \sigma_1$. Then τ has the form $\hat{\sigma}\hat{\sigma}_1 \dots \hat{\sigma}_q v_0 \dots v_r$, as desired.

(b) Let $\tau = \hat{\sigma}_1 \dots \hat{\sigma}_q v_0 \dots v_p$. If $v_0 \dots v_p$ is not missing from this expression, then τ intersects $|K_0|$ in $v_0 \dots v_p$ at least. Conversely, if the set $\tau \cap |K_0|$ is nonempty, then it contains a face of τ and hence a vertex of τ. Since none of the points $\hat{\sigma}_1, \dots, \hat{\sigma}_q$ is in $|K_0|$, the term $v_0 \dots v_p$ cannot be missing. □

To prove the general simplicial approximation theorem, we need to show that given an arbitrary continuous map $h : |K| \to |L|$, there is a subdivision K' of K such that h satisfies the star condition relative to K' and L. This is equivalent to the statement that if $\mathcal{A}$ is the open covering of K defined by

$$\mathcal{A} = \{h^{-1}(\mathrm{St}(w,L)) \mid w \text{ a vertex of } L\},$$

then there is a subdivision of K' of K such that the collection of open stars

$$\mathcal{B} = \{\mathrm{St}(v,K') \mid v \text{ a vertex of } K'\}$$

refines $\mathcal{A}$. (Recall that a collection $\mathcal{B}$ *refines* a collection $\mathcal{A}$ if for each element B of $\mathcal{B}$, there is an element A of $\mathcal{A}$ that contains B.)

To make the proof work, we actually need to prove something slightly stronger than this. We shall construct a subdivision K' of K fine enough that the collection $\{\overline{\mathrm{St}}(v,K')\}$ of *closed* stars in K' refines the collection $\mathcal{A}$.

The following lemma gives the crux of the argument; it will enable us to carry out the induction step of the proof.

Lemma 16.3. *Let $K = p * B$ be a cone over the finite complex B. Let $\mathcal{A}$ be an open covering of $|K|$. Suppose there is a function assigning to each vertex v of the complex B, an element A_v of $\mathcal{A}$ such that*

$$\overline{\mathrm{St}}(v,B) \subset A_v.$$

Then there is an N such that the collection of closed stars of the subdivision $\mathrm{sd}^N(K/B)$ *refines $\mathcal{A}$, and furthermore such that for each vertex v of B,*

$$\overline{\mathrm{St}}(v,\mathrm{sd}^N(K/B)) \subset A_v.$$

Proof. We assume that $|B|$ lies in $\mathbf{R}^m \times 0$ for some m, and that $p = (0,\dots,0,1)$ in $\mathbf{R}^m \times \mathbf{R}$. Let $n = \dim K$.

Step 1. In general, as N increases, the maximum diameter of the simplices of $\mathrm{sd}^N(K/B)$ does not go to zero. For if σ has a positive-dimensional face in B, that face never gets subdivided. However, it is true that as N increases, the simplices that intersect the plane $\mathbf{R}^m \times 0$ lie closer and closer to this plane. More generally, we show that if K' is any subdivision of K that keeps B fixed, and if the simplices of K' that intersect $\mathbf{R}^m \times 0$ lie in the strip $\mathbf{R}^m \times [0,\epsilon]$, then any simplex τ of $\operatorname{sd}(K'/B)$ that intersects $\mathbf{R}^m \times 0$ lies in the strip $\mathbf{R}^m \times [0,n\epsilon/(n+1)]$.

The simplex τ is of the form $\hat{\sigma}_1 \dots \hat{\sigma}_q v_0 \dots v_p$, as in Lemma 16.2; assuming τ intersects $\mathbf{R}^m \times 0$ but does not lie in it, neither $\hat{\sigma}_1 \dots \hat{\sigma}_q$ nor $v_0 \dots v_p$ is missing from this expression. Each vertex v_i lies in $\mathbf{R}^m \times 0$.

Consider the vertex $\hat{\sigma}_j$ of τ. The simplex σ_j of K' intersects $\mathbf{R}^m \times 0$ because σ_j has $v_0 \dots v_p$ as a face; therefore $\sigma_j \subset \mathbf{R}^m \times [0,\epsilon]$. Let $w_0, \dots, w_k$ be the vertices of σ_j, and let $\pi : \mathbf{R}^m \times \mathbf{R} \to \mathbf{R}$ be projection on the last coordinate. Then $\pi(w_i) \leq \epsilon$ for $i = 0, \dots, k$; and $\pi(w_i) = 0$ for at least one i. We compute

$$\pi(\hat{\sigma}_j) = \sum_{i=0}^{k} \left(\frac{1}{k+1}\right) \pi(w_i) \leq \left(\frac{k}{k+1}\right) \epsilon.$$

Thus each vertex of τ lies in the set $\mathbf{R}^m \times [0,(n/(n+1))\epsilon]$. Because this set is convex, all of τ lies in it.

Step 2. For convenience, let K_N denote the complex $\operatorname{sd}^N(K/B)$. We show there is an integer N_0 such that if $N \geq N_0$, then for each vertex v of B,

$$(*) \qquad \overline{\operatorname{St}}(v,K_N) \subset A_v.$$

This is part of what we want to prove.

We are given that $\overline{\operatorname{St}}(v,B) \subset A_v$. We assert that we can choose $\delta > 0$ so that

$$\overline{\operatorname{St}}(v,K) \cap (\mathbf{R}^m \times [0,\delta]) \subset A_v$$

for each $v \in B$. See Figure 16.2. To prove this assertion, consider the continuous map $\rho : |B| \times I \to |K|$ defined by $\rho(x,t) = (1-t)x + tp$. The map ρ carries $\overline{\operatorname{St}}(v,B) \times I$ onto $\overline{\operatorname{St}}(v,K)$, because K is a cone over B. Furthermore, ρ preserves the last coordinate, since

$$\begin{aligned} \pi\rho(x,t) &= (1-t)\pi(x) + t\pi(p) \\ &= (1-t)\cdot 0 + t \cdot 1 = \pi(x,t). \end{aligned}$$

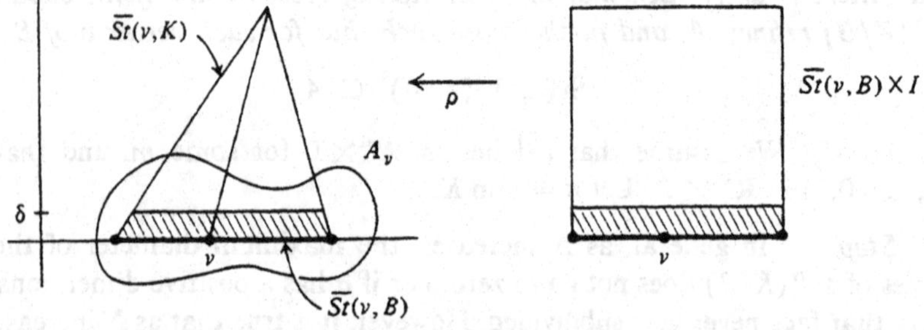

Figure 16.2

Now $\overline{\mathrm{St}}(v,B)$ is a compact set. The "tube lemma" of general topology enables us to choose δ such that

$$\overline{\mathrm{St}}(v,B) \times [0,\delta] \subset \rho^{-1}(A_v).$$

(More directly, one can cover $\overline{\mathrm{St}}(v,B) \times 0$ with finitely many sets of the form $U_i \times [0,\delta_i]$, each lying in $\rho^{-1}(A_v)$; then choose $\delta = \min \delta_i$.) It follows that the set

$$\rho(\overline{\mathrm{St}}(v,B) \times [0,\delta]) = \overline{\mathrm{St}}(v,K) \cap (\mathbf{R}^m \times [0,\delta])$$

lies in A_v, as desired.

Now we can choose N_0. Applying Step 1, we choose N_0 so that for $N \geq N_0$, each simplex of K_N that intersects $\mathbf{R}^m \times 0$ lies in $\mathbf{R}^m \times [0,\delta]$. Then if v is a vertex of B, the set $\overline{\mathrm{St}}(v,K_N)$ lies in $\mathbf{R}^m \times [0,\delta]$. Since this set also lies in $\overline{\mathrm{St}}(v,K)$, it lies in A_v, as desired.

Step 3. The integer N_0 is now fixed. Consider the complex K_{N_0+1}. Let P be the union of all simplices of K_{N_0+1} that intersect B. Let Q be the union of all simplices of K_{N_0+1} that do not intersect B. See Figure 16.3. We prove the following: If $N \geq N_0 + 1$, then for each vertex w of K_N lying in P but not in $|B|$, there is an element A of $\mathcal{A}$ such that

$$(**) \qquad \overline{\mathrm{St}}(w,K_N) \subset A.$$

We prove (**) first in the case $N = N_0 + 1$. Now P is the polytope of a subcomplex of K_{N_0+1} by definition. If w is a vertex of K_{N_0+1} lying in P but not in B, then $w = \hat{\sigma}$ for some simplex σ of K_{N_0} that intersects B but does not lie in B, by Lemma 16.2. See Figure 16.4. Let v be a vertex of σ lying in B. Because w lies in Int σ, we have

$$\mathrm{St}(w,K_{N_0+1}) \subset \mathrm{St}(v,K_{N_0})$$

by Lemma 15.1. Then

$$\overline{\mathrm{St}}(w,K_{N_0+1}) \subset \overline{\mathrm{St}}(v,K_{N_0}) \subset A_v$$

by (*) of Step 2.

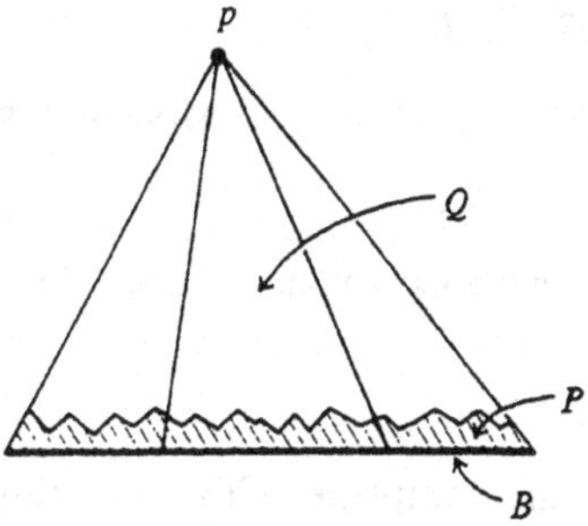

Figure 16.3

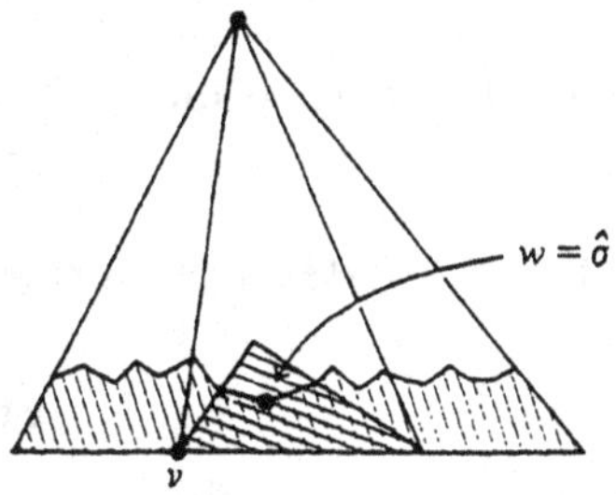

Figure 16.4

Now we prove (**) in the case $N > N_0 + 1$. If w' is a vertex of K_N lying in P, then $w' \in \mathrm{St}(w, K_{N_0+1})$ for some vertex w of K_{N_0+1} lying in P. Then $\mathrm{St}(w', K_N) \subset \mathrm{St}(w, K_{N_0+1})$, by Lemma 15.1. Then (**) follows from the result of the preceding paragraph.

Step 4. We now complete the proof. Let λ be a Lebesgue number for the open covering $\mathcal{A}$ of $|K|$. Consider the space Q. It is the polytope of a subcomplex J of K_{N_0+1}. In forming the subdivision K_{N_0+2}, each simplex of J is subdivided barycentrically, by the preceding lemma. Thus K_{N_0+2} has $\mathrm{sd}\, J$ as a subcomplex. Repeating the argument, we see that in general K_{N_0+1+M} has $\mathrm{sd}^M J$ as a subcomplex.

Choose M large enough that each simplex of $\mathrm{sd}^M J$ has diameter less than $\lambda/2$. Then if $N \geq N_0 + 1 + M$, and if w is a vertex of K_N not in P, we show that there is an element A of $\mathcal{A}$ such that

$$(***) \qquad \overline{\mathrm{St}}(w, K_N) \subset A.$$

For since w is not in P, each simplex of K_N having w as a vertex must lie in Q and hence must be a simplex of $\mathrm{sd}^M J$. Therefore $\overline{\mathrm{St}}(w, K_N)$ has diameter less than λ, so it lies in an element of $\mathcal{A}$.

The combination of (*), (**), and (***) proves the lemma. □

Theorem 16.4. *Let K be a complex; let $\mathcal{A}$ be an open covering of $|K|$. There exists a generalized barycentric subdivision K' of K such that the collection of closed stars $\{\overline{\mathrm{St}}(w, K')\}$, for w a vertex of K', refines $\mathcal{A}$.*

Proof. We proceed step-by-step. Initially, we let $L_0 = K^{(0)}$, and for each point v of $K^{(0)}$, we let A_v denote an element of $\mathcal{A}$ that contains v.

In general, we assume that a subdivision L_p of $K^{(p)}$ is given, and that a function f_p is given assigning to each vertex v of L_p an element A_v of $\mathcal{A}$ such that

$$\overline{\mathrm{St}}(v, L_p) \subset A_v.$$

We extend L_p to a subdivision L_{p+1} of the $p+1$ skeleton of K and we extend f_p to a function f_{p+1}, in a manner we now describe.

We proceed as follows: For each $p+1$ simplex σ of K, the space $\mathrm{Bd}\,\sigma$ is the polytope of a subcomplex L_σ of L_p. Consider the cone $\hat{\sigma} * L_\sigma$. By the preceding lemma, there is an integer $N(\sigma)$ such that if we set

$$C(\sigma) = \mathrm{sd}^{N(\sigma)}(\hat{\sigma} * L_\sigma / L_\sigma),$$

then the following conditions hold: For each vertex v of $C(\sigma)$belonging to L_σ,

$$\overline{\mathrm{St}}(v, C(\sigma)) \subset A_v,$$

and for each vertex w of $C(\sigma)$ not in L_σ, there *exists* an element A of $\mathcal{A}$ such that

$$\overline{\mathrm{St}}(w, C(\sigma)) \subset A.$$

We define L_{p+1} to be the union of L_p and the complexes $C(\sigma)$, as σ ranges over the $p+1$ simplices of K.

If v is a vertex of L_p, then $\overline{\text{St}}(v,L_{p+1})$ is the union of the sets $\overline{\text{St}}(v,L_p)$ and $\overline{\text{St}}(v,C(\sigma))$, as σ ranges over the $p+1$ simplices of K containing v. Each of these sets lies in A_v, by construction.

On the other hand, if w is a vertex of L_{p+1} not in L_p, then w lies interior to some $p+1$ simplex σ of K, so that

$$\overline{\text{St}}(w,L_{p+1}) = \overline{\text{St}}(w,C(\sigma)).$$

The latter set is contained in some element A of $\mathcal{A}$; we define $f_{p+1}(w)$ to be such an element A_w of $\mathcal{A}$. Then the induction step is complete.

The theorem follows. The complex K' is defined to be the union of the complexes L_p, and the function $f(v) = A_v$ from the vertices of K' to $\mathcal{A}$ is defined to be the union of the functions f_p. The function f satisfies the requirements of the theorem. For let v be a vertex of K'. Then v is a vertex of L_p for some p, and

$$\overline{\text{St}}(v,L_{p+k}) \subset f_{p+k}(v) = f(v) = A_v$$

for all $k \geq 0$, from which it follows that $\overline{\text{St}}(v,K') \subset A_v$. □

Theorem 16.5 (The general simplicial approximation theorem). *Let K and L be complexes; let $h : |K| \to |L|$ be a continuous map. There exists a subdivision K' of K such that h has a simplicial approximation $f : K' \to L$.*

Proof. Let $\mathcal{A}$ be the covering of $|K|$ by the open sets $h^{-1}(\text{St}(w,L))$, as w ranges over the vertices of L. Choose a subdivision K' of K whose closed stars refine $\mathcal{A}$. Then h satisfies the star condition relative to K' and L. □

EXERCISES

1. (a) Using Theorem 16.1, show that if K and L are finite and $\dim K = m$, then any continuous map $h : |K| \to |L|$ is homotopic to a map carrying K into $L^{(m)}$, where $L^{(m)}$ is the m-skeleton of L. [*Hint:* See Exercise 2 of §14.]
   We shall consider the non-finite case later.
   (b) Show that if $h : S^m \to S^n$ and $m < n$, then h is homotopic to a constant map. [*Hint:* Any map $f : X \to S^n - p$ is homotopic to a constant.]
2. Let $h : |K| \to |L|$ be a continuous map. Given $\epsilon > 0$, show that there are subdivisions K' and L' of K and L, respectively, and a simplicial map $f : K' \to L'$, such that $|f(x) - h(x)| < \epsilon$ for all x in $|K|$.
3. Show that if K is a complex of dimension m, then $|K|$ has covering dimension at most m. (See the exercises of §15.)

§17. THE ALGEBRA OF SUBDIVISION

Now we explore some of the algebraic consequences of subdivision, determining what subdivision does to the homology groups. We prove that if K is any complex and if K' is a subdivision of K, then there is a uniquely defined chain map

$\lambda : C_p(K) \to C_p(K')$ called the *subdivision operator* that induces an isomorphism of homology groups. Furthermore, if $g : K' \to K$ is any simplicial approximation to the identity map of $|K|$, then λ and $g_\#$ are chain-homotopy inverse to each other, so g_* is also an isomorphism.

Lemma 17.1. *Let K' be a subdivision of K. Then the identity map $i : |K| \to |K|$ has a simplicial approximation*

$$g : K' \to K.$$

Let τ be a simplex of K' and let σ be a simplex of K; if $\tau \subset \sigma$, then $g(\tau) \subset \sigma$.

Proof. By Lemma 15.1, the map i has a simplicial approximation g. Given $\tau \subset \sigma$, let w be a vertex of τ. Then w lies interior to σ or to a face of σ. Then g maps w to a vertex of σ, by Lemma 14.1. □

Definition. Let K' be a subdivision of K. If σ is a simplex of K, let $K(\sigma)$ denote the subcomplex of K consisting of σ and its faces, and let $K'(\sigma)$ denote the subcomplex of K' whose polytope is σ.

Theorem 17.2 (The algebraic subdivision theorem). *Let K' be a subdivision of K. There is a unique augmentation-preserving chain map*

$$\lambda : \mathcal{C}(K) \to \mathcal{C}(K')$$

such that $\lambda(\sigma)$ is carried by $K'(\sigma)$ for each σ. If $g : K' \to K$ is a simplicial approximation to the identity, then λ and $g_\#$ are chain-homotopy inverses, so λ_* and g_* are isomorphisms.

We call λ the **subdivision operator.**

Proof. Step 1. We show first that the theorem holds if K' satisfies the condition that for each $\sigma \in K$ the induced subdivision $K'(\sigma)$ of σ is acyclic. We shall use Theorem 13.3, the geometric version of the acyclic carrier theorem. We define acyclic carriers as follows:

$$\Psi \circlearrowright K \underset{\Theta}{\overset{\Lambda}{\rightleftarrows}} K' \circlearrowleft \Phi$$

The carriers Ψ and Λ are easy to define; we set

$$\Psi(\sigma) = K(\sigma),$$
$$\Lambda(\sigma) = K'(\sigma),$$

for each $\sigma \in K$. The complex $K(\sigma)$ is acyclic because it consists of a simplex and its faces, and the complex $K'(\sigma)$ is acyclic by hypothesis. The inclusion condition for an acyclic carrier is immediate; if $s \prec \sigma$, then $K(s) \subset K(\sigma)$ and $K'(s) \subset K'(\sigma)$.

To define Θ and Φ, we proceed as follows: For each simplex $\tau \in K'$, let σ_τ be

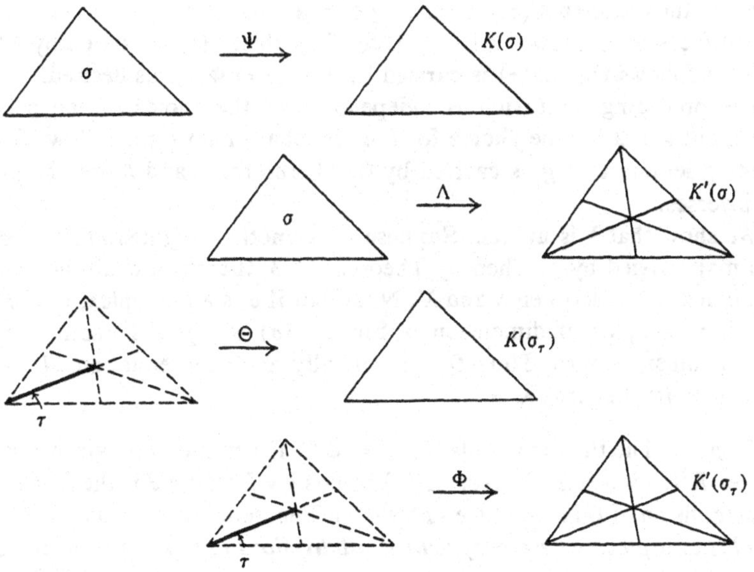

Figure 17.1

the simplex of K *of smallest dimension* such that $\tau \subset \sigma_\tau$. Then if t is a face of τ, we have $\sigma_t \subset \sigma_\tau$. For since both σ_t and σ_τ contain t, their intersection also contains t; because σ_t has minimal dimension, it must equal this intersection. We define

$$\Theta(\tau) = K(\sigma_\tau),$$
$$\Phi(\tau) = K'(\sigma_\tau);$$

both complexes are acyclic. The inclusion condition follows from the fact that if $t \prec \tau$, then $\sigma_t \subset \sigma_\tau$. See Figure 17.1, which illustrates these carriers in the case where $K' = \text{sd } K$.

By Theorem 13.3, there exist chain maps λ and θ

$$C_p(K) \underset{\theta}{\overset{\lambda}{\rightleftarrows}} C_p(K')$$

preserving augmentation and carried by Λ and Θ, respectively.

Now the identity map $C_p(K) \to C_p(K)$ is carried by Ψ (trivially). We show that $\theta \circ \lambda$ is also carried by Ψ; whence it follows that $\theta \circ \lambda$ is chain homotopic to the identity. If σ is a simplex of K, then $\lambda(\sigma)$ is a chain of $K'(\sigma)$. Now each simplex τ in the subdivision $K'(\sigma)$ of σ is contained in σ, whence σ_τ equals σ or a face of σ. In any case, if τ appears in the chain $\lambda(\sigma)$, then $\theta(\tau)$ is carried by $K(\sigma_\tau) \subset K(\sigma)$. Thus $\theta(\lambda(\sigma))$ is a chain of $K(\sigma)$, so $\theta \circ \lambda$ is carried by Ψ.

The identity map $C_p(K') \to C_p(K')$ is carried by Φ; for τ is contained in σ_τ by definition, so that τ is a simplex of $K'(\sigma_\tau)$. We show that $\lambda \circ \theta$ is also carried by Φ, whence $\lambda \circ \theta$ is chain homotopic to the identity. If $\tau \in K'$, then $\theta(\tau)$ is carried by the complex $K(\sigma_\tau)$ consisting of σ_τ and its faces, so it equals a sum of oriented faces of σ_τ. Now if s is any face of σ_τ, then $\lambda(s)$ is carried by $K'(s) \subset K'(\sigma_\tau)$. It follows that $\lambda\theta(\tau)$ is carried by $K'(\sigma_\tau) = \Phi(\tau)$, as desired.

The preceding discussion is independent of the choice of the particular chain maps θ and λ. One choice for θ is the chain map $g_\#$; it follows from the preceding lemma that $g_\#$ is carried by Θ. Therefore, $g_\#$ and λ are chain-homotopy inverses.

We show that λ is unique. Suppose λ' is another augmentation-preserving chain map carried by Λ. Then by Theorem 13.3, there is a chain homotopy D, also carried by Λ, between λ and λ'. Note that if σ is a p-simplex, then $\Lambda(\sigma) = K'(\sigma)$ is a complex of dimension p. Since $D(\sigma)$ is a $p + 1$ chain carried by $K'(\sigma)$, it must be zero. Thus D is identically zero; the equation $\partial D + D\partial = \lambda - \lambda'$ now implies that $\lambda = \lambda'$.

Step 2. The theorem holds if $K' = \operatorname{sd} K$. For in this case, given $\sigma \in K$, the complex $K'(\sigma)$ is a *cone*. In fact, $K'(\sigma)$ equals $\hat{\sigma} * J$, where J is the first barycentric subdivision of $\operatorname{Bd} \sigma$. And we know from Theorem 8.2 that cones are acyclic.

This is a place where barycentric subdivisions are essential to the proof.

Step 3. The theorem is true if $K' = \operatorname{sd}^N K$. In view of Step 1, it suffices to prove that for any simplex σ of K, the complex $\operatorname{sd}^N K(\sigma)$ is acyclic. This follows from Step 2, which implies that for any complex L,

$$\tilde{H}_i(L) \simeq \tilde{H}_i(\operatorname{sd} L) \simeq \tilde{H}_i(\operatorname{sd}^2 L) \simeq \cdots.$$

In particular, if L is acyclic, so is $\operatorname{sd}^N L$.

Step 4. The theorem holds in general. In view of Step 1, it suffices to prove that if $\sigma \in K$ and if K' is any subdivision of K, then $K'(\sigma)$ is acyclic.

Let $L = K(\sigma)$ and $L' = K'(\sigma)$. Then L is acyclic, and we wish to prove that L' is acyclic. We proceed as follows: Let $g : L' \to L$ be a simplicial approximation to the identity. Choose N so that the identity map of $|L|$ to itself has a simplicial approximation $f : \operatorname{sd}^N L \to L'$; for this we need the (finite) simplicial approximation theorem. Proceeding similarly, choose M so the identity has a simplicial approximation $k : \operatorname{sd}^M(L') \to \operatorname{sd}^N L$.

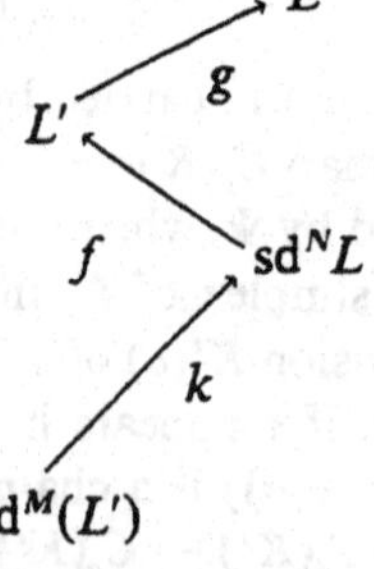

Now we note that $g \circ f$ is a simplicial approximation to the identity, so that by Step 3, $(g \circ f)_* = g_* \circ f_*$ is an isomorphism. For the same reason, $(f \circ k)_* =$

$f_* \circ k_*$ is an isomorphism. The first fact implies f_* is injective and the second implies f_* is surjective. Thus f_* is an isomorphism. Finally, because $g_* \circ f_*$ and f_* are isomorphisms, so is g_*. Thus L' is acyclic. □

Definition. In the special case where K' is the first barycentric subdivision of K, we denote the subdivision operator λ by

$$\mathrm{sd} : C_p(K) \to C_p(\mathrm{sd}\, K)$$

and call it the **barycentric subdivision operator.** (Here we abuse notation, letting sd denote both the "algebraic" and "geometric" subdivision operators.)

There is an inductive formula for the operator sd. It is the following:

$$\mathrm{sd}(v) = v,$$
$$\mathrm{sd}(\sigma) = [\hat{\sigma}, \mathrm{sd}(\partial\sigma)],$$

where the bracket notation has the meaning we gave it in §8. We leave this formula for you to check.

Application to relative homology

We can generalize the preceding theorem to relative homology with no difficulty:

Theorem 17.3. *Let K_0 be a subcomplex of K. Given the subdivision K' of K, let K_0' denote the induced subdivision of K_0. The subdivision operator λ induces a chain map*

$$\lambda : C_p(K,K_0) \to C_p(K',K_0').$$

If $g : (K',K_0') \to (K,K_0)$ is any simplicial approximation to the identity, then λ and $g_\#$ are chain-homotopy inverse to each other.

Proof. We check that each of the acyclic carriers defined in Step 1 of the preceding proof preserves the subcomplexes involved. Certainly if $\sigma \in K_0$, then $\Psi(\sigma) = K(\sigma)$ and $\Lambda(\sigma) = K'(\sigma)$ are subcomplexes of K_0 and K_0', respectively. On the other hand, if $\tau \in K_0'$, then τ is contained in some simplex σ of K_0; it follows that the simplex σ_τ of K of smallest dimension containing τ must belong to K_0. Thus if $\tau \in K_0'$, then $\Theta(\tau) = K(\sigma_\tau)$ and $\Phi(\tau) = K'(\sigma_\tau)$ are subcomplexes of K_0 and K_0', respectively.

It follows that the chain maps λ and θ carried by Λ and Θ, respectively, induce chain maps on the relative level, and that the chain homotopies of $\theta \circ \lambda$ and $\lambda \circ \theta$ to the respective identity maps induce chain homotopies on the relative level as well. Since one choice for θ is the map $g_\#$, it follows that $g_\#$ and λ are chain-homotopy inverses on the relative level. □

EXERCISES

1. Let K' be a subdivision of K; let $\lambda : C_p(K) \rightarrow C_p(K')$ be the subdivision operator. Let $g : K' \rightarrow K$ be a simplicial approximation to the identity. Show that $g_\# \circ \lambda$ equals the identity map of $C_p(K)$.
2. Let K and K' be the complexes pictured in Figure 17.2, whose common underlying space is a square.
   (a) Find a formula for the subdivision operator $\lambda : C_p(K) \rightarrow C_p(K')$.
   (b) Find two different simplicial approximations $g, g' : K' \rightarrow K$ to the identity. Conclude that the chain equivalence θ constructed in the proof of Theorem 17.2 is not unique.
   (c) Check that $\lambda \circ g_\#$ does not equal the identity on the chains of K'.
3. (a) Show that the inductive formula for the barycentric subdivision operator sd defines an augmentation-preserving chain map that is carried by Λ.
   (b) Compute sd σ for the case of a 1-simplex and a 2-simplex.

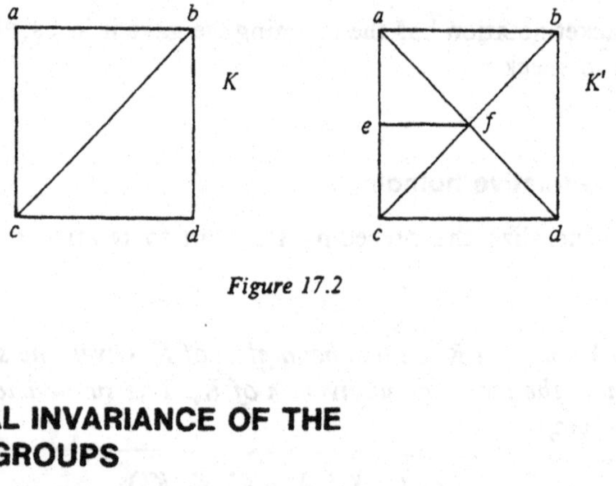

Figure 17.2

§18. TOPOLOGICAL INVARIANCE OF THE HOMOLOGY GROUPS

In this section, we achieve the basic goal of this chapter, to prove the topological invariance of the simplicial homology groups.

Definition. Let K and L be simplicial complexes; let $h : |K| \rightarrow |L|$ be a continuous map. Choose a subdivision K' of K such that h has a simplicial approximation $f : K' \rightarrow L$. Let $\lambda : \mathcal{C}(K) \rightarrow \mathcal{C}(K')$ be the subdivision operator. We define the **homomorphism induced by** h,

$$h_* : H_p(K) \rightarrow H_p(L),$$

by the equation $h_* = f_* \circ \lambda_*$.

Note that once K' has been chosen, the homomorphism h_* is independent of the particular choice of the simplicial approximation $f : K' \rightarrow L$ to h. For any two such simplicial approximations are contiguous.

Note also that if $g : K' \rightarrow K$ is a simplicial approximation to the identity map $i_{|K|}$ of $|K|$ with itself, then λ_* and g_* are inverse to one another. Therefore, one could just as well define

$$h_* = f_* \circ (g_*)^{-1}.$$

We use this fact to show that h_* is independent of the choice of the subdivision K'. Suppose K'' is another subdivision of K such that h has a simplicial approximation mapping K'' into L. We show that if h_* is defined using the subdivision K'', the result is the same as if one uses K'.

The proof is especially easy in the case where the identity map of $|K|$ has a simplicial approximation $k : K'' \to K'$, as in the following diagram:

$$K'' \xrightarrow{k} K' \begin{array}{l} \xrightarrow{\;g\;} K \\ \xrightarrow[\;f\;]{} L \end{array}$$

Then since $g \circ k$ and $f \circ k$ are simplicial approximations to the identity and to h, respectively, the homomorphism h_*, defined using the subdivision K'', equals the composite

$$(f \circ k)_* \circ (g \circ k)_*^{-1} = (f_* \circ k_*) \circ (g_* \circ k_*)^{-1} = f_* \circ g_*^{-1}.$$

The result is thus the same as when h_* is defined using the subdivision K'.

The general case is proved by choosing a subdivision K''' of K such that the identity map has simplicial approximations

$$k_1 : K''' \to K' \qquad \text{and} \qquad k_2 : K''' \to K''.$$

Then using K''' to define h_* gives the same result as using K' or K''.

We should remark that, properly speaking, the homomorphism h_* depends not only on the spaces $X = |K|$ and $Y = |L|$ and the continuous map $h : X \to Y$, but also on the particular complexes K and L whose polytopes are X and Y, respectively. If M and N are other complexes whose polytopes are X and Y, respectively, then h also induces a homomorphism

$$h_* : H_p(M) \to H_p(N).$$

One should really use a notation such as $(h_{KL})_*$ and $(h_{MN})_*$ to distinguish between these homomorphisms. We shall abuse terminology, however, and use the simple notation h_*, relying on the context to make the meaning clear. We return to a further discussion of this point later, when we define the homology of a triangulable space (§27).

Theorem 18.1 (The functorial properties). *The identity map $i : |K| \to |K|$ induces the identity homomorphism $i_* : H_p(K) \to H_p(K)$. If $h : |K| \to |L|$ and $k : |L| \to |M|$ are continuous maps, then $(k \circ h)_* = k_* \circ h_*$. The same results hold for reduced homology.*

Proof. That i_* is the identity is immediate from the definition. To check the second statement, choose $f_0 : L' \to M$ and $g_0 : L' \to L$ as simplicial approximations to k and $i_{|L|}$, respectively. Then choose $f_1 : K' \to L'$ and $g_1 : K' \to K$ as simplicial approximations to h and $i_{|K|}$, respectively. We have the following diagram of continuous maps and simplicial maps:

$$|K| \xrightarrow{h} |L| \xrightarrow{k} |M|$$

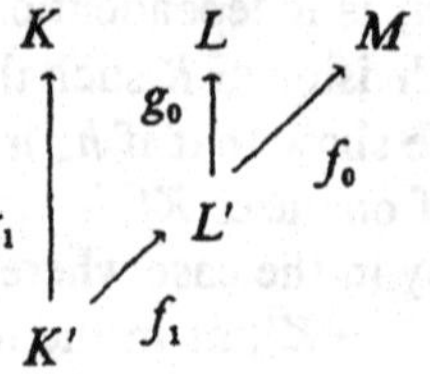

Now $f_0 \circ f_1$ is a simplicial approximation to $k \circ h$; therefore,

$$(k \circ h)_* = (f_0 \circ f_1)_* \circ (g_1)_*^{-1},$$

by definition. Since $g_0 \circ f_1$ is a simplicial approximation to h, we have

$$h_* = (g_0 \circ f_1)_* \circ (g_1)_*^{-1} \quad \text{and} \quad k_* = (f_0)_* \circ (g_0)_*^{-1},$$

again by definition. Combining these equations and applying Theorem 12.2, we obtain the desired result,

$$(k \circ h)_* = k_* \circ h_*. \quad \square$$

Corollary 18.2 (Topological invariance of homology groups). *If $h : |K| \to |L|$ is a homeomorphism, then $h_* : H_p(K) \to H_p(L)$ is an isomorphism. The same result holds for reduced homology.*

Proof. Let $k : |L| \to |K|$ be the inverse of h. Then $h_* \circ k_*$ equals $(i_{|L|})_*$ and $k_* \circ h_*$ equals $(i_{|K|})_*$. Thus $h_* \circ k_*$ and $k_* \circ h_*$ are isomorphisms, so h_* is an isomorphism. $\square$

Application to relative homology

We have proved the topological invariance of the (absolute) homology groups. Can we do the same for the relative homology groups? Yes. Everything we have done in this section goes through for relative homology with no difficulty. One simply replaces each occurrence of a complex by the appropriate pair consisting of a complex and a subcomplex, and applies Theorems 12.6, 14.4, and 17.3 freely. We restate the preceding theorems in this situation.

Theorem 18.3. *The identity map i of $(|K|,|K_0|)$ with itself induces the identity homomorphism in relative homology. If*

$$(|K|,|K_0|) \xrightarrow{h} (|L|,|L_0|) \xrightarrow{k} (|M|,|M_0|)$$

are continuous maps, then $(k \circ h)_* = k_* \circ h_*$ in relative homology. If h is a homeomorphism of $|K|$ with $|L|$ that maps $|K_0|$ onto $|L_0|$, then h_* is an isomorphism in relative homology. $\square$

EXERCISES

1. If $A \subset X$, a **retraction** $r: X \to A$ is a continuous map such that $r(a) = a$ for each $a \in A$.
   (a) If $r: X \to A$ is a retraction and X is Hausdorff, show A is closed in X.
   (b) Let $r: |K| \to |K_0|$ be a retraction, where K_0 is a subcomplex of K. Show $r_*: H_p(K) \to H_p(K_0)$ is surjective and the homomorphism j_* induced by inclusion $j: |K_0| \to |K|$ is injective.
   (c) Show there is no retraction $r: B^n \to S^{n-1}$.
2. (a) Show there is a retraction of the Klein bottle S onto the imbedded circle A pictured in Figure 18.1, but no retraction of S onto the circle C.
   (b) Show there is no retraction of the projective plane onto the imbedded circle C pictured in Figure 18.2.
3. Determine whether there are retractions of the torus onto the tube A pictured in Figure 18.3, onto the disc B, and onto the circle C.

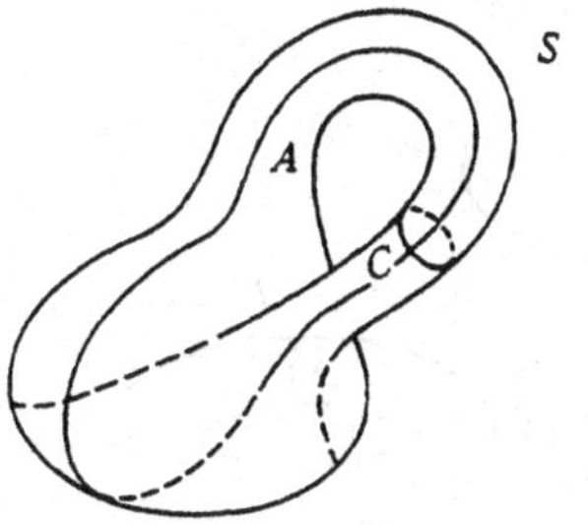

Figure 18.1

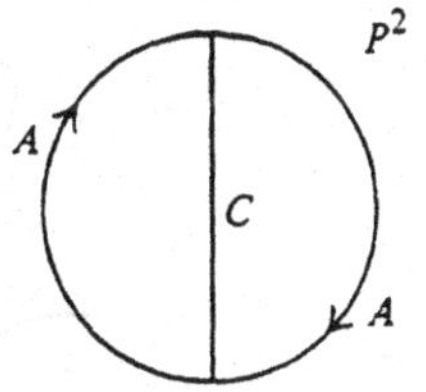

Figure 18.2

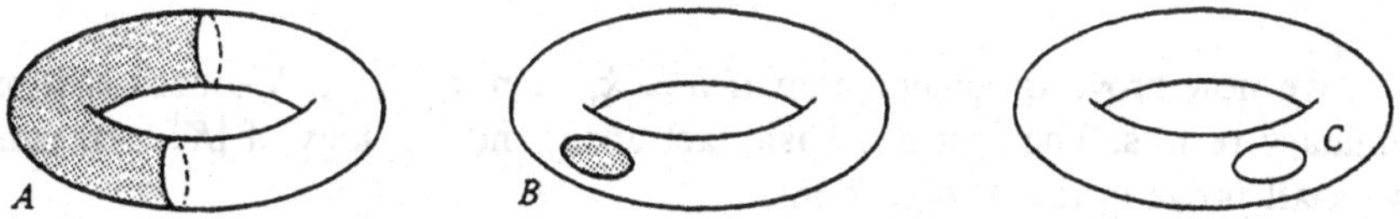

Figure 18.3

§19. HOMOMORPHISMS INDUCED BY HOMOTOPIC MAPS

We now introduce the important concept of homotopy, which was mentioned earlier in the exercises. Throughout, let I denote the closed unit interval $[0,1]$.

Definition. If X and Y are topological spaces, two continuous maps $h, k: X \to Y$ are said to be **homotopic** if there is a continuous map

$$F: X \times I \to Y,$$

such that $F(x,0) = h(x)$ and $F(x,1) = k(x)$ for all $x \in X$. If h and k are homotopic, we write $h \simeq k$. The map F is called a **homotopy** of h to k. We think of F as a way of "deforming" h continuously to k, as t varies from 0 to 1.

We shall prove that if $h, k : |K| \to |L|$ are homotopic, then the homology homomorphisms h_*, k_* they induce are the same. This leads to the important result that the homology groups are invariants of the "homotopy type" of a space.

Example 1. Let $X = S^1$. Then $H_1(X)$ is infinite cyclic; choose the cycle z generating $H_1(X)$ indicated by the arrow in Figure 19.1. Let T denote the torus; let $h, k : X \to T$ be the maps indicated in Figure 19.1. These maps are clearly homotopic, for one can "push h around the ring" until it coincides with k. It is geometrically clear also that the cycles $h_\#(z)$ and $k_\#(z)$ are homologous. Indeed, the 2-chain d obtained by chopping the right half of the torus into triangles and orienting them appropriately satisfies the equation $\partial d = h_\#(z) - k_\#(z)$. Since z represents a generator of $H_1(X)$, it follows that in this case at least, $h_* = k_*$.

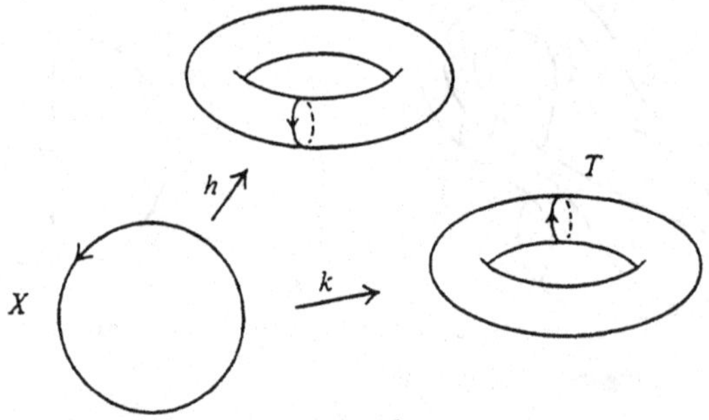

Figure 19.1

We now prove in general that if $h \simeq k$, then $h_* = k_*$. We need two preliminary results. The first is a basic fact about the topology of $|K| \times I$, which we shall prove in the next section:

The topology of the product space $|K| \times I$ is coherent with the subspaces $\sigma \times I$, for $\sigma \in K$.

The second concerns the fact that $|K| \times I$ is a polyhedron:

Lemma 19.1. *If K is a complex, then $|K| \times I$ is the polytope of a complex M, such that each set $\sigma \times I$ is the polytope of a subcomplex of M, and the sets $\sigma \times 0$ and $\sigma \times 1$ are simplices of M, for each simplex σ of K.*

Proof. We have $|K| \subset \mathbf{E}^J$ for some J. Then $|K| \times I \subset \mathbf{E}^J \times \mathbf{R}$. We shall subdivide $|K| \times I$ into simplices by a variant of the starring procedure used in defining barycentric subdivision.

For $p \geq 0$, let us define

$$X_p = (|K| \times 0) \cup (|K| \times 1) \cup (|K^{(p)}| \times I).$$

We proceed inductively to subdivide X_p into simplices. Consider the case $p = 0$. The space $(|K| \times 0) \cup (|K| \times 1)$ is the polytope of a complex consisting of all simplices of the form $\sigma \times 0$ and $\sigma \times 1$, for $\sigma \in K$. The space $|K^{(0)}| \times I$ is the polytope of the complex consisting of all 1-simplices of the form $v \times I$, for $v \in K^{(0)}$, and their vertices. Their union is a complex M_0 whose polytope is X_0.

In general, suppose M_{p-1} is a complex whose polytope is X_{p-1}, such that each set $s \times I$, for s a simplex of K of dimension less than p, is the polytope of a subcomplex of M_{p-1}. Let $\dim \sigma = p$, and consider the set $\sigma \times I$. Now let

$$\begin{aligned} \text{Bd}(\sigma \times I) &= (\sigma \times I) - (\text{Int}\,\sigma \times \text{Int}\,I) \\ &= ((\text{Bd}\,\sigma) \times I) \cup (\sigma \times 0) \cup (\sigma \times 1). \end{aligned}$$

Since Bd σ is the union of simplices s of K of dimension $p - 1$, $\text{Bd}(\sigma \times I)$ is the polytope of a subcomplex M_σ of M_{p-1}. It is finite because $\text{Bd}(\sigma \times I)$ is compact. Let w_σ denote the point $(\hat{\sigma}, 1/2) \in \sigma \times I$. Then the cone $w_\sigma * M_\sigma$ is a complex whose polytope is $\sigma \times I$. The intersection of $|w_\sigma * M_\sigma|$ and $|M_{p-1}|$ is the polytope of a subcomplex of each of them.

Define M_p to be the union of M_{p-1} and the cones $w_\sigma * M_\sigma$, as σ ranges over all p-simplices of K. Finally, define M to be the union of the complexes M_p for all p.

Now M is a complex whose underlying space consists precisely of the points of the space $|K| \times I$. However, it is not at all obvious that the spaces $|M|$ and $|K| \times I$ are equal *as topological spaces*. To prove that result, we need the fact about the topology of $|K| \times I$ that was just quoted.

We know that the topology of $|K| \times I$ is coherent with the subspaces $\sigma \times I$, for $\sigma \in K$. On the other hand, the topology of $|M|$ is coherent with the subspaces s, for $s \in M$. Now if C is closed in $|K| \times I$, then $C \cap (\sigma \times I)$ is closed in $\sigma \times I$. If s is a simplex of M lying in $\sigma \times I$, then s is a subspace of $\sigma \times I$ (both are subspaces of $\mathbf{E}^J \times \mathbf{R}$, being compact). Hence $C \cap s$ is closed in s. It follows that C is closed in $|M|$.

Conversely, if C is closed in $|M|$, then $C \cap s$ is closed in s for each $s \in M$. Because $\sigma \times I$ is a finite union of simplices s of M, the set $C \cap (\sigma \times I)$ is closed in $\sigma \times I$. Thus σ is closed in $|K| \times I$. □

Example 2. If K is the complex consisting of a 1-simplex and its faces, then $|K| \times I$ is by the procedure of the preceding lemma subdivided into the complex pictured in Figure 19.2. If it is the complex consisting of a 2-simplex and its faces, then $|K| \times I$ is subdivided into the complex pictured in Figure 19.3.

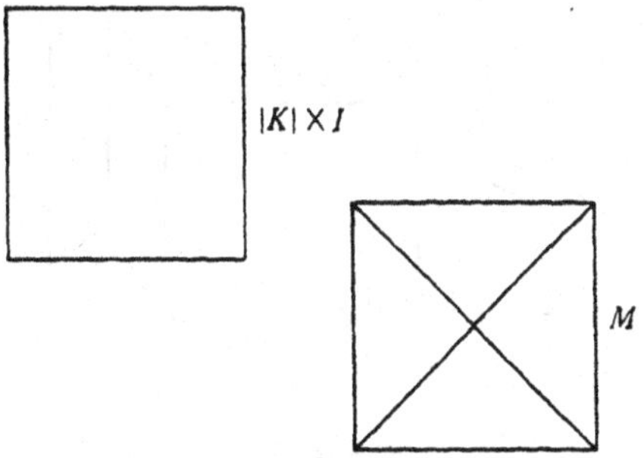

Figure 19.2

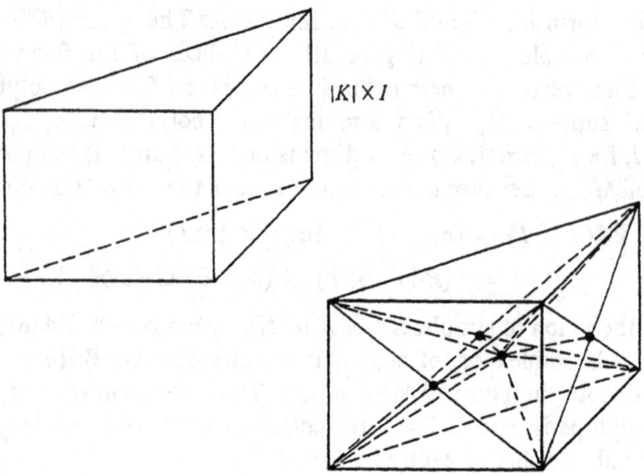

Figure 19.3

Theorem 19.2. *If $h, k : |K| \to |L|$ are homotopic, then $h_*, k_* : H_p(K) \to H_p(L)$ are equal. The same holds for reduced homology.*

Proof. Let K be a complex. Let M be a complex whose underlying space is $|K| \times I$, such that for each $\sigma \in K$, both $\sigma \times 0$ and $\sigma \times 1$ are simplices of M, and $\sigma \times I$ is the polytope of a subcomplex of M.

Let $F : |K| \times I \to |L|$ be the homotopy of h to k. Let $i, j : |K| \to |K| \times I$ be the maps $i(x) = (x,0)$ and $j(x) = (x,1)$, as pictured in Figure 19.4. Then i and j are simplicial maps of K into the complex M; furthermore,

$$F \circ i = h \qquad \text{and} \qquad F \circ j = k.$$

We assert that the chain maps $i_\#$ and $j_\#$ induced by i and j are chain homotopic. Consider the function Φ assigning, to each simplex σ of K, the subcomplex of M whose polytope is $\sigma \times I$. Now the space $\sigma \times I$ is acyclic because it is homeomorphic to a closed ball. And if $s \prec \sigma$, then $s \times I \subset \sigma \times I$, so $\Phi(s)$ is

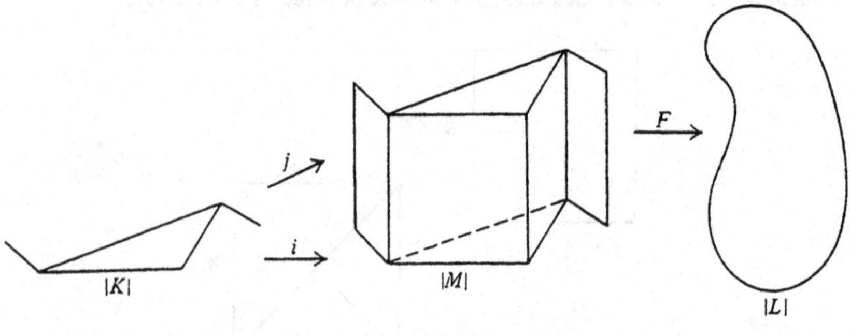

Figure 19.4

a subcomplex of $\Phi(\sigma)$. Therefore, Φ is an acyclic carrier from K to M. Furthermore, it carries both $i_{\#}$ and $j_{\#}$, for both $i(\sigma) = \sigma \times 0$ and $j(\sigma) = \sigma \times 1$ belong to $\Phi(\sigma)$. It follows from Theorem 13.3 that $i_{\#}$ and $j_{\#}$ are chain homotopic. We conclude that $i_* = j_*$. Then

$$h_* = F_* \circ i_* = F_* \circ j_* = k_*,$$

as desired. □

This result carries over readily to the relative homology groups. Given maps $h, k : (|K|,|K_0|) \to (|L|,|L_0|)$, we say they are homotopic (as maps of pairs of spaces) if there is a homotopy $H : |K| \times I \to |L|$ of h to k such that H maps $|K_0| \times I$ into $|L_0|$. We have the following theorem:

Theorem 19.3. *If h and k are homotopic as maps of pairs of spaces, then $h_* = k_*$ as maps of relative homology groups.*

Proof. The proof of Theorem 19.2 goes through without difficulty. Both i and j carry $|K_0|$ into $|K_0| \times I$, and so does the chain homotopy connecting $i_{\#}$ and $j_{\#}$. Then $i_* = j_*$ as maps of relative homology, and the proof proceeds as before. □

Here is another result which follows from our knowledge of the topology of $|K| \times I$; we shall use it later on.

Theorem 19.4. *If $f : K \to L$ is a simplicial approximation to the continuous map $h : |K| \to |L|$, then f is homotopic to h.*

Proof. For each x in $|K|$, we know from Lemma 14.2 that $f(x)$ and $h(x)$ lie in a single simplex of L. Therefore, the "straight-line homotopy" given by

$$F(x,t) = (1 - t)f(x) + th(x)$$

maps $|K| \times I$ into $|L|$. If L is finite, then F is automatically continuous, because it is continuous as a map into euclidean space, of which $|L|$ is a subspace. To show F continuous in general, we show that its restriction to $\sigma \times I$ is continuous, for each $\sigma \in K$. Since the topology of $|K| \times I$ is coherent with the subspaces $\sigma \times I$, this will suffice.

For each $x \in \sigma$, let τ_x denote the simplex of L whose interior contains $h(x)$. Because $h(\sigma)$ is compact, the collection of simplices τ_x, for $x \in \sigma$, is finite. Let L_σ be the subcomplex of L consisting of these simplices and their faces. By Lemma 14.2, the point $f(x)$ lies in τ_x. Therefore, F carries the set $x \times I$ into τ_x. Thus F maps $\sigma \times I$ into $|L_\sigma|$; since L_σ is a finite complex, its space is a subspace of euclidean space. Hence $F : \sigma \times I \to |L_\sigma|$ is continuous. Because inclusion $|L_\sigma| \to |L|$ is continuous, the map $F : \sigma \times I \to |L|$ is also continuous, as desired. □

We know that if two spaces are homeomorphic, they have isomorphic homology groups. There is a weaker relation than homeomorphism that implies the same result. It is the relation of homotopy equivalence, which we now introduce.

Definition. Two spaces X and Y are said to be **homotopy equivalent**, or to have the same **homotopy type**, if there are maps

$$f : X \to Y \quad \text{and} \quad g : Y \to X$$

such that $g \circ f \simeq i_X$ and $f \circ g \simeq i_Y$. The maps f and g are often called **homotopy equivalences**, and g is said to be a **homotopy inverse** to f.

Symmetry and reflexivity of this relation are trivial. Transitivity is left as an exercise.

If X has the homotopy type of a single point, then X is said to be **contractible.** This is equivalent to the statement that the identity map $i_X : X \to X$ is homotopic to a constant map. For example, the unit ball is contractible, because the map $F(x,t) = (1 - t)x$ is a homotopy between the identity and a constant.

Theorem 19.5. *If $f : |K| \to |L|$ is a homotopy equivalence, then f_* is an isomorphism. In particular, if $|K|$ is contractible, then K is acyclic.*

Proof. The proof is immediate. If g is a homotopy inverse for f, then g_* is an inverse for f_*. □

Homotopy equivalences are hard to visualize in general. There is a special kind of homotopy equivalence that is geometrically easier to understand:

Definition. Let $A \subset X$. A **retraction** of X onto A is a continuous map $r : X \to A$ such that $r(a) = a$ for each $a \in A$. If there is a retraction of X onto A, we say A is a **retract** of X. A **deformation retraction** of X onto A is a continuous map $F : X \times I \to X$ such that

$$F(x,0) = x \text{ for } x \in X,$$
$$F(x,1) \in A \text{ for } x \in X,$$
$$F(a,t) = a \text{ for } a \in A.$$

If such an F exists, then A is called a **deformation retract** of X.

If F is a deformation retraction of X onto A, then the map $r(x) = F(x,1)$ is a retraction of X onto A. The latter fact is equivalent to the statement that the composite

$$A \xrightarrow{j} X \xrightarrow{r} A,$$

(where j is inclusion) equals the identity map i_A. On the other hand, the map F is a homotopy between the identity map i_X and the composite

$$X \xrightarrow{r} A \xrightarrow{j} X$$

(and in fact each point of A remains fixed during the homotopy). It follows that r and j are homotopy inverse to each other.

One can visualize a deformation retraction as a gradual collapsing of the space X onto the subspace A, such that each point of A remains fixed during the shrinking process. This type of homotopy equivalence can thus be visualized geometrically. It is intuitively clear that if A is a deformation retract of X, and B is a deformation retract of A, then B is a deformation retract of X. (It is also easy to prove.)

We now consider some special cases.

Theorem 19.6. *The unit sphere S^{n-1} is a deformation retract of punctured euclidean space $\mathbf{R}^n - \mathbf{0}$.*

Proof. Let $X = \mathbf{R}^n - \mathbf{0}$. We define $F : X \times I \to X$ by the equation

$$F(x,t) = (1-t)x + tx/\|x\|.$$

The map F gradually shrinks each open ray emanating from the origin to the point where it intersects the unit sphere. It is a deformation retraction of $\mathbf{R}^n - \mathbf{0}$ onto S^{n-1}. □

Corollary 19.7. *The euclidean spaces $\mathbf{R}^n$ and $\mathbf{R}^m$ are not homeomorphic if $n \neq m$.*

Proof. Suppose that h is a homeomorphism of $\mathbf{R}^n$ with $\mathbf{R}^m$. Then h is a homeomorphism of $\mathbf{R}^n - \mathbf{0}$ with $\mathbf{R}^m - p$ for some $p \in \mathbf{R}^m$. The latter space is homeomorphic with $\mathbf{R}^m - \mathbf{0}$. It follows from the preceding theorem that S^{n-1} and S^{m-1} are homotopy equivalent. This cannot be true if $m \neq n$, for in that case $\tilde{H}_{n-1}(S^{n-1}) \simeq \mathbf{Z}$ and $\tilde{H}_{n-1}(S^{m-1}) = 0$. □

Example 3. The wedge of two circles has the same homotopy type as the letter θ. Let X be a doubly punctured elliptical region in the plane, as pictured in Figure 19.5. The sequence of arrows on the left side of this figure indicates how one can collapse X to the wedge of two circles. The arrows on the right side indicate how to collapse X to the letter θ. Since each of these spaces is homotopy equivalent to X, they are homotopy equivalent to each other.

The situation that occurs in Example 3 is more general than might be supposed. It is an interesting fact that *every* homotopy equivalence may be expressed, as in this example, in terms of deformation retractions. Specifically, there is a theorem to the effect that two spaces X and Y have the same homotopy type if and only if there is a space Z and imbeddings $h : X \to Z$ and $k : Y \to Z$ such that both $h(X)$ and $k(Y)$ are deformation retracts of Z. (See [Wh], [F].) This fact helps in visualizing what the notion of homotopy equivalence really means, geometrically.

Homotopy equivalences are a powerful tool for computing homology groups. Given a complex K, it is often much easier to show that its space is homotopy equivalent to a space whose homology is known than it is to calculate the homology of K directly. The exercises following will illustrate this fact.

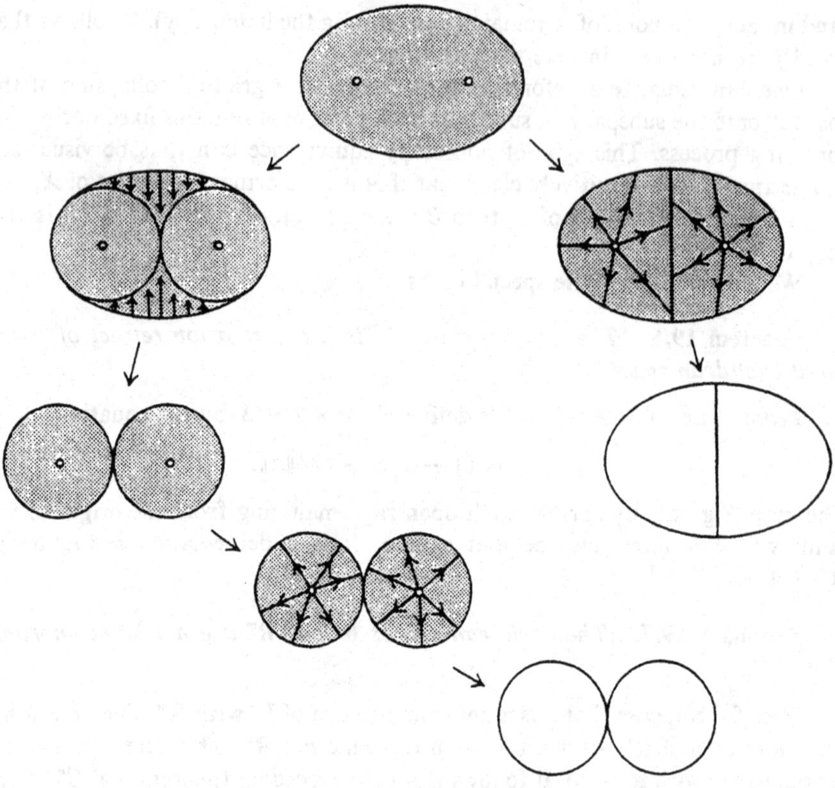

Figure 19.5

EXERCISES

1. Show that, if $\mathcal{A}$ is a collection of spaces, then homotopy equivalence is an equivalence relation on $\mathcal{A}$.
2. Show that if A is a deformation retract of X, and B is a deformation retract of A, then B is a deformation retract of X.
3. Group the following spaces into homotopy equivalence classes. Assuming they are all polytopes of complexes, calculate their homology groups.
   (a) The Möbius band
   (b) The torus
   (c) The solid torus $B^2 \times S^1$
   (d) The torus minus one point
   (e) The torus minus two points
   (f) The Klein bottle minus a point
   (g) $\mathbf{R}^3$ with the z-axis deleted
   (h) $\mathbf{R}^3$ with the circle $\{x^2 + y^2 = 1, z = 0\}$ deleted
   (i) The intersection of the spaces in (g) and (h)
   (j) S^3 with two linked circles deleted
   (k) S^3 with two unlinked circles deleted.

4. *Theorem. If K is a complex of dimension n, then $|K|$ has covering dimension at least n.*

Proof. Let $\mathcal{A}$ be a finite open covering of the n-simplex $\sigma = v_0 \dots v_n$ that refines the open covering $\{\text{St } v_0, \dots, \text{St } v_n\}$. Let $\{\phi_A\}$ be a partition of unity subordinate to $\mathcal{A}$. For each A, let v_A be a vertex of σ such that $A \subset \text{St } v_A$. Define $h : \sigma \to \sigma$ by the rule

$$h(x) = \Sigma \phi_A(x) v_A.$$

If no $x \in X$ belongs to more than n elements of $\mathcal{A}$, then h maps σ into $\text{Bd } \sigma$. Further, h maps each face of σ into itself. Conclude that $h : \text{Bd } \sigma \to \text{Bd } \sigma$ is homotopic to the identity, and derive a contradiction.

5. Prove Theorem 19.2 without using the fact that $|K| \times I$ is the space of a complex, as follows:
   (a) Let K and L be finite; let $h : |K| \to |L|$. Show there is an $\epsilon > 0$ such that if $k : |K| \to |L|$ and $|h(x) - k(x)| < \epsilon$ for all x, then $h_* = k_*$. [*Hint:* Choose K' so that $h(\overline{\text{St}}(v,K')) \subset \text{St}(w,L)$ for some w. If ϵ is small, this same inclusion holds with h replaced by k.]
   (b) Prove the theorem when K and L are finite.
   (c) Prove the theorem in general. [*Hint:* Each cycle of K is carried by a finite subcomplex of K.]

§20. REVIEW OF QUOTIENT SPACES

Here we review some standard definitions and theorems concerning quotient spaces that we shall need.

A surjective map $p : X \to Y$ is called a **quotient map** provided a subset U of Y is open if and only if the set $p^{-1}(U)$ is open in X. It is equivalent to require that A be closed in Y if and only if $p^{-1}(A)$ is closed in X.

A subset C of X is **saturated** (with respect to p) if it equals the complete inverse image $p^{-1}(A)$ of some subset A of Y. To say that p is a quotient map is equivalent to saying that p is continuous and that p maps saturated open sets of X to open sets of Y (or saturated closed sets of X to closed sets of Y).

Let $p : X \to Y$ be a surjective continuous map. If p is either a **closed map** or an **open map** (i.e., if p maps closed sets of X to closed sets of Y, or open sets of X to open sets of Y), then p is a quotient map. In particular, if X is compact and Y is Hausdorff, then p is a closed map and hence a quotient map.

First, we list some elementary facts about quotient maps, whose proofs are straightforward and will be left to the reader.

A *one-to-one quotient map* is a homeomorphism.

A *composite* of quotient maps is a quotient map; if $p : X \to Y$ and $q : Y \to Z$ are quotient maps, so is $q \circ p : X \to Z$.

A *restriction* of a quotient map is sometimes a quotient map: If $p : X \to Y$ is a quotient map, and if A is a saturated subspace of X that is either open or closed in X, then the map $p|A : A \to p(A)$ is a quotient map.

A relation between *continuous maps* and quotient maps is the following: If $p : X \to Y$ is a quotient map, and if $f : X \to Z$ is a continuous map that is constant on each set $p^{-1}(y)$, then there is a unique continuous map $g : Y \to Z$ such that $g \circ p = f$. See the following diagram; we say g is **induced by** f.

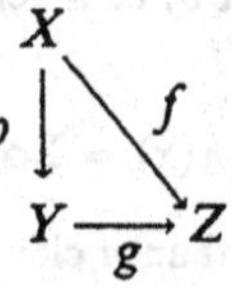

We define the notion of *quotient space* as follows: Let X^* be a partition of the space X into disjoint subsets whose union is X. Let $\pi : X \to X^*$ map each point to the set containing it. If we topologize X^* by declaring the subset U of X^* to be open in X^* if and only if $\pi^{-1}(U)$ is open in X, then π is a quotient map. The space X^* is called a **quotient space** of X. We often say it is obtained by "identifying each element of the partition to a point."

If $p : X \to Y$ is a quotient map, we can always "consider" Y to be a quotient space of X by the following device: Given p, let X^* denote the partition of X into the disjoint sets $p^{-1}(y)$, for $y \in Y$. Then X^* is homeomorphic to Y. We need only apply the preceding remark twice, to obtain continuous maps $g : X^* \to Y$ and $h : Y \to X^*$, as in the following diagram:

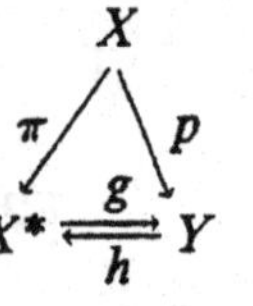

These maps are readily seen to be inverse to each other.

Example 1. Let X be the subspace of $\mathbf{R}^3$ obtained by rotating the unit circle in the x-z plane centered at (2,0,0) about the z-axis. Using cylindrical coordinates (r,θ,z) in $\mathbf{R}^3$, we can express X as the set of points satisfying the equation $(r-2)^2 + z^2 = 1$. Letting $I^2 = I \times I$, we define $p : I^2 \to X$ by setting, for $(s,t) \in I^2$,

$$r - 2 = \cos 2\pi t, \qquad z = \sin 2\pi t, \qquad \theta = 2\pi s.$$

You can check that p maps I^2 onto X and is a closed quotient map.

By use of this map, we can consider X to be the quotient space obtained from I^2 by identifying $(s,0)$ with $(s,1)$, and $(0,t)$ with $(1,t)$, for $s,t \in I$. See Figure 20.1. This

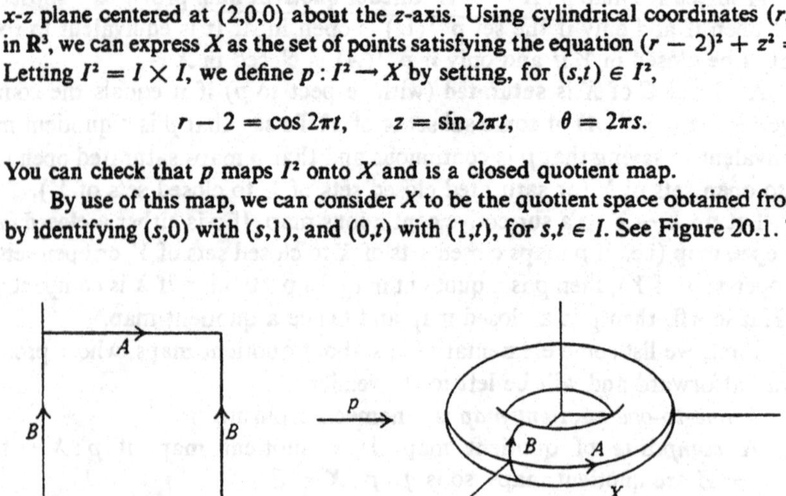

Figure 20.1

is the topological version of the method by which we constructed the torus in §3 by pasting together the edges of a rectangle.

The *separation axioms* do not behave well for quotient spaces. For instance, a quotient space of a Hausdorff space need not be Hausdorff. All one can say in general is the following: If $p: X \to Y$ is a quotient map, and if each set $p^{-1}(y)$ is closed in X, then Y is a T_1-space; that is, one-point sets are closed in Y. This means that if X^* is a partition of X into closed sets, then the quotient space X^* is a T_1-space.

We will return to the matter of separation axioms and quotient spaces in a later section (§37).

Now we consider one final question: Under what conditions is the *cartesian product* of two quotient maps a quotient map? We prove two results in this direction:

Theorem 20.1. *Let $p: X \to Y$ be a quotient map. If C is a locally compact Hausdorff space, then*

$$p \times i_C : X \times C \to Y \times C$$

is a quotient map.

Proof. Let $\pi = p \times i_C$. Let A be a subset of $Y \times C$ such that $\pi^{-1}(A)$ is open in $X \times C$. We show A is open in $Y \times C$. That is, given (y_0, c_0) in A, we find an open set about (y_0, c_0) lying in A.

Choose x_0 so that $p(x_0) = y_0$; then $\pi(x_0, c_0) = (y_0, c_0)$. Since $\pi^{-1}(A)$ is open, we can choose neighborhoods U_1 of x_0 and W of c_0 such that $U_1 \times W \subset \pi^{-1}(A)$. Because C is locally compact Hausdorff, we can choose a neighborhood V of c_0 so that $\bar{V}$ is compact and $\bar{V} \subset W$. Then $U_1 \times V$ is a neighborhood of (x_0, c_0) such that $\bar{V}$ is compact and

$$U_1 \times \bar{V} \subset \pi^{-1}(A).$$

In general, suppose U_i is a neighborhood of x_0 such that $U_i \times \bar{V} \subset \pi^{-1}(A)$. Now $p^{-1}p(U_i)$ is not necessarily open in X, but it contains U_i. We construct an open set U_{i+1} of X such that

$$p^{-1}p(U_i) \times \bar{V} \subset U_{i+1} \times \bar{V} \subset \pi^{-1}(A),$$

as follows: For each point x of $p^{-1}p(U_i)$, the space $\{x\} \times \bar{V}$ lies in $\pi^{-1}(A)$. Using compactness of $\bar{V}$, we choose a neighborhood W_x of x such that $W_x \times \bar{V} \subset \pi^{-1}(A)$. Let U_{i+1} be the union of the open sets W_x; then U_{i+1} is the desired open set of X. See Figure 20.2.

Finally, let U be the union of the open sets $U_1 \subset U_2 \subset \cdots$. Then $U \times V$ is a neighborhood of (x_0, c_0) and $U \times \bar{V} \subset \pi^{-1}(A)$. Furthermore, U is saturated with respect to p, for

$$U \subset p^{-1}p(U) = \bigcup_1^\infty p^{-1}p(U_i) \subset \bigcup_1^\infty U_{i+1} = U.$$

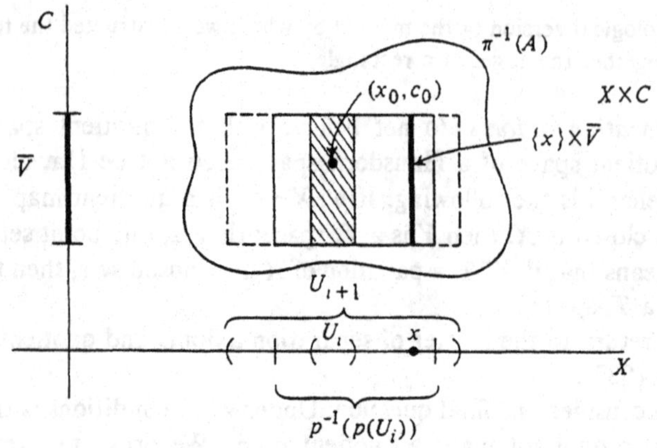

Figure 20.2

Therefore $p(U)$ is open in Y. Then

$$p(U)\times V=\pi(U\times V)\subset A$$

is a neighborhood of (y_0,c_0) lying in A, as desired. □

Corollary 20.2. *If $p:A\to B$ and $q:C\to D$ are quotient maps, and if the domain of p and the range of q are locally compact Hausdorff spaces, then*

$$p\times q:A\times C\to B\times D$$

is a quotient map.

Proof. We can write $p\times q$ as the composite

$$A\times C\xrightarrow{i_A\times q}A\times D\xrightarrow{p\times i_D}B\times D.$$

Since each of these maps is a quotient map, so is $p\times q$. □

There is a close connection between coherent topologies and quotient maps, which can be described as follows:

Definition. Suppose E is a space that is the union of disjoint subspaces E_α, each of which is open (and closed) in E. Then we say E is the **topological sum** of the spaces E_α, and we write $E=\Sigma E_\alpha$. A set U is open in E if and only if $U\cap E_\alpha$ is open in E_α for each α.

More generally, let $\{X_\alpha\}_{\alpha\in J}$ be a family of topological spaces, which may or may not be disjoint. Let E be the *set* that is the union of the disjoint topological spaces

$$E_\alpha=X_\alpha\times\{\alpha\},$$

for $\alpha\in J$. If we topologize E by declaring U to be open in E if and only if

$U \cap E_\alpha$ is open in E_α for each α, then E is the topological sum of the disjoint spaces E_α. One has a natural map $p : E \to \cup X_\alpha$, which projects $X_\alpha \times \{\alpha\}$ onto X_α for each α. (We sometimes abuse terminology and speak of E as the topological sum of the spaces X_α in this situation.)

In this situation, one has the following result, whose proof is immediate:

Lemma 20.3. *Let X be a space which is the union of certain of its subspaces X_α. Let E be the topological sum of the spaces X_α; let $p : E \to X$ be the natural projection. Then the topology of X is coherent with the subspaces X_α if and only if p is a quotient map.* □

In this situation, we often say that X is the **coherent union** of the spaces X_α.

Theorem 20.4. *If the topology of X is coherent with the subspaces X_α, and if Y is a locally compact Hausdorff space, then the topology of $X \times Y$ is coherent with the subspaces $X_\alpha \times Y$.*

Proof. Let $E = \Sigma (X_\alpha \times \{\alpha\})$; let $p : E \to X$ be the projection map. Because Y is locally compact Hausdorff, the map

$$p \times i_Y : E \times Y \to X \times Y$$

is also a quotient map. Now E is the topological sum of the subspaces $E_\alpha = X_\alpha \times \{\alpha\}$. Then $E \times Y$ is the disjoint union of its subspaces $E_\alpha \times Y$, each of which is open in $E \times Y$. Therefore, $E \times Y$ is the topological sum of the spaces $E_\alpha \times Y = X_\alpha \times \{\alpha\} \times Y$. Since $p \times i_Y$ is a quotient map, the topology of $X \times Y$ is coherent with the subspaces $X_\alpha \times Y$. □

Corollary 20.5. *The topology of $|K| \times I$ is coherent with the subspaces $\sigma \times I$, for $\sigma \in K$.*

Proof. By definition, the topology of $|K|$ is coherent with the subspaces σ, for $\sigma \in K$. Since I is locally compact Hausdorff (in fact, compact Hausdorff), the preceding theorem applies. □

Corollary 20.6. *Let $w * K$ be a cone over the complex K. The map $\pi : |K| \times I \to |w * K|$ defined by*

$$\pi(x,t) = (1 - t)x + tw$$

is a quotient map; it collapses $|K| \times 1$ to the point w and is otherwise one-to-one.

Proof. If $\sigma = v_0 \ldots v_n$ is a simplex of K, let $w * \sigma$ denote the simplex $wv_0 \ldots v_n$ of $w * K$. A set B is closed in $|w * K|$ if and only if its intersection with each simplex $w * \sigma$ is closed in that simplex. A set A is closed in $|K| \times I$ if and only if its intersection with each set $\sigma \times I$ is closed in $\sigma \times I$. Therefore, in order that π be a quotient map, it suffices to show that the map

$$\pi' : \sigma \times I \to w * \sigma$$

obtained by restricting π, is a quotient map. But that fact is immediate, since π' is continuous and surjective, and the spaces involved are compact Hausdorff. □

The preceding corollary suggests a way to define a cone over an arbitrary topological space.

Definition. Let X be a space. We define the **cone** over X to be the quotient space obtained from $X \times I$ by identifying the subset $X \times 1$ to a point. This point is called the **vertex** of the cone; the cone itself is denoted by $C(X)$. Formally, we form $C(X)$ by partitioning $X \times I$ into the one-point sets $\{(x,t)\}$ for $t < 1$, and the set $X \times 1$, and passing to the resulting quotient space.

EXERCISES

1. Verify the results about quotient spaces stated without proof in this section.
2. Let X be the space obtained from two copies of $\mathbf{R}^2$, say $\mathbf{R}^2 \times \{a\}$ and $\mathbf{R}^2 \times \{b\}$, by identifying (x,a) and (x,b) whenever $x \neq \mathbf{0}$. Then X is called the **plane with two origins.**
   (a) Show that each point of X has a neighborhood homeomorphic with an open set in $\mathbf{R}^2$.
   (b) Show that X is not Hausdorff.
3. (a) Show that if $p: X \to Y$ is an open quotient map and A is open in X, then $p|A: A \to p(A)$ is an open quotient map.
   (b) Repeat (a) replacing "open" by "closed."
4. Show that if $p: A \to B$ and $q: C \to D$ are open quotient maps, so is $p \times q$.
5. Let X be the coherent union of the subspaces $\{X_\alpha\}$. Show that if Y is a subspace of X that is open or closed in X, then Y is the coherent union of its subspaces $\{Y \cap X_\alpha\}$.
6. Let K and L be complexes. Show that if K is locally finite, then the topology of $|K| \times |L|$ is coherent with the subspaces $\sigma \times \tau$, for $\sigma \in K$ and $\tau \in L$.

*§21. APPLICATION: MAPS OF SPHERES

In this section, we give several applications of homology theory to classical problems of geometry and topology. The theorems we prove here will be generalized in the next section, when we prove the Lefschetz fixed-point theorem.

Definition. Let $n \geq 1$. Let $f: S^n \to S^n$ be a continuous map. If α is one of the two generators of the infinite cyclic group $H_n(S^n)$, then $f_*(\alpha) = d\alpha$ for some d. The integer d is independent of the choice of generator, since $f_*(-\alpha) = d(-\alpha)$. It is called the **degree** of the map f.

Degree has the following properties:

(1) If $f \simeq g$, then $\deg f = \deg g$.

(2) If f extends to a continuous map $h : B^{n+1} \to S^n$, then $\deg f = 0$.

(3) The identity map has degree 1.

(4) $\deg(f \circ g) = (\deg f) \cdot (\deg g)$.

Property (1) follows from Theorem 19.2; while (2) follows from the fact that $f_* : H_n(S^n) \to H_n(S^n)$ equals the composite

$$H_n(S^n) \xrightarrow{j_*} H_n(B^{n+1}) \xrightarrow{h_*} H_n(S^n),$$

where j is inclusion. Since B^{n+1} is acyclic, this composite is the zero homomorphism. Properties (3) and (4) are immediate consequences of Theorem 18.1.

Theorem 21.1. *There is no retraction $r : B^{n+1} \to S^n$.*

Proof. Such a map r would be an extension of the identity map $i : S^n \to S^n$. Since i has degree $1 \neq 0$, there is no such extension. □

Theorem 21.2 (Brouwer fixed-point theorem). *Every continuous map $\phi : B^n \to B^n$ has a fixed point.*

Proof. If $\phi : B^n \to B^n$ has no fixed point, we can define a map $h : B^n \to S^{n-1}$ by the equation

$$h(x) = \frac{x - \phi(x)}{\|x - \phi(x)\|},$$

since $x - \phi(x) \neq 0$. Let $f : S^{n-1} \to S^{n-1}$ denote the restriction of h to S^{n-1}; then f has degree 0.

On the other hand, we show that f has degree 1, giving a contradiction. Define a homotopy $H : S^{n-1} \times I \to S^{n-1}$ by the equation

$$H(u,t) = \frac{u - t\phi(u)}{\|u - t\phi(u)\|}.$$

The denominator is non-zero for $t = 1$ since $u \neq \phi(u)$; and it is non-zero for $0 \leq t < 1$ because $\|u\| = 1$ and $\|t\phi(u)\| = t\|\phi(u)\| \leq t < 1$. The map H is a homotopy between the identity map of S^{n-1} and the map f. Therefore, $\deg f = 1$. □

Definition. The **antipodal map** $a : S^n \to S^n$ is the map defined by the equation $a(x) = -x$, for all x.

In order to make further applications, we need to compute the degree of the antipodal map. We do this here by a direct proof. A second proof is given in the next section, and a third, in §31.

Theorem 21.3. *Let $n \geq 1$. The degree of the antipodal map $a : S^n \to S^n$ is $(-1)^{n+1}$.*

Proof. We show in fact that the reflection map

$$\rho(x_1, \ldots, x_{n+1}) = (x_1, \ldots, x_n, -x_{n+1})$$

has degree -1. It then follows that any reflection map

$$\rho_i(x_1, \ldots, x_i, \ldots, x_{n+1}) = (x_1, \ldots, -x_i, \ldots, x_{n+1})$$

has degree -1. For $\rho_i = h^{-1} \circ \rho \circ h$, where h is the homeomorphism of $\mathbf{R}^{n+1}$ that simply exchanges x_i and x_{n+1}, so that

$$\begin{aligned} \deg \rho_i &= (\deg h^{-1})(\deg \rho)(\deg h) \\ &= \deg(h^{-1} \circ h) \deg \rho = \deg \rho. \end{aligned}$$

Since a equals the composite $\rho_1 \rho_2 \ldots \rho_{n+1}$, we have $\deg a = (-1)^{n+1}$.

Step 1. A **triangulation** of a space X is a complex L and a homeomorphism $h : |L| \to X$. We shall construct a triangulation of S^n by an n-dimensional complex such that the reflection map ρ induces a simplicial map of this complex to itself.

In general, if K is a finite complex in $\mathbf{R}^N \times 0 \subset \mathbf{R}^{N+1}$, let $w_0 = (0, \ldots, 0, 1)$ and $w_1 = (0, \ldots, 0, -1)$ in $\mathbf{R}^{N+1}$, and let

$$S(K) = (w_0 * K) \cup (w_1 * K).$$

Then $S(K)$ is called a *suspension* of K. (See the exercises of §8.) Let $r : S(K) \to S(K)$ be the simplicial map that exchanges w_0 and w_1 and maps each vertex of K to itself. We show that there exists a complex K of dimension $n - 1$, and a triangulation

$$k : |S(K)| \to S^n,$$

such that the following diagram commutes:

$$\begin{array}{ccc} S(K) & \xrightarrow{k} & S^n \\ {\scriptstyle r}\downarrow & & \downarrow{\scriptstyle \rho} \\ S(K) & \xrightarrow{k} & S^n \end{array}$$

Then Step 1 is proved.

Let $h : |K| \to S^{n-1}$ be any triangulation of S^{n-1} by a complex of dimension $n - 1$. Let $y \in |S(K)|$. If $y = (1 - t)x + tw_0$ for some $x \in |K|$, define

$$k(y) = \left(\sqrt{1 - t^2}\, h(x), t\right).$$

If $y = (1 - t)x + tw_1$, define

$$k(y) = \left(\sqrt{1 - t^2}\, h(x), -t\right).$$

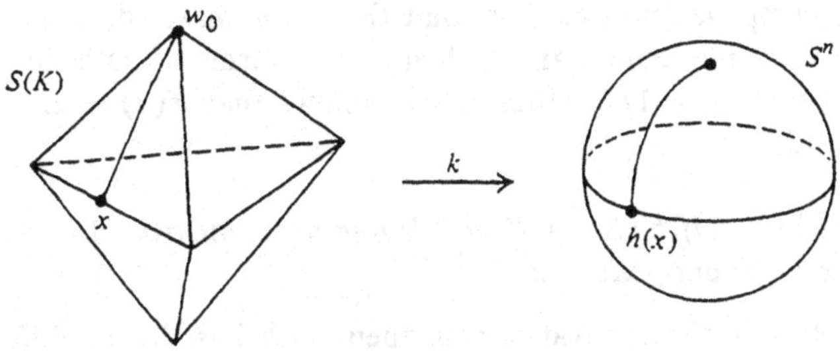

Figure 21.1

See Figure 21.1. It is easy to check that k carries $|S(K)|$ homeomorphically onto S^n. See Exercise 3. The fact that $\rho \circ k = k \circ r$ is immediate, since

$$r((1-t)x + tw_0) = (1-t)x + tw_1.$$

Step 2. In view of Step 1, in order to prove our theorem it suffices to show that $\deg r = -1$.

Let z be an n-cycle of $S(K)$; then z is a chain of the form

$$z = [w_0,c_m] + [w_1,d_m],$$

where c_m and d_m are chains of K, and $m = n - 1$. (Here we use the bracket notation of §8.) Assume $n > 1$. Since z is a cycle,

$$0 = \partial z = c_m - [w_0,\partial c_m] + d_m - [w_1,\partial d_m].$$

Restricting this chain to K, we obtain the equation $c_m + d_m = 0$, whence

$$z = [w_0,c_m] - [w_1,c_m].$$

Since r simply exchanges w_0 and w_1, we have

$$r_\#(z) = [w_1,c_m] - [w_0,c_m] = -z,$$

as desired. A similar computation holds if $n = 1$. □

Theorem 21.4. *If $h : S^n \to S^n$ has degree different from $(-1)^{n+1}$, then h has a fixed point.*

Proof. We shall suppose that $h : S^n \to S^n$ has no fixed point and prove that $h \simeq a$. The theorem follows. Intuitively, we construct the homotopy by simply moving the point $h(x)$ to the point $-x$, along the shorter great circle arc joining these two points; because $h(x)$ and $-x$ are not antipodal, there is a unique such arc, so the homotopy is well-defined. Formally, we define the homotopy $H : S^n \times I \to S^n$ by the equation

$$H(x,t) = \frac{(1-t)h(x) + t(-x)}{\|(1-t)h(x) + t(-x)\|}.$$

The proof is complete once we show that the denominator does not vanish. If $(1 - t)h(x) = tx$ for some x and t, then taking norms of both sides, we conclude that $1 - t = t = 1/2$. From this it follows that $h(x) = x$, contrary to hypothesis. □

Theorem 21.5. *If $h : S^n \to S^n$ has degree different from 1, then h carries some point x to its antipode $-x$.*

Proof. If a is the antipodal map, then $a \circ h$ has degree different from $(-1)^{n+1}$, so it has a fixed point x. Thus $a(h(x)) = x$, so $-h(x) = x$ as desired. □

Corollary 21.6. *S^n has a non-zero tangent vector field if and only if n is odd.*

Proof. If n is odd, let $n = 2k - 1$. Then for $x \in S^n$, we define

$$\bar{v}(x) = (-x_2, x_1, -x_4, x_3, \ldots, -x_{2k}, x_{2k-1}).$$

Note that $\bar{v}(x)$ is perpendicular to x, so that $\bar{v}(x)$ is tangent to S^n at x.

Conversely, suppose $\bar{v}(x)$ is a non-zero vector field defined for $x \in S^n$, such that $\bar{v}(x)$ is tangent to S^n at x. Then $h(x) = \bar{v}(x)/\|\bar{v}(x)\|$ is a map of S^n into S^n. Since $\bar{v}(x)$ is perpendicular to x, for all x, we cannot have $h(x) = x$ or $h(x) = -x$. Thus h has no fixed point and h maps no point to its antipode. We conclude that $\deg h = (-1)^{n+1}$ and $\deg h = 1$, so that n must be odd. □

EXERCISES

1. (a) Let K be the complex pictured in Figure 21.2, whose space is the boundary of a square; let sd K be its first barycentric subdivision, as indicated. Let $f : \text{sd } K \to K$ be the simplicial map specified by $f(a_i) = a_{2i}$, where the subscript $2i$ is reduced modulo 8 if necessary. Show that the map f has degree 2.

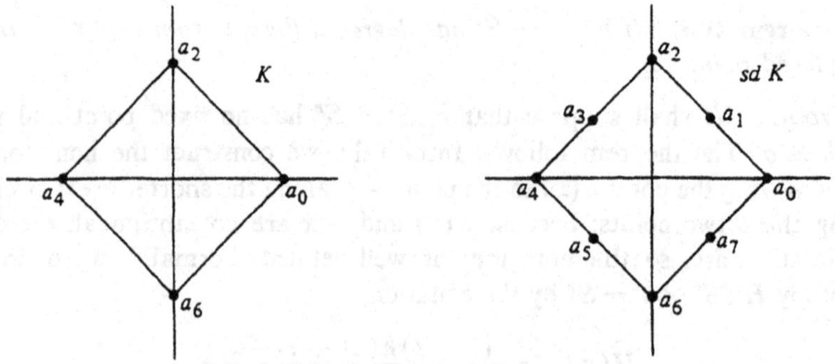

Figure 21.2

(b) Consider S^1 as the set of complex numbers of unit modulus. Show the map $h : S^1 \to S^1$ given by $h(z) = z^2$ has degree 2. [*Hint:* By radial projection, h induces a map of $|K|$ to itself that is homotopic to the simplicial map f of part (a).]

(c) If n is any integer, show that the map $k : S^1 \to S^1$ given by $k(z) = z^n$ has degree n.

2. Using the result of Exercise 1, prove the following:

The Fundamental Theorem of Algebra. Every polynomial

$$z^n + a_{n-1}z^{n-1} + \cdots + a_1 z + a_0$$

with real or complex coefficients has a zero in the complex plane.

Proof. Let S_c be the circle $|z| = c$ of radius c. Suppose the given polynomial has no zero in the ball $|z| \leq c$. Let

$$h : S_c \to \mathbf{R}^2 - \mathbf{0}$$

be defined by $h(z) = z^n + a_{n-1}z^{n-1} + \cdots + a_0$.

(a) Show that h_* is trivial.

(b) Show that if c is sufficiently large, then h is homotopic to the map $k : S_c \to \mathbf{R}^2 - \mathbf{0}$ given by $k(z) = z^n$. [*Hint:* Set $F(z,t) = z^n + t(a_{n-1}z^{n-1} + \cdots + a_0)$.]

(c) Derive a contradiction.

3. Show that the map k defined in Step 1 of Theorem 21.3 is a homeomorphism, as follows:

(a) Show that the maps

$$p : |K| \times [-1,1] \to |S(K)|,$$
$$q : |K| \times [-1,1] \to S^n,$$

given by the equations

$$p(x,t) = \begin{cases} (1-t)x + tw_0 & \text{if} \quad t \geq 0, \\ (1+t)x - tw_1 & \text{if} \quad t \leq 0, \end{cases}$$

$$q(x,t) = \left(\sqrt{1-t^2}\,h(x), t\right),$$

are quotient maps.

(b) Show that the quotient maps p and q induce the homeomorphism k.

*§22. APPLICATION: THE LEFSCHETZ FIXED-POINT THEOREM

The fixed-point theorems proved in the preceding section concern maps of balls and spheres to themselves. There is a far-reaching generalization of these theorems, due to Lefschetz. We shall prove it now.

First we need a few facts from algebra.

If $A = (a_{ij})$ is an $n \times n$ square matrix, then the **trace** of A, denoted $\operatorname{tr} A$, is defined by the equation

$$\operatorname{tr} A = \sum_{i=1}^{n} a_{ii}.$$

If A and B are $n \times n$ matrices, then

$$\operatorname{tr} AB = \sum_{i,j} a_{ij} b_{ji} = \operatorname{tr} BA.$$

If G is a free abelian group with basis $e_1, \ldots, e_n$, and if $\phi : G \to G$ is a homomorphism, we define the **trace** of ϕ to be the number $\operatorname{tr} A$, where A is the matrix of ϕ relative to the given basis. This number is independent of the choice of basis, since the matrix of ϕ relative to another basis equals $B^{-1}AB$ for some square matrix B, and $\operatorname{tr}(B^{-1}(AB)) = \operatorname{tr}((AB)B^{-1}) = \operatorname{tr} A$. The same argument shows that if $i : G \to G'$ is an isomorphism, then $\operatorname{tr}(i \circ \phi \circ i^{-1}) = \operatorname{tr} \phi$.

If K is a finite complex, and if $\phi : C_p(K) \to C_p(K)$ is a chain map, then since $C_p(K)$ is free abelian of finite rank, the trace of ϕ is defined. We denote it by $\operatorname{tr}(\phi, C_p(K))$. The group $H_p(K)$ is not necessarily free abelian, but if $T_p(K)$ is its torsion subgroup, then the group $H_p(K)/T_p(K)$ *is* free abelian. Furthermore, ϕ_* induces a homomorphism of this group with itself. We use the notation $\operatorname{tr}(\phi_*, H_p(K)/T_p(K))$ to denote the trace of this induced homomorphism.

There is no obvious relation between these two numbers; as a result, the following formula is rather striking.

Theorem 22.1 (Hopf trace theorem). *Let K be a finite complex; let $\phi : C_p(K) \to C_p(K)$ be a chain map. Then*

$$\sum_p (-1)^p \operatorname{tr}(\phi, C_p(K)) = \sum_p (-1)^p \operatorname{tr}(\phi_*, H_p(K)/T_p(K)).$$

Proof. *Step 1.* Let G be free abelian of finite rank, let H be a subgroup (necessarily free abelian), and suppose that G/H is free abelian. Let $\phi : G \to G$ be a homomorphism that carries H into itself. We show that

$$\operatorname{tr}(\phi, G) = \operatorname{tr}(\phi', G/H) + \operatorname{tr}(\phi'', H),$$

where ϕ' and ϕ'' denote the homomorphisms induced by ϕ.

Let $\{\alpha_1 + H, \ldots, \alpha_n + H\}$ be a basis for G/H and let $\beta_1, \ldots, \beta_p$ be a basis for H. If A and B are the matrices of ϕ' and ϕ'' relative to these respective bases, then

$$\phi'(\alpha_j + H) = \sum_i a_{ij}(\alpha_i + H);$$

$$\phi''(\beta_j) = \sum_i b_{ij}\beta_i.$$

Now it is easy to check that $\alpha_1,\ldots,\alpha_n,\beta_1,\ldots,\beta_p$ is a basis for G. Furthermore, it follows from the preceding equations that

$$\phi(\alpha_j) = \sum_i a_{ij}\alpha_i + (\text{something in } H);$$

$$\phi(\beta_j) = \sum_i b_{ij}\beta_i.$$

Therefore, the matrix of ϕ relative to this basis for G has the form

$$C = \begin{bmatrix} A & 0 \\ * & B \end{bmatrix}.$$

Obviously, $\operatorname{tr} C = \operatorname{tr} A + \operatorname{tr} B$; our result follows.

Step 2. As usual, let C_p denote the chain group $C_p(K)$. Let Z_p denote the p-cycles, let B_p denote the p-boundaries, and let W_p denote the group of *weak* p-boundaries, which consists of those p-chains some multiple of which bounds. Then

$$B_p \subset W_p \subset Z_p \subset C_p.$$

Since ϕ is a chain map, it carries each of these groups into itself. We shall show that the quotient groups C_p/Z_p and Z_p/W_p are free; then it will follow from Step 1 that

(i) $$\operatorname{tr}(\phi,C_p) = \operatorname{tr}(\phi,C_p/Z_p) + \operatorname{tr}(\phi,Z_p/W_p) + \operatorname{tr}(\phi,W_p).$$

(Here we abandon the use of primes to distinguish among the various induced homomorphisms.) We compute each of these terms.

Step 3. Consider the group C_p/Z_p. The homomorphism $\partial_0 : C_p \to B_{p-1}$ obtained by restricting the range of ∂ is surjective and has kernel equal to Z_p. Therefore, it induces an isomorphism of C_p/Z_p with B_{p-1}, so C_p/Z_p is free. Furthermore, because ϕ commutes with ∂, it commutes with this isomorphism. Therefore (as remarked earlier),

(ii) $$\operatorname{tr}(\phi,C_p/Z_p) = \operatorname{tr}(\phi,B_{p-1}).$$

Step 4. Consider the group Z_p/W_p. Consider the projection mappings

$$Z_p \to Z_p/B_p = H_p \to H_p/T_p.$$

Their composite is surjective and has kernel W_p. Therefore, it induces an isomorphism

$$Z_p/W_p \to H_p/T_p.$$

Thus Z_p/W_p is free. Because the projections commute with ϕ, so does this isomorphism; therefore,

(iii) $$\operatorname{tr}(\phi,Z_p/W_p) = \operatorname{tr}(\phi_*,H_p/T_p).$$

Step 5. Consider the group W_p. We show that

$$\text{(iv)} \qquad \operatorname{tr}(\phi, W_p) = \operatorname{tr}(\phi, B_p).$$

Recall that B_p is a subgroup of W_p. Applying the basic theorem of free abelian groups, we can choose a basis $\alpha_1, \ldots, \alpha_n$ for W_p such that for some integers $m_1, \ldots, m_k \geq 1$, the elements $m_1\alpha_1, \ldots, m_k\alpha_k$ form a basis for B_p. Now W_p/B_p is a torsion group; therefore, $k = n$. We compute the trace of ϕ on W_p and B_p. Let

$$\phi(\alpha_j) = \sum_i a_{ij}\alpha_i$$

$$\phi(m_j\alpha_j) = \sum_i b_{ij}(m_i\alpha_i),$$

where the summations extend from 1 to n. Then $\operatorname{tr}(\phi, W_p) = \Sigma\, a_{ii}$ and $\operatorname{tr}(\phi, B_p) = \Sigma\, b_{ii}$, by definition. Multiplying the first equation above by m_j, we conclude that $m_j a_{ij} = b_{ij} m_i$ for all i and j. In particular, $a_{ii} = b_{ii}$ for all i. Hence $\operatorname{tr}(\phi, W_p) = \operatorname{tr}(\phi, B_p)$.

Step 6. To complete the proof, we substitute formulas (ii), (iii), and (iv) into (i) to obtain the equation

$$\operatorname{tr}(\phi, C_p) = \operatorname{tr}(\phi, B_{p-1}) + \operatorname{tr}(\phi_*, H_p/T_p) + \operatorname{tr}(\phi, B_p).$$

If we multiply this equation by $(-1)^p$ and sum over all p, the first and last terms cancel out in pairs, and our desired formula is proved. □

Definition. The **Euler number** of a finite complex K is defined, classically, by the equation

$$\chi(K) = \Sigma\,(-1)^p \operatorname{rank}(C_p(K)).$$

Said differently, $\chi(K)$ is the alternating sum of the number of simplices of K in each dimension.

We show the Euler number of K is a topological invariant of $|K|$, as follows:

Theorem 22.2. *Let K be a finite complex. Let $\beta_p = \operatorname{rank} H_p(K)/T_p(K)$; it is the betti number of K in dimension p. Then*

$$\chi(K) = \Sigma_p(-1)^p\beta_p.$$

Proof. If $\phi : C_p(K) \to C_p(K)$ is the identity chain map, then the matrix of ϕ relative to any basis is the identity matrix. We conclude that $\operatorname{tr}(\phi, C_p) = \operatorname{rank} C_p$. Similarly, because ϕ_* is the identity map, $\operatorname{tr}(\phi_*, H_p/T_p) = \operatorname{rank} H_p/T_p = \beta_p$. Our formula now follows from the Hopf trace theorem. □

This theorem has a number of consequences. For instance, the fact that $\chi(K) = 2$ if $|K|$ is the boundary of a convex open region in $\mathbf{R}^3$ (since then $|K|$ is homeomorphic to S^2) can be used to show that there are only five such complexes that are regular polyhedra. These give the five classical Platonic solids. See the exercises.

Definition. Let K be a finite complex; let $h : |K| \to |K|$ be a continuous map. The number

$$\Lambda(h) = \Sigma(-1)^p \operatorname{tr}(h_*, H_p(K)/T_p(K))$$

is called the **Lefschetz number** of h.

Note that $\Lambda(h)$ depends only on the homotopy class of h, by Theorem 19.2. Furthermore, it depends only on the topological space $|K|$, not on the particular complex K: If L is another complex with $|L| = |K|$, then the homomorphism

$$j_* : H_p(L) \to H_p(K)$$

induced by the identity is an isomorphism. The fact that $(h_L)_* = j_*^{-1} \circ (h_K)_* \circ j_*$ implies that $(h_L)_*$ and $(h_K)_*$ have the same trace.

Theorem 22.3 (Lefschetz fixed-point theorem). *Let K be a finite complex; let $h : |K| \to |K|$ be a continuous map. If $\Lambda(h) \neq 0$, then h has a fixed point.*

Proof. Assume that h has no fixed point. We prove that $\Lambda(h) = 0$.

Step 1. We shall assume in subsequent steps that K satisfies the condition

$$h(\overline{\operatorname{St}}(v,K)) \cap \overline{\operatorname{St}}(v,K) = \varnothing,$$

for all v. Thus we must show that this assumption is justified.

To begin with, let $\epsilon = \min |x - h(x)|$. Using the uniform continuity of h, choose δ so that whenever $|x - y| < \delta$, we have $|h(x) - h(y)| < \epsilon/3$. Let $\lambda = \min\{\delta, \epsilon/2\}$. Then for any set A of diameter less than λ, both A and $h(A)$ have diameter less than $\epsilon/2$, so they are necessarily disjoint. Replace K by a subdivision of K in which the closed stars have diameter less than λ. As noted earlier, this does not affect the calculation of $\Lambda(h)$. Then our condition holds.

Step 2. Assume that K satisfies the condition of Step 1. Now let us choose a subdivision K' of K such that h has a simplicial approximation $f : K' \to K$. We show that if s and σ are simplices of K' and K, respectively, such that $s \subset \sigma$, then $f(s) \neq \sigma$.

Suppose $f(s) = \sigma$. Let w be a vertex of s, and let $f(w) = v$, a vertex of σ. The fact that $s \subset \sigma$ implies that $w \in \overline{\operatorname{St}}(v,K)$, so that

$$(*) \qquad h(w) \in h(\overline{\operatorname{St}}(v,K)).$$

On the other hand, we have by the definition of simplicial approximation

$$h(\operatorname{St}(w,K')) \subset \operatorname{St}(f(w),K),$$

which implies in particular that

$$(**) \qquad h(w) \in \overline{\operatorname{St}}(v,K).$$

The combination of (*) and (**) contradicts the assumption of Step 1.

Step 3. Now we compute $\Lambda(h)$ by applying the Hopf trace theorem.

Let $f: K' \to K$ be a simplicial approximation to h; let $\lambda : \mathcal{C}(K) \to \mathcal{C}(K')$ be the subdivision operator. Then h_* is induced by the chain map $\phi = f_\# \circ \lambda$, by definition.

We compute the trace of ϕ. Let A be the matrix of ϕ relative to the usual basis for $C_p(K)$, which consists of oriented p-simplices of K. Let σ be a typical basis element. The chain $\lambda(\sigma)$ is a linear combination of oriented simplices s of K' such that $s \subset \sigma$. For any such simplex s, it follows from Step 2 that $f(s) \neq \sigma$. We conclude that the chain $\phi(\sigma) = f_\#(\lambda(\sigma))$ is a linear combination of oriented p-simplices of K different from σ. The matrix A thus has an entry of 0 in the row and column corresponding to σ. It follows that all the diagonal entries of A vanish, so that $\operatorname{tr} A = \operatorname{tr} \phi = 0$.

The Hopf trace theorem tells us that

$$\Lambda(h) = \Sigma(-1)^p \operatorname{tr}(\phi, C_p(K)).$$

Because each of the terms in this summation vanishes, $\Lambda(h) = 0$. □

In order to apply this theorem, we need the following lemma.

Lemma 22.4. *Let K be a finite complex; let $h : |K| \to |K|$ be a continuous map. If $|K|$ is connected, then $h_* : H_0(K) \to H_0(K)$ is the identity.*

Proof. Let $f : K' \to K$ be a simplicial approximation to h. If v is a vertex of K, the subdivision operator λ carries v to a 0-chain carried by the subdivision of v, which is just v itself. Thus $\lambda(v)$ is a multiple of v. Because λ preserves augmentation, $\lambda(v) = v$.

Then $f_\#\lambda(v) = f_\#(v)$, which is a vertex of K. Because $|K|$ is connected, $f_\#(v)$ is homologous to v. Therefore $h_* = f_* \circ \lambda_*$ equals the identity on $H_0(K)$. □

Theorem 22.5. *Let K be a finite complex; let $h : |K| \to |K|$ be a continuous map. If $|K|$ is acyclic, then h has a fixed point.*

Proof. The group $H_0(K)$ is infinite cyclic, and h_* is the identity on $H_0(K)$. Thus $\operatorname{tr}(h_*, H_0(K)) = 1$. Since all the higher dimensional homology vanishes, $\Lambda(h) = 1$. Therefore, h has a fixed point. □

Theorem 22.6. *The antipodal map of S^n has degree $(-1)^{n+1}$.*

Proof. Let $h : S^n \to S^n$ be a map of degree d. We compute $\Lambda(h)$. Now h_* is the identity on 0-dimensional homology. On n-dimensional homology, its matrix is a one by one matrix with single entry $d =$ degree f. Therefore,

$$\Lambda(h) = 1 + (-1)^n d.$$

Now the antipodal map a has no fixed points, so that $\Lambda(a) = 0$. It follows that the degree of a is $(-1)^{n+1}$. □

EXERCISES

1. Show that every map $f : P^2 \to P^2$ has a fixed point.
2. Let K be a finite complex. Show that if $h : |K| \to |K|$ is homotopic to a constant map, then h has a fixed point.
3. (a) Show there is a map f of the Klein bottle S to itself that carries the curve C indicated in Figure 22.1 homeomorphically onto the curve D, and a map g that carries C into the shaded region E.
   (b) Let f and g be as in (a). Show that if $f' \simeq f$ and $g' \simeq g$, then f' and g' have fixed points.
   (c) Find a map $h : S \to S$ that has no fixed point.

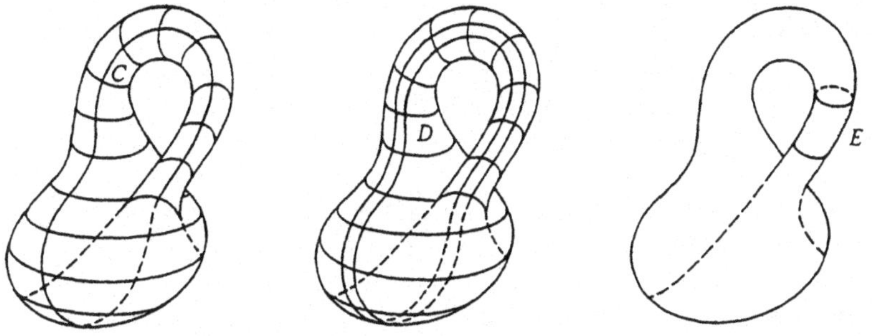

Figure 22.1

4. If M is a compact smooth surface in $\mathbf{R}^n$, and if $\overline{v}(x)$ is a tangent vector field to M, it is a standard theorem of differential geometry that for some $\epsilon > 0$ there is a continuous map

$$F : M \times (-\epsilon,\epsilon) \to M$$

having the property that for each x_0, the curve

$$t \to F(x_0,t)$$

passes through the point x_0 when $t = 0$, with velocity vector $\overline{v}(x_0)$. Furthermore, if $\overline{v}(x) \neq \mathbf{0}$ for all x, then there is a δ such that $F(x_0,t) \neq x_0$ for $0 < |t| < \delta$.
   (a) Using these facts, show that if M has a non-zero tangent vector field, then $\chi(M) = 0$.
   (b) Determine which compact surfaces have non-zero tangent vector fields.
5. Suppose B is a finite collection of polygons in $\mathbf{R}^3$, each two of which intersect in at most a common edge or a common vertex. Each of these polygons is called a face of B; its edges are called edges of B; and its vertices are called vertices of B.
   (a) Let $|B|$ be the union of the elements of B. Show that

$$\chi(|B|) = (\#\text{ faces}) - (\#\text{ edges}) + (\#\text{ vertices}),$$

where # stands for "the number of." [*Hint:* Subdivide B into a simplicial complex by starring from an interior point of each face.]

(b) We say B is "combinatorially regular" if all the faces of B have the same number (say k) of edges, if each edge of B belongs to exactly two faces of B, and if each vertex of B belongs to the same number (say l) of faces of B. The triangles making up the boundary of a tetrahedron form one such collection; the squares making up the boundary of a cube form another.

Show there are only five combinatorially regular collections B such that $|B|$ is homeomorphic to S^2. [*Hint:* Geometric considerations imply that $l \geq 3$ and $k \geq 3$.]

(c) There are many combinatorially regular collections B such that $|B|$ is homeomorphic to the torus. But there are only a limited number of possibilities for k and l. What are these?

3

Relative Homology and the Eilenberg-Steenrod Axioms

Until now we have concentrated mainly on studying the "absolute" simplicial homology groups, although we have defined the relative groups and have proved their topological invariance. Now we study the relative groups in more detail. Their uses are many. For one thing, they arise naturally in many of the applications of topology. For another, they are involved in an essential way in expressing those fundamental properties of homology that are called the Eilenberg-Steenrod axioms, as we shall see.

§23. THE EXACT HOMOLOGY SEQUENCE

One of the many ways that relative homology groups can be useful is for giving information about the absolute homology groups. There are relationships between the relative groups $H_p(K,K_0)$ and the absolute groups $H_p(K)$ and $H_p(K_0)$. For example, the vanishing of $H_p(K,K_0)$ and $H_{p+1}(K,K_0)$ implies that $H_p(K)$ and $H_p(K_0)$ are isomorphic, a fact that is not at all obvious at first glance. Formulating this relationship in a precise and general manner is a rather subtle problem.

In the early days of algebraic topology, the theorems proved along these lines were often awkward and wordy. The right language for formulating them had not been found. A remarkable algebraic idea due to Eilenberg clarified the matter immensely. Really just a new and convenient notation, it is called an "exact sequence" (of groups, or rings, or what have you). The usefulness of this concept, in algebra as well as topology, is hard to overestimate. Obscure algebraic arguments often become beautifully transparent once they are formulated in terms of exact sequences. Other arguments that were difficult even for professional algebraists, become so straightforward that they can be safely left to the reader.

The relationship between the absolute and relative homology groups is expressed by an exact sequence called the "exact homology sequence of a pair." In this section, we shall study exact sequences of groups (usually, abelian groups), and we shall define the exact homology sequence of a pair.

Definition. Consider a sequence (finite or infinite) of groups and homomorphisms

$$\cdots \to A_1 \xrightarrow{\phi_1} A_2 \xrightarrow{\phi_2} A_3 \to \cdots.$$

This sequence is said to be **exact at A_2** if

$$\text{image } \phi_1 = \text{kernel } \phi_2.$$

If it is everywhere exact, it is said simply to be an **exact sequence.** Of course, exactness is not defined at the first or last group of a sequence, if such exist.

We list here several basic facts about exact sequences; you should have them at your fingertips. Proofs are left as exercises. Because the groups we consider are abelian, we shall let 0 denote the trivial (one-element) group.

(1) $A_1 \xrightarrow{\phi} A_2 \to 0$ is exact if and only if ϕ is an epimorphism.

(2) $0 \to A_1 \xrightarrow{\phi} A_2$ is exact if and only if ϕ is a monomorphism.

(3) Suppose the sequence

$$0 \to A_1 \xrightarrow{\phi} A_2 \xrightarrow{\psi} A_3 \to 0$$

is exact; such a sequence is called a **short exact sequence.** Then $A_2/\phi(A_1)$ is isomorphic to A_3; the isomorphism is induced by ψ. Conversely if $\psi: A \to B$ is an epimorphism with kernel K, then the sequence

$$0 \to K \xrightarrow{i} A \xrightarrow{\psi} B \to 0$$

is exact, where i is inclusion.

(4) Suppose the sequence

$$A_1 \xrightarrow{\alpha} A_2 \xrightarrow{\phi} A_3 \xrightarrow{\beta} A_4$$

is exact. Then the following are equivalent:

(i) α is an epimorphism.
(ii) β is a monomorphism.
(iii) ϕ is the zero homomorphism.

(5) Suppose the sequence

$$A_1 \xrightarrow{\alpha} A_2 \to A_3 \to A_4 \xrightarrow{\beta} A_5$$

is exact. Then so is the induced sequence

$$0 \to \operatorname{cok} \alpha \to A_3 \to \ker \beta \to 0.$$

Definition. Consider two sequences of groups and homomorphisms having the same index set,

$$\cdots \to A_1 \to A_2 \to \cdots,$$
$$\cdots \to B_1 \to B_2 \to \cdots.$$

A **homomorphism** of the first sequence into the second is a family of homomorphisms $\alpha_i : A_i \to B_i$ such that each square of maps

$$\begin{array}{ccc} A_i & \longrightarrow & A_{i+1} \\ {\scriptstyle \alpha_i}\downarrow & & \downarrow{\scriptstyle \alpha_{i+1}} \\ B_i & \longrightarrow & B_{i+1} \end{array}$$

commutes. It is an **isomorphism** of sequences if each α_i is an isomorphism.

For example, if a chain complex $\mathcal{C}$ is looked at as a sequence of groups C_i and homomorphisms ∂_i, then a homomorphism of one such sequence $\mathcal{C}$ into another $\mathcal{D}$ is just what we have called a chain map of $\mathcal{C}$ into $\mathcal{D}$.

Definition. Consider a short exact sequence

$$0 \to A_1 \xrightarrow{\phi} A_2 \xrightarrow{\psi} A_3 \to 0.$$

This sequence is said to **split** if the group $\phi(A_1)$ is a direct summand in A_2.

This means that A_2 is the internal direct sum of $\phi(A_1)$ and some other subgroup B; the group B is not uniquely determined, of course. In this case, the sequence becomes

$$0 \to A_1 \xrightarrow{\phi} \phi(A_1) \oplus B \xrightarrow{\psi} A_3 \to 0,$$

where ϕ defines an isomorphism of A_1 with $\phi(A_1)$, and ψ defines an isomorphism of B with A_3. An equivalent formulation is to state that there is an isomorphism θ such that the following diagram commutes:

$$\begin{array}{ccccccccc} 0 & \longrightarrow & A_1 & \longrightarrow & A_2 & \longrightarrow & A_3 & \longrightarrow & 0 \\ & & \downarrow{=} & & \downarrow{\theta} & & \downarrow{=} & & \\ 0 & \longrightarrow & A_1 & \xrightarrow[i]{} & A_1 \oplus A_3 & \xrightarrow[\pi]{} & A_3 & \longrightarrow & 0. \end{array}$$

In this case, $\oplus$ denotes external direct sum; i is inclusion and π is projection. The map θ is defined by writing $A_2 = \phi(A_1) \oplus B$ and letting θ equal ϕ^{-1} on the first summand and ψ on the second.

Theorem 23.1. *Let* $0 \to A_1 \xrightarrow{\phi} A_2 \xrightarrow{\psi} A_3 \to 0$ *be exact. The following are equivalent:*

(1) *The sequence splits.*

(2) *There is a map* $p : A_2 \to A_1$ *such that* $p \circ \phi = i_{A_1}$.

(3) *There is a map* $j : A_3 \to A_2$ *such that* $\psi \circ j = i_{A_3}$.

$$0 \to A_1 \underset{p}{\overset{\phi}{\rightleftarrows}} A_2 \underset{j}{\overset{\psi}{\rightleftarrows}} A_3 \to 0.$$

Proof. We show that (1) implies (2) and (3). It suffices to prove (2) and (3) for the sequence

$$0 \to A_1 \xrightarrow{i} A_1 \oplus A_3 \xrightarrow{\pi} A_3 \to 0.$$

And this is easy; we define $p : A_1 \oplus A_3 \to A_1$ as projection, and $j : A_3 \to A_1 \oplus A_3$ as inclusion.

(2) $\Rightarrow$ (1). We show $A_2 = \phi(A_1) \oplus (\ker p)$. First, for $x \in A_2$ we can write $x = \phi p(x) + (x - \phi p(x))$. The first term is in $\phi(A_1)$ and the second is in $\ker p$ because $p(x) - p(\phi p(x)) = p(x) - (p\phi)p(x) = 0$. Second, if $x \in \phi(A_1) \cap \ker p$, then $x = \phi(y)$ for some y, whence $p(x) = p\phi(y) = y$. Since $x \in \ker p$, the element y vanishes, so $x = \phi(y)$ vanishes also.

(3) $\Rightarrow$ (1). We show $A_2 = (\ker \psi) \oplus j(A_3)$. Since $\ker \psi = \operatorname{im} \phi$, this will suffice. First, for $x \in A_2$ we can write $x = (x - j\psi(x)) + j\psi(x)$. The first term is in $\ker \psi$, since $\psi(x) - (\psi j)\psi(x) = \psi(x) - \psi(x) = 0$; the second term is in $j(A_3)$. Second, if $x \in (\ker \psi) \cap j(A_3)$, then $x = j(z)$ for some z, whence $\psi(x) = \psi j(z) = z$; since $x \in \ker \psi$, the element z vanishes, so $x = j(z)$ vanishes also. □

Corollary 23.2. *Let* $0 \to A_1 \xrightarrow{\phi} A_2 \xrightarrow{\psi} A_3 \to 0$ *be exact. If* A_3 *is free abelian, the sequence splits.*

Proof. We choose a basis for A_3, and define the value of $j : A_3 \to A_2$ on the basis element e to be any element of the nonempty set $\psi^{-1}(e)$. □

With these basic facts about exact sequences at our disposal, we can now describe the exact homology sequence of a pair.

First we need to define a certain homomorphism

$$\partial_* : H_p(K,K_0) \to H_{p-1}(K_0)$$

that is induced by the boundary operator and is called the **homology boundary homomorphism.** It is constructed as follows: Given a cycle z in $C_p(K,K_0)$, it is the coset modulo $C_p(K_0)$ of a chain d of K whose boundary is carried by K_0. The chain ∂d is automatically a cycle of K_0. We define $\partial_* \{z\} = \{\partial d\}$, where $\{\ \}$ means "homology class of"; we prove later that ∂_* is a well-defined homomorphism.

Algebraically, the construction of ∂_* can be described as follows: Consider the following diagram, where $i : K_0 \to K$ and $\pi : (K, \varnothing) \to (K,K_0)$ are inclusions:

$$\begin{array}{ccc} & & C_p(K) \xrightarrow{\pi_\#} C_p(K,K_0) \\ & & \big\downarrow \partial \\ C_{p-1}(K_0) & \xrightarrow{i_\#} & C_{p-1}(K) \end{array}$$

Now $i_\#$ is inclusion, and $\pi_\#$ is projection of $C_p(K)$ onto $C_p(K)/C_p(K_0)$. Given a cycle z of $C_p(K,K_0)$, the chain d of $C_p(K)$ representing it is a chain such that $\pi_\#(d) = z$. We take ∂d, and note that since it is carried by K_0, it equals $i_\#(c)$ for some $p - 1$ chain c of K_0. Now c is actually a cycle; its homology class is defined to be $\partial_*(\{z\})$. Thus ∂_* is defined by a certain "zig-zag" process: Pull back via $\pi_\#$; apply ∂; pull back via $i_\#$. This description of ∂_* will be useful in the next section.

Now we can state our basic theorem relating the homology of K, K_0, and (K,K_0).

Definition. A **long exact sequence** is an exact sequence whose index set is the set of integers. That is, it is a sequence that is infinite in both directions. It may, however, begin or end with an infinite string of trivial groups.

Theorem 23.3 (The exact homology sequence of a pair). *Let K be a complex; let K_0 be a subcomplex. Then there is a long exact sequence*

$$\cdots \longrightarrow H_p(K_0) \xrightarrow{i_*} H_p(K) \xrightarrow{\pi_*} H_p(K,K_0) \xrightarrow{\partial_*} H_{p-1}(K_0) \longrightarrow \cdots,$$

where $i : K_0 \to K$ and $\pi : (K, \varnothing) \to (K,K_0)$ are inclusions, and ∂_* is induced by the boundary operator ∂. There is a similar exact sequence in reduced homology:

$$\cdots \longrightarrow \tilde{H}_p(K_0) \longrightarrow \tilde{H}_p(K) \longrightarrow H_p(K,K_0) \longrightarrow \tilde{H}_{p-1}(K_0) \longrightarrow \cdots.$$

The proof of this theorem is basically algebraic in nature. We shall formulate it in a purely algebraic fashion and prove it in the next section. For the present, we apply this sequence to some specific examples.

Example 1. Let K be the complex pictured in Figure 23.1, whose polytope is a square. Let K_0 be the subcomplex whose polytope is the boundary of the square. We know from Example 3 of §9 that $H_2(K,K_0)$ is infinite cyclic and is generated by the 2-chain γ that is the sum of the 2-simplices, oriented counterclockwise. Furthermore, $H_1(K_0)$ is infinite cyclic and is generated by the 1-chain $s_1 + s_2 + s_3 + s_4$,

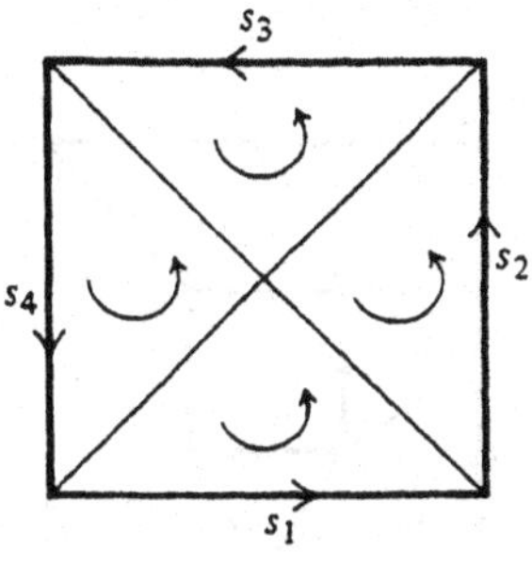

Figure 23.1

which happens to equal $\partial\gamma$. Thus in this particular case, the boundary homomorphism

$$\partial_* : H_2(K,K_0) \to H_1(K_0)$$

is an isomorphism.

This fact can also be proved by considering the exact homology sequence of the pair (K,K_0). A portion of this sequence is

$$H_2(K) \to H_2(K,K_0) \xrightarrow{\partial_*} H_1(K_0) \to H_1(K).$$

The end groups vanish because $|K|$ is contractible; therefore ∂_* is an isomorphism.

Example 2. Let K be the complex pictured in Figure 23.2, whose underlying space is an annulus. Let K_0 be the subcomplex of K whose polytope equals the union of the inner and outer edges of the square. In Example 4 of §9, we computed the homology of (K,K_0). We recompute it here, using our knowledge of the homology of K and K_0. Consider the following portion of the exact sequence in reduced homology:

$$\begin{array}{ccccccccccccc} 0 \to & H_2(K,K_0) & \xrightarrow{\partial_*} & H_1(K_0) & \xrightarrow{i_*} & H_1(K) & \xrightarrow{\pi_*} & H_1(K,K_0) & \xrightarrow{\partial_*} & \tilde{H}_0(K_0) & \to 0, \\ 0 \to & (?) & \to & \mathbf{Z} \oplus \mathbf{Z} & \to & \mathbf{Z} & \to & (?) & \to & \mathbf{Z} & \to 0. \end{array}$$

The zeros at the ends come from the fact that $|K|$ has the homotopy type of a circle, so $H_2(K) = \tilde{H}_0(K) = 0$. Furthermore, $H_1(K)$ is infinite cyclic. It is generated by the cycle z_1 indicated in Figure 23.2 running around the outer edge of K, or by the cycle z_2 running around the inner edge; these cycles are homologous. Because $|K_0|$ is topologically the disjoint union of two circles, $H_1(K_0) \simeq \mathbf{Z} \oplus \mathbf{Z}$ and has as basis the cosets of z_1 and z_2, while $\tilde{H}_0(K_0) \simeq \mathbf{Z}$ and is generated by $\{v_1 - v_0\}$.

If $i : K_0 \to K$ is inclusion, then i_* maps both $\{z_1\}$ and $\{z_2\}$ to the same generator of $H_1(K)$. It follows that (1) i_* is an epimorphism, and (2) its kernel is infinite cyclic and is generated by $\{z_1\} - \{z_2\}$.

From the first fact, it follows that π_* is the zero homomorphism, whence

$$\partial_* : H_1(K,K_0) \to \tilde{H}_0(K_0)$$

is an isomorphism. Thus $H_1(K,K_0)$ is infinite cyclic and is generated by the 1-chain e_0, whose boundary is $v_1 - v_0$.

From the second fact, it follows that since

$$\partial_* : H_2(K,K_0) \to \ker i_*$$

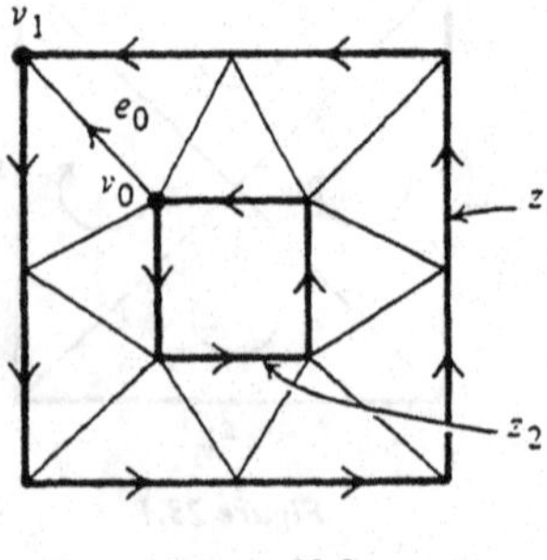

Figure 23.2

is an isomorphism, the group $H_2(K,K_0)$ is infinite cyclic, and is generated by the chain γ that is the sum of the 2-simplices of K, oriented counterclockwise. For $\{\partial\gamma\} = \{z_1 - z_2\}$, and $z_1 - z_2$ generates ker i_*.

Example 3. We consider the next two examples together. Let (K,K_0) denote either the cylinder and its top edge, or the Möbius band and its edge. In each case, $|K_0|$ is a circle. Furthermore, $|K|$ has the homotopy type of a circle; the central circle C indicated in Figure 23.3 is a deformation retract of $|K|$. Thus we know the homology of K and K_0; we compute the homology of (K,K_0) from the exact homology sequence, a portion of which is as follows:

$$\begin{array}{ccccccccccc} H_2(K) & \to & H_2(K,K_0) & \to & H_1(K_0) & \xrightarrow{i_*} & H_1(K) & \to & H_1(K,K_0) & \to & \tilde{H}_0(K_0), \\ 0 & \to & (?) & \to & \mathbf{Z} & \to & \mathbf{Z} & \to & (?) & \to & 0. \end{array}$$

Everything depends on computing the homomorphism i_*. Since the retraction $r:|K| \to C$ is a homotopy equivalence, it suffices to compute the homomorphism induced by the composite map $r \circ i:|K_0| \to C$, which collapses the edge $|K_0|$ to the central circle C. In the case of the cylinder, this map has degree 1; while in the case of the Möbius band the induced homomorphism clearly equals multiplication by 2.

Thus $H_2(K,K_0) = H_1(K,K_0) = 0$ in the case of the cylinder. In the case of the Möbius band,

$$H_2(K,K_0) = 0 \quad \text{and} \quad H_1(K,K_0) \simeq \mathbf{Z}/2.$$

The central circle C represents the non-zero element of $H_1(K,K_0)$ in this case. So does the chain D, as you may prove.

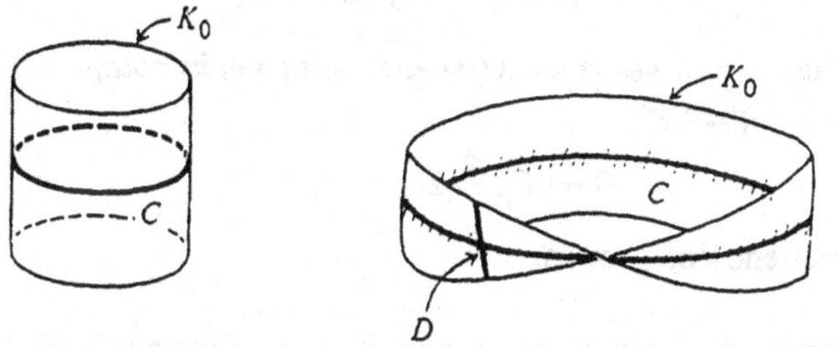

Figure 23.3

EXERCISES

1. Check statements (1)–(5) of this section concerning exact sequences.
2. (a) Suppose

$$A_1 \xrightarrow{\alpha_1} B_1 \xrightarrow{\beta_1} C_1 \quad \text{and} \quad A_2 \xrightarrow{\alpha_2} B_2 \xrightarrow{\beta_2} C_2$$

are exact. Show that

$$A_1 \times A_2 \xrightarrow{\alpha_1 \times \alpha_2} B_1 \times B_2 \xrightarrow{\beta_1 \times \beta_2} C_1 \times C_2$$

is exact.

(b) Generalize to arbitrary direct products.

(c) Generalize to arbitrary direct sums.

3. Show that if K_0 is acyclic, $H_i(K,K_0) \simeq \tilde{H}_i(K)$.
4. Suppose inclusion $|K_0| \to |K|$ is a homotopy equivalence. Show $H_p(K,K_0) = 0$ for all p.
5. Let (K,K_0) denote the Möbius band and its edge, as in Example 3. Show that the non-zero element of $H_1(K,K_0)$ is represented by the chain D indicated in Figure 23.3.
6. Let S denote the Klein bottle. Let A and C be the usual simple closed curves in S. (See Exercise 2 of §18.) Compute the exact sequences of the pairs (S,A) and (S,C).

§24. THE ZIG-ZAG LEMMA

Now we prove exactness of the homology sequence of a pair. We shall reformulate this result as a theorem about chain complexes and prove it in that form. First we need a definition.

Definition. Let $\mathcal{C}$, $\mathcal{D}$, and $\mathcal{E}$ be chain complexes. Let 0 denote the trivial chain complex whose groups vanish in every dimension. Let $\phi : \mathcal{C} \to \mathcal{D}$ and $\psi : \mathcal{D} \to \mathcal{E}$ be chain maps. We say the sequence

$$0 \to \mathcal{C} \xrightarrow{\phi} \mathcal{D} \xrightarrow{\psi} \mathcal{E} \to 0$$

is exact, or that it is a **short exact sequence of chain complexes,** if in each dimension p, the sequence

$$0 \to C_p \xrightarrow{\phi} D_p \xrightarrow{\psi} E_p \to 0$$

is an exact sequence of groups.

For example, if K is a complex and K_0 is a subcomplex of K, the sequence

$$0 \to \mathcal{C}(K_0) \to \mathcal{C}(K) \to \mathcal{C}(K,K_0) \to 0$$

is exact, because $C_p(K,K_0) = C_p(K)/C_p(K_0)$ by definition.

Lemma 24.1 (The zig-zag lemma). *Suppose one is given chain complexes $\mathcal{C} = \{C_p,\partial_C\}$, $\mathcal{D} = \{D_p,\partial_D\}$, and $\mathcal{E} = \{E_p,\partial_E\}$, and chain maps ϕ, ψ such that the sequence*

$$0 \to \mathcal{C} \xrightarrow{\phi} \mathcal{D} \xrightarrow{\psi} \mathcal{E} \to 0$$

is exact. Then there is a long exact homology sequence

$$\cdots \to H_p(\mathcal{C}) \xrightarrow{\phi_*} H_p(\mathcal{D}) \xrightarrow{\psi_*} H_p(\mathcal{E}) \xrightarrow{\partial_*} H_{p-1}(\mathcal{C}) \xrightarrow{\phi_*} H_{p-1}(\mathcal{D}) \to \cdots,$$

where ∂_* is induced by the boundary operator in $\mathcal{D}$.

Proof. The proof is of a type now commonly known as "diagram-chasing." *Master this proof*; in the future we shall leave all proofs of this sort as exercises.

We shall use the following commutative diagram:

$$
\begin{array}{ccccccccc}
0 & \longrightarrow & C_{p+1} & \xrightarrow{\phi} & D_{p+1} & \xrightarrow{\psi} & E_{p+1} & \longrightarrow & 0 \\
& & \downarrow{\scriptstyle\partial_C} & & \downarrow{\scriptstyle\partial_D} & & \downarrow{\scriptstyle\partial_E} & & \\
0 & \longrightarrow & C_p & \xrightarrow{\phi} & D_p & \xrightarrow{\psi} & E_p & \longrightarrow & 0 \\
& & \downarrow{\scriptstyle\partial_C} & & \downarrow{\scriptstyle\partial_D} & & \downarrow{\scriptstyle\partial_E} & & \\
0 & \longrightarrow & C_{p-1} & \xrightarrow{\phi} & D_{p-1} & \xrightarrow{\psi} & E_{p-1} & \longrightarrow & 0.
\end{array}
$$

with boxes labeled d_{p+1}, e_{p+1}, c_p, d_p, e_p, c_{p-1}.

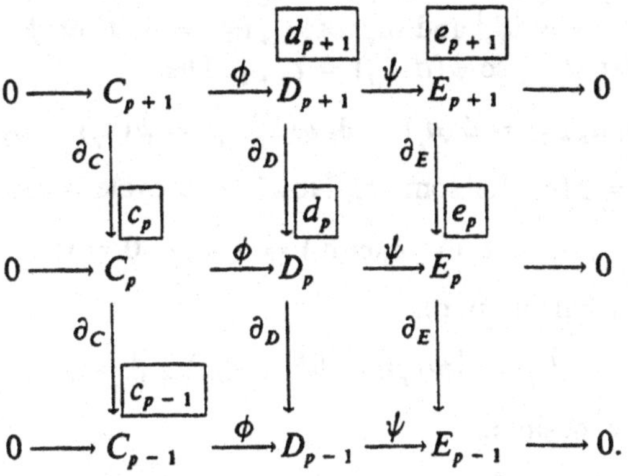

Step 1. First, we define ∂_*. Given a *cycle* e_p in E_p (that is, an element of ker ∂_E), *choose* $d_p \in D_p$ so that $\psi(d_p) = e_p$. (Recall ψ is surjective.) The element $\partial_D d_p$ of D_{p-1} lies in ker ψ, since $\psi(\partial_D d_p) = \partial_E \psi(d_p) = \partial_E e_p = 0$. Therefore, there exists an element c_{p-1} in C_{p-1} such that $\phi(c_{p-1}) = \partial_D d_p$, since ker $\psi = $ im ϕ. This element is unique because ϕ is injective. Furthermore, c_{p-1} is a cycle. For

$$\phi(\partial_C c_{p-1}) = \partial_D \phi(c_{p-1}) = \partial_D \partial_D d_p = 0;$$

then $\partial_C c_{p-1} = 0$ because ϕ is injective. Define

$$\partial_* \{e_p\} = \{c_{p-1}\},$$

where $\{\ \}$ means "homology class of."

Step 2. We show ∂_* is a well-defined homomorphism. *Notation:* Let e_p and e_p' be two elements in the kernel of $\partial_E : E_p \to E_{p-1}$. Choose d_p and d_p' so that $\psi(d_p) = e_p$ and $\psi(d_p') = e_p'$. Then choose c_{p-1} and c_{p-1}' so $\phi(c_{p-1}) = \partial_D d_p$ and $\phi(c_{p-1}') = \partial_D d_p'$.

To show ∂_* well defined, we suppose e_p and e_p' are homologous, and show that c_{p-1} and c_{p-1}' are homologous. So suppose $e_p - e_p' = \partial_E e_{p+1}$. Choose d_{p+1} so that $\psi(d_{p+1}) = e_{p+1}$. Then note that

$$\psi(d_p - d_p' - \partial_D d_{p+1}) = e_p - e_p' - \partial_E \psi(d_{p+1})$$
$$= e_p - e_p' - \partial_E e_{p+1} = 0.$$

Therefore, we can choose c_p so $\phi(c_p) = d_p - d_p' - \partial_D d_{p+1}$. Then

$$\phi(\partial_C c_p) = \partial_D \phi(c_p) = \partial_D(d_p - d_p') - 0 = \phi(c_{p-1} - c_{p-1}').$$

Because ϕ is injective, $\partial_C c_p = c_{p-1} - c_{p-1}'$, as desired.

To show ∂_* is a homomorphism, note that $\psi(d_p + d_p') = e_p + e_p'$, and

$\phi(c_{p-1} + c'_{p-1}) = \partial_D(d_p + d'_p)$. Thus $\partial_* \{e_p + e'_p\} = \{c_{p-1} + c'_{p-1}\}$ by definition. The latter equals, of course, $\partial_* \{e_p\} + \partial_* \{e'_p\}$.

Step 3. We prove exactness at $H_p(\mathcal{D})$. Let $\gamma \in H_p(\mathcal{D})$. Because $\psi \circ \phi = 0$, we have $\psi_* \circ \phi_* = 0$. Hence if $\gamma \in$ image ϕ_*, then $\psi_*(\gamma) = 0$.

Conversely, let $\gamma = \{d_p\}$ and suppose $\psi_*(\gamma) = 0$. Then $\psi(d_p) = \partial_E e_{p+1}$ for some e_{p+1}. Choose d_{p+1} so $\psi(d_{p+1}) = e_{p+1}$. Then

$$\psi(d_p - \partial_D d_{p+1}) = \psi(d_p) - \partial_E \psi(d_{p+1}) = \psi(d_p) - \partial_E e_{p+1} = 0$$

so $d_p - \partial_D d_{p+1} = \phi(c_p)$ for some c_p. Now c_p is a cycle because

$$\phi(\partial_C c_p) = \partial_D \phi(c_p) = \partial_D d_p - 0 = 0$$

and ϕ is injective. Furthermore,

$$\phi_*\{c_p\} = \{\phi(c_p)\} = \{d_p - \partial_D d_{p+1}\} = \{d_p\},$$

so $\{d_p\} \in \text{im}\, \phi_*$, as desired.

Step 4. We prove exactness at $H_p(\mathcal{E})$. *Notation:* Let $\alpha = \{e_p\}$ be an element of $H_p(\mathcal{E})$. Choose d_p so $\psi(d_p) = e_p$; then choose c_{p-1} so $\phi(c_{p-1}) = \partial_D d_p$. Then $\partial_* \alpha = \{c_{p-1}\}$ by definition.

If $\alpha \in \text{im}\, \psi_*$, then $\alpha = \{\psi(d_p)\}$, where d_p is a *cycle* of $\mathcal{D}$. Then $\phi(c_{p-1}) = 0$; whence $c_{p-1} = 0$. Thus $\partial_* \alpha = 0$.

Conversely, suppose $\partial_* \alpha = 0$. Then $c_{p-1} = \partial_C c_p$ for some c_p. We assert that $d_p - \phi(c_p)$ is a cycle, and $\alpha = \psi_*\{d_p - \phi(c_p)\}$, so $\alpha \in \text{im}\, \psi_*$. By direct computation,

$$\partial_D(d_p - \phi(c_p)) = \partial_D d_p - \phi(\partial_C c_p) = \partial_D d_p - \phi(c_{p-1}) = 0,$$
$$\psi_*\{d_p - \phi(c_p)\} = \{\psi(d_p) - 0\} = \{e_p\} = \alpha.$$

Step 5. We prove exactness at $H_{p-1}(\mathcal{C})$. Let $\beta \in H_{p-1}(\mathcal{C})$. If $\beta \in \text{im}\, \partial_*$, then $\beta = \{c_{p-1}\}$, where $\phi(c_{p-1}) = \partial_D d_p$ for some d_p, by definition. Then

$$\phi_*(\beta) = \{\phi(c_{p-1})\} = \{\partial_D d_p\} = 0.$$

Conversely, suppose $\phi_*(\beta) = 0$. Let $\beta = \{c_{p-1}\}$; then $\{\phi(c_{p-1})\} = 0$, so $\phi(c_{p-1}) = \partial_D d_p$ for some d_p. Define $e_p = \psi(d_p)$; then e_p is a cycle because $\partial_E e_p = \psi(\partial_D d_p) = \psi\phi(c_{p-1}) = 0$. And $\beta = \partial_* \{e_p\}$ by definition. Thus $\beta \in \text{im}\, \partial_*$. □

Note that nowhere in the proof of this lemma did we assume that the chain complexes involved were free or that they were non-negative. This lemma thus has a much broader range of application than just to chain groups of simplicial complexes. However, applied to the simplicial chain groups $\mathcal{C}(K_0)$ and $\mathcal{C}(K)$ and $\mathcal{C}(K,K_0)$, we obtain Theorem 23.3 for non-reduced homology as an immediate consequence.

To obtain the theorem for reduced homology, we let $\mathcal{E} = \mathcal{C}(K,K_0)$, as before; but we replace $\mathcal{C}$ and $\mathcal{D}$ by the augmented chain complexes $\{\mathcal{C},\epsilon\}$ and $\{\mathcal{D},\epsilon\}$, respectively. Note that exactness and commutativity hold in dimension -1, since inclusion $C_0(K_0) \to C_0(K)$ preserves augmentation:

$$\begin{array}{ccccccccc}
0 & \longrightarrow & C_0(K_0) & \longrightarrow & C_0(K) & \longrightarrow & C_0(K,K_0) & \longrightarrow & 0 \\
 & & \downarrow \epsilon & & \downarrow \epsilon & & \downarrow & & \\
0 & \longrightarrow & \mathbf{Z} & \xrightarrow[i_Z]{} & \mathbf{Z} & \longrightarrow & 0 & \longrightarrow & 0.
\end{array}$$

Application of the zig-zag lemma now gives us Theorem 23.3 for reduced homology.

Before considering further applications, let us extract some additional information from the proof of the zig-zag lemma. We have constructed a function assigning to each short exact sequence of chain complexes, a long exact sequence of their homology groups. Now we point out that this assignment is "natural," in the sense that a homomorphism of short exact sequences of chain complexes gives rise to a homomorphism of the corresponding long exact homology sequences.

Theorem 24.2. *Suppose one is given the commutative diagram*

$$\begin{array}{ccccccccc}
0 & \longrightarrow & \mathcal{C} & \xrightarrow{\phi} & \mathcal{D} & \xrightarrow{\psi} & \mathcal{E} & \longrightarrow & 0 \\
 & & \downarrow \alpha & & \downarrow \beta & & \downarrow \gamma & & \\
0 & \longrightarrow & \mathcal{C}' & \xrightarrow[\phi']{} & \mathcal{D}' & \xrightarrow[\psi']{} & \mathcal{E}' & \longrightarrow & 0
\end{array}$$

where the horizontal sequences are exact sequences of chain complexes, and α, β, γ are chain maps. Then the following diagram commutes as well:

$$\begin{array}{ccccccccc}
\longrightarrow & H_p(\mathcal{C}) & \xrightarrow{\phi_*} & H_p(\mathcal{D}) & \xrightarrow{\psi_*} & H_p(\mathcal{E}) & \xrightarrow{\partial_*} & H_{p-1}(\mathcal{C}) & \longrightarrow \\
 & \downarrow \alpha_* & & \downarrow \beta_* & & \downarrow \gamma_* & & \downarrow \alpha_* & \\
\longrightarrow & H_p(\mathcal{C}') & \xrightarrow{\phi'_*} & H_p(\mathcal{D}') & \xrightarrow{\psi'_*} & H_p(\mathcal{E}') & \xrightarrow{\partial'_*} & H_{p-1}(\mathcal{C}') & \longrightarrow .
\end{array}$$

Proof. Commutativity of the first two squares is immediate, because commutativity holds already on the *chain* level. Commutativity of the last square involves examining the definitions of ∂_* and ∂'_*.

Given $\{e_p\} \in H_p(\mathcal{E})$, choose d_p so $\psi(d_p) = e_p$, and choose c_{p-1} so $\phi(c_{p-1}) = \partial_D d_p$. Then $\partial_* \{e_p\} = \{c_{p-1}\}$, by definition. Let $e'_p = \gamma(e_p)$; we wish to show that $\partial'_* \{e'_p\} = \alpha_* \{c_{p-1}\}$. Roughly speaking, this follows because each step in the definition of ∂_* commutes: The chain $\beta(d_p)$ is a suitable "pull-back" for e'_p, since $\psi'\beta(d_p) = \gamma\psi(d_p) = \gamma(e_p) = e'_p$. And then the chain $\alpha(c_{p-1})$ is the pull-back of $\partial'_D \beta(d_p)$, since $\phi'\alpha(c_{p-1}) = \beta\phi(c_{p-1}) = \beta(\partial_D d_p) = \partial'_D \beta(d_p)$. Thus $\partial'_* \{e'_p\} = \{\alpha(c_{p-1})\}$ by definition. $\square$

Naturality of the long exact homology sequence is extremely useful. We give one application now. It makes use of the following lemma, whose proof is simple "diagram-chasing." As promised, we leave the proof to you. The lemma itself is of great usefulness.

Lemma 24.3 (The Steenrod five-lemma). *Suppose one is given a commutative diagram of abelian groups and homomorphisms:*

$$\begin{array}{ccccccccc} A_1 & \longrightarrow & A_2 & \longrightarrow & A_3 & \longrightarrow & A_4 & \longrightarrow & A_5 \\ \downarrow f_1 & & \downarrow f_2 & & \downarrow f_3 & & \downarrow f_4 & & \downarrow f_5 \\ B_1 & \longrightarrow & B_2 & \longrightarrow & B_3 & \longrightarrow & B_4 & \longrightarrow & B_5 \end{array}$$

where the horizontal sequences are exact. If $f_1, f_2, f_4,$ and f_5 are isomorphisms, so is f_3. □

Lemma 24.4. *Let $h : (K,K_0) \to (L,L_0)$ be a simplicial map.*

(a) *The induced homology homomorphisms h_* give a homomorphism of the exact homology sequence of (K,K_0) with that of (L,L_0).*

(b) *If $h_* : H_i(K) \to H_i(L)$ and $h_* : H_i(K_0) \to H_i(L_0)$ are isomorphisms for $i = p$ and $i = p - 1$, then*

$$h_* : H_p(K,K_0) \to H_p(L,L_0)$$

is an isomorphism.

(c) *Both these results hold if absolute homology is replaced throughout by reduced homology.*

Proof. We know $h_\#$ is a chain map, and the following diagram commutes:

$$\begin{array}{ccccccccc} 0 & \longrightarrow & C_p(K_0) & \xrightarrow{i_\#} & C_p(K) & \xrightarrow{\pi_\#} & C_p(K)/C_p(K_0) & \longrightarrow & 0 \\ & & \downarrow h_\# & & \downarrow h_\# & & \downarrow h_\# & & \\ 0 & \longrightarrow & C_p(L_0) & \xrightarrow{i_\#} & C_p(L) & \xrightarrow{\pi_\#} & C_p(L)/C_p(L_0) & \longrightarrow & 0. \end{array}$$

Then (a) follows. To deal with the case of reduced homology, we recall that $h_\#$ is augmentation-preserving, so it gives the desired chain map of the augmented chain complexes, provided we define $h_\# : \mathbf{Z} \to \mathbf{Z}$ as the identity on the chain groups in dimension -1. Thus (a) holds for reduced homology.

Result (b) follows immediately from the Five-lemma. □

The following is an immediate consequence of this lemma and the results of Chapter 2.

Theorem 24.5. *The preceding lemma holds for an arbitrary continuous map $h : (|K|,|K_0|) \to (|L|,|L_0|)$.* □

EXERCISES

1. Let X be a complex; let A be a subcomplex of X; let B be a subcomplex of A. Prove the existence and naturality of the following sequence; it is called the **exact homology sequence of the triple (X,A,B)**:
$$\cdots \to H_i(A,B) \to H_i(X,B) \to H_i(X,A) \to H_{i-1}(A,B) \to \cdots.$$
2. Prove the following:
Lemma (The serpent lemma). Given a homomorphism of short exact sequences of abelian groups
$$\begin{array}{ccccccccc} 0 & \longrightarrow & A & \longrightarrow & B & \longrightarrow & C & \longrightarrow & 0 \\ & & \downarrow{\scriptstyle\alpha} & & \downarrow{\scriptstyle\beta} & & \downarrow{\scriptstyle\gamma} & & \\ 0 & \longrightarrow & D & \longrightarrow & E & \longrightarrow & F & \longrightarrow & 0, \end{array}$$
there is an exact sequence
$$0 \to \ker\alpha \to \ker\beta \to \ker\gamma \to \operatorname{cok}\alpha \to \operatorname{cok}\beta \to \operatorname{cok}\gamma \to 0.$$
3. (a) Prove the Five-lemma.
(b) Suppose one is given a commutative diagram of abelian groups, as in the Five-lemma. Consider the following eight hypotheses:
$$f_i \text{ is a monomorphism, for } i = 1, 2, 4, 5,$$
$$f_i \text{ is an epimorphism, for } i = 1, 2, 4, 5.$$
Which of these hypotheses will suffice to prove that f_3 is a monomorphism? Which will suffice to prove that f_3 is an epimorphism?
4. Show by example that one can have $H_i(K_0) \simeq H_i(L_0)$ and $H_i(K) \simeq H_i(L)$ for all i, without having $H_i(K,K_0) \simeq H_i(L,L_0)$.
5. Let $w * K$ be a cone over K. Show that
$$H_i(w * K, K) \simeq \tilde{H}_{i-1}(K).$$
6. Let K_0 be a subcomplex of K.
(a) If there is a retraction $r : |K| \to |K_0|$, show that
$$H_p(K) \simeq H_p(K,K_0) \oplus H_p(K_0).$$
(b) If the identity map $i : |K| \to |K|$ is homotopic to a map carrying $|K|$ into $|K_0|$, show that
$$H_p(K_0) \simeq H_p(K) \oplus H_{p+1}(K,K_0).$$
(c) If the inclusion $j : |K_0| \to |K|$ is homotopic to a constant, show that
$$H_p(K,K_0) \simeq \tilde{H}_p(K) \oplus \tilde{H}_{p-1}(K_0).$$
[*Hint:* Show that the map j extends to a map $f : |w * K_0| \to |K|$. (See Corollary 20.6.) Then apply Exercise 5.]
*(d) Show by example that the conclusion of (c) does not hold if one assumes only that j_* is the zero homomorphism in reduced homology.

7. Given a complex K and a short exact sequence of abelian groups

$$0 \to G \to G' \to G'' \to 0,$$

one has a short exact sequence of chain complexes

$$0 \to C_p(K;G) \to C_p(K;G') \to C_p(K;G'') \to 0,$$

and hence a long exact sequence in homology. The zig-zag homomorphism in this sequence is commonly denoted

$$\beta_* : H_p(K;G'') \to H_{p-1}(K;G)$$

and called the **Bockstein homomorphism** associated with the given coefficient sequence.

(a) Compute β_* for the coefficient sequences

$$0 \to \mathbf{Z} \to \mathbf{Z} \to \mathbf{Z}/2 \to 0,$$
$$0 \to \mathbf{Z}/2 \to \mathbf{Z}/4 \to \mathbf{Z}/2 \to 0,$$

when $|K|$ equals P^2.

(b) Repeat (a) when $|K|$ is the Klein bottle.

§25. MAYER-VIETORIS SEQUENCES

The homology exact sequence is one useful device for computing homology groups. Another is the Mayer-Vietoris sequence, which we now construct. It is another consequence of the zig-zag lemma.

Theorem 25.1. *Let K be a complex; let K_0 and K_1 be subcomplexes such that $K = K_0 \cup K_1$. Let $A = K_0 \cap K_1$. Then there is an exact sequence*

$$\cdots \to H_p(A) \to H_p(K_0) \oplus H_p(K_1) \to H_p(K) \to H_{p-1}(A) \to \cdots,$$

called the **Mayer-Vietoris sequence** of (K_0,K_1). *There is a similar exact sequence in reduced homology if A is nonempty.*

Proof. The proof consists of constructing a short exact sequence of chain complexes

$$0 \to \mathcal{C}(A) \xrightarrow{\phi} \mathcal{C}(K_0) \oplus \mathcal{C}(K_1) \xrightarrow{\psi} \mathcal{C}(K) \to 0,$$

and applying the zig-zag lemma.

First we need to define the chain complex in the middle. Its chain group in dimension p is

$$C_p(K_0) \oplus C_p(K_1),$$

and its boundary operator ∂' is defined by

$$\partial'(d,e) = (\partial_0 d, \partial_1 e),$$

where ∂_0 and ∂_1 are the boundary operators in $\mathcal{C}(K_0)$ and $\mathcal{C}(K_1)$, respectively.

Second, we need to define the chain maps ϕ and ψ. We do this as follows. Consider the inclusion mappings in the following commutative diagram:

$$\begin{array}{ccccc} & & K_0 & & \\ & {\scriptstyle i}\nearrow & & \searrow{\scriptstyle k} & \\ A & & \xrightarrow{\quad m \quad} & & K. \\ & {\scriptstyle j}\searrow & & \nearrow{\scriptstyle l} & \\ & & K_1 & & \end{array}$$

Define homomorphisms ϕ and ψ by the equations

$$\phi(c) = (i_\#(c), -j_\#(c)),$$
$$\psi(d,e) = k_\#(d) + l_\#(e).$$

It is immediate that ϕ and ψ are chain maps.

Let us check exactness: First, note that ϕ is injective, since $i_\#$ is just inclusion of chains. Second, we check that ψ is surjective. Given $d \in C_p(K)$, write d as a sum of oriented simplices, and let d_0 consist of those terms of d carried by K_0. Then $d - d_0$ is carried by K_1; and $\psi(d_0, d - d_0) = d$.

To check exactness at the middle term, note first that $\psi\phi(c) = m_\#(c) - m_\#(c) = 0$. Conversely, if $\psi(d,e) = 0$, then $d = -e$, when considered as chains of K. Since d is carried by K_0 and e is carried by K_1, they must both be carried by $K_0 \cap K_1 = A$. Then $(d,e) = (d,-d) = \phi(d)$, as desired.

The homology of the middle chain complex in dimension p equals

$$\frac{\ker \partial'}{\operatorname{im} \partial'} = \frac{\ker \partial_0 \oplus \ker \partial_1}{\operatorname{im} \partial_0 \oplus \operatorname{im} \partial_1}$$
$$\simeq H_p(K_0) \oplus H_p(K_1).$$

The Mayer-Vietoris sequence now follows from the zig-zag lemma.

To obtain the Mayer-Vietoris sequence in reduced homology, we replace the chain complexes considered earlier by the corresponding augmented chain complexes. Let ϵ_A, ϵ_0, ϵ_1, and ϵ denote the augmentation maps for $\mathcal{C}(A)$, $\mathcal{C}(K_0)$, $\mathcal{C}(K_1)$, and $\mathcal{C}(K)$, respectively. Consider the diagram

$$\begin{array}{ccccccccc} 0 & \longrightarrow & C_0(A) & \longrightarrow & C_0(K_0) \oplus C_0(K_1) & \longrightarrow & C_0(K) & \longrightarrow & 0 \\ & & \downarrow{\scriptstyle \epsilon_A} & & \downarrow{\scriptstyle \epsilon_0 \oplus \epsilon_1} & & \downarrow{\scriptstyle \epsilon} & & \\ 0 & \longrightarrow & \mathbf{Z} & \xrightarrow[\phi]{} & \mathbf{Z} \oplus \mathbf{Z} & \xrightarrow[\psi]{} & \mathbf{Z} & \longrightarrow & 0. \end{array}$$

Commutativity and exactness hold at the bottom level if we define $\phi(n) = (n,-n)$ and $\psi(m,n) = m + n$. Each map ϵ_A, $\epsilon_0 \oplus \epsilon_1$, and ϵ is surjective (since A is nonempty). Thus the homology of these respective chain complexes vanishes in dimension -1, and in dimension 0 equals the respective groups $\tilde{H}_0(A)$, $\tilde{H}_0(K_0) \oplus \tilde{H}_0(K_1)$ and $\tilde{H}_0(K)$. We now apply the zig-zag lemma. □

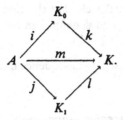

Lemma 25.2. *Let $h:(K,K_0,K_1) \to (L,L_0,L_1)$ be a simplicial map, where $K = K_0 \cup K_1$ and $L = L_0 \cup L_1$. Then h induces a homomorphism of Mayer-Vietoris sequences.*

Proof. One checks immediately that the chain maps $h_\#$ induced by h commute with the chain maps ϕ and ψ defined in the preceding proof. Naturality then follows from Theorem 24.2. □

The following is an immediate consequence of this lemma and the results of Chapter 2.

Theorem 25.3. *The preceding lemma holds for an arbitrary continuous map $h:(|K|,|K_0|,|K_1|) \to (|L|,|L_0|,|L_1|)$.* □

To illustrate how the Mayer-Vietoris sequence is used in practice, we shall compute the homology of a suspension of a complex. We recall the definition from the exercises of §8:

Definition. Let K be a complex; let $w_0 * K$ and $w_1 * K$ be two cones on K whose polytopes intersect in $|K|$ alone. Then

$$S(K) = (w_0 * K) \cup (w_1 * K)$$

is a complex; it is called a **suspension** of K. Given K, the complex $S(K)$ is uniquely defined up to a simplicial isomorphism.

Theorem 25.4. *If K is a complex, then for all p, there is an isomorphism*

$$\tilde{H}_p(S(K)) \to \tilde{H}_{p-1}(K).$$

Proof. Let $K_0 = w_0 * K$ and $K_1 = w_1 * K$. Then $K_0 \cup K_1 = S(K)$ and $K_0 \cap K_1 = K$. In the reduced Mayer-Vietoris sequence

$$\tilde{H}_p(K_0) \oplus \tilde{H}_p(K_1) \to \tilde{H}_p(S(K)) \to \tilde{H}_{p-1}(K) \to \tilde{H}_{p-1}(K_0) \oplus \tilde{H}_{p-1}(K_1)$$

both end terms vanish, because K_0 and K_1 are cones. Therefore the middle map is an isomorphism. □

EXERCISES

1. Let K be the union of the subcomplexes K_0 and K_1, where $|K_0|$ and $|K_1|$ are connected. What can you say about the homology of K in each of the following cases?
   (a) $K_0 \cap K_1$ is nonempty and acyclic.
   (b) $|K_0| \cap |K_1|$ consists of two points.
   (c) $K_0 \cap K_1$ has the homology of S^n, where $n > 0$.
   (d) K_0 and K_1 are acyclic.

2. Let K_0 and K_1 be subcomplexes of K; let L_0 and L_1 be subcomplexes of K_0 and K_1, respectively. Construct an exact sequence

$$\cdots \to H_i(K_0 \cap K_1, L_0 \cap L_1) \to H_i(K_0, L_0) \oplus H_i(K_1, L_1) \to H_i(K_0 \cup K_1, L_0 \cup L_1) \to \cdots.$$

This sequence is called the **relative Mayer-Vietoris sequence.**

3. Show that if d is an integer, and if $n \geq 1$, there is a map $h : S^n \to S^n$ of degree d. [*Hint:* Proceed by induction, using naturality of the Mayer-Vietoris sequence. The case $n = 1$ was considered in Exercise 1 of §21.]

4. Let $\phi : C_p(K) \to C_{p+1}(S(K))$ be the homomorphism defined in Exercise 1 of §8. Show that the isomorphism of Theorem 25.4 is inverse to ϕ_*.

5. Given a sequence $G_0, \ldots, G_n$ of finitely generated abelian groups, with G_0 and G_n free and G_0 non-trivial, show there is a finite complex K of dimension n such that $H_i(K) \simeq G_i$ for $i = 0, \ldots, n$. [*Hint:* See Exercise 8 of §6.]

6. We shall study the homology of $X \times Y$ in Chapter 7. For the present, prove the following, assuming all the spaces involved are polyhedra:
   (a) Show that if $p \in S^n$,

   $$H_q(X \times S^n, X \times p) \simeq H_{q-n}(X).$$

   [*Hint:* Write S^n as the union of its upper and lower hemispheres, and proceed by induction on n.]
   (b) Show that if $p \in Y$, the homology exact sequence of $(X \times Y, X \times p)$ breaks up into short exact sequences that split.
   (c) Prove that

   $$H_q(X \times S^n) \simeq H_{q-n}(X) \oplus H_q(X).$$

   (d) Compute the homology of $S^n \times S^m$.

§26. THE EILENBERG-STEENROD AXIOMS

We have defined homology groups for a particular class of spaces—namely, the polyhedra. Historically, these were the first homology groups to be defined. Later, various generalized definitions of homology were formulated that applied to more general spaces. These various homology theories had many features in common, and they all gave the same results as simplicial homology theory on the class of polyhedra.

This plethora of homology theories led Eilenberg and Steenrod to *axiomatize* the notion of a homology theory. They formulated certain crucial properties these theories have in common, and showed that these properties characterize the homology groups completely on the class of polyhedra.

We shall not try to reproduce the axiomatic approach at this point, nor shall we prove that the axioms characterize homology for polyhedra. For this, the reader is referred to Eilenberg and Steenrod's book [E-S]. We shall simply list the seven axioms here, together with an additional axiom that is needed when one deals with non-compact spaces. In the next section, we verify that simplicial homology theory satisfies the axioms.

Definition. Let $\mathcal{A}$ be a class of pairs (X,A) of topological spaces such that:

(1) If (X,A) belongs to $\mathcal{A}$, so do (X,X), $(X,\varnothing)$, (A,A), and $(A,\varnothing)$.

(2) If (X,A) belongs to $\mathcal{A}$, so does $(X \times I, A \times I)$.

(3) There is a one-point space P such that $(P,\varnothing)$ is in $\mathcal{A}$.

Then we shall call $\mathcal{A}$ an **admissible class of spaces** for a homology theory.

Definition. If $\mathcal{A}$ is admissible, a **homology theory** on $\mathcal{A}$ consists of three functions:

(1) A function H_p defined for each integer p and each pair (X,A) in $\mathcal{A}$, whose value is an abelian group.

(2) A function that, for each integer p, assigns to each continuous map $h:(X,A) \to (Y,B)$ a homomorphism

$$(h_*)_p : H_p(X,A) \to H_p(Y,B).$$

(3) A function that, for each integer p, assigns to each pair (X,A) in $\mathcal{A}$, a homomorphism

$$(\partial_*)_p : H_p(X,A) \to H_{p-1}(A),$$

where A denotes the pair $(A,\varnothing)$.

These functions are to satisfy the following axioms, where all pairs of spaces are in $\mathcal{A}$. As usual, we shall simplify notation and delete the dimensional subscripts on h_* and ∂_*.

Axiom 1. If i is the identity, then i_* is the identity.

Axiom 2. $(k \circ h)_* = k_* \circ h_*$.

Axiom 3. If $f:(X,A) \to (Y,B)$, then the following diagram commutes:

$$\begin{array}{ccc} H_p(X,A) & \xrightarrow{f_*} & H_p(Y,B) \\ \downarrow \partial_* & & \downarrow \partial_* \\ H_{p-1}(A) & \xrightarrow{(f|A)_*} & H_{p-1}(B) \end{array}$$

Axiom 4 (Exactness axiom). The sequence

$$\cdots \rightarrow H_p(A) \xrightarrow{i_*} H_p(X) \xrightarrow{\pi_*} H_p(X,A) \xrightarrow{\partial_*} H_{p-1}(A) \rightarrow \cdots$$

is exact, where $i : A \rightarrow X$ and $\pi : X \rightarrow (X,A)$ are inclusion maps.

Recall that two maps $h,k : (X,A) \rightarrow (Y,B)$ are said to be *homotopic* (written $h \simeq k$) if there is a map

$$F : (X \times I, A \times I) \rightarrow (Y,B)$$

such that $F(x,0) = h(x)$ and $F(x,1) = k(x)$ for all $x \in X$.

Axiom 5 (Homotopy axiom). If h and k are homotopic, then $h_* = k_*$.

Axiom 6 (Excision axiom). Given (X,A), let U be an open subset of X such that $\bar{U} \subset \text{Int}\, A$. If $(X - U, A - U)$ is admissible, then inclusion induces an isomorphism

$$H_p(X - U, A - U) \simeq H_p(X,A).$$

Axiom 7 (Dimension axiom). If P is a one-point space, then $H_p(P) = 0$ for $p \neq 0$ and $H_0(P) \simeq \mathbf{Z}$.

Axiom 8 (Axiom of compact support). If $\alpha \in H_p(X,A)$, there is an admissible pair (X_0,A_0) with X_0 and A_0 compact, such that α is in the image of the homomorphism $H_p(X_0,A_0) \rightarrow H_p(X,A)$ induced by inclusion.

A pair (X_0,A_0) with both X_0 and A_0 compact is called a **compact pair.**

Note that one can modify Axiom 7 by writing $H_0(P) \simeq G$, where G is a fixed abelian group. What one has then is called "homology with coefficients in G." We shall stick to integer coefficients for the present.

Of all the Eilenberg-Steenrod axioms, the dimension axiom seems most innocuous and least interesting. But in some sense, just the opposite is true. Since Eilenberg and Steenrod's axioms were published, several new mathematical theories have been discovered (invented?) that resemble homology theory. Cobordism theories in differential topology and K-theory in vector bundle theory are examples. Although these theories were invented for purposes quite different from those for which the homology groups were invented, they share many formal properties with homology theory. In particular, *they satisfy all the Eilenberg-Steenrod axioms except for the dimension axiom.*

Such a theory is nowadays called an *extraordinary homology theory,* or a *generalized homology theory.* It differs from ordinary homology in that a one-point space may have non-zero homology in many dimensions.

This situation illustrates a phenomenon that occurs over and over again in mathematics; a theory formulated for one purpose turns out to have consequences far removed from what its originators envisaged.

EXERCISES

1. Consider the following commutative diagram of abelian groups and homomorphisms, which is called a **braid:**

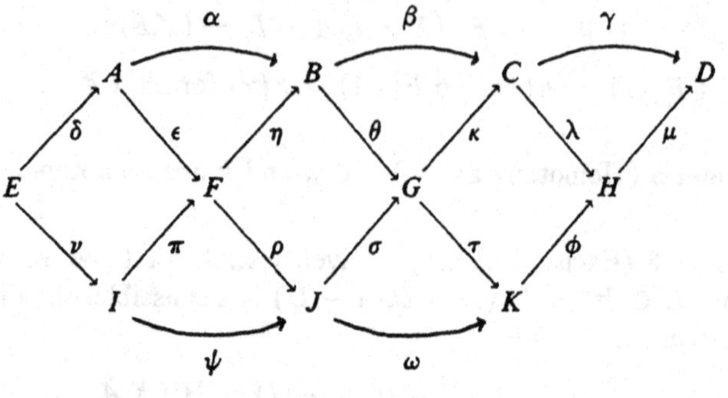

This diagram contains the following four sequences, arranged in the form of overlapping sine curves:

$$E \to A \to B \to G \to K,$$

$$E \to I \to J \to G \to C \to D,$$

$$A \to F \to J \to K \to H \to D,$$

$$I \to F \to B \to C \to H.$$

If all four sequences are exact, this diagram is called an **exact braid.** Prove the following two facts about braids:

(a) If this braid is exact, there is an isomorphism

$$\Delta : \frac{\ker \omega}{\operatorname{im} \psi} \longrightarrow \frac{\ker \beta}{\operatorname{im} \alpha}$$

defined as follows: If $\omega(j) = 0$, choose f so $\rho(f) = j$; then define $\Delta(j) = \{\eta(f)\}$.

(b) *Lemma (The braid lemma). In order that a braid be exact, it suffices if the first three of the preceding sequences are exact, and if the composite $I \to F \to B$ is zero.*

2. Using only the axioms for a homology theory, prove exactness of the homology sequence of a triple:

$$\cdots \to H_p(A,B) \xrightarrow{\pi} H_p(X,B) \xrightarrow{\eta} H_p(X,A) \xrightarrow{\beta} H_{p-1}(A,B) \to \cdots$$

where π and η are induced by inclusion. The map β is the composite

$$H_p(X,A) \xrightarrow{\partial_*} H_{p-1}(A) \xrightarrow{i_*} H_{p-1}(A,B),$$

where ∂_* is given by the axioms and i is inclusion. Assume the pairs involved are admissible. [*Hint:* Prove that $H_p(A,A) = 0$. Show that $\eta \circ \pi = 0$. Then apply Exercise 1.]

3. Using only the axioms for a homology theory, derive the Mayer-Vietoris sequence, as follows:

   Let $X = X_1 \cup X_2$; let $A = X_1 \cap X_2$. We say that (X_1,X_2) is an **excisive couple** for the given homology theory if (X_1,A) and (X,X_2) are admissible and if inclusion $(X_1,A) \to (X,X_2)$ induces a homology isomorphism.

   (a) Consider the following diagram of inclusions, and the corresponding homomorphism of long exact homology sequences:

$$\begin{array}{ccccc} A & \xrightarrow{i} & X_1 & \xrightarrow{\alpha} & (X_1,A) \\ {\scriptstyle j}\downarrow & & {\scriptstyle k}\downarrow & & \downarrow{\scriptstyle \gamma} \\ X_2 & \xrightarrow[l]{} & X & \xrightarrow[\beta]{} & (X,X_2) \end{array}$$

   Given that (X_1,X_2) is excisive, define a sequence

$$\cdots \to H_p(A) \xrightarrow{\phi} H_p(X_1) \oplus H_p(X_2) \xrightarrow{\psi} H_p(X) \xrightarrow{\theta} H_{p-1}(A) \to \cdots$$

   by letting

$$\begin{aligned} \phi(a) &= (i_*(a), -j_*(a)), \\ \psi(x_1,x_2) &= k_*(x_1) + l_*(x_2), \\ \theta(x) &= \partial_*(\gamma_*)^{-1}\beta_*(x). \end{aligned}$$

   Here ∂_* is the boundary homomorphism in the homology exact sequence of (X_1,A). Show that this sequence is exact and is natural with respect to homomorphisms induced by continuous maps. [*Hint:* The proof is a diagram-chase. One begins with a homomorphism of one long exact sequence to another, where every third homomorphism is an isomorphism.]

   (b) Show that if X is the polytope of a complex K, and if X_1 and X_2 are polytopes of subcomplexes of K, then (X_1,X_2) is excisive for simplicial theory, and this sequence is the same as the Mayer-Vietoris sequence of Theorem 25.1.

   (c) Suppose (X_1,A) and (X,X_2) are admissible. Show that if Int X_1 and Int X_2 cover X and if X_1 is closed in X, then (X_1,X_2) is excisive for any homology theory satisfying the axioms.

§27. THE AXIOMS FOR SIMPLICIAL THEORY

Before showing that simplicial homology theory satisfies the Eilenberg-Steenrod axioms, we must treat several points of theory with more care than we have done up to now. The axioms involve homology groups that are defined on an admissible class of spaces. Strictly speaking, we have not defined homology groups for *spaces,* but only for *simplicial complexes.* Given a polyhedron X, there are many different simplicial complexes whose polytopes equal X. Their homology groups are isomorphic to one another in a natural way, but they are nevertheless distinct groups. Similarly, if $h : X \to Y$ is a continuous map, where $X = |K|$ and $Y = |L|$, we have defined an induced homomorphism $h_* : H_p(K) \to H_p(L)$. Of course, if $X = |M|$ and $Y = |N|$, we also have an induced homomorphism $H_p(M) \to H_p(N)$, which we also denote by h_*. We noted this notational ambiguity earlier.

The way out of this difficulty is the following: Given a polyhedron X, we can consider the class of all simplicial complexes whose polytopes equal X, and we can *identify* their homology groups in a natural way. The resulting groups can be called the homology groups of the polyhedron X.

More generally, we can perform this same construction for any space that is homeomorphic to a polyhedron. We give the details now.

Definition. Let A be a subspace of the space X. A **triangulation** of the pair (X,A) is a complex K, a subcomplex K_0 of K, and a homeomorphism

$$h : (|K|,|K_0|) \to (X,A).$$

If such a triangulation exists, we say (X,A) is a **triangulable pair.** If A is empty, we say simply that X is a **triangulable space.**

Now let (X,A) be a triangulable pair. We define the simplicial homology $H_p(X,A)$ of this pair as follows: Consider the collection of all triangulations of (X,A). They are of the form

$$h_\alpha : (|K_\alpha|,|C_\alpha|) \to (X,A),$$

where C_α is a subcomplex of K_α.

Now there is some set-theoretic difficulty with the concept of the "set of all triangulations of a pair," just as there is with the "set of all sets." We avoid such problems by assuming that each K_α lies in some fixed space $\mathbf{E}^J$. This is justified by noting that if J is large enough, each K_α has an isomorphic copy lying in $\mathbf{E}^J$. For instance, we can let J have the cardinality of X itself!

For fixed p, consider the groups $H_p(K_\alpha,C_\alpha)$. We make sure they are disjoint as sets by forming $H_p(K_\alpha,C_\alpha) \times \{\alpha\}$. Then in the disjoint union

$$\cup_\alpha H_p(K_\alpha,C_\alpha) \times \{\alpha\},$$

we introduce an equivalence relation. We define

$$(x, \alpha) \in H_p(K_\alpha,C_\alpha) \times \{\alpha\}$$
$$(y, \beta) \in H_p(K_\beta,C_\beta) \times \{\beta\}$$

to be equivalent if $(h_\beta^{-1} h_\alpha)_*(x) = y$. And we let $H_p(X,A)$ denote the set of equivalence classes.

Now each equivalence class contains exactly one element from each group $H_p(K_\alpha,C_\alpha) \times \{\alpha\}$. That is, the map

$$H_p(K_\alpha,C_\alpha) \times \{\alpha\} \to H_p(X,A)$$

that carries each element to its equivalence class is bijective. We make $H_p(X,A)$ a group by requiring this map to be an isomorphism. This group structure is unambiguous because $(h_\beta^{-1}h_\alpha)_*$ is an isomorphism for each pair of indices α, β.

A continuous map $h:(X,A) \to (Y,B)$ induces a homomorphism in homology as follows: Take any pair of triangulations

$$h_\alpha : (|K_\alpha|,|C_\alpha|) \to (X,A),$$
$$k_\beta : (|L_\beta|,|D_\beta|) \to (Y,B).$$

The map h induces a map $h' : (|K_\alpha|,|C_\alpha|) \to (|L_\beta|,|D_\beta|)$, which in turn gives rise to a homomorphism

$$h'_* : H_p(K_\alpha,C_\alpha) \to H_p(L_\beta,D_\beta)$$

of simplicial homology groups. By passing to equivalence classes, this homomorphism induces a well-defined homomorphism

$$h_* : H_p(X,A) \to H_p(Y,B).$$

In a similar manner, the boundary homomorphism $\partial_* : H_p(K_\alpha,C_\alpha) \to H_{p-1}(C_\alpha)$ induces a boundary homomorphism

$$\partial_* : H_p(X,A) \to H_{p-1}(A).$$

We now have all the constituents for a homology theory.

First, we note that the class of triangulable pairs forms an admissible class of spaces for a homology theory. If (X,A) is triangulable, so are (X,X), $(X,\varnothing)$, (A,A), and $(A,\varnothing)$. Any one-point space is triangulable. Finally, if (X,A) is triangulable, then so is $(X \times I, A \times I)$, by Lemma 19.1.

Theorem 27.1. *Simplicial homology theory on the class of triangulable pairs satisfies the Eilenberg-Steenrod axioms.*

Proof. Axioms 1–5 and 7 express familiar properties of the homology of simplicial complexes that carry over at once to the homology of triangulable pairs. Only Axioms 6 and 8, the excision axiom and the axiom of compact support, require comment.

To check the axiom of compact support, it suffices to show that it holds for simplicial complexes. Let α be an element of $H_p(K,K_0)$. Let c be a chain of K representing α; its boundary is carried by K_0. Since c is carried by a finite subcomplex L of K, it can be considered as a cycle of (L,L_0), where $L_0 = K_0 \cap L$. If β is the homology class of c in $H_p(L,L_0)$, then $j_*(\beta) = \alpha$, where j is inclusion of (L,L_0) in (K,K_0). Thus the axiom is verified.

Checking the excision axiom involves a subtlety that may not be apparent at first sight. The problem is that even though both (X,A) and $(X - U, A - U)$ may be triangulable, the two triangulations may be entirely unrelated to one another! If this is *not* the case, then the excision axiom follows readily from Theorem 9.1, as we now show.

Let $U \subset A \subset X$. Suppose there is a triangulation

$$h : (|K|,|K_0|) \to (X,A)$$

of the pair (X,A) that induces a triangulation of the subspace $X - U$. This means that $X - U = h(|L|)$ for some subcomplex L of K. Let $L_0 = L \cap K_0$; then $A - U = h(|L_0|)$. (It follows that U is open and A is closed.) Theorem 9.1 states that inclusion induces an isomorphism

$$H_p(L,L_0) \simeq H_p(K,K_0).$$

Thus our result is proved.

Now we prove excision in the general case. Let $U \subset \text{Int}\,A$; let

$$h : (K,K_0) \to (X,A)$$
$$k : (M,M_0) \to (X - U, A - U)$$

be triangulations of these respective pairs of spaces. Let X_1 denote the closure of $X - A$, and let $A_1 = X_1 \cap A$. We assert that the pair (X_1,A_1) is triangulated by *both* h and k. (See Figure 27.1, where the maps j_0 and j_1 denote inclusions.)

To verify this assertion, note that the space $|K| - |K_0|$ is the union of all open simplices Int σ such that $\sigma \in K$ and $\sigma \notin K_0$. Then its closure C is the polytope of the subcomplex of K consisting of all simplices σ of K that are not in K_0, and their faces. The image of the set C under the homeomorphism h equals the closure of $X - A$, which is X_1. Therefore, X_1 is triangulated by h. Since both X_1 and A are triangulated by h, so is $A_1 = X_1 \cap A$. Similarly, the closure of $|M| - |M_0|$ is the polytope of a subcomplex of M, and its image under k equals the closure of $(X - U) - (A - U) = X - A$. Thus X_1 is triangulated by k, and so is $A_1 = X_1 \cap A$. (Here is where we need the fact that $U \subset \text{Int}\,A$.)

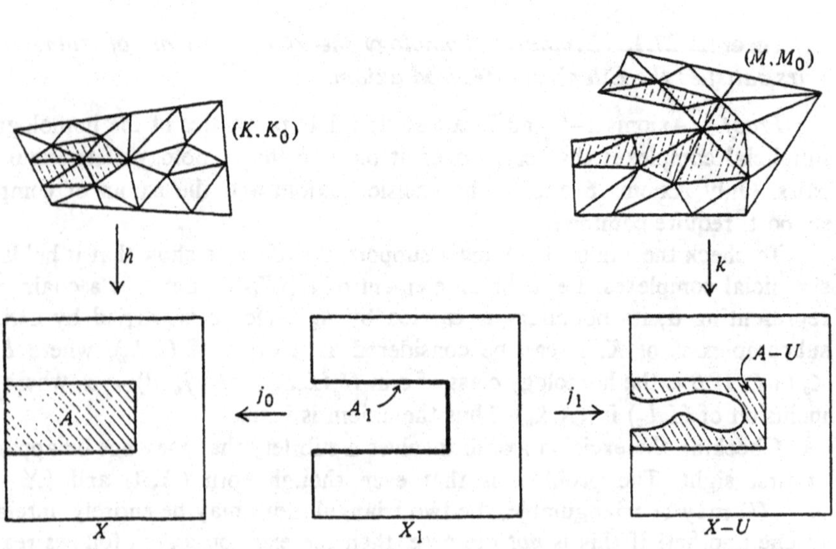

Figure 27.1

It follows from the special case already proved that in the diagram of inclusions

$$\begin{array}{ccc} (X_1,A_1) & \xrightarrow{j_0} & (X,A) \\ & {\scriptstyle j_1}\searrow \quad \nearrow{\scriptstyle j} & \\ & (X-U,A-U) & \end{array}$$

both j_0 and j_1 induce isomorphisms. Because the diagram commutes, j_* is also an isomorphism. $\square$

We remark that in the preceding proof we did not need the full strength of the hypothesis of the excision axiom. We did not use the fact that $\bar{U} \subset \operatorname{Int} A$, nor that U is open. Thus for simplicial theory on the class of triangulable pairs, the following stronger version of the excision axiom holds:

Theorem. 27.2 (Excision in simplicial theory). *Let A be a subspace of X. Let U be a subset of X such that $U \subset \operatorname{Int} A$. If both pairs (X,A) and $(X-U, A-U)$ are triangulable, then inclusion induces an isomorphism*

$$H_p(X-U,A-U) \simeq H_p(X,A). \quad \square$$

The axiom of compact support states, roughly speaking, that every homology class is compactly supported. It is also true that every homology relation between such classes is compactly supported. More precisely, one has the following useful result, which we shall verify directly for simplicial theory. It may also be derived from the axioms; see the exercises.

Theorem 27.3. *Let $i:(X_0,A_0) \to (X,A)$ be an inclusion of triangulable pairs, where (X_0,A_0) is a compact pair. If $\alpha \in H_p(X_0,A_0)$ and $i_*(\alpha) = 0$, then there are a compact triangulable pair (X_1,A_1) and inclusion maps*

$$(X_0,A_0) \xrightarrow{j} (X_1,A_1) \xrightarrow{k} (X,A)$$

such that $j_*(\alpha) = 0$.

Proof. We may assume that (X,A) is the polytope of a simplicial pair (K,C). Because X_0 is compact, it is contained in the polytope of a finite subcomplex K_0 of K. Then A_0 is contained in the polytope of $C \cap K_0 = C_0$. The theorem thus reduces to the case where

$$i:(K_0,C_0) \to (K,C)$$

is an inclusion of subcomplexes and K_0 is finite.

Let $\alpha \in H_p(K_0,C_0)$ and suppose that $i_*(\alpha) = 0$. Let c_p be a chain of K_0 representing α. Since $i_*(\alpha) = \{i_\#(c_p)\} = 0$, there is a chain d_{p+1} of K such that $c_p - \partial d_{p+1}$ is carried by C. Choose a finite subcomplex of C carrying

$c_p - \partial d_{p+1}$; let C_1 be the union of this subcomplex and C_0. Then choose a finite subcomplex of K carrying d_{p+1}; let K_1 be the union of this subcomplex and K_0 and C_1. Then the homomorphism induced by the inclusion

$$(K_0,C_0) \to (K_1,C_1)$$

carries α to zero. □

EXERCISES

1. *Theorem. Given a sequence $G_0, G_1, \ldots$ of finitely generated abelian groups with G_0 free and non-trivial, there is a complex K such that $H_i(K) \simeq G_i$ for each i.*

   [*Hint:* See Exercise 5 of §25.]

2. Let $\mathcal{A}$ be either the class of triangulable pairs, or the class of all topological pairs. Prove Theorem 27.3, with the word "triangulable" replaced throughout by the word "admissible," directly from the axioms.

   [*Hint:* Show that in the triangulable case, one can assume that the triangulation of (X,A) triangulates X_0 as well. Examine the diagram

$$\begin{array}{ccccccc} & & H_q(X_0,A_0) & & & & \\ & & \downarrow l_* & \searrow & & & \\ \to H_q(A,A_0) & \to & H_q(X,A_0) & \to & H_q(X,A) & \to & \end{array}$$

   to find a compact A_1 such that $l_*(\alpha)$ is in the image of $H_q(A_1,A_0) \to H_q(X,A_0)$. Then inclusion $(X_0,A_0) \to (X,A_1)$ induces a homomorphism carrying α to zero. Examine the diagram

$$\begin{array}{ccccc} & & H_q(X_0,A_0) & & \\ & & \downarrow m_* & \searrow & \\ H_{q+1}(X,X_0 \cup A_1) & \to & H_q(X_0 \cup A_1, A_1) & \to & H_q(X,A_1) \end{array}$$

   to find a compact X_1 such that $m_*(\alpha)$ is in the image of

$$\partial_* : H_{q+1}(X_1,X_0 \cup A_1) \to H_q(X_0 \cup A_1,A_1).$$

   Complete the proof.]

*§28. CATEGORIES AND FUNCTORS[†]

By now you have seen enough references to "induced homomorphisms" and their "functorial properties," and to the "naturality" of the way one assigns one mathematical object to another, that you may suspect there is some common idea underlying all this language. There is; we study it in this section. It consists

[†]This section will be needed when we study cohomology, in Chapter 5.

mostly of new terminology; what proofs there are, are elementary and are left to the reader.

Definition. A **category** $\mathbf{C}$ consists of three things:

(1) A class of **objects** X.

(2) For every ordered pair (X,Y) of objects, a set $\hom(X,Y)$ of **morphisms** f.

(3) A function, called **composition of morphisms,**

$$\hom(X,Y) \times \hom(Y,Z) \rightarrow \hom(X,Z),$$

which is defined for every triple (X,Y,Z) of objects.

The image of the pair (f,g) under the composition operation is denoted by $g \circ f$. The following two properties must be satisfied:

Axiom 1 (Associativity). If $f \in \hom(W,X)$ and $g \in \hom(X,Y)$ and $h \in \hom(Y,Z)$, then $h \circ (g \circ f) = (h \circ g) \circ f$.

Axiom 2 (Existence of identities). If X is an object, there is an element $1_X \in \hom(X,X)$ such that

$$1_X \circ f = f \quad \text{and} \quad g \circ 1_X = g$$

for every $f \in \hom(W,X)$ and every $g \in \hom(X,Y)$, where W and Y are arbitrary.

One standard example of a category consists of topological spaces and continuous maps, with the usual composition of functions. This example illustrates why we speak of the objects of a category as forming a *class* rather than a *set,* for one cannot speak of the "set of all topological spaces" or the "set of all sets" without becoming involved in logical paradoxes. (Is the set of all sets a member of itself?) A *class* is something larger than a set, to which we do not apply the usual set-theoretic operations (such as taking the set of all subsets).

Let us note the following fact: *The identity morphism 1_X is unique.* For suppose

$$1_X \circ f = f \quad \text{and} \quad g = g \circ 1'_X$$

for every $f \in \hom(W,X)$ and $g \in \hom(X,Y)$. Then setting $f = 1'_X$ and $g = 1_X$, we have

$$1_X \circ 1'_X = 1'_X \quad \text{and} \quad 1_X = 1_X \circ 1'_X,$$

whence $1'_X = 1_X$.

Definition. Let $f \in \hom(X,Y)$ and $g, g' \in \hom(Y,X)$. If $g \circ f = 1_X$, we call g a **left inverse** for f; if $f \circ g' = 1_Y$, we call g' a **right inverse** for f.

We note the following fact: *If f has a left inverse g and a right inverse g', they are equal.* For one computes as follows:

$$(g \circ f) \circ g' = 1_X \circ g' = g',$$
$$g \circ (f \circ g') = g \circ 1_Y = g,$$

whence $g = g'$. The map $g = g'$ is called an **inverse** to f; it is unique.

If f has an inverse, then f is called an **equivalence** in the category in question.

In general, we write $f : X \to Y$ to mean $f \in \hom(X,Y)$; and we call X the **domain object** of f, and Y, the **range object** of f.

Definition. A (covariant) **functor** G from a category $\mathbf{C}$ to a category $\mathbf{D}$ is a function assigning to each object X of $\mathbf{C}$, an object $G(X)$ of $\mathbf{D}$, and to each morphism $f : X \to Y$ of $\mathbf{C}$, a morphism $G(f) : G(X) \to G(Y)$ of $\mathbf{D}$. The following two conditions must be satisfied:

$$G(1_X) = 1_{G(X)} \quad \text{for all} \quad X,$$
$$G(g \circ f) = G(g) \circ G(f).$$

That is, a functor must preserve composition and identities. It is immediate that if f is an equivalence in $\mathbf{C}$, then $G(f)$ is an equivalence in $\mathbf{D}$.

Example 1. We list a number of categories. In all these examples, composition is either the usual composition of functions or is induced by it. Equivalences in some of these categories are given special names; in such cases, the name is listed in parentheses.

(a) Sets and maps (bijective correspondences).

(b) Topological spaces and continuous maps (homeomorphisms).

(c) Topological spaces and homotopy classes of maps (homotopy equivalences).

(d) Simplicial complexes and simplicial maps (simplicial homeomorphisms).

(e) Simplicial complexes and continuous maps of their polytopes.

(f) Simplicial complexes and homotopy classes of continuous maps.

(g) Groups and homomorphisms (isomorphisms).

(h) Chain complexes and chain maps.

(i) Chain complexes and chain-homotopy classes (chain equivalences).

(j) Short exact sequences of abelian groups and homomorphisms of such.

(k) Short exact sequences of chain complexes and homomorphisms of such.

(l) Long exact sequences of abelian groups and homomorphisms of such.

(m) Pairs (X,Y) of topological spaces and pairs (f,g) of continuous maps.

Example 2. Now we list several examples of functors.

(a) The correspondence assigning to a pair (X,Y) of spaces, the space $X \times Y$, and to a pair (f,g) of continuous maps, the map $f \times g$, is a functor from pairs of spaces to spaces.

(b) The correspondence assigning to a space X its underlying set, and to a continuous map, the underlying set map, is a functor from spaces to sets. It is called the **forgetful functor**; it "forgets" the topological structure involved.

(c) The correspondence $K \to \mathcal{C}(K)$ and $f \to f_\#$ is a functor from the category of simplicial complexes and simplicial maps to the category of chain complexes and chain maps.

(d) Given a homology theory, the correspondence $X \to H_p(X)$ and $[h] \to h_*$ is a functor from the category of admissible spaces and homotopy classes of maps to the category of abelian groups and homomorphisms. (Here $[h]$ denotes the homotopy class of h.) This is precisely the substance of the first two Eilenberg-Steenrod axioms and the homotopy axiom.

(e) The zig-zag lemma assigns to each short exact sequence of chain complexes, a long exact sequence of their homology groups. The "naturality" property expressed in Theorem 24.2 is just the statement that this assignment is a functor.

Definition. Let G and H be two functors from category $\mathbf{C}$ to category $\mathbf{D}$. A **natural transformation** T from G to H is a rule assigning to each object X of $\mathbf{C}$, a morphism

$$T_X : G(X) \to H(X)$$

of $\mathbf{D}$, such that the following diagram commutes, for all morphisms $f: X \to Y$ of the category $\mathbf{C}$:

$$\begin{array}{ccc} G(X) & \xrightarrow{T_X} & H(X) \\ \downarrow{\scriptstyle G(f)} & & \downarrow{\scriptstyle H(f)} \\ G(Y) & \xrightarrow{T_Y} & H(Y). \end{array}$$

If for each X, the morphism T_X is an equivalence in the category $\mathbf{D}$, then T is called a **natural equivalence** of functors.

Example 3. Given a homology theory, let p be fixed, and consider the following two functors, defined on admissible pairs:

$$G(X,A) = H_p(X,A); \qquad G(f) = f_*.$$
$$H(X,A) = H_{p-1}(A); \qquad H(f) = (f|A)_*.$$

The commutativity of the diagram

$$\begin{array}{ccc} H_p(X,A) & \xrightarrow{\partial_*} & H_{p-1}(A) \\ \downarrow{\scriptstyle f_*} & & \downarrow{\scriptstyle (f|A)_*} \\ H_p(Y,B) & \xrightarrow{\partial_*} & H_{p-1}(B) \end{array}$$

tells us that ∂_* is a natural transformation of the functor G to the functor H. This is precisely the third of the Eilenberg-Steenrod axioms.

Example 4. Consider the category of pairs of spaces and pairs of maps. Let G and H be the functors

$$G(X,Y) = X \times Y; \qquad G(f,g) = f \times g.$$
$$H(X,Y) = Y \times X; \qquad H(f,g) = g \times f.$$

Given (X,Y), let $T_{(X,Y)}$ be the homeomorphism of $X \times Y$ with $Y \times X$ that switches coordinates. Then T is a natural equivalence of G with H.

We have until now been dealing with what we call *covariant* functors. There is also a notion of *contravariant* functor, which differs, roughly speaking, only inasmuch as "all the arrows are reversed"! Formally, it is defined as follows:

Definition. A **contravariant functor** G from a category **C** to a category **D** is a rule that assigns to each object X of **C**, an object $G(X)$ of **D**, and to each morphism $f : X \to Y$ of **C**, a morphism

$$G(f) : G(Y) \to G(X)$$

of **D**, such that $G(1_X) = 1_{G(X)}$ and

$$G(g \circ f) = G(f) \circ G(g).$$

A natural transformation between contravariant functors is defined in the obvious way.

In this book, we have not yet studied any contravariant functors, but we shall in the future. Here is an example of one such, taken from linear algebra:

Example 5. If V is a vector space over $\mathbf{R}$, consider the space $\mathcal{L}(V,\mathbf{R})$ of linear functionals on V (linear transformations of V into $\mathbf{R}$). It is often called the **dual space** to V. The space $\mathcal{L}(V,\mathbf{R})$ has the structure of vector space, in the obvious way. Now if $f : V \to W$ is a linear transformation, there is a linear transformation

$$f^{tr} : \mathcal{L}(W,\mathbf{R}) \to \mathcal{L}(V,\mathbf{R}),$$

which is often called the **transpose** (or **adjoint**) of f. It is defined as follows: if $\alpha : W \to \mathbf{R}$ is a linear functional on W, then $f^{tr}(\alpha) : V \to \mathbf{R}$ is the linear functional on V which is the composite $V \xrightarrow{f} W \xrightarrow{\alpha} \mathbf{R}$.

The assignment

$$V \to \mathcal{L}(V,\mathbf{R}) \quad \text{and} \quad f \to f^{tr}$$

is a contravariant functor from the category of vector spaces and linear transformations to itself.

EXERCISES

1. Check that each class of objects and morphisms listed in Example 1 is indeed a category, and that the equivalences are as stated. (Only (c) and (i) need any attention.)

2. Functors have appeared already in your study of mathematics, although that terminology may not have been used. Here are some examples; recall their definitions now:
   (a) Consider the category of complexes and equivalence classes of simplicial maps, the equivalence relation being generated by the relation of contiguity. Define a functor to chain complexes and chain-homotopy classes of chain maps.
   (b) Give a functor from abstract complexes to geometric complexes, called the "geometric realization functor." What are the maps involved?
   (c) Give functors from families of abelian groups, indexed with the fixed index set J, to abelian groups, called "direct product" and "direct sum."
   (d) In algebra, there is a functor $G \to G/[G,G]$ called the "abelianization functor." Either recall it or look it up.
   (e) In topology, there is a functor from completely regular spaces to compact Hausdorff spaces, called the "Stone-Čech compactification." Either recall it or look it up. Don't forget to deal with maps as well as spaces.

3. Let G and H be the functors assigning to a complex K its oriented chain complex and its ordered chain complex, respectively. To each simplicial map, they assign the induced chain map.
   (a) There is a natural transformation either of G to H or of H to G. Which?
   (b) Show that if you consider G and H as taking values in the category of chain complexes and chain-homotopy classes of chain maps, then both natural transformations exist and are natural equivalences.

4. Consider the category of pairs (X,Y) of triangulable spaces such that $X \times Y$ is triangulable, and pairs of continuous maps. Define

$$G(X,Y) = H_p(X \times Y) \quad \text{and} \quad G(f,g) = (f \times g)_*;$$
$$H(X,Y) = H_p(X) \times H_p(Y) \quad \text{and} \quad H(f,g) = f_* \times g_*.$$

   Define a natural transformation of G to H; show it is not a natural equivalence.

5. We have not yet studied a category where the morphisms are other than maps in the usual sense or equivalence classes of maps. Here is an example, for those who are familiar with the fundamental group of a topological space.

   Let X be a fixed space. Let **C** be the category whose objects are the points of X; and let $\hom(x_1,x_0)$ consist of path-homotopy classes $[\alpha]$ of paths α from x_0 to x_1. The composition operation

$$\hom(x_2,x_1) \times \hom(x_1,x_0) \to \hom(x_2,x_0)$$

   is induced by the usual composition of paths $(\beta,\alpha) \to \alpha * \beta$.
   (a) Check that **C** is a category in which every morphism is an equivalence. (Such a category is called a **groupoid.**)

(b) Check that the assignment $x_0 \to \pi_1(X,x_0)$ and $[\alpha] \to \hat{\alpha}$ is a contravariant functor from **C** to the category of groups and homomorphisms. (Here $\hat{\alpha}([f]) = [\bar{\alpha} * f * \alpha]$, where $\bar{\alpha}$ is the reverse path to α.)

Note: The reason we must let $[\alpha]$ represent an element of $\hom(x_1,x_0)$ rather than $\hom(x_0,x_1)$ arises from the awkward fact that when we compose *paths* we put the first path on the left, while when we compose *maps* we put the first function on the right!

4

Singular Homology Theory

Now we are in some sense going to begin all over again. We construct a new homology theory. As compared with simplicial theory, it is much more "natural." For one thing, the homology groups are defined for arbitrary topological spaces, not just for triangulable ones. For another, the homomorphism induced by a continuous map is defined directly and its functorial properties are proved easily; no difficult results such as the simplicial approximation theorem are needed. The topological invariance of the singular homology groups follows at once.

However, the singular homology groups are not immediately computable. One must develop a good deal of singular theory before one can compute the homology of even such a simple space as the sphere S^n. Eventually, when we develop the theory of CW complexes, we will see how singular homology can be computed fairly readily. Alternatively, since we shall show that simplicial and singular homology groups are isomorphic for triangulable spaces, we can always go back to simplicial theory if we want to compute something.

In this chapter, we construct the singular homology groups and prove that they satisfy the Eilenberg-Steenrod axioms on the class of all topological spaces. (The homotopy axiom and the excision axiom, it turns out, require some effort.) Then we construct a specific isomorphism between the singular and simplicial theories that will be useful later.

Finally, we give a number of applications. They include the Jordan curve theorem, theorems about manifolds, and the computation of the homology of real and complex projective spaces.

§29. THE SINGULAR HOMOLOGY GROUPS

In this section we define the singular homology groups and derive their elementary properties. First, we introduce some notation.

Let $\mathbf{R}^\infty$ denote the vector space $\mathbf{E}^J$, where J is the set of positive integers. An element of $\mathbf{R}^\infty$ is an infinite sequence of real numbers having only finitely many non-zero components. Let Δ_p denote the p-simplex in $\mathbf{R}^\infty$ having vertices

$$\begin{aligned} \epsilon_0 &= (0,0,\ldots,0,\ldots), \\ \epsilon_1 &= (1,0,\ldots,0,\ldots), \\ &\ldots \\ \epsilon_p &= (0,0,\ldots,1,\ldots). \end{aligned}$$

We call Δ_p the **standard p-simplex.** Note that under this definition, Δ_{p-1} is a face of Δ_p.

If X is a topological space, we define a **singular p-simplex** of X to be a continuous map

$$T : \Delta_p \rightarrow X.$$

(The word "singular" is used to emphasize that T need not be an imbedding. The map T could for instance be a constant map.)

The free abelian group generated by the singular p-simplices of X is denoted $S_p(X)$ and called the **singular chain group** of X in dimension p. By our usual convention for free abelian groups (see §4), we shall denote an element of $S_p(X)$ by a formal linear combination, with integer coefficients, of singular p-simplices.

It is convenient to consider a special type of singular simplex. Given points $a_0,\ldots,a_p$ in some euclidean space $\mathbf{E}^J$, which need not be independent, there is a unique affine map l of Δ_p into $\mathbf{E}^J$ that maps ϵ_i to a_i for $i = 0,\ldots,p$. It is defined by the equation

$$l(x_1,\ldots,x_p,0,\ldots) = a_0 + \sum_{i=1}^{p} x_i(a_i - a_0).$$

We call this map the **linear singular simplex** determined by $a_0,\ldots,a_p$; and we denote it by $l(a_0,\ldots,a_p)$.

For example, the map $l(\epsilon_0,\ldots,\epsilon_p)$ is just the inclusion map of Δ_p into $\mathbf{R}^\infty$.

Similarly, if as usual we use the notation $\hat{v}_i$ to mean that the symbol v_i is to be deleted, then the map

$$l(\epsilon_0,\ldots,\hat{\epsilon}_i,\ldots,\epsilon_p)$$

is a map of Δ_{p-1} into $\mathbf{R}^\infty$ that carries Δ_{p-1} by a linear homeomorphism onto the face $\epsilon_0 \ldots \epsilon_{i-1}\epsilon_{i+1} \ldots \epsilon_p$ of Δ_p. We often consider it as a map of Δ_{p-1} into Δ_p rather than into $\mathbf{R}^\infty$. Then if $T : \Delta_p \rightarrow X$, we can form the composite

$$T \circ l(\epsilon_0,\ldots,\hat{\epsilon}_i,\ldots,\epsilon_p).$$

This is a singular $p-1$ simplex of X, which we think of as the "ith face" of the singular p-simplex T.

We now define a homomorphism $\partial : S_p(X) \to S_{p-1}(X)$. If $T : \Delta_p \to X$ is a singular p-simplex of X, let

$$\partial T = \sum_{i=0}^{p} (-1)^i T \circ l(\epsilon_0, \ldots, \hat{\epsilon}_i, \ldots, \epsilon_p).$$

Then ∂T equals a formal sum of singular simplices of dimension $p-1$, which are the "faces" of T. We shall verify that $\partial^2 = 0$ presently.

If $f : X \to Y$ is a continuous map, we define a homomorphism $f_\# : S_p(X) \to S_p(Y)$ by defining it on singular p-simplices by the equation $f_\#(T) = f \circ T$. That is, $f_\#(T)$ is the composite

$$\Delta_p \xrightarrow{T} X \xrightarrow{f} Y.$$

Theorem 29.1. *The homomorphism $f_\#$ commutes with ∂. Furthermore, $\partial^2 = 0$.*

Proof. The first statement follows by direct computation:

$$\partial f_\#(T) = \sum_{i=0}^{p} (-1)^i (f \circ T) \circ l(\epsilon_0, \ldots, \hat{\epsilon}_i, \ldots, \epsilon_p),$$

$$f_\#(\partial T) = \sum_{i=0}^{p} (-1)^i f \circ (T \circ l(\epsilon_0, \ldots, \hat{\epsilon}_i, \ldots, \epsilon_p)).$$

To prove the second statement, we first compute ∂ for linear singular simplices. We compute:

$$\partial l(a_0, \ldots, a_p) = \sum_{i=0}^{p} (-1)^i l(a_0, \ldots, a_p) \circ l(\epsilon_0, \ldots, \hat{\epsilon}_i, \ldots, \epsilon_p)$$

$$= \sum_{i=0}^{p} (-1)^i l(a_0, \ldots, \hat{a}_i, \ldots, a_p).$$

(The second equality comes from the fact that a composite of linear maps is linear. See §2.) The fact that $\partial\partial(l(a_0, \ldots, a_p)) = 0$ is now immediate; one simply takes the proof that $\partial^2 = 0$ in simplicial theory (Lemma 5.3) and inserts the letter l at appropriate places! The general result then follows from the fact that

$$\partial\partial(T) = \partial\partial(T_\#(l(\epsilon_0, \ldots, \epsilon_p)))$$

and ∂ commutes with $T_\#$. $\square$

Definition. The family of groups $S_p(X)$ and homomorphisms $\partial : S_p(X) \to S_{p-1}(X)$ is called the **singular chain complex** of X, and is denoted $\mathcal{S}(X)$. The homology groups of this chain complex are called the **singular homology groups** of X, and are denoted $H_p(X)$.

(If X is triangulable, this notation overlaps with that introduced in §27 for the simplicial homology of a triangulable space. We shall prove later that the simplicial and singular homology theories are naturally isomorphic, so there is in fact no real ambiguity involved.)

The chain complex $\mathcal{S}(X)$ is augmented by the homomorphism $\epsilon : S_0(X) \to \mathbf{Z}$ defined by setting $\epsilon(T) = 1$ for each singular 0-simplex $T : \Delta_0 \to X$. It is immediate that if T is a singular 1-simplex, then $\epsilon(\partial T) = 0$. The homology groups of $\{\mathcal{S}(X), \epsilon\}$ are called the **reduced singular homology groups** of X, and are denoted $\tilde{H}_p(X)$. If $f : X \to Y$ is a continuous map, then $f_\# : S_p(X) \to S_p(Y)$ is an augmentation-preserving chain map, since $f_\#(T)$ is a singular 0-simplex if T is. Thus $f_\#$ induces a homomorphism f_* in both ordinary and reduced singular homology.

If the reduced homology of X vanishes in all dimensions, we say that X is **acyclic** (in singular homology).

Theorem 29.2. *If $i : X \to X$ is the identity, then $i_* : H_p(X) \to H_p(X)$ is the identity. If $f : X \to Y$ and $g : Y \to Z$, then $(g \circ f)_* = g_* \circ f_*$. The same holds in reduced homology.*

Proof. Both equations in fact hold on the chain level. For $i_\#(T) = i \circ T = T$. And $(g \circ f)_\#(T) = (g \circ f) \circ T = g \circ (f \circ T) = g_\#(f_\#(T))$. □

Corollary 29.3. *If $h : X \to Y$ is a homeomorphism, then h_* is an isomorphism.* □

The reader will note how quickly we have proved that the singular homology groups are topological invariants. This is in contrast to simplicial theory, where it took us most of Chapter 2 to do the same thing.

Following the pattern of simplicial theory, we next compute the zero-dimensional homology groups.

Theorem 29.4. *Let X be a topological space. Then $H_0(X)$ is free abelian. If $\{X_\alpha\}$ is the collection of path components of X, and if T_α is a singular 0-simplex with image in X_α, for each α, then the homology classes of the chains T_α form a basis for $H_0(X)$.*

The group $\tilde{H}_0(X)$ is free abelian; it vanishes if X is path connected. Otherwise, let α_0 be a fixed index; then the homology classes of the chains $T_\alpha - T_{\alpha_0}$, for $\alpha \neq \alpha_0$, form a basis for $\tilde{H}_0(X)$.

Proof. Let x_α be the point $T_\alpha(\Delta_0)$. If $T : \Delta_0 \to X$ is any singular 0-simplex of X, then there is a path $f : [0,1] \to X$ from the point $T(\Delta_0)$ to some point x_α. Then f is a singular 1-simplex and $\partial f = T_\alpha - T$. We conclude that an arbitrary singular 0-chain on X is homologous to a chain of the form $\Sigma\, n_\alpha T_\alpha$.

We show that no such 0-chain bounds. Suppose that $\Sigma n_\alpha T_\alpha = \partial d$ for some d. Now each singular 1-simplex in the expression for d has path-connected image, so its image lies in some one of the path components X_α. Thus we can write $d = \Sigma\, d_\alpha$, where d_α consists of those terms of d carried by X_α. Then ∂d_α lies in X_α as well. It follows that $n_\alpha T_\alpha = \partial d_\alpha$ for each α. Applying ϵ to both sides of this equation, we conclude that $n_\alpha = 0$.

The computation of $\tilde{H}_0(X)$ proceeds as in the proof of Theorem 7.2. □

Still following the pattern of simplicial theory, we compute the homology of a cone-like space. Actually, it is more convenient here to deal with an analogous notion, that of a "star-convex" set.

Definition. A set X in E^J is said to be **star convex** relative to the point w of X, if for each x in X different from w, the line segment from x to w lies in X.

Definition. Suppose $X \subset E^J$ is star convex relative to w. We define a **bracket operation** on singular chains of X. Let $T : \Delta_p \to X$ be a singular p-simplex of X. Define a singular $p + 1$ simplex $[T,w] : \Delta_{p+1} \to X$ by letting it carry the line segment from x to ϵ_{p+1}, for x in Δ_p, linearly onto the line segment from $T(x)$ to w. See Figure 29.1. We extend the definition to arbitrary p-chains as follows: if $c = \Sigma\, n_i T_i$ is a singular p-chain on X, let

$$[c,w] = \Sigma\, n_i [T_i,w].$$

This operation is similar to the one we introduced in §8 for cones, except here we have put the vertex w last instead of first.

Note that when restricted to the face Δ_p of Δ_{p+1}, the map $[T,w]$ equals the map T. Note also that if T is the linear singular simplex $l(a_0, \ldots, a_p)$, then $[T,w]$ equals the linear singular simplex $l(a_0, \ldots, a_p, w)$.

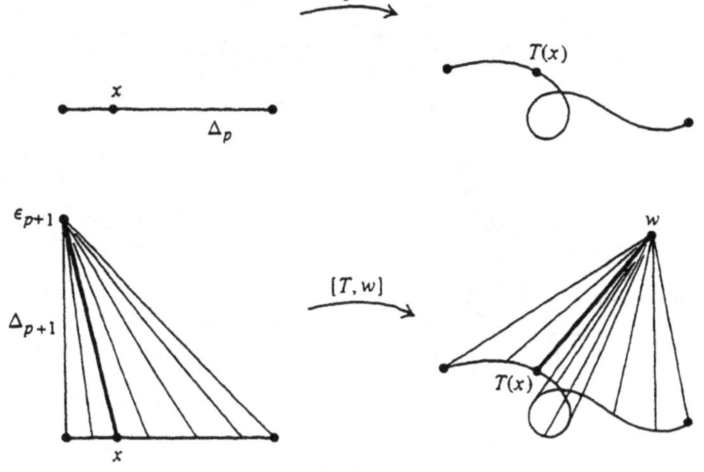

Figure 29.1

We must show that the map $[T,w]$ is continuous. First, we note that the map

$$\pi : \Delta_p \times I \to \Delta_{p+1}$$

defined by the equation $\pi(x,t) = (1-t)x + t\epsilon_{p+1}$ is a quotient map that collapses $\Delta_p \times 1$ to the vertex ϵ_{p+1} and is otherwise one-to-one. The continuous map

$$f : \Delta_p \times I \to X$$

defined by $f(x,t) = (1-t)T(x) + tw$ is constant on $\Delta_p \times 1$; because π is a quotient map, it induces a continuous map of Δ_{p+1} into X. Since π maps the line segment $x \times I$ linearly onto the line segment from x to ϵ_{p+1}, and f maps $x \times I$ linearly onto the line segment from $T(x)$ to w, this induced map equals the singular $p+1$ simplex $[T,w]$ defined previously. Thus $[T,w]$ is continuous.

We now compute how the bracket operation and the boundary operator interact.

Lemma 29.5. *Let X be star convex with respect to w; let c be a singular p-chain of X. Then*

$$\partial[c,w] = \begin{cases} [\partial c,w] + (-1)^{p+1}c & \text{if} \quad p > 0, \\ \epsilon(c)T_w - c & \text{if} \quad p = 0, \end{cases}$$

where T_w is the singular 0-simplex mapping Δ_0 to w.

Proof. If T is a singular 0-simplex, then $[T,w]$ maps the simplex Δ_1 linearly onto the line segment from $T(\Delta_0)$ to w. Then $\partial[T,w] = T_w - T$. The second formula follows.

Let $p > 0$. It suffices to check the formula when c is a singular p-simplex T. The formula's plausibility when $p = 1$ is illustrated in Figure 29.2.

Using the definition of ∂, we compute

$$(*) \qquad \partial[T,w] = \sum_{i=0}^{p+1} (-1)^i [T,w] \circ l_i,$$

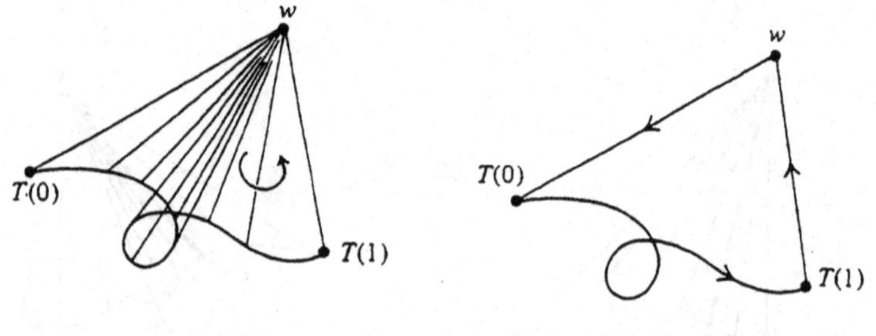

Figure 29.2

where for convenience we use l_i to denote the linear singular simplex

$$l_i = l(\epsilon_0, \dots, \hat{\epsilon}_i, \dots, \epsilon_{p+1})$$

mapping Δ_p into Δ_{p+1}. Now l_{p+1} equals the inclusion map of Δ_p into Δ_{p+1}; since the restriction of $[T,w]$ to Δ_p equals the map T, the last term of (*) equals $(-1)^{p+1}T$.

To complete the proof, we consider the singular simplex $[T,w] \circ l_i$ for $i < p + 1$. Now the map l_i carries Δ_p homeomorphically onto the ith face of Δ_{p+1}; it carries $\epsilon_0, \dots, \epsilon_p$ to $\epsilon_0, \dots, \epsilon_{i-1}, \epsilon_{i+1}, \dots, \epsilon_{p+1}$, respectively. Therefore the restriction of l_i to $\Delta_{p-1} = \epsilon_0 \dots \epsilon_{p-1}$ carries this simplex by a linear map onto the simplex spanned by $\epsilon_0, \dots, \epsilon_{i-1}, \epsilon_{i+1}, \dots, \epsilon_p$. (Recall that $i < p + 1$.) Thus

$$(**) \qquad l_i \mid \Delta_{p-1} = l(\epsilon_0, \dots, \hat{\epsilon}_i, \dots, \epsilon_p).$$

Now we can compute $[T,w] \circ l_i : \Delta_p \to X$. Let x be the general point of Δ_{p-1}. Since $l_i : \Delta_p \to \Delta_{p+1}$ is a linear map, it carries the line segment from x to ϵ_p linearly onto the line segment from $l_i(x)$ to ϵ_{p+1}. Since $l_i(x) \in \Delta_p$, the map $[T,w] : \Delta_{p+1} \to X$ by definition carries this line segment linearly onto the line segment from $T(l_i(x))$ to w. Therefore, by definition,

$$[T,w] \circ l_i = [T \circ (l_i \mid \Delta_{p-1}), w].$$

Substituting this formula into (*) and using (**), we obtain the equation

$$\begin{aligned} \partial[T,w] &= \sum_{i=0}^{p} (-1)^i [T \circ l(\epsilon_0, \dots, \hat{\epsilon}_i, \dots, \epsilon_p), w] + (-1)^{p+1}T \\ &= [\partial T, w] + (-1)^{p+1}T. \quad \square \end{aligned}$$

Theorem 29.6. *Let X be a subspace of E^J that is star convex relative to w. Then X is acyclic in singular homology.*

Proof. To show that $\tilde{H}_0(X) = 0$, let c be a singular 0-chain on X such that $\epsilon(c) = 0$. Then by the preceding lemma,

$$\partial[c,w] = \epsilon(c)T_w - c = -c,$$

so c bounds a 1-chain.

To show $H_p(X) = 0$ for $p > 0$, let z be a singular p-cycle on X. By the preceding lemma,

$$\partial[z,w] = [\partial z, w] + (-1)^{p+1}z = (-1)^{p+1}z.$$

Thus z bounds a $p + 1$ chain. $\square$

Corollary 29.7. *Any simplex is acyclic in singular homology.* $\square$

EXERCISES

1. (a) Let $\{X_\alpha\}$ be the set of path components of X. Show that $H_p(X) \simeq \bigoplus_\alpha H_p(X_\alpha)$.
   (b) The **topologist's sine curve** is the subspace of $\mathbf{R}^2$ consisting of all points $(x, \sin 1/x)$ for $0 < x \leq 1$, and all points $(0,y)$ for $-1 \leq y \leq 1$. See Figure 29.3. Compute the singular homology of this space.
2. Let X be a compact subspace of $\mathbf{R}^N$; let $f: X \to Y$.
   (a) Let w be the point $(0,\ldots,0,1)$ in $\mathbf{R}^N \times \mathbf{R}$; let C be the union of all line segments joining points of X to w. Show that C is a quotient space of $X \times I$.
   (b) Show that if f is homotopic to a constant map, then f_* is the zero homomorphism in reduced homology. [*Hint:* Show that f extends to C.]
   (c) Show that if X is contractible, then X is acyclic.

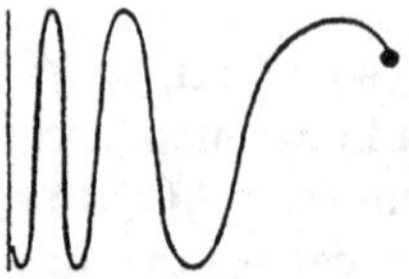

Figure 29.3

§30. THE AXIOMS FOR SINGULAR THEORY

We now introduce the relative singular homology groups. Then we show, in this section and the next, that they satisfy the Eilenberg-Steenrod axioms on the class of all topological pairs.

If X is a space and A is a subspace of X, there is a natural inclusion $S_p(A) \to S_p(X)$. The group of **relative singular chains** is defined by

$$S_p(X,A) = S_p(X)/S_p(A).$$

The boundary operator $\partial : S_p(X) \to S_{p-1}(X)$ restricts to the boundary operator on $S_p(A)$; hence it induces a boundary operator

$$\partial : S_p(X,A) \to S_{p-1}(X,A)$$

on relative chains. The family of groups $S_p(X,A)$ and homomorphisms ∂ is called the **singular chain complex** of the pair (X,A) and denoted $\mathcal{S}(X,A)$. The homology groups of this chain complex are called the **singular homology groups** of the pair (X,A) and are denoted $H_p(X,A)$.

Note that the chain complex $\mathcal{S}(X,A)$ is free; the group $S_p(X,A)$ has as basis all cosets of the form $T + S_p(A)$, where T is a singular p-simplex whose image set does not lie in A.

If $f: (X,A) \to (Y,B)$ is a continuous map, then the homomorphism

$f_{\#}: S_p(X) \to S_p(Y)$ carries chains of A into chains of B, so it induces a homomorphism (also denoted $f_{\#}$)

$$f_{\#}: S_p(X,A) \to S_p(Y,B).$$

This map commutes with ∂, so it in turn induces a homomorphism

$$f_*: H_p(X,A) \to H_p(Y,B).$$

The following theorem is immediate:

Theorem 30.1. *If $i: (X,A) \to (X,A)$ is the identity, then i_* is the identity. If $h: (X,A) \to (Y,B)$ and $k: (Y,B) \to (Z,C)$ are continuous, then $(k \circ h)_* = k_* \circ h_*$.* □

Theorem 30.2. *There is a homomorphism $\partial_*: H_p(X,A) \to H_{p-1}(A)$, defined for $A \subset X$ and all p, such that the sequence*

$$\cdots \to H_p(A) \xrightarrow{i_*} H_p(X) \xrightarrow{\pi_*} H_p(X,A) \xrightarrow{\partial_*} H_{p-1}(A) \to \cdots$$

is exact, where i and π are inclusions. The same holds if reduced homology is used for X and A, provided $A \neq \varnothing$.

A continuous map $f: (X,A) \to (Y,B)$ induces a homomorphism of the corresponding exact sequences in singular homology, either ordinary or reduced.

Proof. For the existence of ∂_* and the exact sequence, we apply the "zig-zag lemma" of §24 to the short exact sequence of chain complexes

$$0 \to S_p(A) \xrightarrow{i_{\#}} S_p(X) \xrightarrow{\pi_{\#}} S_p(X,A) \to 0,$$

where $i: A \to X$ and $\pi: (X,\varnothing) \to (X,A)$ are inclusions. The naturality of ∂_* follows from Theorem 24.2, once we note that $f_{\#}$ commutes with $i_{\#}$ and $\pi_{\#}$. The corresponding results for reduced homology follow by the same methods used for simplicial theory (see §24). □

Theorem 30.3. *If P is a one-point space, then $H_p(P) = 0$ for $p \neq 0$ and $H_0(P) \simeq \mathbf{Z}$.*

Proof. This follows from Theorem 29.6, once one notes that a one-point space in $\mathbf{R}^N$ is star convex! For a more direct proof, we compute the chain complex $\mathcal{S}(P)$. There is exactly one singular simplex $T_p: \Delta_p \to P$ in each non-negative dimension, so $S_p(P)$ is infinite cyclic for $p \geq 0$. Each of the "faces" of T_p,

$$T_p \circ l(\epsilon_0, \ldots, \hat{\epsilon}_i, \ldots, \epsilon_p),$$

equals the singular simplex T_{p-1}. Therefore, $\partial T_p = 0$ if p is odd (the terms cancel in pairs), and $\partial T_p = T_{p-1}$ if p is even (since there is one term left over). The chain complex $\mathcal{S}(P)$ is thus of the form

$$\begin{array}{ccccccccccc} S_{2k}(P) & \to & S_{2k-1}(P) & \to & \cdots & \to & S_1(P) & \to & S_0(P) & \to & 0 \\ \mathbf{Z} & \xrightarrow{\simeq} & \mathbf{Z} & \xrightarrow{0} & \cdots & \to & \mathbf{Z} & \xrightarrow{0} & \mathbf{Z} & \to & 0. \end{array}$$

In dimension $2k-1$, every chain is a boundary; while in dimension $2k$, for $k>0$, no chain is a cycle. In dimension 0, every chain is a cycle and no chain is a boundary, so the homology is infinite cyclic. □

Theorem 30.4. *Given $\alpha \in H_p(X,A)$, there is a compact pair $(X_0,A_0) \subset (X,A)$, such that α is in the image of the homomorphism induced by inclusion*

$$i_* : H_p(X_0,A_0) \to H_p(X,A).$$

Proof. If $T:\Delta_p \to X$ is a singular simplex, its ***minimal carrier*** is defined to be the image set $T(\Delta_p)$. The minimal carrier of a singular p-chain $\Sigma\, n_i T_i$ (where each $n_i \neq 0$) is the union of the minimal carriers of the T_i; since the sum is finite, this set is compact. Let c_p be a chain in $S_p(X)$ whose coset modulo $S_p(A)$ represents α; then ∂c_p is carried by A. Let X_0 be the minimal carrier of c_p and let A_0 be the minimal carrier of ∂c_p. Then c_p can be taken to represent a homology class β in $H_p(X_0,A_0)$; and $i_*(\beta)=\alpha$. □

The preceding theorem shows that singular theory satisfies the compact support axiom. There is an addendum to this theorem, which we prove here. (It can also be derived directly from the axioms; see the exercises of §27.)

Theorem 30.5. *Let $i:(X_0,A_0) \to (X,A)$ be inclusion, where (X_0,A_0) is a compact pair. If $\alpha \in H_p(X_0,A_0)$ and $i_*(\alpha)=0$, then there are a compact pair (X_1,A_1) and inclusions*

$$(X_0,A_0) \xrightarrow{j} (X_1,A_1) \xrightarrow{k} (X,A)$$

such that $j_*(\alpha)=0$.

Proof. Let c_p be a singular p-chain of X_0 representing α; then ∂c_p is carried by A_0. By hypothesis, there is a chain d_{p+1} of X such that $c_p - \partial d_{p+1}$ is carried by A. Let X_1 be the union of X_0 and the minimal carrier of d_{p+1}; let A_1 be the union of A_0 and the minimal carrier of $c_p - \partial d_{p+1}$. □

Thus far everything has been relatively easy; we have verified all but the homotopy and excision axioms. These two require more work. We verify the homotopy axiom now.

In outline, the proof is similar to the proof of the homotopy axiom for simplicial theory (Theorem 19.2). We consider the inclusion maps $i,j:X \to X \times I$ defined by

$$i(x)=(x,0) \qquad \text{and} \qquad j(x)=(x,1),$$

and we construct a chain homotopy D between $i_\#$ and $j_\#$. If F is a homotopy between f and g, then $F_\# \circ D$ is a chain homotopy between $f_\#$ and $g_\#$, and the theorem follows. To construct D in simplicial theory, we used acyclic carriers; here we need something more general. It is a special case of what we shall later call the "method of acyclic models."

It is plausible that a chain homotopy D between $i_\#$ and $j_\#$ should exist. For

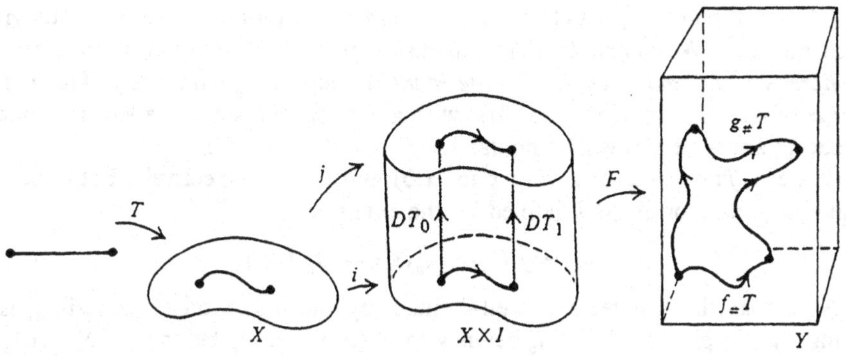

Figure 30.1

example, if T is the singular 1-simplex pictured in Figure 30.1, and if $\partial T = T_1 - T_0$, we can take DT_1 and DT_0 to be the "vertical" singular 1-simplices pictured. Then we can take DT to be a 2-chain that covers the shaded region in the figure, having as boundary $j_\#(T) - DT_1 - i_\#(T) + DT_0$.

We state the existence of D as a separate lemma:

Lemma 30.6. *There exists, for each space X and each non-negative integer p, a homomorphism*

$$D_X : S_p(X) \to S_{p+1}(X \times I)$$

having the following properties:

(a) *If $T : \Delta_p \to X$ is a singular simplex, then*

$$\partial D_X T + D_X \partial T = j_\#(T) - i_\#(T).$$

Here the map $i : X \to X \times I$ carries x to $(x,0)$; and the map $j : X \to X \times I$ carries x to $(x,1)$.

(b) *D_X is natural; that is, if $f : X \to Y$ is a continuous map, then the following diagram commutes:*

$$\begin{array}{ccc} S_p(X) & \xrightarrow{D_X} & S_{p+1}(X \times I) \\ \downarrow f_\# & & \downarrow (f \times i_I)_\# \\ S_p(Y) & \xrightarrow{D_Y} & S_{p+1}(Y \times I). \end{array}$$

Proof. We proceed by induction on p. The case $p = 0$ is easy. Given $T : \Delta_0 \to X$, let x_0 denote the point $T(\Delta_0)$. Define $D_X T : \Delta_1 \to X \times I$ by the equation

$$D_X T(t,0,\ldots) = (x_0,t) \qquad \text{for} \qquad 0 \le t \le 1.$$

(Recall that $\Delta_1 = \epsilon_0\epsilon_1$ consists of all points $(t,0,\ldots)$ of $\mathbf{R}^\infty$ with $0 \le t \le 1$.) Properties (a) and (b) follow at once.

Now suppose D_X is defined in dimensions less than p, for *all* X, satisfying (a) and (b). We define $D_X T$ in dimension p, *by first defining it in the case where $X = \Delta_p$ and T equals i_p, the identity map of Δ_p with itself.* This is the crux of the technique; we deal first with a very special space, called the "model space," and a very special singular simplex on that space.

Let $\hat{i}, \hat{j} : \Delta_p \to \Delta_p \times I$ map x to $(x,0)$ and $(x,1)$, respectively. Let c_p be the singular chain on $\Delta_p \times I$ defined by the equation

$$c_p = \hat{j}_\#(i_p) - \hat{i}_\#(i_p) - D_{\Delta_p}(\partial i_p).$$

(Note that the last term is well-defined, by the induction hypothesis; i_p is a singular simplex on Δ_p, so i_p belongs to $S_p(\Delta_p)$ and ∂i_p belongs to $S_{p-1}(\Delta_p)$.) Now c_p is a cycle, for using the induction hypothesis, we have

$$\partial c_p = \partial \hat{j}_\#(i_p) - \partial \hat{i}_\#(i_p) - [\hat{j}_\#(\partial i_p) - \hat{i}_\#(\partial i_p) - D_{\Delta_p} \partial\partial i_p],$$

and this chain vanishes. *Since $\Delta_p \times I$ is convex*, it is acyclic (by Theorem 29.6). Therefore, we can *choose* an element of $S_{p+1}(\Delta_p \times I)$ whose boundary equals c_p. We denote this element by $D_{\Delta_p}(i_p)$; then formula (a) holds because

$$\partial D_{\Delta_p}(i_p) = c_p.$$

Now, given an arbitrary space X and an arbitrary singular p-simplex $T : \Delta_p \to X$, we define

$$D_X T = (T \times i_I)_\#(D_{\Delta_p} i_p).$$

See Figure 30.2. Intuitively, $D_{\Delta_p} i_p$ is a singular $p + 1$ chain filling up the entire prism $\Delta_p \times I$; its boundary is, roughly speaking, the boundary of the prism. The map $(T \times i_I)_\#$ carries this chain over to a singular chain on $X \times I$.

Checking (a) and (b) is now straightforward. To check (a), we compute

$$(*) \qquad \begin{aligned} \partial D_X T &= (T \times i_I)_\# \partial D_{\Delta_p} i_p \\ &= (T \times i_I)_\#(\hat{j}_\#(i_p) - \hat{i}_\#(i_p) - D_{\Delta_p} \partial i_p) \\ &= j_\#(T) - i_\#(T) - (T \times i_I)_\# D_{\Delta_p}(\partial i_p). \end{aligned}$$

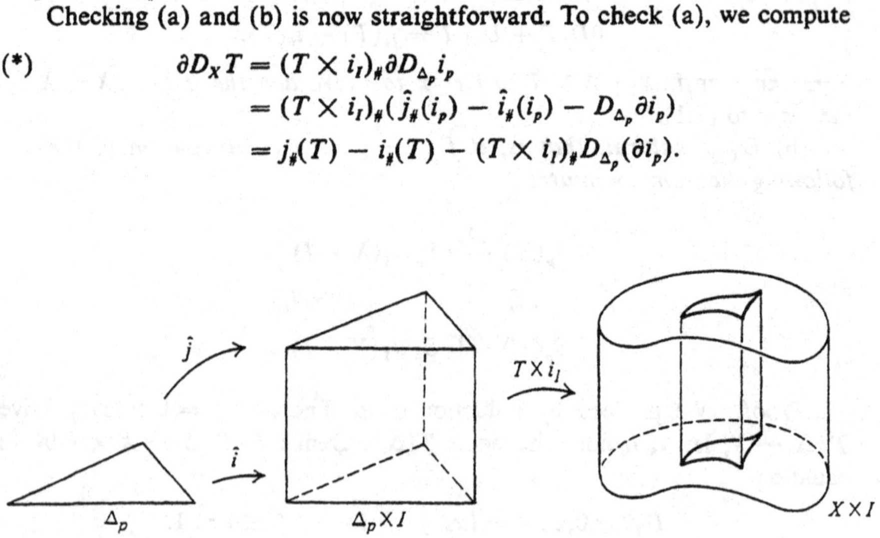

Figure 30.2

Now we use naturality. By the induction hypothesis, we have commutativity in the diagram:

$$\begin{array}{ccc} S_{p-1}(\Delta_p) & \xrightarrow{D_{\Delta_p}} & S_p(\Delta_p \times I) \\ T_{\#}\big\downarrow & & \big\downarrow (T \times i_I)_{\#} \\ S_{p-1}(X) & \xrightarrow{D_X} & S_p(X \times I). \end{array}$$

Therefore, the last term of (*) equals

$$-D_X T_{\#}(\partial i_p) = -D_X \partial T_{\#}(i_p) = -D_X \partial T,$$

as desired.

Checking (b) is easier. If $f: X \to Y$, note that $(f \circ T) \times i_I = (f \times i_I) \circ (T \times i_I)$. Now

$$D_Y(f_{\#}(T)) = D_Y(f \circ T) = ((f \circ T) \times i_I)_{\#} D_{\Delta_p}(i_p),$$

by definition. This in turn equals

$$(f \times i_I)_{\#}(T \times i_I)_{\#} D_{\Delta_p}(i_p) = (f \times i_I)_{\#} D_X T. \quad \square$$

Note that in the proof just given, we gave a direct construction for D_X in the case $p = 0$. Only in the inductive step did we use the "model space" Δ_p. Here is a proof for the case $p = 0$ that looks more like the inductive step; this is the proof we shall later generalize:

Let $p = 0$. We define $D_X T$ first in the case where $X = \Delta_0$, and T equals i_0, the identity map of Δ_0 with itself. We want the formula

$$\partial D_{\Delta_0}(i_0) = \hat{j}_{\#}(i_0) - \hat{i}_{\#}(i_0)$$

to hold. Note that the right side of this formula is a 0-chain carried by $\Delta_0 \times I$, and ϵ of it is zero. *Since the space $\Delta_0 \times I$ is convex, it is acyclic.* Therefore, we can choose $D_{\Delta_0} i_0$ to be an element of $S_1(\Delta_0 \times I)$ whose boundary is $\hat{j}_{\#}(i_0) - \hat{i}_{\#}(i_0)$. Then for general X and T, we define

$$D_X T = (T \times i_I)_{\#} D_{\Delta_0} i_0.$$

Properties (a) and (b) are proved just as they were in the inductive step.

Theorem 30.7. *If $f, g: (X,A) \to (Y,B)$ are homotopic, then $f_* = g_*$. The same holds in reduced homology if $A = B = \varnothing$.*

Proof. Let $F: (X \times I, A \times I) \to (Y,B)$ be the homotopy between $f, g: (X,A) \to (Y,B)$. Let $i, j: (X,A) \to (X \times I, A \times I)$ be given by $i(x) = (x,0)$ and $j(x) = (x,1)$. Let $D_X: S_p(X) \to S_{p+1}(X \times I)$ be the chain homotopy of the preceding lemma. Naturality of D_X with respect to inclusion $A \to X$

shows that the restriction of D_X to $S_p(A)$ equals D_A. Thus D_X carries $S_p(A)$ into $S_{p+1}(A \times I)$, and thus induces a chain homotopy

$$D_{X,A} : S_p(X,A) \to S_{p+1}(X \times I, A \times I)$$

on the relative level. Formula (a) of Lemma 30.6 holds because $D_{X,A}$ is induced by D_X. Define D to be the composite of $D_{X,A}$ and the homomorphism

$$F_\# : S_{p+1}(X \times I, A \times I) \to S_{p+1}(Y,B).$$

Then we compute

$$\begin{aligned} \partial D = \partial F_\# D_{X,A} &= F_\# \partial D_{X,A} \\ &= F_\#(j_\# - i_\# - D_{X,A}\partial) \\ &= (F \circ j)_\# - (F \circ i)_\# - F_\# D_{X,A}\partial \\ &= f_\# - g_\# - D\partial. \quad \square \end{aligned}$$

Definition. Let $f : (X,A) \to (Y,B)$. If there is a map $g : (Y,B) \to (X,A)$ such that $f \circ g$ and $g \circ f$ are homotopic to the appropriate identities *as maps of pairs,* we call f a **homotopy equivalence,** and we call g a **homotopy inverse** for f.

Theorem 30.8. *Let $f : (X,A) \to (Y,B)$.*

(a) *If f is a homotopy equivalence, then f_* is an isomorphism in relative homology.*

(b) *More generally, if $f : X \to Y$ and $f|A : A \to B$ are homotopy equivalences, then f_* is an isomorphism in relative homology.*

Proof. If f is a homotopy equivalence, it is immediate that f_* is an isomorphism. To prove (b), we examine the long exact homology sequences of (X,A) and of (Y,B), and the homomorphism f_* carrying the one exact sequence to the other. The hypotheses of the theorem tell us that

$$f_* : H_p(X) \to H_p(Y) \qquad \text{and} \qquad (f|A)_* : H_p(A) \to H_p(B)$$

are isomorphisms. The theorem then follows from the Five-lemma. $\square$

If $f : (X,A) \to (Y,B)$ is a homotopy equivalence, then so are $f : X \to Y$ and $f|A : A \to B$. The converse does not hold, however, as the following example shows.

Example 1. Consider the inclusion map $j : (B^n, S^{n-1}) \to (\mathbf{R}^n, \mathbf{R}^n - \mathbf{0})$. Since B^n is a deformation retract of $\mathbf{R}^n$, and S^{n-1} is a deformation retract of $\mathbf{R}^n - \mathbf{0}$, the map j_* is an isomorphism in relative homology. Suppose there existed a map $g : (\mathbf{R}^n, \mathbf{R}^n - \mathbf{0}) \to (B^n, S^{n-1})$ that served as a homotopy inverse for j. Then since $\mathbf{0}$ is a limit point of $\mathbf{R}^n - \mathbf{0}$, the map g necessarily carries $\mathbf{0}$ into S^{n-1}. Hence

$$g \circ j : (B^n, S^{n-1}) \to (B^n, S^{n-1})$$

carries all of B^n into S^{n-1}, so it induces the trivial homomorphism in homology. On the other hand, this map is by hypothesis homotopic to the identity, so it induces the identity homomorphism of $H_n(B^n, S^{n-1})$. Since this group is non-trivial (as we shall prove shortly), we have a contradiction.

EXERCISES

1. Construct the exact sequence of a triple in singular homology. (See Exercise 2 of §26.)
2. Show that if $f : X \to Y$ is homotopic to a constant, then f_* is the zero homomorphism in reduced homology. Conclude that if X is contractible, then X is acyclic in singular homology.
3. Suppose inclusion $j : A \to X$ is a homotopy equivalence. Show $H_p(X,A) = 0$.
4. Give an example in which $H_p(X,A) \not\simeq H_p(Y,B)$, although X has the homotopy type of Y and A has the homotopy type of B. Compare with Theorem 30.8.
5. In the proof of Lemma 30.6, one has some freedom in choosing the chain $D_{\Delta_p} i_p$. Show that the formula
$$D_{\Delta_p} i_p = \sum_{i=0}^{p} (-1)^i l((\epsilon_0,0), \ldots, (\epsilon_i,0), (\epsilon_i,1), \ldots, (\epsilon_p,1))$$
gives a $p + 1$ chain on $\Delta_p \times I$ that satisfies the requirements of the lemma. Draw a picture when $p = 1$ and $p = 2$.
6. Show that if $f : (X,A) \to (Y,B)$ is a homotopy equivalence, then both $f : X \to Y$ and $f|A : A \to B$ are homotopy equivalences.
7. Consider the category **C** whose objects are chain complexes; a morphism of **C** (of degree d), mapping $\mathcal{C}$ to $\mathcal{C}'$, is a family of homomorphisms $\phi_p : C_p \to C'_{p+d}$. Consider the following functors from the topological category to **C**:
$$G(X) = \mathcal{S}(X); \qquad G(f) = f_\#,$$
$$H(X) = \mathcal{S}(X \times I); \quad H(f) = (f \times i_I)_\#.$$
Show that the chain homotopy D_X of Lemma 30.6 is a natural transformation of G to H.

§31. EXCISION IN SINGULAR HOMOLOGY

In this section, we verify the excision axiom for singular homology. The techniques involved in the proof will be useful later in other situations.

One of the facts we proved about simplicial complexes was that we could chop up a finite complex barycentrically into simplices that were as small as desired. We need a similar result for singular chains. To be precise, suppose one is given a space X and a collection $\mathcal{A}$ of subsets of X whose interiors cover X. A singular simplex of X is said to be **$\mathcal{A}$-small** if its image set lies in an element of $\mathcal{A}$. Given a singular chain on X, we show how to "chop it up" so that all its simplices are $\mathcal{A}$-small. Not surprisingly, what we do is to introduce something like barycentric subdivision in order to accomplish this. We give the details now.

Definition. Let X be a topological space. We define a homomorphism $\text{sd}_X : S_p(X) \to S_p(X)$ by induction. If $T : \Delta_0 \to X$ is a singular 0-simplex, we define $\text{sd}_X T = T$. Now suppose sd_X is defined in dimensions less than p. If $i_p : \Delta_p \to \Delta_p$ is the identity map, let $\hat{\Delta}_p$ denote the barycenter of Δ_p and define (using the bracket operation of §29),

$$\text{sd}_{\Delta_p} i_p = (-1)^p [\text{sd}_{\Delta_p}(\partial i_p), \hat{\Delta}_p].$$

The definition makes sense since Δ_p is star convex relative to $\hat{\Delta}_p$. Then if $T : \Delta_p \to X$ is any singular p-simplex on X, we define

$$\text{sd}_X T = T_\#(\text{sd}_{\Delta_p} i_p).$$

See Figure 31.1. It is called the **barycentric subdivision operator** in singular theory.

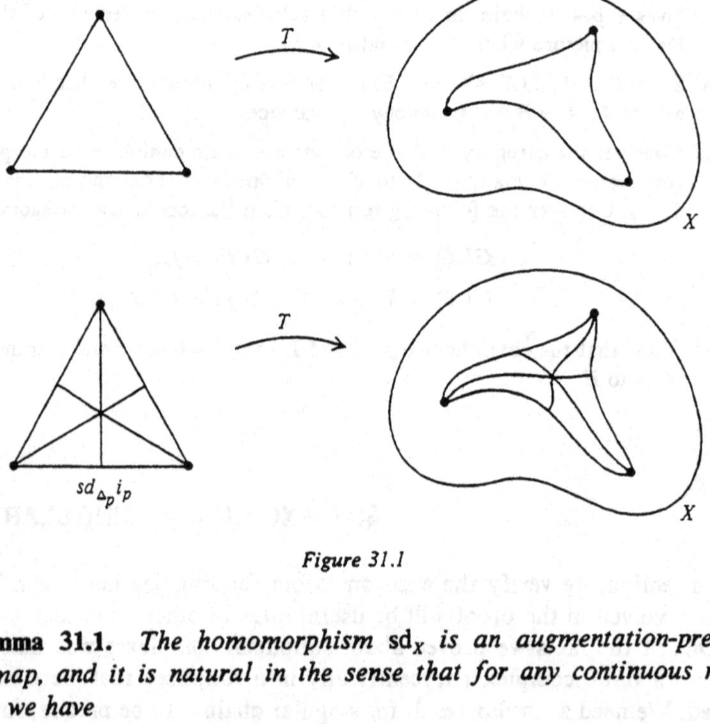

Figure 31.1

Lemma 31.1. *The homomorphism sd_X is an augmentation-preserving chain map, and it is natural in the sense that for any continuous map $f : X \to Y$, we have*

$$f_\# \circ \text{sd}_X = \text{sd}_Y \circ f_\#.$$

Proof. The map sd_X preserves augmentation because it is the identity in dimension 0. Naturality holds in dimension 0 for the same reason. Naturality holds in positive dimensions by direct computation:

$$\begin{aligned} f_\#(\text{sd}_X T) &= f_\# T_\#(\text{sd}_{\Delta_p} i_p) = (f \circ T)_\#(\text{sd}_{\Delta_p} i_p) \\ &= \text{sd}_Y(f \circ T) = \text{sd}_Y(f_\#(T)). \end{aligned}$$

Henceforth, we shall normally omit the subscript on the operator sd_X, relying on the context to make the meaning clear.

To check that sd is a chain map, we proceed by induction on p. The fact that $\text{sd} \circ \partial = \partial \circ \text{sd}$ in dimension 0 is trivial. Assuming the result true in dimensions less than p, we apply Lemma 29.5 to compute

$$\begin{aligned} \partial \,\text{sd}\, i_p &= (-1)^p \partial [\text{sd}\, \partial i_p, \hat{\Delta}_p] \\ &= \begin{cases} (-1)^p [\partial \,\text{sd}\, \partial i_p, \hat{\Delta}_p] + \text{sd}\, \partial i_p & \text{if} \quad p > 1, \\ -\epsilon(\text{sd}\, \partial i_1) T_0 \quad + \text{sd}\, \partial i_1 & \text{if} \quad p = 1, \end{cases} \end{aligned}$$

where T_0 is the 0-simplex whose image point is $\hat{\Delta}_1$. Now if $p > 1$, we have $\partial \,\text{sd}\, \partial i_p = \text{sd}\, \partial\partial i_p = 0$ by the induction hypothesis. If $p = 1$, we have $\epsilon(\text{sd}\, \partial i_1) = \epsilon(\partial i_1) = 0$, because sd preserves augmentation. Hence in either case, $\partial \,\text{sd}\, i_p = \text{sd}\, \partial i_p$. In general, we compute

$$\begin{aligned} \partial \,\text{sd}\, T &= \partial T_\#(\text{sd}\, i_p) && \text{by definition,} \\ &= T_\#(\partial \,\text{sd}\, i_p) && \text{because } T_\# \text{ is a chain map,} \\ &= T_\#(\text{sd}\, \partial i_p) && \text{by the formula just proved,} \\ &= \text{sd}\, T_\#(\partial i_p) && \text{because sd is natural,} \\ &= \text{sd}\, \partial(T_\#(i_p)) && \text{because } T_\# \text{ is a chain map,} \\ &= \text{sd}\, \partial T. \quad \square \end{aligned}$$

Lemma 31.2. *Let $T : \Delta_p \to \sigma$ be a linear homeomorphism of Δ_p with the p-simplex σ. Then each term of* sd T *is a linear homeomorphism of Δ_p with a simplex in the first barycentric subdivision of σ.*

Proof. The lemma is trivial for $p = 0$; suppose it is true in dimensions less than p. Consider first the identity linear homeomorphism $i_p : \Delta_p \to \Delta_p$. Now

$$\text{sd}\, i_p = [\text{sd}\, \partial i_p, \hat{\Delta}_p].$$

Each term in ∂i_p is a linear homeomorphism of Δ_{p-1} with a simplex in Bd Δ_p. By the induction hypothesis, $\text{sd}(\partial i_p) = \Sigma \pm T_i$, where T_i is a linear homeomorphism of Δ_{p-1} with a simplex $\hat{s}_1 \ldots \hat{s}_p$ in the first barycentric subdivision of Bd Δ_p. Then $[T_i, \hat{\Delta}_p]$ is by definition a linear homeomorphism of Δ_p with the simplex $\hat{\Delta}_p \hat{s}_1 \ldots \hat{s}_p$, which belongs to the first barycentric subdivision of Δ_p.

Now consider a general linear homeomorphism $T : \Delta_p \to \sigma$. Note that T defines a linear isomorphism of the first barycentric subdivision of Δ_p with the first barycentric subdivision of σ, because it carries the barycenter of Δ_p to the barycenter of σ. Now $\text{sd } T = T_\#(\text{sd } i_p)$; since the composite of linear homeomorphisms is a linear homeomorphism, each term in $\text{sd } T$ is a linear homeomorphism of Δ_p with a simplex in the first barycentric subdivision of σ. □

Theorem 31.3. *Let $\mathcal{A}$ be a collection of subsets of X whose interiors cover X. Given $T:\Delta_p \to X$, there is an m such that each term of $\text{sd}^m T$ is $\mathcal{A}$-small.*

Proof. It follows from the preceding lemma that if L is a linear homeomorphism of Δ_p with the p-simplex σ, then each term of $\text{sd}^m L$ is a linear homeomorphism of Δ_p with a simplex in the mth barycentric subdivision of σ.

Let us cover Δ_p by the open sets $T^{-1}(\text{Int } A)$, for $A \in \mathcal{A}$. Let λ be a Lebesgue number for this open cover. Choose m so that each simplex in the mth barycentric subdivision of Δ_p has diameter less than λ. By the preceding remark, each term of $\text{sd}^m i_p$ is a linear singular simplex on Δ_p whose image set has diameter less than λ. Then each term of $\text{sd}^m T = T_\#(\text{sd}^m i_p)$ is a singular simplex on X whose image set lies in an element of $\mathcal{A}$. □

Having shown how to chop up singular chains so they are $\mathcal{A}$-small, we now show that these $\mathcal{A}$-small singular chains suffice to generate the homology of X. First, we need a lemma.

Lemma 31.4. *Let m be given. For each space X, there is a homomorphism $D_X : S_p(X) \to S_{p+1}(X)$ such that for each singular p-simplex T of X,*

$$(*) \qquad \partial D_X T + D_X \partial T = \text{sd}^m T - T.$$

Furthermore, D_X is natural. That is, if $f: X \to Y$, then $f_\# \circ D_X = D_Y \circ f_\#$.

Proof. If $T:\Delta_0 \to X$ is a singular 0-simplex, define $D_X T = 0$. Formula (*) and naturality follow trivially. Now let $p > 0$. Suppose D_X is defined, satisfying (*) and naturality, in dimensions less than p. We proceed by a method similar to that used in the proof of Lemma 30.6, which will be formalized in the next section as the "method of acyclic models."

We first define $D_X T$ in the special case $X = \Delta_p$ and $T = i_p$, the identity map of Δ_p with itself. Consider the singular p-chain

$$c_p = \text{sd}^m i_p - i_p - D_{\Delta_p}(\partial i_p).$$

This is by definition a singular p-chain on Δ_p. It is a cycle, by the usual computation (using the hypothesis that (*) holds in dimensions less than p). Since Δ_p is acyclic in singular homology, we can choose $D_{\Delta_p} i_p$ to be an element of $S_{p+1}(\Delta_p)$ whose boundary equals c_p. Then (*) holds for $X = \Delta_p$ and $T = i_p$.

Given a general singular p-simplex $T:\Delta_p \to X$, we define

$$D_X T = T_\#(D_{\Delta_p}(i_p)).$$

Formula (*) holds for D_X by the usual direct computation:

$$\begin{aligned}\partial D_X T &= T_\#(\partial D_{\Delta_p}(i_p))\\ &= T_\#(\mathrm{sd}^m i_p - i_p - D_{\Delta_p}(\partial i_p))\\ &= \mathrm{sd}^m T_\#(i_p) - T_\#(i_p) - D_X T_\#(\partial i_p)\\ &= \mathrm{sd}^m T - T - D_X \partial T,\end{aligned}$$

where the next to last equality uses naturality of D_X for the $p-1$ chain ∂i_p. Naturality of D_X in dimension p follows directly from the definition. □

Note that the naturality of sd^m and D_X shows that if A is a subspace of X, then sd^m and D_X carry $S_p(A)$ into $S_p(A)$ and $S_{p+1}(A)$, respectively. Thus they induce a chain map and a chain homotopy, respectively, on the relative chain complex $\mathcal{S}(X,A)$ as well.

Definition. Let X be a space; let $\mathcal{A}$ be a covering of X. Let $S_p^{\mathcal{A}}(X)$ denote the subgroup of $S_p(X)$ generated by the $\mathcal{A}$-small singular simplices. Let $\mathcal{S}^{\mathcal{A}}(X)$ denote the chain complex whose chain groups are the groups $S_p^{\mathcal{A}}(X)$. It is a subchain complex of $\mathcal{S}(X)$, for if the image set of T lies in the element A of $\mathcal{A}$, so does the image set of each term of ∂T.

Note that each singular 0-chain is automatically $\mathcal{A}$-small; hence $S_0^{\mathcal{A}}(X) = S_0(X)$, and ϵ defines an augmentation for $\mathcal{S}^{\mathcal{A}}(X)$. It follows from the preceding remark that both sd^m and D_X carry $\mathcal{S}^{\mathcal{A}}(X)$ into itself for if the image set of T lies in A, so does the image set of each term in $\mathrm{sd}^m T$ and $D_X T$.

Theorem 31.5. *Let X be a space; let $\mathcal{A}$ be a collection of subsets of X whose interiors cover X. Then the inclusion map $\mathcal{S}^{\mathcal{A}}(X) \to \mathcal{S}(X)$ induces an isomorphism in homology, both ordinary and reduced.*

Proof. The obvious way to proceed is to attempt to define a chain map $\lambda : \mathcal{S}(X) \to \mathcal{S}^{\mathcal{A}}(X)$ that is a chain-homotopy inverse for the inclusion map. This is not as easy as it looks. For any particular singular chain, there is an m such that the map sd^m will work, but as the singular chain changes, one may have to take m larger and larger. We avoid this difficulty by using a different trick (or method, if you prefer).

Consider the short exact sequence of chain complexes

$$0 \to S_p^{\mathcal{A}}(X) \to S_p(X) \to S_p(X)/S_p^{\mathcal{A}}(X) \to 0.$$

It gives rise to a long exact sequence in homology (either ordinary or reduced). To prove our theorem, it will suffice to show that the homology of the chain complex $\{S_p(X)/S_p^{\mathcal{A}}(X), \partial\}$ vanishes in every dimension. This we can do. We need only prove the following:

Suppose c_p is an element of $S_p(X)$ whose boundary belongs to $S_{p-1}^{\mathcal{A}}(X)$.

Then there is an element d_{p+1} of $S_{p+1}(X)$ such that $c_p + \partial d_{p+1}$ belongs to $S_p^{\mathcal{A}}(X)$.

Note that c_p is a *finite* formal linear combination of singular p-simplices. In view of Theorem 31.3, we can choose m so that each singular simplex appearing in the expression for $\mathrm{sd}^m c_p$ is $\mathcal{A}$-small. Once m is chosen, let D_X be the chain homotopy of Lemma 31.4. We show that $c_p + \partial D_X c_p$ belongs to $S_p^{\mathcal{A}}(X)$. Then we are finished.

We know that

$$\partial D_X c_p + D_X \partial c_p = \mathrm{sd}^m c_p - c_p,$$

or

$$c_p + \partial D_X c_p = \mathrm{sd}^m c_p - D_X \partial c_p.$$

The chain $\mathrm{sd}^m c_p$ is in $S_p^{\mathcal{A}}(X)$, by choice of m. And since ∂c_p belongs to $S_{p-1}^{\mathcal{A}}(X)$, the chain $D_X \partial c_p$ belongs to $S_p^{\mathcal{A}}(X)$, as noted earlier. □

Now there does exist a chain map

$$\lambda : \mathcal{S}(X) \to \mathcal{S}^{\mathcal{A}}(X)$$

that is a chain-homotopy inverse for the inclusion map. A specific formula for λ, involving the chain maps sd^m, is given in [V], p. 207. We shall derive the existence of λ shortly from a more general result. (See the exercises of §46.)

Corollary 31.6. *Let X and $\mathcal{A}$ be as in the preceding theorem. If $B \subset X$, let $S_p^{\mathcal{A}}(B)$ be generated by those singular simplices $T : \Delta_p \to B$ whose image sets lie in elements of $\mathcal{A}$. Let $S_p^{\mathcal{A}}(X,B)$ denote $S_p^{\mathcal{A}}(X)/S_p^{\mathcal{A}}(B)$. Then inclusion $S_p^{\mathcal{A}}(X,B) \to S_p(X,B)$ induces a homology isomorphism.*

Proof. The inclusions $\mathcal{S}^{\mathcal{A}}(B) \to \mathcal{S}(B)$ and $\mathcal{S}^{\mathcal{A}}(X) \to \mathcal{S}(X)$ give rise to a homomorphism of the long exact homology sequence derived from

$$0 \to \mathcal{S}(B) \to \mathcal{S}(X) \to \mathcal{S}(X,B) \to 0$$

with the one derived from

$$0 \to \mathcal{S}^{\mathcal{A}}(B) \to \mathcal{S}^{\mathcal{A}}(X) \to \mathcal{S}^{\mathcal{A}}(X,B) \to 0.$$

Since inclusion induces an isomorphism of the absolute homology groups of these respective sequences, the Five-lemma implies that it induces an isomorphism of the relative groups as well. □

Theorem 31.7 (Excision for singular theory). *Let $A \subset X$. If U is a subset of X such that $\bar{U} \subset \operatorname{Int} A$, then inclusion*

$$j : (X - U, A - U) \to (X,A)$$

induces an isomorphism in singular homology.

Proof. Let $\mathcal{A}$ denote the collection $\{X - U, A\}$. Now $X - U$ contains the open set $X - \bar{U}$. Since $\bar{U} \subset \text{Int } A$, the interiors of the sets $X - U$ and A cover X. Consider the homomorphisms

$$\frac{S_p(X - U)}{S_p(A - U)} \to \frac{S_p^{\mathcal{A}}(X)}{S_p^{\mathcal{A}}(A)} \to \frac{S_p(X)}{S_p(A)}$$

induced by inclusion. The second of these homomorphisms induces a homology isomorphism, by the preceding corollary. We show that the first is an isomorphism already on the chain level, and the proof is complete.

Note first that the map

$$\phi : S_p(X - U) \to S_p^{\mathcal{A}}(X)/S_p^{\mathcal{A}}(A)$$

induced by inclusion is surjective. For if c_p is a chain of $S_p^{\mathcal{A}}(X)$, then each term of c_p has image set lying in either $X - U$ or in A. When we form the coset $c_p + S_p^{\mathcal{A}}(A)$ we can discard those terms lying in A. Thus ϕ is surjective. The kernel of ϕ is

$$S_p(X - U) \cap S_p^{\mathcal{A}}(A) = S_p((X - U) \cap A) = S_p(A - U),$$

as desired. □

Note that this theorem is slightly stronger than the excision axiom proper (see §26); in singular theory we do not need to assume U is an *open* subset of X.

As an application, we compute the singular homology of the ball and the sphere.

Theorem 31.8. *Let $n \geq 0$. The group $H_i(B^n, S^{n-1})$ is infinite cyclic for $i = n$ and vanishes otherwise. The group $\tilde{H}_i(S^n)$ is infinite cyclic for $i = n$ and vanishes otherwise. The homomorphism of $\tilde{H}_n(S^n)$ with itself induced by the reflection map*

$$\rho_1(x_1, x_2, \dots, x_{n+1}) = (-x_1, x_2, \dots, x_{n+1})$$

equals multiplication by -1.

Proof. We verify the theorem for $n = 0$. It is trivial that $H_p(B^0, \varnothing)$ is infinite cyclic for $p = 0$ and vanishes otherwise, since B^0 is a single point.

It is similarly easy to see that $H_p(S^0) = 0$ for $p \neq 0$, since S^0 consists of two points a and b. In dimension 0, the singular chain group of S^0 is generated by the constant simplices T_a and T_b. The boundary operator $\partial : S_1(\{a,b\}) \to S_0(\{a,b\})$ is trivial, because any singular 1-simplex on $\{a,b\}$ is constant. It follows that $\tilde{H}_0(S^0)$ is infinite cyclic, and is generated by $T_a - T_b$. As a result, the reflection map, which exchanges a and b, induces the homomorphism of $\tilde{H}_0(S^0)$ that equals multiplication by -1.

Suppose the theorem is true in dimension $n - 1$, where $n \geq 1$. It follows from the long exact homology sequence of (B^n, S^{n-1}) that

$$H_i(B^n, S^{n-1}) \simeq \tilde{H}_{i-1}(S^{n-1}),$$

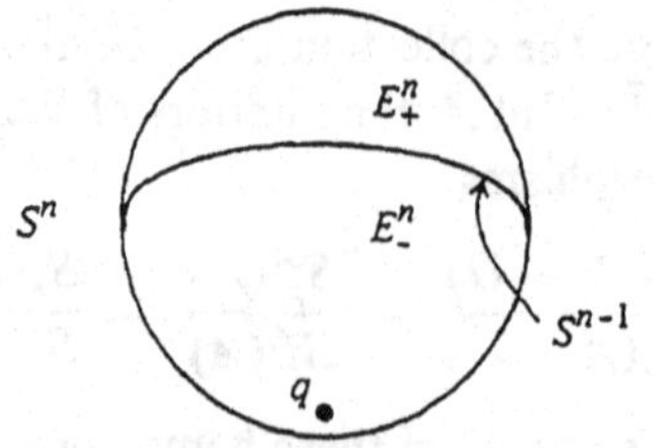

Figure 31.2

since B^n is contractible. Thus $H_i(B^n, S^{n-1})$ is infinite cyclic for $i = n$ and vanishes otherwise.

To compute $\tilde{H}_i(S^n)$, consider the following homomorphisms,

$$\begin{array}{ccc} & H_i(E^n_+,S^{n-1}) & \xrightarrow{\partial_*} \tilde{H}_{i-1}(S^{n-1}) \\ & \downarrow k_* & \\ & H_i(S^n - q, E^n_- - q) & \\ & \downarrow j_* & \\ \tilde{H}_i(S^n) & \xrightarrow{i_*} H_i(S^n,E^n_-) & \end{array}$$

where i, j, k are inclusions, and $q = (0,\ldots,0,-1)$ is the "south pole" of S^n. Recall that E^n_+ and E^n_- are the upper and lower hemispheres of S^n, respectively, and $S^{n-1} = E^n_+ \cap E^n_-$ is the "equator" of S^n. See Figure 31.2. Now both E^n_+ and E^n_- are n-balls; in fact, the projection of $\mathbf{R}^{n+1}$ onto $\mathbf{R}^n \times 0$ carries E^n_+ and E^n_- homeomorphically onto B^n. (It also carries $E^n_- - q$ onto $B^n - \mathbf{0}$.) In particular, E^n_- is contractible, so the long exact homology sequence of (S^n,E^n_-) shows that i_* is an isomorphism. The excision property shows that j_* is an isomorphism, since $q \in \operatorname{Int} E^n_-$. The map k_* is an isomorphism because S^{n-1} is a deformation retract of $E^n_- - q$, and E^n_+ is a deformation retract of $S^n - q$. (Since the pair $(E^n_- - q, S^{n-1})$ is homeomorphic to the pair $(B^n - \mathbf{0}, S^{n-1})$, there is a deformation retraction F_t of $E^n_- - q$ onto S^{n-1}. It extends to a deformation retraction of $S^n - q$ onto E^n_+, by letting F_t equal the identity on E^n_+, for each t.) The fact that ∂_* is an isomorphism follows from the contractibility of E^n_+.

It follows from the induction hypothesis that $\tilde{H}_i(S^n)$ is infinite cyclic for $i = n$ and vanishes otherwise.

Now the reflection map ρ_1 induces a homomorphism of the preceding diagram with itself, since it maps each of p, E^n_-, E^n_+, and S^n into itself. By the induction hypothesis, the homomorphism induced by ρ_1 equals multiplication by -1 on $\tilde{H}_{n-1}(S^{n-1})$; therefore, it equals the same on $\tilde{H}_n(S^n)$. □

Corollary 31.9. *If $a : S^n \to S^n$ is the antipodal map $a(x) = -x$, then a_* equals multiplication by $(-1)^{n+1}$.* □

EXERCISE

1. Express the naturality provisions in Lemmas 31.1 and 31.4 by stating that sd and D_X are natural transformations of certain functors.

*§32. ACYCLIC MODELS†

In the two preceding sections, we constructed certain natural chain homotopies D_X. The method was to define D_X first for a particular singular simplex i_p on a particular space Δ_p. To do this, we needed the acyclicity of another space, either of $\Delta_p \times I$ (in §30) or of Δ_p itself (in §31). This part of the construction involved an arbitrary choice; everything thereafter was forced by naturality. The resemblance to earlier constructions involving acyclic carriers was strong, but nevertheless there were differences.

Now we formalize this method for later use. We state here a theorem, sufficiently strong for our purposes, that we shall call the acyclic model theorem. Its formulation is sufficiently abstract to bother some readers. It may help you keep your feet on the ground (even if your head is in the clouds of abstraction) if you reread the proofs of Lemmas 30.6 and 31.4 before tackling this theorem and its proof.

Throughout this section, *let* **C** *denote an arbitrary category with objects* $X, Y, \ldots$ *and morphisms* $f, g, \ldots$; *and let* **A** *denote the category of augmented chain complexes and chain maps of such*. We will be dealing with functors from **C** to **A**.

For most of the applications we have in mind, **C** will be either the topological category (whose objects are topological spaces and whose morphisms are continuous maps) or the category of pairs of spaces and pairs of continuous maps. So you may think only of those categories if you like.

Definition. Let G be a functor from **C** to **A**; given an object X of **C**, let $G_p(X)$ denote the p-dimensional group of the augmented chain complex $G(X)$. Let $\mathcal{M}$ be a collection of objects of **C** (called **models**, or **model objects**). We say that G is **acyclic** relative to the collection $\mathcal{M}$ if $G(X)$ is acyclic for each $X \in \mathcal{M}$. We say G is **free** relative to the collection $\mathcal{M}$ if for each $p \geq 0$, there are:

(1) An index set J_p.

(2) An indexed family $\{M_\alpha\}_{\alpha \in J_p}$ of objects of $\mathcal{M}$.

(3) An indexed family $\{i_\alpha\}_{\alpha \in J_p}$, where $i_\alpha \in G_p(M_\alpha)$ for each α.

The following condition is to hold: Given X, the elements

$$G(f)(i_\alpha) \in G_p(X)$$

†In this section, we assume familiarity with §28, Categories and Functors. The results of this section will be used when we prove the Eilenberg-Zilber theorem, in §59.

are to be distinct and form a *basis* for $G_p(X)$, as f ranges over all elements of $\hom(M_\alpha, X)$, and α ranges over J_p.

Example 1. Consider the singular chain complex functor, from the topological category to **A**. Let $\mathcal{M}$ be the collection $\{\Delta_p \mid p = 0,1,\ldots\}$. This functor is acyclic relative to $\mathcal{M}$. We show it is free relative to $\mathcal{M}$: for each p, let the index set J_p have only one element; let the corresponding family consist of Δ_p alone; and let the corresponding element of $S_p(\Delta_p)$ be the identity singular simplex i_p. It is immediate that, as T ranges over all continuous maps $\Delta_p \to X$, the elements $T_\#(i_p) = T$ form a basis for $S_p(X)$.

Example 2. Consider the following functor G, defined on the category of topological pairs:

$$(X,Y) \to \mathcal{S}(X \times Y) \qquad \text{and} \qquad (f,g) \to (f \times g)_\#.$$

Let $\mathcal{M} = \{(\Delta_p, \Delta_q) \mid p,q = 0,1,\ldots\}$. Then G is acyclic relative to $\mathcal{M}$, since $\Delta_p \times \Delta_q$ is contractible. We show G is free relative to $\mathcal{M}$: For each index p, let J_p consist of a single element; let the corresponding family consist of (Δ_p, Δ_p) alone; and let the corresponding element of $S_p(\Delta_p \times \Delta_p)$ be the diagonal map $d_p(x) = (x,x)$. As f and g range over all maps from Δ_p into X and Y, respectively, $(f \times g)_\#(d_p)$ ranges over all maps $\Delta_p \to X \times Y$—that is, over a basis for $S_p(X \times Y)$.

Example 3. Let G be the functor

$$X \to \mathcal{S}(X \times I) \qquad \text{and} \qquad f \to (f \times i_I)_\#.$$

Let $\mathcal{M} = \{\Delta_p \mid p = 0,1,\ldots\}$. Then G is acyclic relative to $\mathcal{M}$. It is also true that G is free relative to $\mathcal{M}$, but the proof is not obvious. Let J_p be the set of all continuous functions $\alpha : \Delta_p \to I$. Let the family $\{M_\alpha\}_{\alpha \in J_p}$ be defined by setting $M_\alpha = \Delta_p$ for each α. For each α, let $i_\alpha \in S_p(M_\alpha \times I)$ be the singular simplex

$$i_\alpha : \Delta_p \to \Delta_p \times I$$

defined by

$$i_\alpha(x) = (x, \alpha(x)).$$

As f ranges over all maps of Δ_p into X, and α ranges over the set J_p, the element $(f \times i_I)_\#(i_\alpha)$ ranges over a basis for $S_p(X \times I)$.

Note that if G is free relative to a collection $\mathcal{M}$, then it is automatically free relative to any larger collection, while if it is acyclic relative to $\mathcal{M}$, it is automatically acyclic relative to any smaller collection. Therefore, if we wish G to be both free and acyclic relative to $\mathcal{M}$, we must choose $\mathcal{M}$ to be just the right size, neither too large or too small.

Theorem 32.1 (Acyclic model theorem). *Let G and G' be functors from the category* **C** *to the category* **A** *of augmented chain complexes and chain maps. Let $\mathcal{M}$ be a collection of objects of* **C**.

If G is free relative to $\mathcal{M}$, and G' is acyclic relative to $\mathcal{M}$, then the following hold:

(a) *There is a natural transformation T_X of G to G'.*

(b) *Given two natural transformations T_X, T'_X of G to G', there is a natural chain homotopy D_X between them.*

"Naturality" means the following: For each object X of $\mathbf{C}$,

$$T_X : G_p(X) \to G'_p(X) \qquad \text{and} \qquad D_X : G_p(X) \to G'_{p+1}(X).$$

Naturality of T_X and D_X means that for each $f \in \hom(X,Y)$, we have

$$G'(f) \circ T_X = T_Y \circ G(f),$$
$$G'(f) \circ D_X = D_Y \circ G(f).$$

The proof of Theorem 32.1 is left to the reader!

Actually, this is not as unkind as it might seem. One cannot in fact understand a proof at this level of abstraction simply by reading it. The only way to understand it is to write out the details oneself. If you have labored through the acyclic carrier theorem, and have followed the constructions of D_X in the preceding sections, you should be able to write down the proof of (b). After that, the proof of (a) should not be too difficult.

An immediate corollary is the following.

Theorem 32.2. *Let $\mathbf{C}$ be a category; let G and G' be functors from $\mathbf{C}$ to $\mathbf{A}$. If G and G' are free and acyclic relative to the collection $\mathcal{M}$ of objects of $\mathbf{C}$, then there is a natural transformation $T_X: G(X) \to G'(X)$; any such transformation is a chain equivalence.*

Proof. We apply the preceding theorem four times. Because G is free and G' is acyclic, T_X exists. Because G' is free and G is acyclic, there is a natural transformation S_X of G' to G. Now $S_X \circ T_X$ and the identity transformation are two natural transformations of G to G; because G is free and acyclic, there is a natural chain homotopy of $S_X \circ T_X$ to the identity. Similarly, because G' is free and acyclic, there is a natural chain homotopy of $T_X \circ S_X$ to the identity. □

EXERCISES

1. Prove the acyclic model theorem.
2. Consider the following functors:

$$G : X \to \mathcal{S}(X) \qquad \text{and} \qquad f \to f_\#,$$
$$G' : X \to \mathcal{S}(X \times I) \qquad \text{and} \qquad f \to (f \times i_I)_\#.$$

   (a) Show that the maps T_X, $T'_X: S_p(X) \to S_p(X \times I)$ defined by $T_X(T) = i_\#(T)$ and $T'_X(T) = j_\#(T)$ are natural transformations. Derive Lemma 30.6 as a consequence of the acyclic model theorem.
   (b) Derive Lemma 31.4 from the acyclic model theorem by showing that sd^m and $(i_X)_\#$ are natural transformations of the functor G to itself.

3. Let K and L be simplicial complexes; suppose Φ is an acyclic carrier from K to L.

Consider the category **C** whose objects are subcomplexes of K and whose morphisms are inclusion maps j of subcomplexes of K. If K_0 is a subcomplex of K, let $\Phi(K_0)$ be the union of the subcomplexes $\Phi(\sigma)$ of L, as σ ranges over all simplices of K_0.

(a) If K_0 and K_1 are subcomplexes of K, and $j : K_0 \to K_1$ is inclusion, consider the functors from **C** to **A** given by

$$G : K_0 \to \mathcal{C}(K_0) \quad \text{and} \quad j \to j_\#,$$

$$G' : K_0 \to \mathcal{C}(\Phi(K_0)) \quad \text{and} \quad j \to l_\#,$$

where $l : \Phi(K_0) \to \Phi(K_1)$ is the inclusion map. Show that G and G' are indeed functors.

(b) Derive the acyclic carrier theorem (geometric version) from the acyclic model theorem. [*Hint:* Let $\mathcal{M}$ consist of those subcomplexes of K whose polytopes are simplices of K.]

§33. MAYER-VIETORIS SEQUENCES

If X is the union of two subspaces X_1 and X_2, under suitable hypotheses there is an exact sequence relating the homology of X with that of X_1 and X_2. It is called the Mayer-Vietoris sequence of the pair X_1, X_2. We constructed such a sequence in simplicial theory under the assumption that X_1 and X_2 were polytopes of subcomplexes of a complex. In singular theory, we need an analogous condition:

Definition. Let $X = X_1 \cup X_2$. Let $\mathcal{S}(X_1) + \mathcal{S}(X_2)$ denote the chain complex $\mathcal{S}^{\mathcal{A}}(X)$, where $\mathcal{A} = \{X_1, X_2\}$. Its pth chain group is the sum $S_p(X_1) + S_p(X_2)$, which is not a direct sum unless X_1 and X_2 are disjoint. We say $\{X_1, X_2\}$ is an **excisive couple** if the inclusion

$$\mathcal{S}(X_1) + \mathcal{S}(X_2) \to \mathcal{S}(X)$$

induces an isomorphism in homology.

For singular homology, this definition is equivalent to the one given in the exercises of §26. Proof is left to the reader (see Exercise 2).

In view of Theorem 31.5, one situation in which $\{X_1, X_2\}$ is excisive occurs when the sets Int X_1 and Int X_2 cover X.

Theorem 33.1. *Let $X = X_1 \cup X_2$; suppose $\{X_1, X_2\}$ is an excisive couple. Let $A = X_1 \cap X_2$. Then there is an exact sequence*

$$\cdots \to H_p(A) \xrightarrow{\phi_*} H_p(X_1) \oplus H_p(X_2) \xrightarrow{\psi_*} H_p(X) \to H_{p-1}(A) \to \cdots$$

called the **Mayer-Vietoris sequence** *of* $\{X_1, X_2\}$. *The homomorphisms are defined by*

$$\phi_*(a) = (i_*(a), -j_*(a)),$$
$$\psi_*(x_1,x_2) = k_*(x_1) + l_*(x_2),$$

where the maps

$$\begin{array}{ccc} A & \xrightarrow{i} & X_1 \\ j\downarrow & & \downarrow k \\ X_2 & \xrightarrow{l} & X \end{array}$$

are inclusions. A similar sequence exists in reduced homology if A is nonempty. Both sequences are natural with respect to homomorphisms induced by continuous maps.

Proof. We define a short exact sequence of chain complexes

(*) $$0 \to S_p(A) \xrightarrow{\phi} S_p(X_1) \oplus S_p(X_2) \xrightarrow{\psi} S_p(X_1) + S_p(X_2) \to 0$$

by the equations

$$\phi(c) = (i_\#(c), -j_\#(c)),$$
$$\psi(c_1, c_2) = k_\#(c_1) + l_\#(c_2).$$

The map ϕ is injective, while ψ is surjective and its kernel consists of all chains of the form $(c, -c)$, where $c \in S_p(X_1)$ and $-c \in S_p(X_2)$. Exactness follows. We obtain from the zig-zag lemma a long exact sequence in homology. Since the hypotheses of the theorem guarantee that

(**) $$H_p(\mathcal{S}(X_1) + \mathcal{S}(X_2)) \simeq H_p(X),$$

the proof is complete. Exactness of the Mayer-Vietoris sequence in reduced homology when $A \neq \varnothing$ follows by a similar argument.

Now suppose $f: (X,X_1,X_2) \to (Y,Y_1,Y_2)$ is a continuous map; where $X = X_1 \cup X_2$ and $Y = Y_1 \cup Y_2$ and both $\{X_1,X_2\}$ and $\{Y_1,Y_2\}$ are excisive couples. Since $f_\#$ commutes with inclusions, it commutes with the chain maps ϕ and ψ. Thus $f_\#$ gives a homomorphism of short exact sequences of chain complexes, so that f_* is a homomorphism of the corresponding homology sequences. Finally, we note that the isomorphism (**) commutes with f_*, since it is induced by inclusion. The naturality of the Mayer-Vietoris sequence follows. □

The Mayer-Vietoris sequence has many applications. We give one here, as an illustration. We shall not, however, have occasion to use this result later in the book.

Recall that if K is a complex and if $w_0 * K$ and $w_1 * K$ are cones on K whose polytopes intersect in $|K|$ alone, then their union is a complex denoted $S(K)$ and called a *suspension* of K. (See §25.) It is easy to show that the map

$$\pi : |K| \times [-1,1] \to |S(K)|$$

defined by

$$\pi(x,t) = \begin{cases} (1-t)x + tw_0 & \text{if } t \geq 0, \\ (1+t)x - tw_1 & \text{if } t \leq 0, \end{cases}$$

is a quotient map that collapses $|K| \times 1$ to the point w_0, and $|K| \times (-1)$ to w_1, and is otherwise one-to-one. The proof is similar to the corresponding result for cones (see Corollary 20.6). This fact motivates the definition of suspension for an arbitrary topological space.

Definition. Let X be a space. We define the **suspension** of X to be the quotient space of $X \times [-1,1]$ obtained by identifying the subset $X \times 1$ to a point, and the subset $X \times (-1)$ to a point. It is denoted $S(X)$.

Just as in the case of simplicial theory, one can compute the homology of a suspension by using a Mayer-Vietoris sequence. One has the following theorem:

Theorem 33.2. *Let X be a space. There is for all p an isomorphism*

$$\tilde{H}_p(S(X)) \to \tilde{H}_{p-1}(X).$$

Proof. Let $\pi : X \times [-1,1] \to S(X)$ be the quotient map. Let $v = \pi(X \times 1)$ and $w = \pi(X \times (-1))$; these points are called the "suspension points." Let $X_1 = S(X) - w$ and $X_2 = S(X) - v$; since both X_1 and X_2 are open in $S(X)$, the pair $\{X_1, X_2\}$ is excisive. We show that X_1 and X_2 are acyclic. The Mayer-Vietoris sequence then implies that there is an isomorphism

$$\tilde{H}_p(S(X)) \to \tilde{H}_{p-1}(X_1 \cap X_2).$$

We also show there is a homotopy equivalence of $X_1 \cap X_2$ with X, so there is an isomorphism

$$\tilde{H}_{p-1}(X_1 \cap X_2) \to \tilde{H}_{p-1}(X).$$

The theorem follows.

Now $X \times (-1,1]$ is open in $X \times [-1,1]$ and is saturated with respect to π. Therefore, the restricted map

$$\pi' : X \times (-1,1] \to X_1$$

is a quotient map. Now $X \times 1$ is a deformation retract of $X \times (-1,1]$; the map

$$F : X \times (-1,1] \times I \to X \times (-1,1]$$

defined by

$$F(x,s,t) = (x, (1-t)s + t)$$

is the desired deformation retraction. Since the map $\pi' \circ F$ in the following diagram

$$\begin{array}{ccc} X \times (-1,1] \times I & \xrightarrow{F} & X \times (-1,1] \\ \pi' \times i_I \downarrow & & \downarrow \pi' \\ X_1 \times I & \xrightarrow{G} & X_1 \end{array}$$

is constant on $X \times 1 \times I$, it induces a continuous map G that is a deformation retraction of X_1 to the point v. (Here we use the fact that $\pi' \times i_I$ is a quotient map, which follows from Theorem 20.1.) Thus X_1 is acyclic.

A similar proof shows that X_2 is acyclic. Finally, we note that the restricted map

$$\pi'' : X \times (-1,1) \to X_1 \cap X_2$$

is a one-to-one quotient map, and hence a homeomorphism. Since $X \times 0$ is a deformation retract of $X \times (-1,1)$, there is a homotopy equivalence of X with $X_1 \cap X_2$. $\square$

EXERCISES

1. Consider the **closed topologist's sine curve** X, pictured in Figure 33.1. It is the union of the topologist's sine curve Y (defined in the exercises of §29) and an arc that intersects Y only in the points $(0,-1)$ and $(1,\sin 1)$. Compute the singular homology of X, using a suitable Mayer-Vietoris sequence.

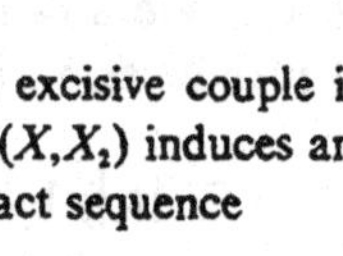

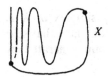

Figure 33.1

2. Show that $\{X_1, X_2\}$ is an excisive couple in singular theory if and only if inclusion $(X_1, X_1 \cap X_2) \to (X, X_2)$ induces an isomorphism in singular homology. [*Hint:* Construct an exact sequence

$$0 \to \mathcal{S}(X_2) \to \mathcal{S}(X_1) + \mathcal{S}(X_2) \to \mathcal{S}(X_1)/\mathcal{S}(X_1 \cap X_2) \to 0.$$

Compare its derived homology sequence with the homology sequence of $\mathcal{S}(X,X_2)$.]

3. Show that $S^n \approx S(S^{n-1})$ for $n \geq 1$. Use this fact to compute the singular homology of S^n.

4. State and prove a relative version of the Mayer-Vietoris sequence in singular homology, where the middle group is $H_p(X_1,B_1) \oplus H_p(X_2,B_2)$, assuming $\{X_1,X_2\}$ and $\{B_1,B_2\}$ are excisive. [*Hint:* Include the sequence

$$0 \to \mathcal{S}(B_1 \cap B_2) \to \mathcal{S}(B_1) \oplus \mathcal{S}(B_2) \to \mathcal{S}(B_1) + \mathcal{S}(B_2) \to 0$$

into the sequence (*) of the proof of Theorem 33.1 and use the serpent lemma.]

5. Let $X = X_1 \cup X_2$ and $A = X_1 \cap X_2$. Suppose X_1 and X_2 are closed in X, and A is a deformation retract of an open set U of X_2. Show that X_1 is a deformation retract of $U \cup X_1$; conclude that $\{X_1,X_2\}$ is an excisive couple.

§34. THE ISOMORPHISM BETWEEN SIMPLICIAL AND SINGULAR HOMOLOGY

In this section, we show that if K is a simplicial complex, the simplicial homology groups of K are isomorphic with the singular homology groups of $|K|$. In fact, we show that the isomorphism commutes with induced homomorphisms and with the boundary homomorphism ∂_*, so it is an isomorphism between the two homology *theories*.

The proof involves a notion we introduced in §13, that of the ordered chain complex $\{C'_p(K),\partial'\}$ of a simplicial complex K. We proved there that the homology groups of this chain complex, called the ordered homology groups of K, are isomorphic to the usual (oriented) homology groups of K. In this section, we show that the ordered homology groups of K are in turn isomorphic with the singular homology groups of $|K|$, thus showing that simplicial and singular homology agree for polyhedra.

Recall that an ordered p-simplex of K is a $p+1$ tuple $(v_0,\dots,v_p)$ of vertices of K (not necessarily distinct) that span a simplex of K, and the group $C'_p(K)$ is the free abelian group generated by the ordered p-simplices. Also recall that

$$\partial'(v_0,\dots,v_p) = \Sigma\,(-1)^i(v_0,\dots,\hat{v}_i,\dots,v_p)$$

and that $\epsilon'(v) = 1$ for each vertex v. We now construct a chain map carrying $C'_p(K)$ to $S_p(|K|)$ and prove that it induces an isomorphism in homology.

Definition. Define $\theta : C'_p(K) \to S_p(|K|)$ by the equation

$$\theta((v_0,\dots,v_p)) = l(v_0,\dots,v_p).$$

Then θ assigns, to the ordered simplex $(v_0,\dots,v_p)$ of K, the linear singular simplex mapping Δ_p into $|K|$ and carrying ϵ_i to v_i for $i = 0,\dots,p$.

It is immediate from the definitions that θ is a chain map and that it preserves augmentation. If K_0 is a subcomplex of K, then θ commutes with inclusion, so it maps $C'_p(K_0)$ into $S_p(|K_0|)$, and thus induces a chain map

$$\theta : C'_p(K,K_0) \to S_p(|K|,|K_0|).$$

To show θ_* is an isomorphism, the following lemma will be useful:

Lemma 34.1. *Let $\psi : \mathcal{C} \to \mathcal{C}'$ be a chain map of augmented chain complexes. Then ψ_* is an isomorphism in reduced homology if and only if it is an isomorphism in ordinary homology.*

Proof. We recall the proof that $H_0(\mathcal{C}) \simeq \tilde{H}_0(\mathcal{C}) \oplus \mathbf{Z}$. (See the exercises of §7.) Begin with the exact sequences

$$\begin{array}{ccccccccc} 0 & \longrightarrow & \ker \epsilon & \longrightarrow & C_0 & \xrightarrow{\epsilon} & \mathbf{Z} & \longrightarrow & 0 \\ & & \downarrow \psi & & \downarrow \psi & & \downarrow = & & \\ 0 & \longrightarrow & \ker \epsilon' & \longrightarrow & C'_0 & \xrightarrow{\epsilon'} & \mathbf{Z} & \longrightarrow & 0. \end{array}$$

Choose $j : \mathbf{Z} \to C_0$ so that $j \circ \epsilon$ is the identity. Define $j' : \mathbf{Z} \to C'_0$ by setting $j' = \psi \circ j$. Then j and j' split the two sequences, so

$$C_0 = \ker \epsilon \oplus \operatorname{im} j \qquad \text{and} \qquad C'_0 = \ker \epsilon' \oplus \operatorname{im} j'.$$

Note that ψ defines an isomorphism of $\operatorname{im} j$ with $\operatorname{im} j'$. Now

$$H_0(\mathcal{C}) \simeq \frac{\ker \epsilon}{\partial C_1} \oplus \operatorname{im} j \qquad \text{and} \qquad H_0(\mathcal{C}') \simeq \frac{\ker \epsilon'}{\partial' C'_1} \oplus \operatorname{im} j'.$$

It follows that $\psi_* : H_0(\mathcal{C}) \to H_0(\mathcal{C}')$ is an isomorphism if and only if ψ induces an isomorphism of $\ker \epsilon/\partial C_1$ with $\ker \epsilon'/\partial' C'_1$. □

Lemma 34.2. *Let K_0 be a subcomplex of K. For all p, the chain map θ induces isomorphisms*

(1) $\theta_* : \tilde{H}_p(\mathcal{C}'(K)) \to \tilde{H}_p(|K|)$,

(2) $\theta_* : H_p(\mathcal{C}'(K)) \to H_p(|K|)$,

(3) $\theta_* : H_p(\mathcal{C}'(K,K_0)) \to H_p(|K|,|K_0|)$.

Proof. We assume $K_0 \neq K$, since otherwise (3) is trivial.

Step 1. We prove the theorem first when K is finite, by induction on the number of simplices in K. If $n = 1$, then K consists of a single vertex v. In each dimension p, there is exactly one ordered p-simplex $(v, \ldots, v)$ of K, and exactly one singular p-simplex $T : \Delta_p \to v$ of $|K|$. Furthermore, $\theta(v, \ldots, v) = l(v, \ldots, v) = T$. Then $\theta : \mathcal{C}'(K) \to \mathcal{S}(|K|)$ is an isomorphism already on the chain level. Hence (1), (2), and (3) hold.

Now suppose the lemma holds for any complex having fewer than n simplices. Let K have n simplices. We note that it suffices to prove (1) for K. For then (2) follows by Lemma 34.1. And since $\theta_* : H_p(\mathcal{C}'(K_0)) \to H_p(|K_0|)$ is an isomorphism by the induction hypothesis, (3) follows from the long exact homology sequences and the Five-lemma.

To prove (1), let σ be a simplex of K of maximal dimension. Then σ is not a face of any other simplex, so that the collection of all simplices of K different from σ is a subcomplex K_1 of K having $n-1$ simplices. Let Σ denote the complex consisting of σ and its faces; let Bd Σ be the collection of proper faces of σ. Consider the following commutative diagram, where i and j are inclusions:

$$\begin{array}{ccc}
\tilde{H}_p(\mathcal{C}'(K)) & \xrightarrow{\theta_*} & \tilde{H}_p(|K|) \\
\downarrow i_* & & \downarrow i_* \\
H_p(\mathcal{C}'(K,\Sigma)) & \xrightarrow{\theta_*} & H_p(|K|,\sigma) \\
\uparrow j_* & & \uparrow j_* \\
H_p(\mathcal{C}'(K_1,\text{Bd}\,\Sigma)) & \xrightarrow{\theta_*} & H_p(|K_1|,\text{Bd}\,\sigma).
\end{array}$$

(If $\dim \sigma = 0$, then Bd Σ and Bd σ are empty.) It follows from the induction hypothesis that the map θ_* on the bottom line of the diagram is an isomorphism. We shall show that the vertical maps are isomorphisms; it then follows that the map θ_* on the top line is an isomorphism and the proof is complete.

First we consider the homomorphisms i_*. Because σ is acyclic in singular homology, the long exact homology sequence of $(|K|,\sigma)$ shows that the right-hand map i_* is an isomorphism. The same argument applies to the map i_* in ordered simplicial homology, since Σ is acyclic in ordered homology, by Lemma 13.5.

The map j_* is an isomorphism in ordered simplicial homology, because $j_\#$ is an isomorphism already on the chain level; the inclusion map

$$\mathcal{C}'(K_1) \to \mathcal{C}'(K) \to \mathcal{C}'(K)/\mathcal{C}'(\Sigma)$$

is surjective and carries to zero precisely those chains carried by $K_1 \cap \Sigma =$ Bd Σ. (In fact, j is just an excision map.)

It is tempting to assert that j is also an excision map for singular homology, so j_* is an isomorphism. But this is not true. The domain of j is formed by "excising away" from $|K|$ and σ the set $U = \text{Int}\,\sigma$. Since $\bar{U} = \sigma$, Theorem 31.7 does not apply; we are excising away something too large. We can however excise something smaller—namely, the barycenter $\hat{\sigma}$ of σ. This we now do.

Consider the diagram

$$\begin{array}{ccc}
H_p(|K|,\sigma) & & \\
\uparrow j_* & \overset{l_*}{\nwarrow} & \\
 & & H_p(|K|-\hat{\sigma},\sigma-\hat{\sigma}) \\
 & \underset{k_*}{\nearrow} & \\
H_p(|K_1|,\text{Bd}\,\sigma) & &
\end{array}$$

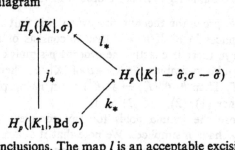

where k and l are inclusions. The map l is an acceptable excision map, since $\hat{\sigma}$ is closed and is contained in Int σ. (Note that Int σ is open in $|K|$, because $|K_1|$ is closed. Therefore, the interior of σ in $|K|$, in the sense of point-set topology, equals the "open simplex" Int σ.) Therefore, l_* is an isomorphism. Why is k_* an isomorphism? If $\dim \sigma = 0$, then $\sigma - \hat{\sigma}$ and Bd σ are empty, and $|K| - \sigma = |K_1|$; the map k is the identity in this case. If $\dim \sigma > 0$, then we use the fact that Bd σ is a deformation retract of $\sigma - \hat{\sigma}$. This deformation retraction extends to a deformation retraction F_t of $|K| - \hat{\sigma}$ onto $|K_1|$, by letting F_t equal the identity on $|K_1|$. See Figure 34.1. We conclude from Theorem 30.8 that k_* is an isomorphism. Then j_* is an isomorphism in singular homology, as desired.

Step 2. Having proved the theorem when K is finite, we prove it in general. As before, it suffices to prove (1) for all K. For then (2) follows from the preceding lemma; and (3) follows from (2) and the Five-lemma.

First, we show θ_* is surjective. Given $\{z\} \in \tilde{H}_p(|K|)$, there is a compact subset A of $|K|$ such that the chain z is carried by A. Let L be a finite subcomplex of K such that A is contained in $|L|$. Consider the commutative diagram

$$\begin{array}{ccc} \tilde{H}_p(\mathcal{C}'(L)) & \xrightarrow[\simeq]{\theta_*} & \tilde{H}_p(|L|) \\ {\scriptstyle i_*}\downarrow & & \downarrow{\scriptstyle j_*} \\ \tilde{H}_p(\mathcal{C}'(K)) & \xrightarrow{\theta_*} & \tilde{H}_p(|K|) \end{array}$$

where the vertical maps are induced by inclusion. Then $\{z\}$ lies in the image of j_*. The map θ_* on the top line is an isomorphism by Step 2. Hence $\{z\}$ lies in the image of the map θ_* on the bottom line.

We show θ_* has trivial kernel. Suppose $\{z\} \in \tilde{H}_p(\mathcal{C}'(K))$ and $\theta_*(\{z\}) = 0$. Then $\theta(z) = \partial d$ for some singular $p + 1$ chain d of $|K|$. The chain d is carried by a compact subset of $|K|$; choose a finite subcomplex L of K such that z is carried by L and d is carried by $|L|$. Consider the same commutative diagram as before. Let α denote the homology class of z in $\tilde{H}_p(\mathcal{C}'(L))$. Because $\theta(z) = \partial d$, where d is carried by $|L|$, $\theta_* : H_p(\mathcal{C}'(L)) \to H_p(|L|)$ carries α to zero. Because this map is an isomorphism, $\alpha = 0$. Hence $\{z\} = i_*(\alpha) = 0$ as well. □

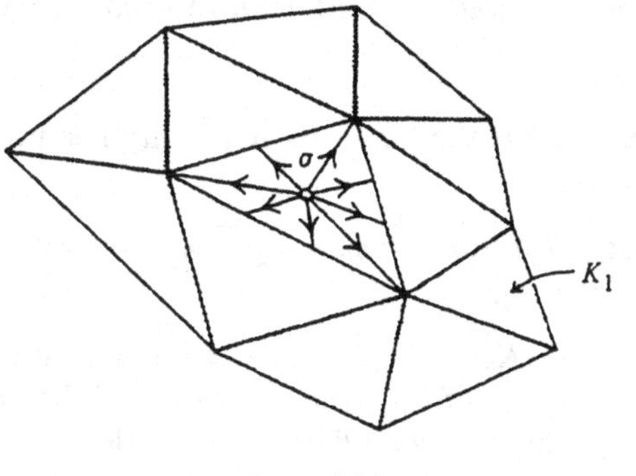

Figure 34.1

The argument given in Step 2 is of the general kind known as a "direct limit argument." It is a standard way of carrying over results that hold for the homology of finite complexes, to complexes in general.

We now combine the results of §13 on ordered homology with the lemma just proved.

Definition. Let K be a complex. We define

$$\eta : \mathcal{C}(K) \to \mathcal{S}(|K|)$$

as follows: Choose a partial ordering of the vertices of K that induces a linear ordering on the vertices of each simplex of K. Orient the simplices of K by using this ordering, and define

$$\eta([v_0,\ldots,v_p]) = l(v_0,\ldots,v_p),$$

where $v_0 < \cdots < v_p$ in the given ordering. It is immediate that η is a chain map, that it preserves augmentation, and that it commutes with inclusions. Thus it induces a chain map on the relative level as well.

In fact, η is just the composite

$$\mathcal{C}(K) \xrightarrow{\phi} \mathcal{C}'(K) \xrightarrow{\theta} \mathcal{S}(|K|),$$

where θ is the chain map of the preceding lemma, and ϕ is the chain equivalence of Theorem 13.6. Since both ϕ and θ induce homology isomorphisms, so does η. This result holds for relative homology and reduced homology as well. Although η depends on the chosen ordering of vertices of K, the homomorphism η_* does not. For $\eta_* = \theta_* \circ \phi_*$, where θ obviously does not depend on the ordering, and ϕ_* does not depend on the ordering, by Theorem 13.6.

Finally, we note that since the chain map η commutes with inclusions, the naturality of the zig-zag lemma implies that η_* commutes with ∂_*.

We summarize these facts as follows.

Theorem 34.3. *The map η_* is a well-defined isomorphism of simplicial with singular homology that commutes with the boundary homomorphism ∂_*.* □

Furthermore, one has the following naturality result.

Theorem 34.4. *The isomorphism η_* commutes with homomorphisms induced by simplicial maps.*

Proof. Let $f : (K,K_0) \to (L,L_0)$ be a simplicial map. We have already proven that f_* commutes with ϕ_*. (See Theorem 13.7.) We show it commutes with θ_*. In fact, $f_\#$ commutes with θ on the chain level.

Recall that in ordered homology, the chain map induced by f is defined by the equation

$$f'_{\#}((w_0,\ldots,w_p)) = (f(w_0),\ldots,f(w_p)),$$

where $(w_0,\ldots,w_p)$ is an ordered p-simplex of K. We compute directly

$$f_{\#}\theta((w_0,\ldots,w_p)) = f \circ l(w_0,\ldots,w_p).$$

This map equals the linear map of Δ_p into $|K|$ that carries ϵ_i to w_i for each i, followed by the map f, which carries the simplex spanned by $w_0,\ldots,w_p$ onto a simplex of L. Since the composite of linear maps is linear, this map equals

$$\begin{aligned} l(f(w_0),\ldots,f(w_p)) &= \theta((f(w_0),\ldots,f(w_p))) \\ &= \theta f'_{\#}((w_0,\ldots,w_p)). \quad \square \end{aligned}$$

It is *possible* to define the homomorphism of simplicial homology induced by a continuous map h as the composite $\eta_*^{-1} \circ h_* \circ \eta_*$, where h_* denotes the induced homomorphism in singular homology. This would give us the same homomorphism of simplicial homology as we defined in Chapter 2 by use of simplicial approximations. We prove this result as follows:

Theorem 34.5. *The isomorphism η_* commutes with homomorphisms induced by continuous maps.*

Proof. Let

$$h : (|K|,|K_0|) \to (|L|,|L_0|)$$

be a continuous map. Let

$$f : (K',K'_0) \to (L,L_0) \qquad \text{and} \qquad g : (K',K'_0) \to (K,K_0)$$

be simplicial approximations to h and to the identity, respectively. In simplicial homology, we have

$$h_* = f_* \circ (g_*)^{-1},$$

by definition. Now in singular homology, the map g_* equals the identity isomorphism, because g is homotopic to the identity map (Theorem 19.4). Similarly, $f_* = h_*$ in singular homology, because f is homotopic to h. Thus in singular homology, we also have the equation $h_* = f_* \circ (g_*)^{-1}$.

Our theorem now follows by applying the preceding theorem to the maps f and g. $\square$

Although we have proven that the chain map η induces an isomorphism in homology, we have not found a chain-homotopy inverse λ for η. The fact that such a λ exists is a consequence of results we shall prove later. (See the exercises of §46.) A specific formula for λ can be derived using the theory of "regular neighborhoods." (See [E-S].)

EXERCISE

1. Show that if K_1 and K_2 are subcomplexes of a complex K, then $\{|K_1|, |K_2|\}$ is an excisive couple in singular homology.

*§35. APPLICATION: LOCAL HOMOLOGY GROUPS AND MANIFOLDS[†]

In this section, we define the local homology groups of a space X at a point x of X, and we use these groups to prove several non-trivial facts about manifolds. *Throughout this section, let X denote a Hausdorff space.*

Definition. If X is a space and if $x \in X$, then the **local homology groups** of X at x are the singular homology groups

$$H_p(X, X - x).$$

The reason for the term "local" comes from the following lemma.

Lemma 35.1. *Let $A \subset X$. If A contains a neighborhood of the point x, then*

$$H_p(X, X - x) \simeq H_p(A, A - x).$$

Therefore, if $x \in X$ and $y \in Y$ have neighborhoods U, V, respectively, such that $(U,x) \approx (V,y)$, then the local homology groups of X at x and of Y at y are isomorphic.

Proof. Let U denote the set $X - A$. Because A contains a neighborhood of x,

$$\bar{U} \subset X - x = \text{Int}(X - x).$$

It follows from the excision property that

$$H_p(X, X - x) \simeq H_p(X - U, X - x - U) = H_p(A, A - x). \quad \square$$

Let us compute some local homology groups.

Example 1. If $x \in \mathbf{R}^m$, we show that $H_i(\mathbf{R}^m, \mathbf{R}^m - x)$ is infinite cyclic for $i = m$ and vanishes otherwise.

Let B denote a ball centered at x. By the preceding lemma,

$$H_i(\mathbf{R}^m, \mathbf{R}^m - x) \simeq H_i(B, B - x) \simeq H_i(B^m, B^m - \mathbf{0}).$$

[†]This section will be assumed in Chapter 8. It is also used in treating one of the examples in §38.

Now S^{m-1} is a deformation retract of $B^m - \mathbf{0}$. See Figure 35.1; the formula for the deformation is given in the proof of Theorem 19.6. Therefore,

$$H_i(B^m, B^m - \mathbf{0}) \simeq H_i(B^m, S^{m-1}).$$

This group is infinite cyclic for $i = m$ and vanishes otherwise.

Example 2. Let $\mathbf{H}^m$ denote **euclidean half-space**

$$\mathbf{H}^m = \{(x_1, \ldots, x_m) \mid x_m \geq 0\}.$$

Let Bd $\mathbf{H}^m$ denote the set $\mathbf{R}^{m-1} \times 0$. If $x \in$ Bd $\mathbf{H}^m$, then the group $H_i(\mathbf{H}^m, \mathbf{H}^m - x)$ vanishes for all i. If $x \in \mathbf{H}^m$ and $x \notin$ Bd $\mathbf{H}^m$, then this group is infinite cyclic for $i = m$ and vanishes otherwise. We prove these facts as follows.

If $x \notin$ Bd $\mathbf{H}^m$, this result follows from the preceding example, once we note that x has a neighborhood that is an open set of $\mathbf{R}^m$. So suppose $x \in$ Bd $\mathbf{H}^m$; we can assume without loss of generality that $x = \mathbf{0}$. Let B^m be the unit ball in $\mathbf{R}^m$, centered at $\mathbf{0}$. The set $B^m \cap \mathbf{H}^m$ contains a neighborhood of $\mathbf{0}$ in $\mathbf{H}^m$. Letting D^m denote the half-ball $B^m \cap \mathbf{H}^m$, we have

$$H_i(\mathbf{H}^m, \mathbf{H}^m - \mathbf{0}) \simeq H_i(D^m, D^m - \mathbf{0}).$$

See Figure 35.2. Now there is a deformation retraction of $B^m - \mathbf{0}$ onto S^{m-1}. It restricts to a deformation retraction of the punctured half-ball $D^m - \mathbf{0}$ onto the set

$$S^{m-1} \cap \mathbf{H}^m = E_+^{m-1}.$$

Therefore,

$$H_i(D^m, D^m - \mathbf{0}) \simeq H_i(D^m, E_+^{m-1}).$$

Now D^m is acyclic, being a convex set in $\mathbf{R}^m$; and E_+^{m-1} is acyclic, being homeomorphic to B^{m-1}. The long exact homology sequence shows that their relative homology vanishes.

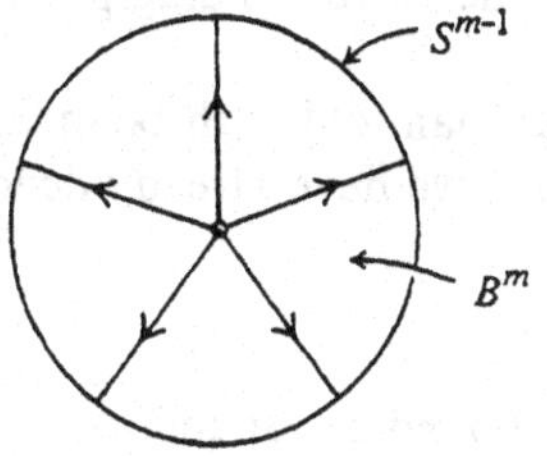

Figure 35.1

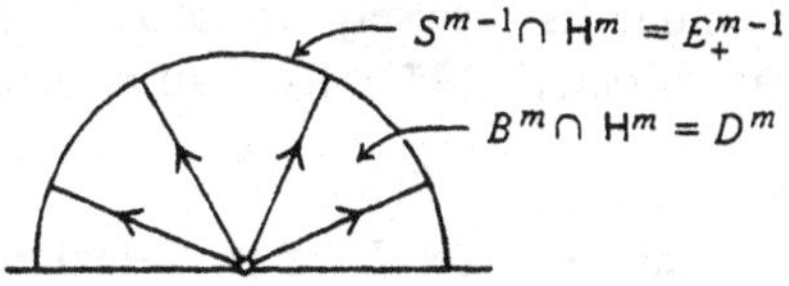

Figure 35.2

Definition. A nonempty Hausdorff space X is called an ***m*-manifold** if each point of X has a neighborhood homeomorphic with an open subset of euclidean space $\mathbf{R}^m$. It is called an ***m*-manifold with boundary** if each point has a neighborhood homeomorphic with an open set of euclidean half-space $\mathbf{H}^m$.

Note that an m-manifold is automatically an m-manifold with boundary. For if x has a neighborhood homeomorphic with an open set in $\mathbf{R}^m$, it has a neighborhood U homeomorphic with an open ball in $\mathbf{R}^m$. Then U is homeomorphic with the open unit ball in $\mathbf{R}^m$ centered at $(0, \ldots, 0,1)$, which is an open set of $\mathbf{H}^m$.

One often includes in the definition the requirement that X have a countable basis, or at least that X be metrizable. We shall not make either assumption.

Manifolds and manifolds with boundary are among the most familiar and important of geometric objects; they are the main objects of study in differential geometry and differential topology.

If X is a manifold with boundary and if $h : U \to V$ is a homeomorphism of an open set U in X onto an open set V in $\mathbf{H}^m$, then h is called a **coordinate patch** on X. A point x of X may be mapped by h either into the open upper half space $\mathbf{R}^{m-1} \times \mathbf{R}_+$ of $\mathbf{H}^m$, or onto the "edge" $\mathbf{R}^{m-1} \times 0$. The local homology groups distinguish between these two possibilities. For it follows from Example 2 that if x is mapped into the open upper half space of $\mathbf{H}^m$, then $H_m(X,X - x)$ is infinite cyclic, while if x is mapped into Bd $\mathbf{H}^m$, then $H_m(X,X - x)$ vanishes. This fact leads to the following definition:

Definition. Let X be an m-manifold with boundary. If the point x of X maps to a point of Bd $\mathbf{H}^m$ under one coordinate patch about x, it maps to a point of Bd $\mathbf{H}^m$ under every such coordinate patch. Such a point is called a **boundary point** of X. The set of all such points x is called the **boundary** of X, and is denoted Bd X. The space $X -$ Bd X is called the **interior** of X and denoted Int X.

Note that there is nothing in the definition requiring that X have any boundary points. If it does not, then Bd X is empty, and X is an m-manifold. While Bd X may be empty, the set Int X cannot be. For if $h : U \to V$ is a coordinate patch about a point x of X, then V is open in $\mathbf{H}^m$ and hence contains at least one point of the open upper half space. The corresponding point y of X lies in Int X by definition.

We remark that the space $\mathbf{H}^m$ is itself an m-manifold with boundary, and its boundary is precisely the set $\mathbf{R}^{m-1} \times 0$, which we have already denoted by Bd $\mathbf{H}^m$. Similarly, $\mathbf{R}^m$ is itself an m-manifold.

Definition. Let X be an m-manifold with boundary. It follows from Example 2 that m is uniquely determined by X, for it is the unique integer such that the group $H_m(X,X - x)$ is non-trivial for at least one x in X. The number m is called (obviously) the **dimension** of the manifold with boundary X.

Example 3. The unit ball B^n in $\mathbf{R}^n$ is an n-manifold with boundary, and Bd $B^n = S^{n-1}$. We prove this fact as follows.

If $p \in B^n - S^{n-1}$, then the set of all x with $\|x\| < 1$ is an open set of $\mathbf{R}^n$; thus there is a coordinate patch about p. Now let $p \in S^{n-1}$; we find a coordinate patch about p. Some coordinate of p is non-zero; suppose for convenience that $p_n < 0$.

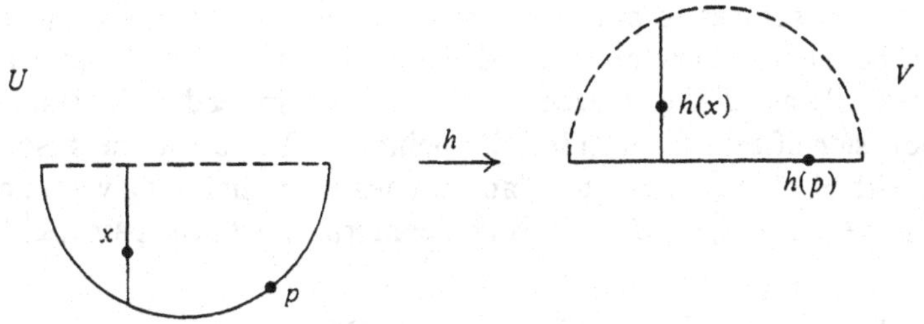

Figure 35.3

Let U be the open set in B^n consisting of all points x of B^n with $x_n < 0$. Define $h : U \to \mathbf{H}^n$ by the equation

$$h(x) = (x_1, \ldots, x_{n-1}, x_n + f(x)),$$

where $f(x) = [1 - x_1^2 - \cdots - x_{n-1}^2]^{1/2}$. Then you can check that h is a homeomorphism of U onto the open set V of $\mathbf{H}^m$ consisting of all points y with $\|y\| < 1$ and $y_n \geq 0$; and it carries p to a point of Bd $\mathbf{H}^n$. See Figure 35.3.

Example 4. Let σ be an n-simplex. Then σ is an n-manifold with boundary, because there is a homeomorphism of σ with B^n. This homeomorphism carries the union Y of the proper faces of σ onto S^{n-1}. Thus the set Y, which we have been denoting by Bd σ, is just the boundary of σ when it is considered as an n-manifold with boundary! And the set $\sigma - Y$, which we have been denoting by Int σ, is just the interior of σ as a manifold with boundary.

There is a certain overlapping of terminology here, which we should clarify. In general topology, if A is a subset of a space X, then the *interior* of A, denoted Int A, is the union of all open sets of X contained in A. And the *boundary* of A, denoted Bd A, is the set $\bar{A} \cap \overline{X - A}$. In the special case where A is an open set of X, it turns out that Bd $A = \bar{A} - A$. We have used this terminology from general topology earlier in this book. For instance, the notion Bd U appeared in Lemma 1.1; and the notion Int A was used in formulating the excision axiom.

The concepts of boundary and interior for a manifold with boundary are entirely different; it is unfortunate that the same terminology is commonly used in two different ways. Some authors use ∂X to denote the boundary of a manifold with boundary. But that can lead to difficulty when one wishes to distinguish the boundary of the space σ from the simplicial chain $\partial\sigma$! We will simply endure the ambiguity, relying on the context to make the meaning clear.

It happens that for the subset B^n of the topological space $\mathbf{R}^n$, its boundary in the sense of general topology is the same as its boundary in the sense of manifolds with boundary. The same remark applies to the subspace $\mathbf{H}^n$ of $\mathbf{R}^n$. But these cases are the exception rather than the rule.

Now we prove a result about triangulations of manifolds.

Lemma 35.2. *Let s be a simplex of the complex K. If x and y are points of* Int s, *then the local homology groups of $|K|$ at x and at y are isomorphic.*

Proof. It suffices to prove the theorem when $x = \hat{s}$, the barycenter of s. Let sd K be the first barycentric subdivision of K. Let K' be a subdivision of K defined exactly as sd K was, except that y is used instead of $\hat{s}$ when "starring" the subdivision of Bd s from an interior point of s. There is a linear isomorphism of sd K with K' that carries $\hat{s}$ to y, and carries each remaining vertex of sd K to itself. Then the pair $(|K|, |K| - \hat{s})$ is homeomorphic with the pair $(|K|, |K| - y)$. □

Recall that a homeomorphism $h : |K| \to M$ is called a *triangulation* of M. It is not known whether an arbitrary manifold with boundary has a triangulation.

Theorem 35.3. *Let M be an m-manifold with boundary; suppose K is a complex and $h : |K| \to M$ is a homeomorphism. Then $h^{-1}(\text{Bd } M)$ is the polytope of a subcomplex of K.*

Proof. If an open simplex Int s of K intersects the set $h^{-1}(\text{Bd } M)$, it lies in this set, by the preceding lemma; since this set is closed, it must contain s. □

Now we give a final application of local homology groups. We show that the dimension of a finite-dimensional simplicial complex K is a topological invariant of $|K|$. (A different proof, using the notion of "covering dimension," was outlined in the exercises of §16 and §19.)

Recall that if v is a vertex of K, then St v is the union of the interiors of all simplices that have v as a vertex, and Lk $v = \overline{\text{St}}\, v - \text{St}\, v$.

Lemma 35.4. *Let v be a vertex of the simplicial complex K. Then*

$$H_i(|K|, |K| - v) \simeq H_i(\overline{\text{St}}\, v, \text{Lk}\, v).$$

Proof. The set $\overline{\text{St}}\, v$ contains a neighborhood of v; therefore, it follows from Lemma 35.1 that

$$H_i(|K|, |K| - v) \simeq H_i(\overline{\text{St}}\, v, \overline{\text{St}}\, v - v).$$

Let L denote the subcomplex of K whose polytope is Lk v. Then $\overline{\text{St}}\, v$ is the polytope of the cone $v * L$, by definition. See Figure 35.4. The following lemma implies that Lk v is a deformation retract of $\overline{\text{St}}\, v - v$; then our proof is complete. □

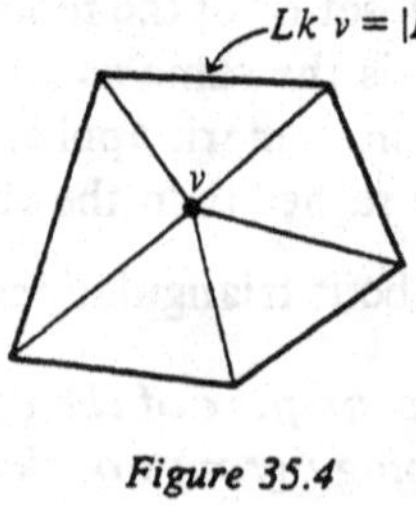

Figure 35.4

Lemma 35.5. *Let $v * L$ be a cone over L. Then $|L|$ is a deformation retract of $|v * L| - v$.*

Proof. Consider the quotient map

$$\pi : |L| \times I \rightarrow |v * L|$$

defined by $\pi(x,t) = (1-t)x + tv$. (See Corollary 20.6.) Since $|L| \times [0,1)$ is open in $|L| \times I$ and is saturated with respect to π, the restriction of π,

$$\pi' : |L| \times [0,1) \rightarrow |v * L| - v,$$

is a quotient map; being one-to-one, it is a homeomorphism. Since $|L| \times 0$ is a deformation retract of $|L| \times [0,1)$, the space $|L|$ is a deformation retract of $|v * L| - v$. □

Theorem 35.6. *Let K be a complex of dimension n; let $X = |K|$. For $p > n$ the local homology groups $H_p(X,X - x)$ vanish, while for $p = n$ at least one of the groups $H_n(X,X - x)$ is non-trivial.*

Proof. Let σ be an n-simplex of K. Then σ is a face of no other simplex of K, so the set Int σ is in fact an open set of $|K|$. (Its complement is the union of all simplices of K different from σ.) If x is the barycenter $\hat{\sigma}$ of σ, it follows from Lemma 35.1 that

$$\begin{aligned} H_n(X,X - x) &\simeq H_n(\sigma,\sigma - \hat{\sigma}) \\ &\simeq H_n(B^n,B^n - \mathbf{0}). \end{aligned}$$

By Example 1, this group is infinite cyclic.

Now let x be an arbitrary point of X. We wish to show that $H_p(X,X - x) = 0$ for $p > n$. In view of Lemma 35.2, it suffices to consider the case where x is the barycenter of a simplex of K. Then x is a vertex of the complex $L = \text{sd } K$, and Lemma 35.4 applies. We have

$$H_p(X,X - x) \simeq H_p(\overline{\text{St}}(x,L), \text{Lk}(x,L)).$$

Since L is a complex of dimension n, $\overline{\text{St}}(x,L)$ is the polytope of a complex of dimension at most n. Therefore, this group vanishes in simplicial homology for $p > n$, so it vanishes in singular homology as well. □

EXERCISES

1. Check the details of Example 3.
2. Show that if M and N are manifolds with boundary of dimensions m and n, respectively, then $M \times N$ is a manifold with boundary of dimension $m + n$, and

$$\text{Bd}(M \times N) = (M \times (\text{Bd } N)) \cup ((\text{Bd } M) \times N).$$

3. A space X is **homogeneous** if given x and y, there is a homeomorphism of X with itself carrying x to y. Show that a connected manifold is homogeneous. [*Hint:* Define $x \sim y$ if there is such a homeomorphism; show the equivalence classes are open.]
4. Let M be an m-manifold with boundary; suppose $h : |K| \to M$ is a triangulation of M.
   (a) Show that if v is a vertex of K, then $\text{Lk}\, v$ is a homology $m-1$ sphere or ball, according as $h(x) \in \text{Int}\, M$ or $h(x) \in \text{Bd}\, M$.
   (b) Show that every simplex of K either has dimension m or is the face of a simplex of dimension m.
   (c) Show that an $m-1$ simplex s of K is a face of precisely one m-simplex of K if $s \subset h^{-1}(\text{Bd}\, M)$, and it is a face of precisely two m-simplices of K otherwise.
5. A **solid torus** is a space homeomorphic to $S^1 \times B^2$; it is a 3-manifold with boundary whose boundary is homeomorphic to the torus. Use the fact that S^3 is homeomorphic to $\text{Bd}(B^2 \times B^2)$ to write S^3 as the union of two solid tori T_1, T_2 that intersect in their common boundary. Compute the Mayer-Vietoris sequence of T_1, T_2.

*§36. APPLICATION: THE JORDAN CURVE THEOREM

Using the basic properties of singular homology, we now prove several classical theorems of topology, including the generalized Jordan curve theorem and the Brouwer theorem on invariance of domain.

Definition. If A is a subspace of X, we say that A **separates** X if the space $X - A$ is not connected.

We are going to be concerned with the case where X is $\mathbf{R}^n$ or S^n, and A is closed in X. Since $X - A$ is then locally path connected, its components and path components are identical. In particular, the group $\tilde{H}_0(X - A)$ vanishes if and only if $X - A$ is connected, and in general its rank is one less than the number of components of $X - A$. We shall use this fact freely in what follows.

Definition. A space homeomorphic to the unit k-ball B^k is called a ***k*-cell.**

Theorem 36.1. *Let B be a k-cell in S^n. Then $S^n - B$ is acyclic. In particular, B does not separate S^n.*

Proof. Let n be fixed. We proceed by induction on k. First take the case $k = 0$. Then B is a single point. The space $S^n - B$ is a single point if $n = 0$, while if $n > 0$, it is homeomorphic to $\mathbf{R}^n$. In either case, $S^n - B$ is acyclic.

We now suppose the theorem holds for a $k-1$ cell in S^n. Let B be a k-cell in S^n; let $h : I^k \to B$ be a homeomorphism.

Step 1. Let B_1 and B_2 be the two k-cells

$$B_1 = h(I^{k-1} \times [0,\tfrac{1}{2}]) \qquad \text{and} \qquad B_2 = h(I^{k-1} \times [\tfrac{1}{2},1])$$

in S^n. Let C be the $k-1$ cell $h(I^{k-1} \times (\tfrac{1}{2}))$ in S^n. See Figure 36.1. We show that if α is a non-zero element of $\tilde{H}_i(S^n - B)$, then its image is non-zero under at least one of the homomorphisms induced by the inclusion mappings

$$i : (S^n - B) \to (S^n - B_1) \qquad \text{and} \qquad j : (S^n - B) \to (S^n - B_2).$$

To prove this fact, let $X = S^n - C$. By the induction hypothesis, X is acyclic. We write X as the union of the two subspaces

$$X_1 = S^n - B_1 \qquad \text{and} \qquad X_2 = S^n - B_2.$$

Since X_1 and X_2 are open in S^n, they are open in X; so we have an exact Mayer-Vietoris sequence

$$\tilde{H}_{i+1}(X) \to \tilde{H}_i(A) \to \tilde{H}_i(X_1) \oplus \tilde{H}_i(X_2) \to \tilde{H}_i(X),$$

where $A = X_1 \cap X_2 = S^n - B$. Since X is acyclic, the map in the middle is an isomorphism. By Theorem 33.1, it carries $\alpha \in \tilde{H}_i(A)$ to $(i_*(\alpha), -j_*(\alpha))$. Therefore, at least one of the elements $i_*(\alpha)$ and $j_*(\alpha)$ is non-trivial.

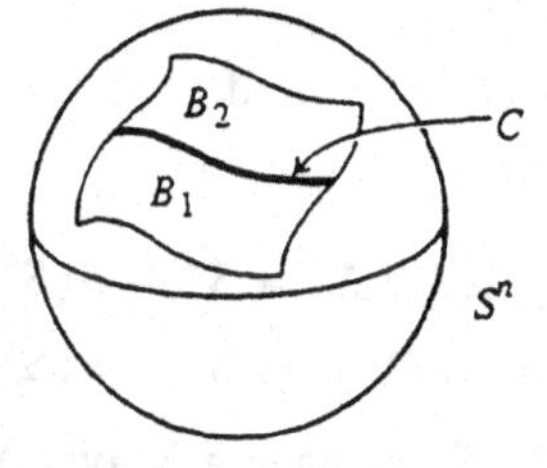

Figure 36.1

Step 2. We suppose there exists a non-zero element α in $\tilde{H}_i(S^n - B)$ and derive a contradiction. Let B_1 and B_2 be as in Step 1. Then the image of α in either $\tilde{H}_i(S^n - B_1)$ or $\tilde{H}_i(S^n - B_2)$ is non-zero. Suppose the former. Then write B_1 as the union of the k-cells

$$B_{11} = h(I^{k-1} \times [0,\tfrac{1}{4}]) \qquad \text{and} \qquad B_{12} = h(I^{k-1} \times [\tfrac{1}{4},\tfrac{1}{2}]).$$

Applying Step 1 again, we conclude that the image of α in either $\tilde{H}_i(S^n - B_{11})$ or $\tilde{H}_i(S^n - B_{12})$ is non-zero.

Continuing similarly, we obtain a sequence of closed intervals $[a_1,b_1] \supset [a_2,b_2] \supset \cdots$, each half the length of the preceding one; furthermore (letting D_m denote the k-cell

$$D_m = h(I^{k-1} \times [a_m,b_m])$$

for convenience) the image of α in $\tilde{H}_i(S^n - D_m)$ is non-zero for all m. Let e be the unique point in the intersection of the intervals $[a_m, b_m]$; then the set $E = h(I^{k-1} \times \{e\})$ is a $k - 1$ cell in S^n that equals the intersection of the nested sequence of k-cells $D_1 \supset D_2 \supset \cdots$. By the induction hypothesis, the group $\tilde{H}_i(S^n - E)$ vanishes. In particular, the image of α in this group vanishes. By Theorem 30.5, there is a compact subset A of $S^n - E$ such that the image of α in $\tilde{H}_i(A)$ vanishes. Since $S_n - E$ is the union of the open sets

$$S^n - D_1 \subset S^n - D_2 \subset \cdots,$$

the set A lies in one of them, say in $S^n - D_m$. But this means that the image of α in $\tilde{H}_i(S^n - D_m)$ vanishes, contrary to construction. □

Theorem 36.2. *Let $n > k \geq 0$. Let $h : S^k \to S^n$ be an imbedding. Then*

$$\tilde{H}_i(S^n - h(S^k)) \simeq \begin{cases} \mathbf{Z} & \text{if } i = n - k - 1, \\ 0 & \text{otherwise.} \end{cases}$$

Proof. Let n be fixed. We prove the theorem by induction on k. First take the case $k = 0$. Then $h(S^0)$ consists of two points p and q. Since $S^n - p - q \approx \mathbf{R}^n - \mathbf{0}$, and $\mathbf{R}^n - \mathbf{0}$ has the homotopy type of S^{n-1}, we see that $\tilde{H}_i(S^n - p - q)$ is infinite cyclic for $i = n - 1$ and vanishes otherwise.

Now suppose the theorem is true in dimension $k - 1$. Let $h : S^k \to S^n$ be an imbedding. We construct a certain Mayer-Vietoris sequence. Let X_1 and X_2 be the following open sets of S^n:

$$X_1 = S^n - h(E^k_+) \qquad \text{and} \qquad X_2 = S^n - h(E^k_-).$$

See Figure 36.2. Then let

$$X = X_1 \cup X_2 = S^n - h(S^{k-1}),$$
$$A = X_1 \cap X_2 = S^n - h(S^k).$$

Since X_1 and X_2 are open in X, we have a Mayer-Vietoris sequence

$$\tilde{H}_{i+1}(X_1) \oplus \tilde{H}_{i+1}(X_2) \to \tilde{H}_{i+1}(X) \to \tilde{H}_i(A) \to \tilde{H}_i(X_1) \oplus \tilde{H}_i(X_2).$$

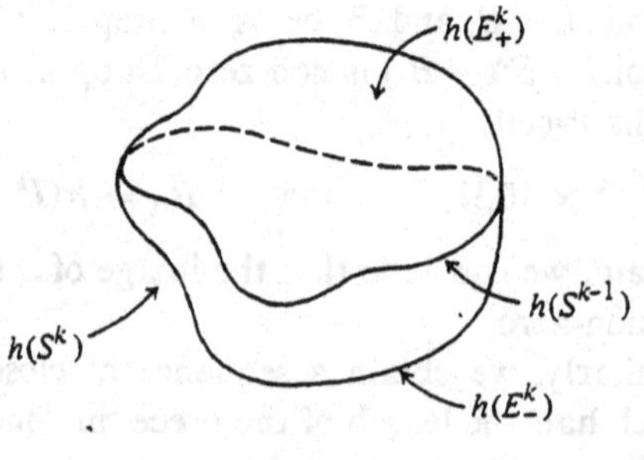

Figure 36.2

Both X_1 and X_2 are acyclic by the preceding theorem; therefore, the middle map is an isomorphism. That is,

$$\tilde{H}_{i+1}(S^n - h(S^{k-1})) \simeq \tilde{H}_i(S^n - h(S^k))$$

for all i. By the induction assumption the group on the left is infinite cyclic for $i + 1 = n - (k - 1) - 1$ and vanishes otherwise. Hence the group on the right is infinite cyclic for $i = n - k - 1$ and vanishes otherwise. □

Theorem 36.3 (The generalized Jordan curve theorem). *Let $n > 0$. Let C be a subset of S^n homeomorphic to the $n - 1$ sphere. Then $S^n - C$ has precisely two components, of which C is the common (topological) boundary.*

Proof. Applying the preceding theorem to the case $k = n - 1$, we see that $\tilde{H}_0(S^n - C) \simeq \mathbf{Z}$. Thus $S^n - C$ has precisely two path components (which are the same as its components, as noted earlier). Let W_1 and W_2 be these path components; because S^n is locally path connected, they are open in S^n. Then the (topological) boundary of W_i is the set $\bar{W}_i - W_i$. We need to show that

$$\bar{W}_1 - W_1 = C = \bar{W}_2 - W_2.$$

It suffices to prove the first of these equations. Since W_2 is open, no point of W_2 is a limit point of W_1; therefore $\bar{W}_1 - W_1 \subset C$. We show that $C \subset \bar{W}_1 - W_1$, whence equality holds.

Given $x \in C$ and given a neighborhood U of x, we show that U intersects the closed set $\bar{W}_1 - W_1$. This will suffice. Since C is homeomorphic to S^{n-1}, we can write C as the union of two $n - 1$ cells, C_1 and C_2, such that C_1 is small enough to lie in U. Figure 36.3 illustrates the case $n = 2$.

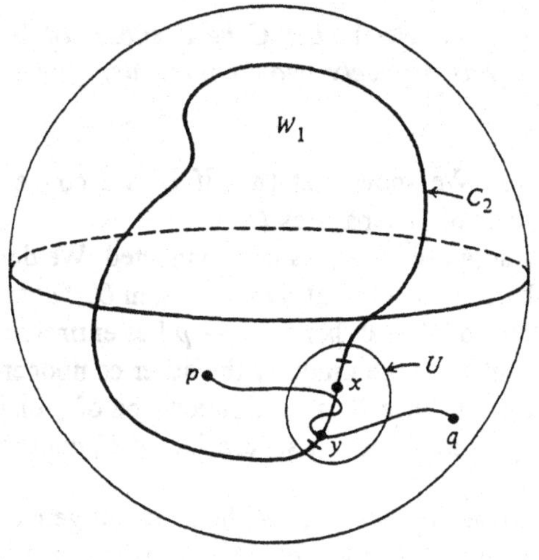

Figure 36.3

Now C_2 does not separate S^n, so we can choose a path α in $S^n - C_2$ joining a point p of W_1 to a point q of W_2. Now α must contain a point of $\overline{W}_1 - W_1$, since otherwise α would lie in the union of the disjoint open sets W_1 and $S^n - \overline{W}_1$ and contain a point of each of them, contrary to the fact that the image set of α is connected. Let y be a point of $\overline{W}_1 - W_1$ lying on the path α; then y lies in C. Since it cannot lie in C_2, it must lie in C_1 and hence in U. Then U intersects $\overline{W}_1 - W_1$ in the point y, as desired. □

We remark that, under the hypotheses of this theorem, it seems likely that if W_1 and W_2 are the components of $S^n - C$, where C is an $n - 1$ sphere, then the sets $\overline{W}_1$ and $\overline{W}_2$ should be n-cells. But actually this is not true. It is not even true in general that W_1 and W_2 are open balls. There is a famous imbedding of S^2 in S^3, called the *Alexander horned sphere*, for which one of the sets W_i is not even simply connected! (See [H-Y], p. 176.)

What *can* one prove about the sets W_i? In the case $n = 2$, the answer has been known for a long time. If C is a simple closed curve in S^2, then C separates S^2 into two components W_1 and W_2, and both $\overline{W}_1$ and $\overline{W}_2$ *are* 2-cells. This result is called the *Schoenflies Theorem*. A proof may be found in [N].

More recently (1960), results have been proved in higher dimensions, assuming additional hypotheses about the imbedding. Suppose the map $h : S^{n-1} \to S^n$ can be "collared," which means there is an imbedding $H : S^{n-1} \times I \to S^n$ such that $H(x, \frac{1}{2}) = h(x)$ for each x. (This hypothesis is satisfied, for instance, if h is differentiable with Jacobian of maximal rank.) In this case both $\overline{W}_1$ and $\overline{W}_2$ are n-cells; this result is known as the ***Brown-Mazur Theorem*** [B].

Classically, the Jordan curve theorem is usually stated for spheres imbedded in $\mathbf{R}^n$ rather than in S^n. We prove that version of the theorem now.

Corollary 36.4. *Let $n > 1$. Let C be a subset of $\mathbf{R}^n$ homeomorphic to S^{n-1}. Then $\mathbf{R}^n - C$ has precisely two components, of which C is the common boundary.*

Proof. Step 1. We show first that if U is a connected open set in S^n, where $n > 1$, no point of U separates U.

Let $p \in U$ and suppose $U - p$ is not connected. We derive a contradiction. Choose an open ϵ-ball B_ϵ centered at p and lying in U. Then $B_\epsilon - p$ is connected, being homeomorphic to $\mathbf{R}^n - \mathbf{0}$; hence $B_\epsilon - p$ lies entirely in one of the components C of $U - p$. Let D be the union of the other components of $U - p$. Now p is not a limit point of D, since B_ϵ is a neighborhood of p disjoint from D. Hence the two sets $C \cup \{p\}$ and D form a separation of U, contrary to hypothesis.

Step 2. We prove the theorem. Without loss of generality, we can replace $\mathbf{R}^n$ in the statement of the theorem by $S^n - p$, where p is the north pole of S^n. Now $S^n - C$ has two components W_1 and W_2. Suppose $p \in W_1$. By the result of

Step 1, $W_1 - p$ is connected. Then $W_1 - p$ and W_2 are the components of $S^n - p - C$, and C equals the boundary of W_2 and $W_1 - p$. □

Theorem 36.5 (Invariance of domain). *Let U be open in $\mathbf{R}^n$; let $f : U \to \mathbf{R}^n$ be continuous and injective. Then $f(U)$ is open in $\mathbf{R}^n$ and f is an imbedding.*

Proof. Without loss of generality, we can replace $\mathbf{R}^n$ by S^n.

Step 1. Given a point y of $f(U)$, we show that $f(U)$ contains a neighborhood of y. This proves that $f(U)$ is open in S^n.

Let x be the point of U such that $f(x) = y$. Choose an open ball B_ϵ of radius ϵ, centered at x, whose closure lies in U. Let $S_\epsilon = \bar{B}_\epsilon - B_\epsilon$. See Figure 36.4. Now the set $f(S_\epsilon)$ is homeomorphic to S^{n-1}; therefore, $f(S_\epsilon)$ separates S^n into two components W_1 and W_2, each open in S^n. The set $f(B_\epsilon)$ is connected and disjoint from $f(S_\epsilon)$. Therefore, it lies in either W_1 or W_2; suppose it lies in W_1. Now it must equal all of W_1, for otherwise the set

$$S_n - f(B_\epsilon) - f(S_\epsilon) = S^n - f(\bar{B}_\epsilon)$$

would consist of all of W_2 along with points of W_1 as well, contradicting the fact that the cell $f(\bar{B}_\epsilon)$ does not separate S^n. Therefore, $f(B_\epsilon) = W_1$. We conclude that $f(U)$ contains the neighborhood W_1 of y, as desired.

Step 2. In view of Step 1, f carries any set open in U to a set that is open in S^n and hence open in $f(U)$. Thus f is a homeomorphism of U with $f(U)$. □

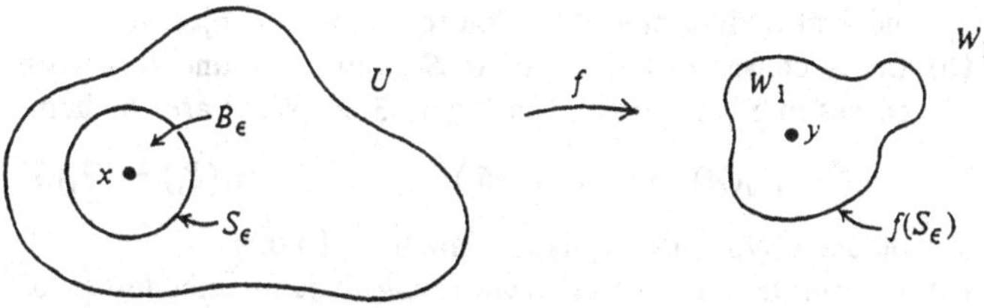

Figure 36.4

The theorem on invariance of domain is much easier to prove *if one assumes that f is continuously differentiable with non-singular Jacobian.* In this case the theorem follows from the inverse function theorem of analysis. The true profundity of invariance of domain is that it has nothing to do with differentiability or Jacobians, but depends only on continuity and injectivity of the map f.

Similarly, the Jordan curve theorem is much easier to prove in the case that the imbedding $f : S^{n-1} \to \mathbf{R}^n$ is a simplicial map (of some complex whose space is homeomorphic to S^{n-1}), or a differentiable map with Jacobian of maximal rank. The true difficulties appear only when one assumes no more than continuity and injectivity.

EXERCISES

1. Let M be an m-manifold with boundary, as defined in the preceding section. Use invariance of domain to prove the following.
   (a) Show that if a point x of M maps to a point of $\mathbf{R}^{m-1} \times 0$ under one coordinate patch about x, it does so under every such coordinate patch.
   (b) Let U be open in $\mathbf{R}^m$ and let V be open in $\mathbf{R}^n$. Show that if U and V are homeomorphic, then $m = n$.
   (c) Show that the number m is uniquely determined by M.
2. (a) Let Y be the topologist's sine curve (see the exercises of §29). Show that if $h : Y \to S^n$ is an imbedding, then $S^n - h(Y)$ is acyclic.
   (b) Let X be the closed topologist's sine curve (see the exercises of §33). Show that if $h : X \to S^n$ is an imbedding, then $\tilde{H}_i(S^n - h(X))$ is infinite cyclic if $i = n - 2$, and vanishes otherwise.
   (c) Show that if $h : X \to S^2$ is an imbedding, then $S^2 - h(X)$ has precisely two components, of which $h(X)$ is the common boundary.
3. (a) Consider S^3 as $\mathbf{R}^3$ with a point at infinity adjoined. Let A, B_1, B_2, and B_3 be the simple closed curves in S^3 pictured in Figure 36.5. The map
   $$H_1(B_i) \to H_1(S^3 - A)$$
   induced by inclusion is a homomorphism of infinite cyclic groups, so it equals multiplication by d_i, where d_i is well-defined up to sign. The integer d_i measures how many times B_i *links* A. What is it in each case? Similarly, determine the integer corresponding to the homomorphism
   $$H_1(A) \to H_1(S^3 - B_i)$$
   induced by inclusion. Can you formulate a conjecture?
   (b) Let A consist of two points of S^2, and let B_1 and B_2 be two simple closed curves in S^2, as pictured in Figure 36.6. What are the homomorphisms
   $$\tilde{H}_0(A) \to \tilde{H}_0(S^2 - B_i) \quad \text{and} \quad H_1(B_i) \to H_1(S^2 - A)$$
   induced by inclusion? Formulate a conjecture.
   (c) Formulate a conjecture concerning disjoint imbeddings of S^p and S^q in S^{p+q+1}.

   We shall return to this conjecture later on, after we prove the Alexander duality theorem. (See the exercises of §72.)

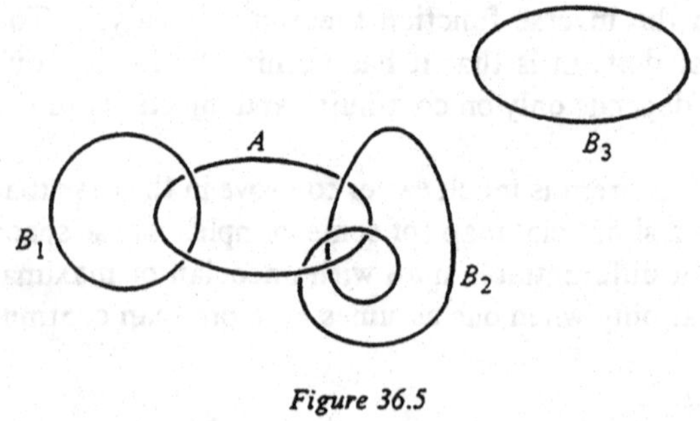

Figure 36.5

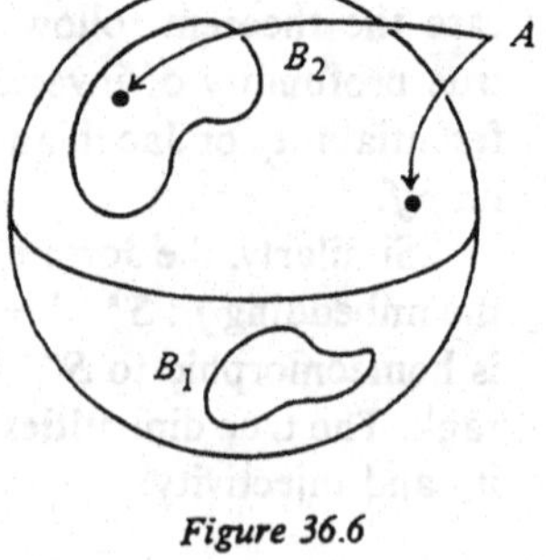

Figure 36.6

§37. MORE ON QUOTIENT SPACES

We reviewed some aspects of the theory of quotient spaces in §20. We now discuss the topic further; in particular, we consider separation axioms for quotient spaces.

We have already noted that the separation axioms do not behave well for quotient spaces. In general, it is difficult to ensure that a quotient space satisfies any stronger separation axiom than the T_1-axiom. Even the Hausdorff axiom is often hard to verify. We give here three situations in which we can be sure that the quotient space is Hausdorff (in fact, normal).

Theorem 37.1. *Let $p : X \to Y$ be a quotient map. If p is a closed map, and if X is normal, then Y is normal.*

Proof. If x is a point of X, then x is closed in X, so the one-point set $p(x)$ is closed in Y (because p is a closed map). Thus Y is a T_1-space.

Let A and B be disjoint closed sets in Y. Then $p^{-1}(A)$ and $p^{-1}(B)$ are disjoint closed sets in X. Choose disjoint open sets U and V in X about $p^{-1}(A)$ and $p^{-1}(B)$, respectively. The sets $p(U)$ and $p(V)$ need not be disjoint nor open in Y. See Figure 37.1. However, the sets $C = X - U$ and $D = X - V$ are closed in X, so $p(C)$ and $p(D)$ are closed in Y. The sets $Y - p(C)$ and $Y - p(D)$ are then disjoint open sets about A and B, respectively.

To show these sets are disjoint, we begin by noting that $U \cap V = \varnothing$. Taking complements, we have

$$C \cup D = (X - U) \cup (X - V) = X.$$

Then $p(C) \cup p(D) = Y$; taking complements again,

$$(Y - p(C)) \cap (Y - p(D)) = \varnothing,$$

as desired.

To show these sets contain A and B, respectively, note first that because C

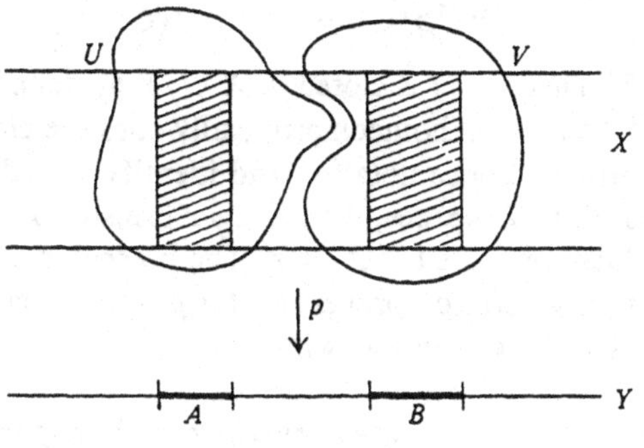

Figure 37.1

is disjoint from $p^{-1}(A)$, the set $p(C)$ is disjoint from A. Thus $p(C) \subset Y - A$; taking complements, we have $Y - p(C) \supset A$. Similarly, $Y - p(D)$ contains B. □

If X^* is a partition of X into closed sets, and if the quotient map $p: X \to X^*$ is closed, then X^* is called, classically, an **upper semicontinuous decomposition** of X. Specifically, this means that for each closed set A of X, the set $p^{-1}p(A)$, which is called the **saturation** of A, is also closed in X. In such a case, normality of X implies normality of X^*.

Definition. Let X and Y be disjoint topological spaces; let A be a closed subset of X; let $f: A \to Y$ be a continuous map. We define a certain quotient space as follows: Topologize $X \cup Y$ as the topological sum. Form a quotient space by identifying each set

$$\{y\} \cup f^{-1}(y)$$

for $y \in Y$, to a point. That is, partition $X \cup Y$ into these sets, along with the one-point sets $\{x\}$, for $x \in X - A$. We denote this quotient space by $X \cup_f Y$, and call it the **adjunction space determined by** f. See Figure 37.2.

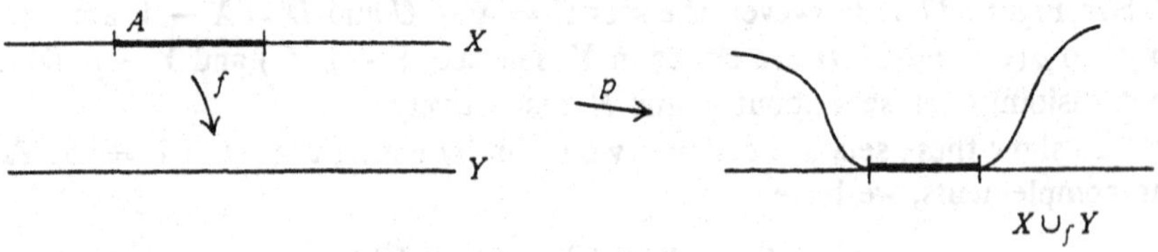

Figure 37.2

Let $p: X \cup Y \to X \cup_f Y$ be the quotient map. We show first that p defines a homeomorphism of Y with a closed subspace of $X \cup_f Y$. Obviously $p|Y$ is continuous and injective. Furthermore, if C is closed in Y, then $f^{-1}(C)$ is closed in X, because $f: A \to Y$ is continuous and A is closed in X. It follows that

$$p^{-1}p(C) = C \cup f^{-1}(C)$$

is closed in $X \cup Y$. Then $p(C)$ is closed in $X \cup_f Y$ by definition of the quotient topology. Thus $p|Y$ carries Y homeomorphically onto the closed subspace $p(Y)$ of $X \cup_f Y$. We normally abuse notation and identify Y with $p(Y)$.

Now if X and Y are T_1-spaces, then $X \cup_f Y$ is also a T_1-space, since each of the equivalence classes is closed in $X \cup Y$. The preceding theorem does not apply to give us further separation properties, for p is not in general a closed map. However, we can still prove the following.

Theorem 37.2. *If X and Y are normal, then the adjunction space $X \cup_f Y$ is normal.*

Proof. As usual, A is closed in X and $f: A \to Y$ is continuous. Let B and C be disjoint closed sets in $X \cup_f Y$. Let

$$B_X = p^{-1}(B) \cap X; \qquad C_X = p^{-1}(C) \cap X;$$
$$B_Y = p^{-1}(B) \cap Y; \qquad C_Y = p^{-1}(C) \cap Y.$$

See Figure 37.3. By the Urysohn lemma, we can choose a continuous function $g: Y \to [0,1]$ that maps B_Y to 0 and C_Y to 1. Then the function $g \circ f: A \to [0,1]$ equals 0 on $A \cap B_X$ and 1 on $A \cap C_X$. We define a function

$$h: A \cup B_X \cup C_X \to [0,1]$$

by letting it equal $g \circ f$ on A, and 0 on B_X, and 1 on C_X. Since A, B_X, and C_X are closed in X, the map h is continuous. By the Tietze theorem, we can now extend h to a continuous function k defined on all of X.

The function $X \cup Y \to [0,1]$ that equals k on X and g on Y is continuous. It is constant on each equivalence class, because $k(f^{-1}(y)) = g(y)$ if $y \in f(A)$. Therefore, it induces a continuous map of $X \cup_f Y$ into $[0,1]$ that equals 0 on B and equals 1 on C. □

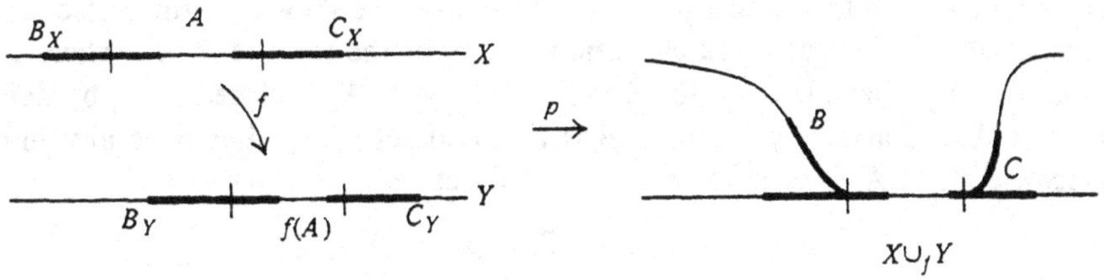

Figure 37.3

We have already defined what it means for the topology of a given space X to be coherent with a collection of subspaces of X. We wish to deal with the separation axioms in this context.

First, we extend the notion of coherent topology slightly. We suppose that we have a collection of topological spaces $\{X_\alpha\}$ whose union X has no topology. We seek to find conditions under which there *exists* a topology on X, of which the X_α are subspaces, such that this topology is coherent with the subspaces X_α.

Lemma 37.3. *Let X be a set which is the union of the topological spaces $\{X_\alpha\}$.*

(a) *If there is a topological space X_T having X as its underlying set, and each X_α is a subspace of X_T, then X has a topology, of which the X_α are subspaces, that is coherent with the X_α. This topology is in general finer than the topology of X_T.*

(b) *If for each pair α, β of indices, the set $X_\alpha \cap X_\beta$ is closed in both X_α and X_β, and inherits the same subspace topology from each of them, then X has a topology coherent with the subspaces X_α. Each X_α is a closed set in this topology.*

Proof. (a) Let us define a topological space X_C whose underlying set is X by declaring a set A to be closed in X_C if and only if its intersection with each X_α is a closed set of X_α. The collection of such sets contains arbitrary intersections and finite unions of its elements, so it does define a topology on the set X.

If A is closed in X_T, then because X_α is a subspace of X_T, the set $A \cap X_\alpha$ is closed in X_α for each α. It follows that A is closed in X_C. Thus the topology of X_C is finer than that of X_T.

We show that each X_α is a subspace of X_C; it then follows from its definition that X_C is coherent with the subspaces X_α. For this purpose, we show that the collection of closed sets of X_α equals the collection of sets of the form $A \cap X_\alpha$, where A is closed in X_C. First, note that if A is closed in X_C, then $A \cap X_\alpha$ is closed in X_α by definition of X_C. Conversely, suppose B is closed in X_α. Because X_α is a subspace of X_T, we have $B = A \cap X_\alpha$ for some set A closed in X_T. Because the topology of X_C is finer than that of X_T, the set A is also closed in X_C. Thus $B = A \cap X_\alpha$ for some A closed in X_C, as desired.

(b) As before, we define a topology X_C on the set X by declaring A to be closed in X_C if $A \cap X_\alpha$ is closed in X_α for each α.

We show that each X_α is a subspace of X_C; it then follows immediately that X_C is coherent with the subspaces X_α. As before, we show that the collection of closed sets of X_α equals the collection of sets of the form $A \cap X_\alpha$, where A is closed in X_C. First, if A is closed in X_C, then $A \cap X_\alpha$ is closed in X_α by definition of X_C. Conversely, suppose B is a closed set of X_α. Let β be any index. Because $X_\alpha \cap X_\beta$ is a closed set in X_α, the set

$$B \cap X_\beta = B \cap (X_\alpha \cap X_\beta)$$

is a closed set of X_α. Because $X_\alpha \cap X_\beta$ is a subspace of X_α, it is also a closed set in $X_\alpha \cap X_\beta$; because $X_\alpha \cap X_\beta$ is a closed subspace of X_β, it is a closed set in X_β. Since β is arbitrary, B is by definition a closed set of X_C. Thus $B = B \cap X_\alpha$, where B is closed in X_C, as desired.

Incidentally, we have shown that every closed set of X_α is also closed in X_C. In particular, X_α is itself closed in X_C. □

Example 1. Let K be a complex in $\mathbf{E}^J$. Since $\mathbf{E}^J$ is a space, $|K|$ has a topology inherited from $\mathbf{E}^J$; each simplex σ of K is a subspace in this topology. By (a) of the preceding theorem, $|K|$ has a topology coherent with the subspaces σ, which is in general finer than the topology $|K|$ inherits from $\mathbf{E}^J$. This is a fact we proved directly in §2.

Suppose X is a space whose topology is coherent with its subspaces X_α. In general, even if each of the spaces X_α has nice separation properties, X need not have these properties. In the special case of a countable union of closed subspaces, however, one can prove the following.

Theorem 37.4. *Let X be a space that is the countable union of certain closed subspaces X_n. Suppose the topology of X is coherent with the spaces X_n. Then if each X_i is normal, so is X.*

Proof. If p is a point of X, then $\{p\} \cap X_i$ is closed in X_i for each i, so $\{p\}$ is closed in X. Thus X is a T_1-space.

Let A and B be disjoint closed sets in X. Define $Y_0 = A \cup B$, and for $n > 0$, define

$$Y_n = X_1 \cup \cdots \cup X_n \cup A \cup B.$$

We define a continuous function $f_0 : Y_0 \to I$ by letting f_0 equal 0 on A and 1 on B. In general, suppose we are given a continuous function $f_n : Y_n \to I$. The space X_{n+1} is normal and $Y_n \cap X_{n+1}$ is closed in X_{n+1}; we use the Tietze theorem to extend the function $f_n|(Y_n \cap X_{n+1})$ to a continuous function $g : X_{n+1} \to I$. Because Y_n and X_{n+1} are closed subsets of Y_{n+1}, the functions f_n and g combine to define a continuous function $f_{n+1} : Y_{n+1} \to I$ that extends f_n. The functions f_n in turn combine to define a function $f : X \to I$ that equals 0 on A and 1 on B. Because X has the topology coherent with the subspaces X_n, the map f is continuous. □

EXERCISES

1. Let X be a set which is the union of the topological spaces $\{X_\alpha\}$. Suppose that each set $X_\alpha \cap X_\beta$ inherits the same topology from each of X_α and X_β.
   (a) Show that if $X_\alpha \cap X_\beta$ is open in both X_α and X_β, for each pair α,β, then there is a topology on X of which the X_α are subspaces.
   (b) Show that in general there is no topology on X of which each X_α is a subspace. [*Hint:* Let A, B, and C be three disjoint subsets of $\mathbf{R}$, each of which is dense in $\mathbf{R}$. Let A, B, $X_1 = \mathbf{R} - A$, and $X_2 = \mathbf{R} - B$ be topologized as subspaces of $\mathbf{R}$; let $X_3 = A \cup B$ be topologized as the topological sum of A and B. Let $X = X_1 \cup X_2 \cup X_3$. Compute $\bar{A}$.]
2. Recall that if $J = \mathbf{Z}_+$, we denote the space $\mathbf{E}^J$ by $\mathbf{R}^\infty$. Each space $\mathbf{R}^n \times \mathbf{0}$ is a subspace of $\mathbf{R}^\infty$. Show that the function $f : \mathbf{R}^\infty \to \mathbf{R}$ given by
   $$f(x) = \sum_{i=1}^{\infty} i\, x_i$$
   is not continuous in the usual (metric) topology of $\mathbf{R}^\infty$, but is continuous in the topology coherent with the subspaces $\mathbf{R}^n \times \mathbf{0}$.
3. Let X be a space. Let X_C denote the set X in the topology coherent with the collection of compact subspaces of X.
   (a) Show that a subset B of X is compact in the topology it inherits from X if and only if it is compact in the topology it inherits from X_C. [*Hint:* Show also that these two distinct subspace topologies on the set B are in fact the same topology.
   (b) A space is said to be **compactly generated** if its topology is coherent with the collection of its compact subspaces. Show that X_C is compactly generated. Conclude that $(X_C)_C = X_C$.
   (c) Show that inclusion $X_C \to X$ induces an isomorphism in singular homology.

(d) In general, let $X \times_C Y$ denote the compactly generated topology $(X \times Y)_C$ derived from $X \times Y$. Show that if K and L are complexes, then the topology of $|K| \times_C |L|$ is coherent with the subspaces $\sigma \times \tau$, for $\sigma \in K$ and $\tau \in L$. Compare Exercise 6 of §20. [*Hint:* If $D \subset X \times Y$ is compact, then $D \subset A \times B \subset X \times Y$, where A and B are compact.]

4. Show that Theorem 37.4 does not hold for uncountable collections. [*Hint:* Let X be an uncountable well-ordered set with a smallest element 0 and a largest element Ω, such that $[0,\alpha]$ is countable for each $\alpha < \Omega$. Then $[0,\Omega) \times [0,\Omega]$ is not normal. See [Mu], p. 201, or [K], p. 131.]

§38. CW COMPLEXES

We have stated that one of the advantages of simplicial homology theory is its effective computability. But in fact this statement is somewhat misleading. The amount of labor involved in a straightforward calculation in all but the simplest cases is too large to carry out in practice. Even when we calculated the homology of such simple spaces as the torus and the Klein bottle, in §6, we did not proceed straightforwardly. Instead, we used geometric arguments (of a rather ad hoc nature) to reduce the computations to simpler ones.

We now refine these ad hoc techniques into a systematic method for computing homology groups. This method will apply not only to simplicial homology, but to singular homology as well. We introduce in this section a notion of complex more general than that of simplicial complex. It was invented by J. H. C. Whitehead, and is called a "CW complex." In the next section, we show how to assign to each CW complex a certain chain complex, called its "cellular chain complex," which can be used to compute the homology of the underlying space. This chain complex is much simpler and easier to deal with than the singular or simplicial chain complexes. In a final section, we apply these methods to compute, among other things, the homology of real and complex projective spaces.

Definition. Recall that a space is called a **cell** of dimension m if it is homeomorphic with B^m. It is called an **open cell** of dimension m if it is homeomorphic with Int B^m. In each case the integer m is uniquely determined by the space in question.

Definition. A **CW complex** is a space X and a collection of disjoint open cells e_α whose union is X such that:

(1) X is Hausdorff.

(2) For each open m-cell e_α of the collection, there exists a continuous map $f_\alpha : B^m \to X$ that maps Int B^m homeomorphically onto e_α and carries Bd B^m into a finite union of open cells, each of dimension less than m.

(3) A set A is closed in X if $A \cap \bar{e}_\alpha$ is closed in $\bar{e}_\alpha$ for each α.

The finiteness part of condition (2) was called "closure-finiteness" by J. H. C. Whitehead. Condition (3) expresses the fact that X has what he called the "weak topology" relative to the collection $\{\bar{e}_\alpha\}$. These terms are the origin of the letters C and W in the phrase "CW complex."

We commonly denote $\bar{e}_\alpha - e_\alpha$ by $\dot{e}_\alpha$. We remark that conditions (1) and (2) imply that f_α carries B^m onto $\bar{e}_\alpha$ and Bd B^m onto $\dot{e}_\alpha$: Because f_α is continuous, it carries B^m, which is the closure of Int B^m, *into* the closure of $f_\alpha(\text{Int } B^m)$, which is $\bar{e}_\alpha$. Because $f_\alpha(B^m)$ is compact, it is closed (since X is Hausdorff); because it contains e_α, it must contain $\bar{e}_\alpha$. Thus $f_\alpha(B^m) = \bar{e}_\alpha$. Finally, because $f_\alpha(\text{Bd } B^m)$ is disjoint from e_α, it must equal $\dot{e}_\alpha$.

We remark also that the converse of (3) holds trivially; if A is closed in X, then $A \cap \bar{e}_\alpha$ is closed in $\bar{e}_\alpha$ for each α.

The map f_α is called a "characteristic map" for the open cell e_α. Note that the maps f_α are not uniquely specified in the definition of a CW complex. Only the space X and the collection $\{e_\alpha\}$ are specified. We customarily abuse notation and use the symbol X to refer both to the CW complex and to the underlying space.

A **finite CW complex** X is a CW complex for which the collection of open cells is finite. If X has only finitely many open cells, then the finiteness part of (2) is automatic, and condition (3) is implied by the other conditions: If the set $A \cap \bar{e}_\alpha$ is closed in $\bar{e}_\alpha$, it is closed in X; then since A is a finite union of such sets, A is also closed in X.

A finite CW complex is of course compact. Conversely, any compact subset A of a CW complex X can intersect only finitely many open cells of X; we leave the proof to the exercises. One needs the finiteness part of condition (2) in the proof.

The following lemma is an immediate consequence of our general results about coherent topologies (see §20).

Lemma 38.1. *Let X be a CW complex with open cells e_α. A function $f: X \to Y$ is continuous if and only if $f|\bar{e}_\alpha$ is continuous for each α. A function $F: X \times I \to Y$ is continuous if and only if $F|(\bar{e}_\alpha \times I)$ is continuous for each α.* □

Example 1. Consider the torus as a quotient space of a rectangle, as usual. See Figure 38.1. We can express T as a CW complex having a single open 2-cell (the image under π of the interior of the rectangle), two open 1-cells (the images of the open edges), and one 0-cell (the image of the vertices). Conditions (1)–(3) hold at once.

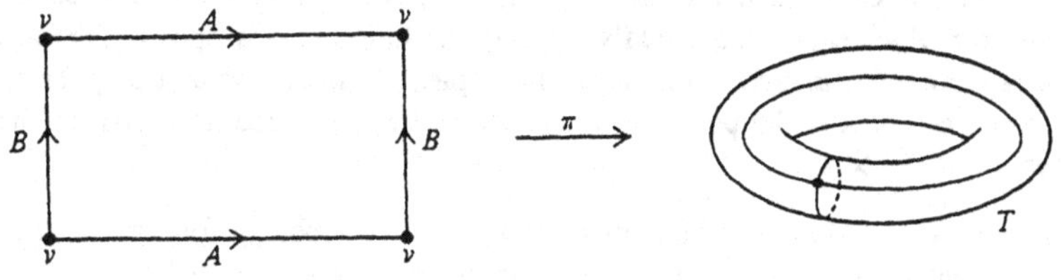

Figure 38.1

Similarly, one can express the Klein bottle as a CW complex having the same number of cells in each dimension as the torus. The projective plane can be expressed as a CW complex having one open cell in each dimension 0, 1, 2.

More generally, the discussion in the exercises of §6 shows how one can express the n-fold connected sum $T \# \cdots \# T$ (or $P^2 \# \cdots \# P^2$, respectively) as a CW complex having one open cell in dimension 2, one cell in dimension 0, and $2n$ (or n, respectively) open cells in dimension 1.

Similarly, the k-fold dunce cap can be expressed as a CW complex with one open cell in each dimension 0, 1, 2.

Example 2. The quotient space formed from B^n by collapsing $\text{Bd } B^n$ to a point is homeomorphic to S^n. (We leave the proof to you.) Therefore, S^n can be expressed as a CW complex having one open n-cell and one 0-cell, and *no other cells at all.* See Figure 38.2.

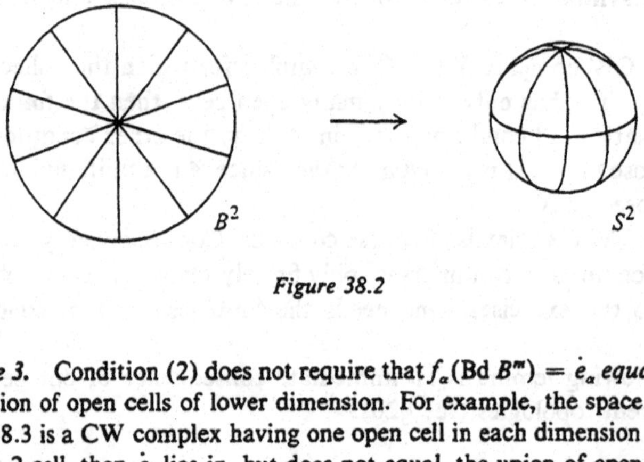

Figure 38.2

Example 3. Condition (2) does not require that $f_\alpha(\text{Bd } B^m) = \dot{e}_\alpha$ *equal* the union of a collection of open cells of lower dimension. For example, the space X pictured in Figure 38.3 is a CW complex having one open cell in each dimension 0, 1, 2; if e_2 is the open 2-cell, then $\dot{e}_2$ lies in, but does not equal, the union of open cells of lower dimensions.

Example 4. Let K and L be simplicial complexes; suppose K is locally finite. The space $X = |K| \times |L|$ can be expressed as a CW complex by taking the sets $(\text{Int } \sigma) \times (\text{Int } \tau)$ as its cells, for $\sigma \in K$ and $\tau \in L$. In this case the characteristic maps

$$f_\alpha : B^m \to \sigma \times \tau$$

can be taken to be *homeomorphisms*. Furthermore, in this case $\text{Bd}(\sigma \times \tau)$ *equals* a union of open cells of lower dimension. Condition (3) is a consequence of Exercise 6 of §20; it depends on the local finiteness of K. Thus $|K| \times |L|$ is a special kind of CW complex. We give the formal definition now.

A CW complex X for which the maps f_α can be taken to be homeomorphisms, and for which each set $\dot{e}_\alpha$ equals the union of finitely many open cells of X, is called a **regular cell complex.** A regular cell complex X can always be triangulated so each closed cell of X is the polytope of a subcomplex. The proof is similar to the one we gave for $|K| \times I$.

Example 5. Many of the constructions used in topology for forming new spaces from old ones, when applied to CW complexes, give rise to CW complexes. The

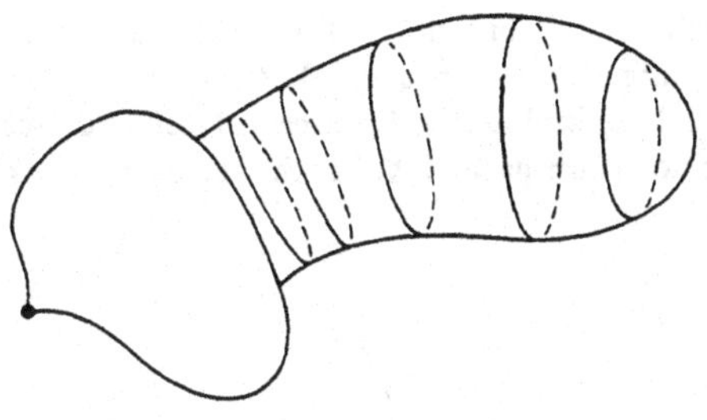

Figure 38.3

product $X \times Y$ of two CW complexes, for instance, is a CW complex, provided Y is locally compact. Similarly, adjunction spaces can often be made into CW complexes. If A is a subcomplex of X, a *cellular map* $f: A \to Y$ is a map carrying each p-cell of A into the union of the open cells of Y of dimension at most p. In this case, one can show that the adjunction space $X \cup_f Y$ is a CW complex.

Definition. Let X be a CW complex. Let Y be a subspace of X that equals a union of open cells of X. Suppose that for each open cell e_α of X contained in Y, its closure is also contained in Y. Then we shall show that Y is a closed set in X, and that Y is a CW complex in its own right. It is called a **subcomplex** of X. In particular, the subspace X^p of X that is the union of the open cells of X of dimension at most p satisfies these conditions. It is thus a subcomplex of X, which is called the **p-skeleton** of X.

Clearly, Y is Hausdorff. If e_α is an open m-cell of X contained in Y, then its characteristic map $f_\alpha : B^m \to X$ carries B^m onto $\bar{e}_\alpha$, which is contained in Y by hypothesis. The open cells of X that intersect $f_\alpha(\text{Bd } B^m) = \dot{e}_\alpha$ must lie in Y; thus f_α carries Bd B^m into the union of finitely many open cells of Y. It only remains to show that Y has the topology specified by (3).

Let $B \subset Y$; suppose $B \cap \bar{e}_\alpha$ is closed in $\bar{e}_\alpha$ for each e_α contained in Y. If e_β is a cell of X not contained in Y, then e_β is disjoint from Y. Hence $Y \cap \bar{e}_\beta \subset \dot{e}_\beta$, so $Y \cap \bar{e}_\beta$ lies in the union of finitely many open cells of Y, say $e_1, \ldots, e_k$. Then

$$B \cap \bar{e}_\beta = [(B \cap \bar{e}_1) \cup \cdots \cup (B \cap \bar{e}_k)] \cap \bar{e}_\beta.$$

By hypothesis, $B \cap \bar{e}_i$ is closed in $\bar{e}_i$ and hence in X. Therefore, $B \cap \bar{e}_\beta$ is closed in X, and in particular is closed in $\bar{e}_\beta$. Since β is arbitrary, it follows that B is closed in X, and in particular, is closed in Y, as desired.

It follows that Y has the topology specified by condition (3). It also follows that if B is closed in Y, then B is closed in X. In particular, Y itself is closed in X.

The finiteness part of condition (2) is crucial for what we have just proved, as the following example shows.

Example 6. Let X be a 2-simplex σ in the plane, in its usual topology. Break X up into a single open 2-cell Int σ, and infinitely many open 1-cells and 0-cells, as indicated in Figure 38.4. Then X satisfies all the conditions for a CW complex except the finiteness part of (2). (Condition (3) is trivial, since the closure of Int σ equals X.) Let Y be the union of the 1-cells and 0-cells of X, topologized by declaring C to be closed in Y if $C \cap \bar{e}_\alpha$ is closed in $\bar{e}_\alpha$ for each 1-cell and 0-cell e_α. Then Y is a CW complex, but it is not a subspace of X. For the subspace Bd σ of X is compact, and Y is not.

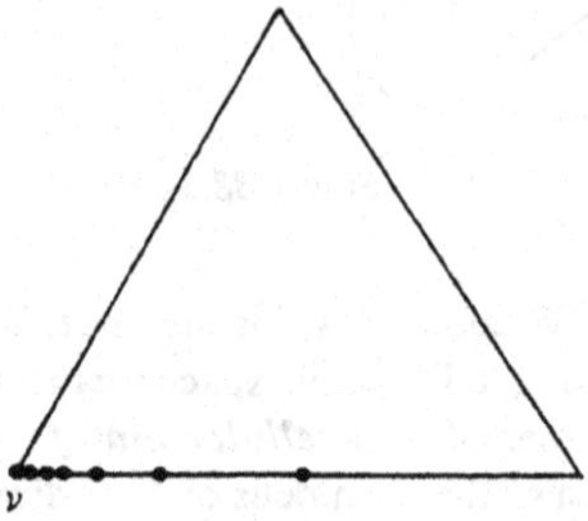

Figure 38.4

Definition. If a CW complex X can be triangulated by a complex K in such a way that each skeleton X^p of X is triangulated by a subcomplex of K of dimension at most p, then we say that X is a **triangulable CW complex.**[†]

Each of the CW complexes mentioned in the earlier examples is a triangulable CW complex. We now give an example of one that is not. The proof uses results about local homology groups from §35.

Example 7. Let A be a subspace of $\mathbf{R}^3$ that is the union of a square and a triangle, such that one of the edges of the triangle coincides with the diagonal D of the square. See Figure 38.5. Then A is the space of a complex consisting of three triangles with an edge in common. Now draw a wiggly 1-cell C in the square, intersecting the diagonal in an infinite, totally disconnected set. (An $x\sin(1/x)$ curve will do.) Take the 3-ball B^3 and attach it to A by a map $f : \text{Bd } B^3 \to C$ that maps each great circle arc in S^2 from the south pole to the north pole homeomorphically onto the 1-cell C. The resulting adjunction space X is easily seen to be a CW complex; one takes the open simplices of A as the open cells of dimensions 0, 1, and 2, and Int B^3 as the open 3-cell e_3. We show that the space X cannot be triangulated; hence in particular, it is not triangulable as a CW complex.

Suppose $h : |K| \to X$ is a triangulation. First we write X as the disjoint union

$$X = (A - C) \cup C \cup e_3.$$

Now if $x \in e_3$, then $H_3(X, X - x)$ is infinite cyclic, because x has a neighborhood homeomorphic to an open 3-ball. On the other hand, if $x \in A - C$, then $H_3(X, X - x)$ vanishes, because x has a neighborhood lying entirely in the 2-dimensional complex A. (See Lemma 35.6.) It follows from Lemma 35.2 that if σ is a simplex of K, then $h(\text{Int } \sigma)$ cannot intersect both $A - C$ and e_3. It follows that $h(\sigma)$ lies either in A or in $\bar{e}_3$. Thus A and $\bar{e}_3$ are triangulated by h, and thus so is $A \cap \bar{e}_3 = C$.

[†]The dimensional condition is in fact redundant. See Exercise 2.

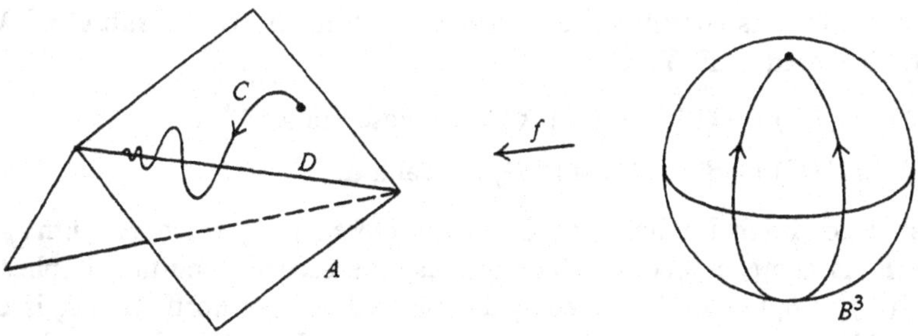

Figure 38.5

On the other hand, similar reasoning shows that D is triangulated by h: Consider the local homology groups $H_i(A,A - x)$ of A. At each point x interior to D, $H_2(A,A - x) \simeq \mathbf{Z} \oplus \mathbf{Z}$. (For A can be expressed as a cone with vertex x and base homeomorphic to a "θ curve"; by the long exact homology sequence, $H_2(A,A - x) \simeq H_1(\theta) \simeq \mathbf{Z} \oplus \mathbf{Z}$.) At each point x of A not in D, either $H_2(A,A - x) \simeq \mathbf{Z}$ (if x is interior to one of the triangles) or $H_2(A,A - x) = 0$. We conclude that if $h(\text{Int}\,\sigma)$ intersects the interior of D, then it lies in D. Then D is triangulated by h.

It follows that $C \cap D$ is triangulated by h. This is impossible, since $C \cap D$ has infinitely many components and K is a finite complex.

It is often helpful to view a CW complex as a space built up from a collection of closed balls by forming appropriate quotient spaces, or to construct a CW complex in this way, as we did in the preceding example. The following two theorems show how this is done.

The **dimension** of a CW complex X is the largest dimension of a cell of X, if such exists; otherwise it is said to be infinite.

Theorem 38.2. (a) *Suppose X is a CW complex of dimension p. Then X is homeomorphic to an adjunction space formed from X^{p-1} and a topological sum ΣB_α of closed p-balls, by means of a continuous map $g : \Sigma\, \text{Bd}\, B_\alpha \to X^{p-1}$. It follows that X is normal.*

(b) *Conversely, let Y be a CW complex of dimension at most $p - 1$, let ΣB_α be a topological sum of closed p-balls, and let $g : \Sigma\, \text{Bd}\, B_\alpha \to Y$ be a continuous map. Then the adjunction space X formed from Y and ΣB_α by means of g is a CW complex, and Y is its $p - 1$ skeleton.*

Proof. (a) For each cell e_α of X of dimension p, one is given the characteristic map $f_\alpha : B^p \to \bar{e}_\alpha$. Let $B_\alpha = B^p \times \{\alpha\}$, and let ΣB_α be the topological sum of these disjoint p-balls. Form the topological sum

$$E = X^{p-1} \cup (\Sigma B_\alpha),$$

and define $\pi : E \to X$ by letting π equal inclusion on X^{p-1} and the composite

$$B_\alpha = B^p \times \{\alpha\} \to B^p \xrightarrow{f_\alpha} X$$

on B_α. To prove (a), it will suffice to show that π is a quotient map.

Obviously π is continuous and surjective. Suppose C is a subset of X and $\pi^{-1}(C)$ is closed in E. Then:

(1) $\pi^{-1}(C) \cap X^{p-1} = C \cap X^{p-1}$ is closed in X^{p-1}.

(2) $\pi^{-1}(C) \cap B_\alpha$ is closed in B_α for each α.

The first condition implies that $C \cap \bar{e}_\beta$ is closed in $\bar{e}_\beta$ whenever $\dim e_\beta < p$. Since B_α is compact and π is continuous, the second condition implies that $\pi(\pi^{-1}(C) \cap B_\alpha) = C \cap \bar{e}_\alpha$ is compact. Since X is Hausdorff, $C \cap \bar{e}_\alpha$ is closed in X and hence closed in $\bar{e}_\alpha$, whenever $\dim e_\alpha = p$. Thus C is closed in X, so π is a quotient map, as desired.

Normality of X follows from Theorem 37.2; one proceeds by induction on p.

(b) Let $f : Y \cup (\Sigma B_\alpha) \rightarrow X$ be the hypothesized quotient map. Now Y is normal, by part (a), and ΣB_α is normal. It follows from Theorem 37.2 that X is normal (and in particular, Hausdorff). As usual, we consider Y to be a subspace of the adjunction space X; then f equals inclusion on Y, and f equals g on $\Sigma \operatorname{Bd} B_\alpha$. We define the open cells of X to be the cells $\{e_\beta\}$ of Y (having dimension less than p), and the cells $e_\alpha = f(\operatorname{Int} B_\alpha)$, (having dimension p). Since $\operatorname{Int} B_\alpha$ is open in the topological sum $Y \cup (\Sigma B_\alpha)$, and it is saturated relative to f, the restriction of f to $\operatorname{Int} B_\alpha$ is a quotient map. Being one-to-one, it is a homeomorphism. Thus f maps $\operatorname{Int} B_\alpha$ homeomorphically onto e_α, so e_α is an open p-cell, as desired.

We check condition (2) for a CW complex. We have already noted that the map $f_\alpha = f|B_\alpha$ maps $\operatorname{Int} B_\alpha$ homeomorphically onto the set e_α. By construction, f_α maps $\operatorname{Bd} B_\alpha$ into Y, which is the union of cells of dimension less than p. Because $\operatorname{Bd} B_\alpha$ is compact, the set $f(\operatorname{Bd} B_\alpha)$ is a compact subset of Y; because Y is a CW complex, it intersects only finitely many open cells of Y. Thus condition (2) is satisfied.

Condition (3) follows readily. Suppose C is a subset of X and $C \cap \bar{e}_\alpha$ is closed in $\bar{e}_\alpha$ for each open cell e_α. We show $f^{-1}(C)$ is closed in $Y \cup (\Sigma B_\alpha)$; from this it follows that C is closed in X.

First note that $f^{-1}(C) \cap Y = C \cap Y$; because $C \cap \bar{e}_\beta$ is closed in $\bar{e}_\beta$ for each cell of dimension less than p, $C \cap Y$ is closed in Y. Similarly,

$$f^{-1}(C) \cap B_\alpha = f^{-1}(C \cap \bar{e}_\alpha) \cap B_\alpha.$$

We use here the fact that $f(B_\alpha) = \bar{e}_\alpha$. Now $C \cap \bar{e}_\alpha$ is closed in $\bar{e}_\alpha$ by hypothesis, and hence closed in X. We apply continuity of f to see that $f^{-1}(C \cap \bar{e}_\alpha) \cap B_\alpha$ is closed in B_α. Thus $f^{-1}(C)$ is closed, as desired. □

Theorem 38.3. (a) *Let X be a CW complex. Then X^p is a closed subspace of X^{p+1} for each p, and X is the coherent union of the spaces $X^0 \subset X^1 \subset \cdots$. It follows that X is normal.*

(b) *Conversely, suppose X_p is a CW complex for each p, and X_p equals the*

p-skeleton of X_{p+1} for each p. If X is the coherent union of the spaces X_p, then X is a CW complex having X_p as its p-skeleton.

Proof. (a) Suppose $C \cap X^p$ is closed in X^p for each p. Then $C \cap \bar{e}_\alpha$ is closed in $\bar{e}_\alpha$ for each cell e_α of dimension at most p. Since p is arbitrary, we conclude that C is closed in X. Thus X has the topology coherent with the subspaces X^p. Normality follows from Theorem 37.4.

(b) If $p < q$, then $X_p \cap X_q = X_p$ is a closed subspace of both X_p and X_q. Therefore by (b) of Lemma 37.3, there is a topology on X coherent with the subspaces X_p, and each X_p is closed in X. By the preceding theorem, each space X_p is normal; therefore, by Theorem 37.4, X is normal (and in particular, Hausdorff). Condition (2) for a CW complex is trivial. We check condition (3). Suppose $C \cap \bar{e}_\alpha$ is closed in $\bar{e}_\alpha$ for each cell e_α. Then $C \cap X_p$ is closed in X_p, because X_p is a CW complex. Then C is closed in X because the topology of X is coherent with the spaces X_p. □

Sometimes one begins with a space X and seeks to give it the structure of CW complex. This we did in Examples 1–3. On the other hand, sometimes one seeks to construct *new* spaces that are CW complexes by the process of pasting balls together. This is what we did in Example 7. The general construction is described in the two theorems we just proved. This construction significantly enlarges the class of spaces about which algebraic topology has something interesting to say.

EXERCISES

1. Let X be a CW complex; let A be a compact subset of X.
   (a) Show that A intersects only finitely many open cells of X. Where do you use "closure-finiteness" in the proof?
   (b) Show that A lies in a finite subcomplex of X.
2. Let X be a CW complex. Suppose $h : |K| \to X$ is a triangulation of X that induces a triangulation of each skeleton X^p.
   (a) Show that h induces a triangulation of each set $\bar{e}_\alpha$.
   (b) Show that if K_p is the subcomplex of K triangulating X^p, then K_p has dimension at most p. [*Hint:* Use local homology groups.]
3. Let X be a CW complex. Show that the topology of X is compactly generated. (See Exercise 3 of §37.)
4. Let X and Y be CW complexes. Then $X \times Y$ is the union of open cells $e_\alpha \times e_\beta$, for e_α a cell of X and e_β a cell of Y.
   (a) Show that if Y is locally compact, then $X \times Y$ is a CW complex.
   (b) Show that $X \times_c Y$ is a CW complex in general. (See Exercise 3 of §37.)
5. Verify that a regular cell complex can be triangulated so that each closed cell is the polytope of a subcomplex.

§39. THE HOMOLOGY OF CW COMPLEXES

We now show how to compute the singular homology of a CW complex.

Throughout this section, X will denote a CW complex with open cells e_α and characteristic maps f_α. The symbol H_p will denote singular homology in general, but if it happens that X is a triangulable CW complex, then H_p can also be taken to denote simplicial homology, since there is a natural isomorphism between singular and simplicial theory.

Definition. If X is a CW complex, let

$$D_p(X) = H_p(X^p, X^{p-1}).$$

Let $\partial : D_p(X) \to D_{p-1}(X)$ be defined to be the composite

$$H_p(X^p, X^{p-1}) \xrightarrow{\partial_*} H_{p-1}(X^{p-1}) \xrightarrow{j_*} H_{p-1}(X^{p-1}, X^{p-2}),$$

where j is inclusion. The fact that $\partial^2 = 0$ follows from the fact that

$$H_{p-1}(X^{p-1}) \xrightarrow{j_*} H_{p-1}(X^{p-1}, X^{p-2}) \xrightarrow{\partial_*} H_{p-2}(X^{p-2})$$

is exact. The chain complex $\mathcal{D}(X) = \{D_p(X), \partial\}$ is called the **cellular chain complex** of X.

Example 1. Consider the case where X is the space of a simplicial complex K, and the open cells of X are the open simplices of K. Let H_p denote ordinary simplicial homology. We compute $H_p(X^p, X^{p-1})$. The simplicial chain group $C_i(K^{(p)}, K^{(p-1)})$ vanishes if $i \neq p$, and it equals the chain group $C_p(K^{(p)}) = C_p(K)$ when $i = p$. Therefore,

$$H_p(X^p, X^{p-1}) = H_p(K^{(p)}, K^{(p-1)}) = C_p(K).$$

Furthermore, the boundary operator in the cellular chain complex is just the ordinary simplicial boundary operator of K. It follows that in this case at least, the cellular chain complex can be used to compute the homology of X.

Our goal is to prove in general that if X is a CW complex, the cellular chain complex $\mathcal{D}(X)$ behaves very much like the simplicial chain complex $\mathcal{C}(K)$. In particular, we show that the group $D_p(X)$ is free abelian with a basis consisting of, roughly speaking, the oriented p-cells. And we show the chain complex $\mathcal{D}(X)$ can be used to compute the singular homology of X.

We begin with a sequence of lemmas.

Lemma 39.1. *Given an open p-cell e_α of X, any characteristic map for e_α,*

$$f_\alpha : (B^p, S^{p-1}) \to (\bar{e}_\alpha, \dot{e}_\alpha),$$

induces an isomorphism in relative homology.

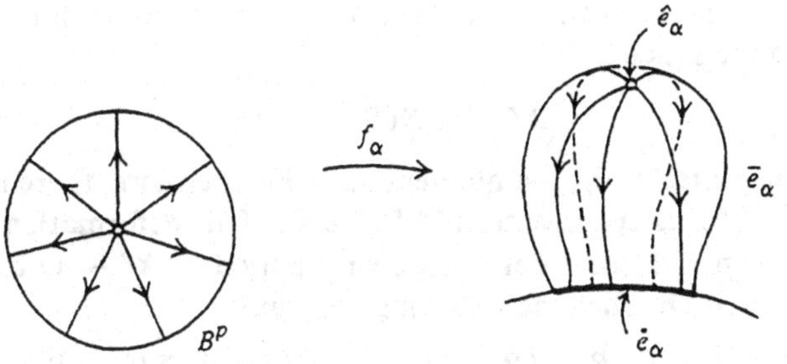

Figure 39.1

Proof. If $p = 0$, the result is trivial. Let $p > 0$. The point $\mathbf{0}$ is the center of B^p; let $\hat{e}_\alpha$ denote $f_\alpha(\mathbf{0})$. Note that because f_α is a quotient map, so is its restriction

$$f'_\alpha : (B^p - \mathbf{0}) \to (\bar{e}_\alpha - \hat{e}_\alpha)$$

to the saturated open set $B^p - \mathbf{0}$. Now S^{p-1} is a deformation retract of $B^p - \mathbf{0}$; this deformation retraction induces, via the quotient map

$$f'_\alpha \times i_I : (B^p - \mathbf{0}) \times I \to (\bar{e}_\alpha - \hat{e}_\alpha) \times I,$$

a deformation retraction of $\bar{e}_\alpha - \hat{e}_\alpha$ onto $\dot{e}_\alpha$. See Figure 39.1.

It follows that the horizontal inclusion maps on the left side of the following diagram,

$$\begin{array}{ccccc} (B^p,S^{p-1}) & \to & (B^p,B^p - \mathbf{0}) & \leftarrow & (\text{Int } B^p, \text{Int } B^p - \mathbf{0}) \\ \downarrow f_\alpha & & \downarrow f_\alpha & & \downarrow f_\alpha \\ (\bar{e}_\alpha,\dot{e}_\alpha) & \to & (\bar{e}_\alpha,\bar{e}_\alpha - \hat{e}_\alpha) & \leftarrow & (e_\alpha, e_\alpha - \hat{e}_\alpha) \end{array}$$

induce isomorphisms in homology, by Theorem 30.8. Since the horizontal inclusion maps on the right side of the diagram are excision maps, they also induce homology isomorphisms. (On the top line, one excises S^{p-1}; on the second line, one excises $\dot{e}_\alpha$.) Now the map $f_\alpha : \text{Int } B^p \to e_\alpha$ is a homeomorphism that carries $\mathbf{0}$ to $\hat{e}_\alpha$. Therefore, the vertical map f_α at the right of the diagram induces a homology isomorphism. Our result follows. □

Lemma 39.2. *Let the map*

$$f : X^{p-1} \cup \Sigma B_\alpha \to X^p$$

express X^p as the adjunction space obtained from X^{p-1} and a topological sum of p-balls ΣB_α via a map $g : \Sigma S_\alpha \to X^{p-1}$, where $S_\alpha =$ Bd B_α. *Then f induces a homology isomorphism*

$$H_i(\Sigma B_\alpha, \Sigma S_\alpha) \simeq H_i(X^p, X^{p-1}).$$

Proof. The proof is similar to that of the preceding lemma. The restriction f' of f to the space

$$X^{p-1} \cup \Sigma(B_\alpha - \mathbf{0}_\alpha),$$

where $\mathbf{0}_\alpha$ is the center of B_α, is a quotient map. Furthermore, there is a deformation retraction of this space onto $X^{p-1} \cup \Sigma S_\alpha$. This deformation induces, via the quotient map $f' \times i_I$, a deformation retraction of $X^p - \cup \hat{e}_\alpha$ onto X^{p-1}, where $\hat{e}_\alpha = f(\mathbf{0}_\alpha)$. One has the following diagram:

$$\begin{array}{ccccc} (\Sigma B_\alpha, \Sigma S_\alpha) & \longrightarrow & (\Sigma B_\alpha, \Sigma(B_\alpha - \mathbf{0}_\alpha)) & \longleftarrow & (\Sigma \operatorname{Int} B_\alpha, \Sigma((\operatorname{Int} B_\alpha) - \mathbf{0}_\alpha)) \\ \downarrow f & & \downarrow f & & \downarrow f \\ (X^p, X^{p-1}) & \longrightarrow & (X^p, X^p - \cup \hat{e}_\alpha) & \longleftarrow & (\cup e_\alpha, \cup (e_\alpha - \hat{e}_\alpha)) \end{array}$$

The map f on the right is a homeomorphism, being a one-to-one quotient map. The horizontal maps are inclusions, and they induce homology isomorphisms for the same reasons as before. □

Theorem 39.3. *The group $H_i(X^p, X^{p-1})$ vanishes for $i \neq p$, and is free abelian for $i = p$. If γ generates $H_p(B^p, S^{p-1})$, then the elements $(f_\alpha)_*(\gamma)$ form a basis for $H_p(X^p, X^{p-1})$, as f_α ranges over a set of characteristic maps for the p-cells of X.*

Proof. The preceding lemma tells us that

$$H_i(X^p, X^{p-1}) \simeq H_i(\Sigma B_\alpha, \Sigma S_\alpha),$$

where ΣB_α is a topological sum of p-balls and $S_\alpha = \operatorname{Bd} B_\alpha$. Because the sets B_α are disjoint open sets in ΣB_α, this group is isomorphic to the direct sum $\bigoplus H_i(B_\alpha, S_\alpha)$. The theorem follows. □

Definition. Given a triple $X \supset A \supset B$ of spaces, one has a short exact sequence of chain complexes

$$0 \longrightarrow \frac{S_p(A)}{S_p(B)} \longrightarrow \frac{S_p(X)}{S_p(B)} \longrightarrow \frac{S_p(X)}{S_p(A)} \longrightarrow 0.$$

It gives rise to the following sequence, which is called the **exact homology sequence of a triple:**

$$\cdots \longrightarrow H_p(A,B) \longrightarrow H_p(X,B) \longrightarrow H_p(X,A) \longrightarrow H_{p-1}(A,B) \longrightarrow \cdots.$$

This sequence was mentioned earlier in the exercises. As usual, a continuous map $f:(X,A,B) \to (Y,C,D)$ induces a homomorphism of the corresponding exact homology sequences.

In the case $(X,A,B) = (X^p, X^{p-1}, X^{p-2})$, the boundary operator ∂_* in the above sequence equals the boundary operator ∂ of the cellular chain complex

$\mathcal{D}(X)$. This follows from the fact that ∂_* commutes with the homomorphism j_* induced by inclusion:

$$\begin{array}{ccc} \to H_p(X^p,X^{p-1}) & \xrightarrow{\partial_*} & H_{p-1}(X^{p-1},\varnothing) \to \\ = \downarrow & & \downarrow j_* \\ \to H_p(X^p,X^{p-1}) & \xrightarrow{\partial_*} & H_{p-1}(X^{p-1},X^{p-2}) \to \end{array}$$

Using this fact, we now prove that the cellular chain complex of X can be used to compute the homology of X. For later purposes, we are going to prove this theorem in a somewhat more general form. We shall assume that we have a space X that is written as the union of a sequence of subspaces

$$X_0 \subset X_1 \subset X_2 \subset \cdots .$$

Then we form the chain complex whose p-dimensional chain group is $H_p(X_p,X_{p-1})$ and whose boundary operator is the boundary homomorphism in the exact sequence of the triple (X_p,X_{p-1},X_{p-2}). We shall show that under suitable hypotheses (which are satisfied in the case of a CW complex) this chain complex gives the homology of X.

Definition. If X is a space, a **filtration** of X is a sequence $X_0 \subset X_1 \subset \cdots$ of subspaces of X whose union is X. A space X together with a filtration of X is called a **filtered space.** If X and Y are filtered spaces, a continuous map $f: X \to Y$ such that $f(X_p) \subset Y_p$ for all p is said to be **filtration-preserving.**

Theorem 39.4. *Let X be filtered by the subspaces $X_0 \subset X_1 \subset \cdots$; let $X_i = \varnothing$ for $i < 0$. Assume that $H_i(X_p,X_{p-1}) = 0$ for $i \neq p$. Suppose also that given any compact set C in X, there is an n such that $C \subset X_n$. Let $\mathcal{D}(X)$ be the chain complex defined by setting $D_p(X) = H_p(X_p,X_{p-1})$ and letting the boundary operator be the boundary homomorphism ∂_* in the exact sequence of a triple. Then there is an isomorphism*

$$\lambda: H_p(\mathcal{D}(X)) \to H_p(X).$$

It is natural with respect to homomorphisms induced by filtration-preserving continuous maps.

Proof. As motivation for the proof, let us consider the situation of Example 1, where X_p is the p-skeleton $K^{(p)}$ of a simplicial complex and H_p denotes simplicial homology. Then $\mathcal{D}(X) = \mathcal{C}(K)$ in this case, and the theorem holds. It is also true in this case that

$$H_p(K) = H_p(K^{(p+1)},K^{(p-2)}),$$

because only chains of dimensions $p+1$, p, and $p-1$ are used in defining $H_p(K)$. We shall show in Steps 1 and 2 that an analogous result holds in the present situation.

Step 1. We show that the homomorphism

$$i_*: H_p(X_{p+1}) \to H_p(X)$$

induced by inclusion is an isomorphism. For this purpose, we first note that the homomorphisms

$$H_p(X_{p+1}) \to H_p(X_{p+2}) \to H_p(X_{p+3}) \to \cdots$$

induced by inclusion are isomorphisms. This follows by examining the exact sequence

$$H_{p+1}(X_{p+i+1}, X_{p+i}) \to H_p(X_{p+i}) \to H_p(X_{p+i+1}) \to H_p(X_{p+i+1}, X_{p+i})$$

and noting that both end groups vanish if $i \geq 1$, by hypothesis.

Our result now follows from the compact support properties of homology. To show that i_* is surjective, let β be an element of $H_p(X)$. Choose a compact set C in X such that β is in the image of $H_p(C) \to H_p(X)$ under the homomorphism induced by inclusion. Since C is compact, $C \subset X_{p+k}$ for some k. Then β is the image of an element of $H_p(X_{p+k})$ in the diagram

$$H_p(X_{p+1}) \to H_p(X_{p+k}) \to H_p(X).$$

Because the first of these homomorphisms is an isomorphism, β is the image of an element of $H_p(X_{p+1})$, as desired.

To show that i_* has kernel 0, suppose $\beta \in H_p(X_{p+1})$ maps to zero in $H_p(X)$. There is a compact set C such that β maps to zero in $H_p(C)$. Again, C lies in X_{p+k} for some k. Then

$$H_p(X_{p+1}) \to H_p(X_{p+k})$$

carries β to zero; because this map is an isomorphism, $\beta = 0$.

Step 2. We show that the homomorphism

$$j_*: H_p(X_{p+1}) \to H_p(X_{p+1}, X_{p-2})$$

induced by inclusion is an isomorphism.

This result will follow once we show that the homomorphisms

$$H_p(X_{p+1}, \varnothing) \to H_p(X_{p+1}, X_0) \to \cdots \to H_p(X_{p+1}, X_{p-2})$$

induced by inclusion are isomorphisms. To prove this fact, consider the exact sequence of the triple (X_{p+1}, X_i, X_{i-1}):

$$H_p(X_i, X_{i-1}) \to H_p(X_{p+1}, X_{i-1}) \to H_p(X_{p+1}, X_i) \to H_{p-1}(X_i, X_{i-1}).$$

Both end groups vanish for $i \leq p-2$, by hypothesis; therefore, the middle homomorphism is an isomorphism.

Step 3. We now prove the theorem. Given a quadruple $X \supset A \supset B \supset C$ of spaces, one has four "exact sequences of a triple" associated with this quadruple. They are most conveniently arranged in the form of four overlapping sine curves. We shall consider the special case

$$(X,A,B,C) = (X_{p+1},X_p,X_{p-1},X_{p-2}).$$

In this case, the groups in the upper left and right corners of the diagram, $H_p(B,C)$ and $H_p(X,A)$, vanish, by hypothesis.

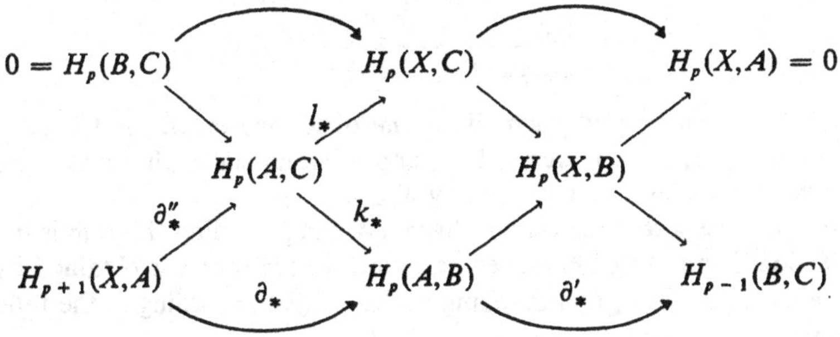

Now k_* carries $H_p(A,C)$ isomorphically onto $\ker \partial'_*$. Furthermore, the map $l_* \circ k_*^{-1}$ carries $\ker \partial'_*$ *onto* $H_p(X,C)$ (because l_* is surjective); its kernel is just $\operatorname{im} \partial_*$ (because the kernel of l_* equals $\operatorname{im} \partial''_*$). Thus $l_* \circ k_*^{-1}$ induces an isomorphism

$$\frac{\ker \partial'_*}{\operatorname{im} \partial_*} \simeq H_p(X,C) = H_p(X_{p+1},X_{p-2}).$$

(A more general result concerning this diagram was given in Exercise 1 of §26.)

Combining this result with those of Steps 1 and 2, we obtain our desired isomorphism

$$H_p(X) \simeq H_p(X_{p+1}) \simeq H_p(X_{p+1},X_{p-2}) \simeq \frac{\ker \partial'_*}{\operatorname{im} \partial_*}.$$

The latter group equals $H_p(\mathcal{D}(X))$.

Naturality of the isomorphism is easy to check. If $f: X \to Y$ preserves filtrations, the first two of the preceding isomorphisms obviously commute with f_*. Furthermore, f_* carries the diagram of Step 3 for X into the corresponding diagram for Y. It follows that the third isomorphism commutes with f_* as well. □

We now prove an addendum to this theorem in the case where X is triangulable.

Theorem 39.5. *Let X be filtered by the subspaces $X_0 \subset X_1 \subset \cdots$; suppose that X is the space of a simplicial complex K, and each subspace X_p is the space of a subcomplex of K of dimension at most p. Let H_i denote simplicial homology. Suppose $H_i(X_p,X_{p-1}) = 0$ for $i \neq p$. Then $H_p(X_p,X_{p-1})$ equals a subgroup of $C_p(K)$, and the isomorphism λ of the preceding theorem is induced by inclusion.*

Indeed, $H_p(X_p,X_{p-1})$ is the subgroup of $C_p(K)$ consisting of all p-chains of K carried by X_p whose boundaries are carried by X_{p-1}.

Proof. Any compact set in X lies in a finite subcomplex of K, so it lies in X_i for some i. Therefore, the hypotheses of the preceding theorem are satisfied. Since X_p contains no $p+1$ simplices, the homology group $H_p(X_p,X_{p-1})$ equals the group of relative p-cycles, which is the kernel of the homomorphism

$$\frac{C_p(X_p)}{C_p(X_{p-1})} \xrightarrow{\partial} \frac{C_{p-1}(X_p)}{C_{p-1}(X_{p-1})}.$$

Because X_{p-1} contains no p-simplices, the denominator on the left side vanishes. Thus $H_p(X_p,X_{p-1})$ equals the group of simplicial p-chains of K carried by X_p whose boundaries are carried by X_{p-1}.

We must check that the isomorphism λ of the preceding theorem is induced by inclusion. Examining the preceding proof, we see that λ is obtained by taking an element of $\ker \partial'_*$ and mapping it into $H_p(X)$ according to the following diagram:

$$\begin{array}{ccccccc} H_p(X_p,X_{p-2}) & \xrightarrow{l_*} & H_p(X_{p+1},X_{p-2}) & \xleftarrow[\simeq]{j_*} & H_p(X_{p+1}) & \xrightarrow[\simeq]{i_*} & H_p(X). \\ \simeq \big\downarrow k_* & & & & & & \\ \ker \partial'_* & & & & & & \end{array}$$

Since each map is induced by inclusion, our theorem follows. □

We now see how strong the analogy is between the homology of simplicial complexes and the homology of CW complexes. Let us introduce some terminology that will make the analogy even stronger.

For each open p-cell e_α of the CW complex X, the group $H_p(\bar{e}_\alpha, \dot{e}_\alpha)$ is infinite cyclic. The two generators of this group will be called the two **orientations** of e_α. An **oriented *p*-cell** of X is an open p-cell e_α together with an orientation of e_α.

The cellular chain group $D_p(X) = H_p(X^p,X^{p-1})$ is a free abelian group. One obtains a basis for it by orienting each open p-cell e_α of X and passing to the corresponding element of $H_p(X^p,X^{p-1})$. [That is, by taking the image of the orientation under the homomorphism induced by inclusion

$$H_p(\bar{e}_\alpha, \dot{e}_\alpha) \to H_p(X^p,X^{p-1}).]$$

The homology of the chain complex $\mathcal{D}(X)$ is isomorphic, by our theorem, with the singular homology of X.

In the special case where X is a triangulable CW complex triangulated by a complex K, and H_p denotes simplicial homology, we interpret these comments as follows: The fact that X^p and X^{p-1} are subcomplexes of K implies that each open p-cell e_α is a union of open simplices of K, so that $\bar{e}_\alpha$ is the polytope of a subcomplex of K. The group $H_p(\bar{e}_\alpha, \dot{e}_\alpha)$ equals the group of p-chains carried by $\bar{e}_\alpha$ whose boundaries are carried by $\dot{e}_\alpha$. This group is infinite cyclic; either generator of this group is called a **fundamental cycle** for $(\bar{e}_\alpha, \dot{e}_\alpha)$.

The cellular chain group $D_p(X)$ equals the group of all simplicial p-chains of X carried by X^p whose boundaries are carried by X^{p-1}. Any such p-chain can be written uniquely as a finite linear combination of fundamental cycles for those pairs $(\bar{e}_\alpha, \dot{e}_\alpha)$ for which $\dim e_\alpha = p$.

Let us interpret these results in some familiar situations.

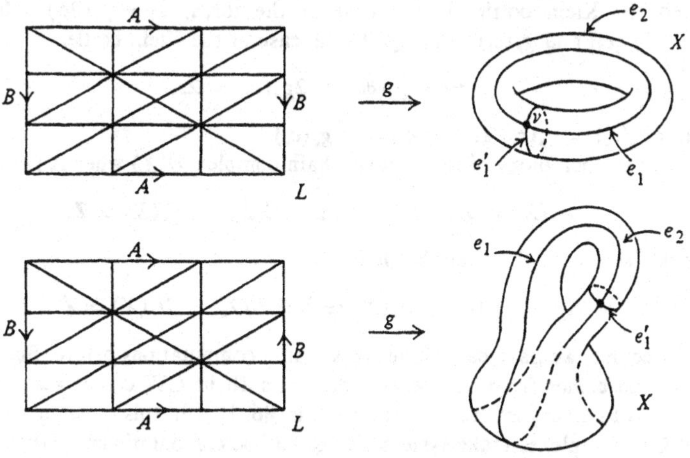

Figure 39.2

Example 2. Let X denote either the torus or the Klein bottle, expressed as a quotient space of the rectangle L in the usual way. See Figure 39.2.

Then X is a triangulable CW complex, having one open 2-cell e_2, two open 1-cells e_1 and e_1' (which are the images of A and B, respectively), and one 0-cell e_0. Then

$$D_2(X) \simeq \mathbf{Z}, \qquad D_1(X) \simeq \mathbf{Z} \oplus \mathbf{Z}, \qquad D_0(X) \simeq \mathbf{Z}.$$

Let us find specific generators for these chain groups. The 2-chain d of L that is the sum of all the 2-simplices of L, oriented counterclockwise, is by inspection a cycle of $(L, \text{Bd } L)$. Because d is a multiple of no other cycle, it is a fundamental cycle for $(L, \text{Bd } L)$. By Lemma 39.1, $\gamma = g_\#(d)$ is a fundamental cycle for $(\bar{e}_2, \dot{e}_2)$.

Let c_1 be the sum of the 1-simplices along the top of L, oriented as indicated in Figure 39.3. Let c_2, c_3, and c_4 denote chains along the other edges of L, as indicated. Now $w_1 = g_\#(c_1)$ is a fundamental cycle for $(\bar{e}_1, \dot{e}_1)$. So is $g_\#(c_3)$, of course. Similarly, $z_1 = g_\#(c_2)$ is a fundamental cycle for $(\bar{e}_1', \dot{e}_1')$, as is $g_\#(c_4)$.

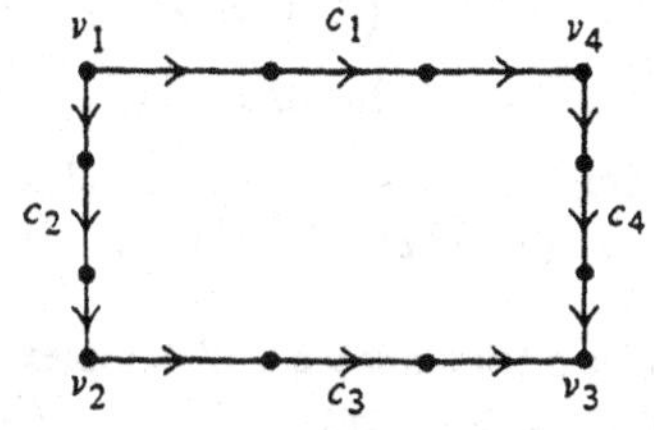

Figure 39.3

In terms of these basis elements, it is easy to compute the boundary operator in the cellular chain complex $\mathcal{D}(X)$. We first compute ∂ in the complex L as follows:

$$\partial c_1 = v_4 - v_1, \qquad \partial c_2 = v_2 - v_1,$$
$$\partial d = -c_1 + c_2 + c_3 - c_4.$$

Applying $g_\#$, we see that $\partial w_1 = g_\#(\partial c_1) = 0$ and $\partial z_1 = g_\#(\partial c_2) = 0$ for both the torus and the Klein bottle. In the case of the torus, $\partial\gamma = g_\#(\partial d) = 0$ because $g_\#(c_1) = g_\#(c_3)$ and $g_\#(c_2) = g_\#(c_4)$. In the case of the Klein bottle,

$$\partial\gamma = g_\#(\partial d) = 2g_\#(c_2) = 2z_1$$

because $g_\#(c_1) = g_\#(c_3)$ and $g_\#(c_2) = -g_\#(c_4)$.

Thus the homology of the cellular chain complex $\mathcal{D}(X)$ when X is the torus is

$$H_2(X) \simeq \mathbf{Z}, \qquad H_1(X) \simeq \mathbf{Z} \oplus \mathbf{Z}, \qquad H_0(X) \simeq \mathbf{Z},$$

while in the case of the Klein bottle it is

$$H_2(X) = 0, \qquad H_1(X) \simeq \mathbf{Z} \oplus \mathbf{Z}/2, \qquad H_0(X) \simeq \mathbf{Z}.$$

Of course, we have carried out these same computations before. But now their justification comes from our general theorems about CW complexes, rather than from the ad hoc arguments we used back in §6. It is in this sense that our results about CW complexes make systematic the ad hoc computational methods we studied there.

Example 3. Let S^n be an n-sphere. Assume $n > 1$ for convenience. We can make S^n into a CW complex having one open cell in dimension n and one cell in dimension 0. It follows that the cellular chain complex of S^n is infinite cyclic in dimensions n and 0, and vanishes otherwise. Therefore, $H_n(S^n) \simeq \mathbf{Z}$ and $H_0(S^n) \simeq \mathbf{Z}$, while $H_i(S^n) = 0$ for $i \neq 0, n$. A similar computation applies when $n = 1$.

EXERCISES

1. Recompute the homology of the n-fold connected sums

$$T \# \cdots \# T \quad \text{and} \quad P^2 \# \cdots \# P^2$$

by expressing them as triangulable CW complexes and finding the corresponding cellular chain complexes.

2. Let A be a closed subset of X; suppose A is a deformation retract of an open set in X. If X/A is the space obtained by collapsing A to a point, show that

$$H_p(X,A) \simeq \tilde{H}_p(X/A).$$

[*Hint:* Examine the proof of Lemma 39.1.]

3. Let X be a CW complex; let A be a subcomplex. Show that inclusion induces a monomorphism $D_p(A) \to D_p(X)$. The quotient $D_p(X)/D_p(A)$ is denoted $D_p(X,A)$.
   (a) Show that if X is a triangulable CW complex, $\mathcal{D}(X,A)$ can be used to compute the simplicial homology of (X,A). [*Hint:* Use the long exact se-

quences to show that inclusion $D_p(X,A) \to C_p(X,A)$ induces a homology isomorphism.]

(b) Show that in general, $\mathcal{D}(X,A)$ can be used to compute the singular homology of (X,A). [*Hint:* Show that $\mathcal{D}(X,A)$ is isomorphic to the chain complex whose pth chain group is $H_p(X^p \cup A, X^{p-1} \cup A)$ and whose boundary operator comes from the exact sequence of a triple. Let $X_p = X^p \cup A$ for all integers p. Repeat the proof of Theorem 39.4, replacing $H_i(X_p)$ by $H_i(X_p,A)$ and $H_i(X)$ by $H_i(X,A)$.]

*4. *Theorem.* *Let $\mathcal{C} = \{C_p,\partial\}$ be a non-negative free chain complex such that $H_0(\mathcal{C})$ is free and non-trivial. Then there is a CW complex X whose cellular chain complex is isomorphic to $\mathcal{C}$.*

Proof. (a) Show that if $n \geq 1$, given an n-simplex σ and a homomorphism

$$\phi : H_n(\sigma, \text{Bd}\,\sigma) \to H_n(S^n,x_0),$$

ϕ is the homomorphism induced by some continuous map. (See Exercise 3 of §25.)

(b) Show that if X is a CW complex consisting of a collection of n-spheres with a point p in common ($n > 0$), and if $\alpha \in H_n(X,p)$, there is a map $f : (S^n,x_0) \to (X,p)$ whose induced homomorphism carries a generator of $H_n(S^n,x_0)$ to α.

(c) Show that for $p > 0$ one can write $C_p = U_p \oplus Z_p$, where Z_p is the group of p-cycles, and for $p = 0$ one can write $C_0 = \partial U_1 \oplus A$, where A is non-trivial.

(d) Complete the proof.

5. Let $G_0, G_1, \ldots$ be a sequence of abelian groups with G_0 free and non-trivial. Assuming Exercise 4, show that there is a CW complex X such that $H_i(X) \simeq G_i$ for all i.

*§40. APPLICATION: PROJECTIVE SPACES AND LENS SPACES[†]

We now apply the theory of CW complexes to compute the homology of certain spaces that are of particular importance in topology and geometry—the projective spaces. We also study those classical 3-manifolds called the lens spaces.

Definition. Let us introduce an equivalence relation on the n-sphere S^n by defining $x \sim -x$ for each $x \in S^n$. The resulting quotient space is called (real) **projective n-space** and denoted P^n.

The quotient map $p : S^n \to P^n$ is a closed map. For if A is closed in S^n, then the saturation $p^{-1}(p(A))$ of A equals the set $A \cup a(A)$, where $a : S^n \to S^n$ is the antipodal map. Because a is a homeomorphism, $A \cup a(A)$ is closed in S^n, so (by definition of quotient space) the set $p(A)$ is closed in P^n.

Therefore, P^n is Hausdorff (in fact, normal).

[†]The results of this section will be used when we compute the cohomology rings of these spaces, in §68 and §69.

If we consider $\mathbf{R}^n$ to be the set of all real sequences $(x_1, x_2, \ldots)$ such that $x_i = 0$ for $i > n$, then $\mathbf{R}^n \subset \mathbf{R}^{n+1}$. As a result, $S^{n-1} \subset S^n$; in fact, S^{n-1} is the intersection of S^n with the plane $x_{n+1} = 0$. Now the equivalence relation $x \sim -x$ is the same in S^{n-1} as it is in S^n; therefore, $P^{n-1} \subset P^n$. In fact, P^{n-1} is a closed *subspace* of P^n. For if C is a subset of P^{n-1}, then $p^{-1}(C)$ is closed in S^{n-1} if and only if it is closed in S^n. Hence C is closed in P^{n-1} if and only if it is closed in P^n.

Theorem 40.1. *The space P^n is a CW complex having one cell in each dimension $0 \leq j \leq n$; its j-skeleton is P^j.*

Proof. The space P^0 is obtained from the 2-point space S^0 by identifying these two points. Thus P^0 consists of a single point.

We proceed by induction. Suppose we restrict the map $p: S^n \to P^n$ to the closed upper hemisphere E_+^n of S^n. Because S^n is compact and P^n is Hausdorff, the map $p' = p \mid E_+^n$ is a quotient map; and it maps E_+^n *onto* P^n because each equivalence class $\{x, -x\}$ contains at least one point of E_+^n. Its restriction to the open upper hemisphere Int E_+^n is also a quotient map; being one-to-one, it is a homeomorphism of Int E_+^n with $P^n - P^{n-1}$. Thus $P^n - P^{n-1}$ is an open n-cell. Call it e_n.

Now the map p' carries Bd $E_+^n = S^{n-1}$ onto P^{n-1}, which by the induction hypothesis is the union of finitely many open cells of dimensions less than n. Thus when we identify E_+^n with B^n, the map p' becomes a characteristic map for e_n. It follows that P^n is a CW complex with one open cell in each dimension $0 \leq j \leq n$. □

Note that P^1, having one open 1-cell and one 0-cell, is homeomorphic to S^1.

Definition. Consider the increasing sequence $P^0 \subset P^1 \subset \cdots$ of projective spaces. Their coherent union is denoted by P^∞, and called **infinite-dimensional** (real) **projective space.** It follows from Theorem 38.3(b) that P^∞ is a CW complex having one open cell in each dimension $j \geq 0$; and its n-skeleton is P^n.

Now we perform an analogous construction with complex numbers replacing real numbers. Let $\mathbf{C}^n$ be the space of all complex sequences $z = (z_1, z_2, \ldots)$ such that $z_i = 0$ for $i > n$. Then $\mathbf{C}^n \subset \mathbf{C}^{n+1}$ for all n. There is an obvious homeomorphism $\rho: \mathbf{C}^{n+1} \to \mathbf{R}^{2n+2}$, which we call the "realification operator," defined by

$$\rho(z_1, z_2, \ldots) = (\text{Re } z_1, \text{Im } z_1, \text{Re } z_2, \text{Im } z_2, \ldots),$$

where Re z_i and Im z_i are the real and imaginary parts of z_i, respectively. Let us define

$$|z| = \|\rho(z)\| = [\Sigma((\text{Re } z_i)^2 + (\text{Im } z_i)^2)]^{1/2} = [\Sigma z_i \bar{z}_i]^{1/2},$$

where $\bar{z}_i$ is the complex conjugate of z_i. The subspace of $\mathbf{C}^{n+1}$ consisting of all points z with $|z| = 1$ is called the **complex n-sphere.** It corresponds to the sphere

S^{2n+1} under the operator ρ, so we use the symbol S^{2n+1} to denote it, whether we consider it in $\mathbf{C}^{n+1}$ or in $\mathbf{R}^{2n+2}$.

Definition. Let us introduce an equivalence relation in the complex n-sphere $S^{2n+1} \subset \mathbf{C}^{n+1}$ by defining

$$(z_1,\ldots,z_{n+1},0,\ldots) \sim (\lambda z_1,\ldots,\lambda z_{n+1},0,\ldots)$$

for each complex number λ with $|\lambda| = 1$. The resulting quotient space is called **complex projective n-space** and is denoted CP^n.

The quotient map $p : S^{2n+1} \to CP^n$ is a closed map. For if A is a closed set in S^{2n+1}, then $p^{-1}p(A)$ is the image of $S^1 \times A$ under the scalar multiplication map $(\lambda,z) \to \lambda z$. Since this map is continuous and $S^1 \times A$ is compact, its image is compact and therefore closed in S^{2n+1}. Then $p(A)$ is closed in CP^n, by definition.

It follows that CP^n is Hausdorff (in fact, normal).

As in the real case, $\mathbf{C}^n \subset \mathbf{C}^{n+1}$ for all n, so that $S^{2n-1} \subset S^{2n+1}$. Then, passing to quotient spaces, we have $CP^{n-1} \subset CP^n$. In fact, CP^{n-1} is a closed subspace of CP^n, by the same argument as before.

Said differently, the elements of CP^n are equivalence classes of sequences $(z_1,z_2,\ldots)$ of complex numbers such that $z_i = 0$ for $i > n + 1$. If one sequence in an equivalence class satisfies the equation $z_{n+1} = 0$, so does every member of the equivalence class; and the class in question belongs to CP^{n-1}, by definition.

Theorem 40.2. *The space CP^n is a CW complex of dimension $2n$. It has one open cell in each even dimension $2j$ for $0 \leq 2j \leq 2n$, and CP^j is its $2j$-skeleton.*

Proof. The space CP^0 is a single point. In general, we show that $CP^n - CP^{n-1}$ is an open $2n$-cell, which we denote by e_{2n}. Consider the subset of S^{2n+1} consisting of all points $z = (z_1,\ldots,z_{n+1},0,\ldots)$ *with z_{n+1} real.* Under the operator ρ, this corresponds to the set of all points of $\mathbf{R}^{2n+2}$ of the form

$$(x_1,y_1,\ldots,x_n,y_n,x_{n+1},0,0,\ldots)$$

having euclidean norm 1. *This is just the unit sphere S^{2n} in $\mathbf{R}^{2n+1}$*; it is the equator of S^{2n+1}. If we further restrict the set by requiring z_{n+1} to be real *and* non-negative, then $x_{n+1} \geq 0$ and we obtain the upper hemisphere E_+^{2n} of this equator sphere. The restriction p' of p to E_+^{2n} is of course a quotient map, because the domain is compact and the range is Hausdorff. The boundary of E_+^{2n} is the sphere S^{2n-1} obtained by setting $z_{n+1} = 0$; the map p' carries S^{2n-1} *onto* CP^{n-1}. We shall show that p' maps $\operatorname{Int} E_+^{2n}$ bijectively onto $e_{2n} = CP^n - CP^{n-1}$; then since it is a one-to-one quotient map, it is a homeomorphism. It follows that e_{2n} is an open $2n$-cell, and that p' is a characteristic map for e_{2n}.

The map $p' : \operatorname{Int} E_+^{2n} \to e_{2n}$ is surjective. Given a point of $e_{2n} = CP^n - CP^{n-1}$, it equals $p(z)$ for some point $z = (z_1,\ldots,z_{n+1},0,\ldots)$ of

$S^{2n+1} - S^{2n-1}$. Then $|z| = 1$ and $z_{n+1} \neq 0$. Write $z_{n+1} = re^{i\theta}$, where $r > 0$. Let $\lambda = e^{-i\theta}$; then

$$\lambda z = (\lambda z_1, \ldots, \lambda z_n, r, 0, \ldots) \in \text{Int } E_+^{2n}.$$

Now $|\lambda| = 1$, so $p(\lambda z) = p(z)$. Thus $p(z) \in p(\text{Int } E_+^{2n})$.

The map $p' : \text{Int } E_+^{2n} \to e_{2n}$ is injective. Suppose $p(z) = p(w)$, where z and w are in $\text{Int } E_+^{2n}$. Then $w = \lambda z$, so in particular $w_{n+1} = \lambda z_{n+1}$. Since w_{n+1} and z_{n+1} are real and positive, so is λ. Since $|\lambda| = 1$, we conclude that $\lambda = 1$. Thus $w = z$, as desired. □

We have noted that CP^0 is a single point. The space CP^1 is obtained by attaching a 2-cell to CP^0; therefore, CP^1 is homeomorphic to S^2.

Definition. Consider the increasing sequence of spaces $CP^0 \subset CP^1 \subset \cdots$. Their coherent union is denoted CP^∞ and called **infinite-dimensional complex projective space.** By Theorem 38.3, it is a CW complex with one cell in each non-negative even dimension, and CP^n is its $2n$-skeleton.

The homology of complex projective space is exceedingly easy to compute:

Theorem 40.3. *The group $H_i(CP^n)$ is infinite cyclic if i is even and $0 \leq i \leq 2n$; it vanishes otherwise. The group $H_i(CP^\infty)$ is infinite cyclic if i is even and $i \geq 0$; it vanishes otherwise.*

Proof. The cellular chain group $D_i(CP^n)$ is infinite cyclic if i is even and $0 \leq i \leq 2n$; otherwise, it vanishes. Therefore, every chain of this chain complex is a cycle, and no chain bounds. A similar computation applies to CP^∞. □

The computations for P^n require more work. The cellular chain group $D_k(P^n)$ is infinite cyclic for $0 \leq k \leq n$; we shall compute the boundary operator in the cellular chain complex. Since the open k-cell e_k of P^n equals $P^k - P^{k-1}$, and $\dot{e}_k = P^{k-1}$, we have $D_k(P^n) = H_k(P^k, P^{k-1})$. Thus we must compute the boundary operator

$$\partial_* : H_{k+1}(P^{k+1}, P^k) \to H_k(P^k, P^{k-1}).$$

First, we prove a lemma.

Lemma 40.4. *Let $p : S^n \to P^n$ be the quotient map ($n \geq 1$). Let $j : P^n \to (P^n, P^{n-1})$ be inclusion. The composite homomorphism*

$$H_n(S^n) \xrightarrow{p_*} H_n(P^n) \xrightarrow{j_*} H_n(P^n, P^{n-1})$$

is zero if n is even, and multiplication by 2 if n is odd.

Chain-level proof. We assume that S^n is triangulated so that the antipodal map $a : S^n \to S^n$ is simplicial, and that P^n is triangulated so that $p : S^n \to P^n$ is simplicial. (See Lemma 40.7 following.) We use simplicial homology. Let c_n

be a fundamental cycle for (E^n_+, S^{n-1}). Then $p_\#(c_n)$ is a fundamental cycle for (P^n, P^{n-1}), by Theorem 39.1.

Consider the following chain of S^n:

$$\gamma_n = c_n + (-1)^{n-1} a_\#(c_n).$$

It is a cycle, for its boundary is

$$\begin{aligned}\partial\gamma_n &= \partial c_n + (-1)^{n-1} a_\#(\partial c_n)\\ &= \partial c_n + (-1)^{2n-1}\partial c_n = 0.\end{aligned}$$

This equation follows from the fact that $a : S^{n-1} \to S^{n-1}$ has degree $(-1)^n$, so that $a_\#$, mapping the $n-1$ cycle group of S^{n-1} to itself, equals multiplication by $(-1)^n$. Furthermore, γ_n is a multiple of no other cycle of S^n, for its restriction to E^n_+ is c_n, which is a fundamental cycle for (E^n_+, S^{n-1}). Thus γ_n is a fundamental cycle for S^n.

Finally, we compute

$$p_\#(\gamma_n) = p_\#(c_n + (-1)^{n-1} a_\#(c_n)).$$

Since $p \circ a = p$, we conclude that

$$p_\#(\gamma_n) = [1 + (-1)^{n-1}] p_\#(c_n).$$

Since γ_n is a fundamental cycle for S^n and $p_\#(c_n)$ is a fundamental cycle for (P^n, P^{n-1}), the lemma follows.

Homology-level proof. This proof is similar to the preceding one, except that the computations are carried out on the homology level rather than the chain level. It may seem more complicated, but the ideas are basically the same.

Step 1. Consider the diagram

$$\begin{array}{ccccc}
 & & \alpha \in H_n(E^n_+, S^{n-1}) & & \\
 & & \downarrow i_* & & \\
H_n(S^n) & \xrightarrow{k_*} & H_n(S^n, S^{n-1}) & \xrightarrow{\partial_*} & H_{n-1}(S^{n-1}) \\
 & \searrow m_* & \downarrow l_* \quad \circlearrowleft a_* & & \circlearrowleft (a|S^{n-1})_* \\
 & & H_n(S^n, E^n_-) & &
\end{array}$$

where i, k, l, m are inclusions and a is the antipodal map. Let α be a generator of $H_n(E^n_+, S^{n-1})$. Consider the element

$$\gamma = i_*(\alpha) + (-1)^{n-1} a_* i_*(\alpha)$$

of $H_n(S^n, S^{n-1})$. We show there is a generator β of $H_n(S^n)$ such that $k_*(\beta) = \gamma$.

First, we show that $\gamma = k_*(\beta)$ for some β. Note that

$$\partial_*\gamma = \partial_* i_*(\alpha) + (-1)^{n-1}(a|S^{n-1})_*(\partial_* i_*(\alpha))$$

because ∂_* is natural. This homology class vanishes, because $(a|S^{n-1})_*$ equals multiplication by $(-1)^n$. By horizontal exactness, there exists *some* $\beta \in H_n(S^n)$ such that $k_*(\beta) = \gamma$.

Second, we show that β generates $H_n(S^n)$. Now m_* is an isomorphism, by the long exact reduced homology sequence. (Recall $n \geq 1$.) Therefore, it suffices to show that $m_*(\beta)$ generates $H_n(S^n,E^n_-)$. Since $m_*(\beta) = l_*(\gamma)$, we shall compute

$$l_*(\gamma) = l_*i_*(\alpha) + (-1)^{n-1}l_*a_*i_*(\alpha).$$

For this purpose, we note first that $l_*i_*(\alpha)$ generates $H_n(S^n,E^n_-)$, because the inclusion map $l \circ i$ induces a homology isomorphism. (See the proof of Theorem 31.8.) Then we note that $l_*a_*i_*(\alpha)$ is trivial; this fact is a consequence of the following commutative diagram, where the unlabelled maps are inclusions.

$$\begin{array}{ccc}
(E^n_+,S^{n-1}) & \xrightarrow{\ i\ } & (S^n,S^{n-1}) \\
\big\downarrow{\scriptstyle a|E^n_+} & & \big\downarrow{\scriptstyle a} \\
(E^n_-,S^{n-1}) & \longrightarrow & (S^n,S^{n-1}) \\
\big\downarrow & & \big\downarrow{\scriptstyle l} \\
(E^n_-,E^n_-) & \longrightarrow & (S^n,E^n_-)
\end{array}$$

It follows that $l_*(\gamma)$ generates $H_n(S^n,E^n_-)$, as desired.

Step 2. We prove the lemma. Consider the following commutative diagram:

$$\begin{array}{ccccc}
 & & \alpha \in H_n(E^n_+,S^{n-1}) & & \\
 & & \big\downarrow{\scriptstyle i_*} & & \\
\beta \in H_n(S^n) & \xrightarrow{\ k_*\ } & H_n(S^n,S^{n-1}) & \xrightarrow{\ a_*\ } & H_n(S^n,S^{n-1}) \\
\big\downarrow{\scriptstyle p_*} & & \big\downarrow{\scriptstyle p_*} & \swarrow{\scriptstyle p_*} & \\
H_n(P^n) & \xrightarrow[j_*]{} & H_n(P^n,P^{n-1}) & &
\end{array}$$

Choose generators α and β as in Step 1. We wish to compute $j_*p_*(\beta)$. The map

$$p \circ i : (E^n_+,S^{n-1}) \to (P^n,P^{n-1})$$

is a characteristic map for the n-cell of the CW complex P^n, so it induces an isomorphism in homology. Thus $p_*i_*(\alpha)$ is a generator of $H_n(P^n,P^{n-1})$. We compute

$$\begin{aligned}
j_*p_*(\beta) = p_*k_*(\beta) &= p_*(i_*(\alpha) + (-1)^{n-1}a_*i_*(\alpha)) \\
&= [1 + (-1)^{n-1}]p_*i_*(\alpha).
\end{aligned}$$

Here we use the fact that $p \circ a = p$. The lemma follows. □

Theorem 40.5. *The homomorphism*

$$\partial_* : H_{n+1}(P^{n+1},P^n) \to H_n(P^n,P^{n-1})$$

is zero if n is even, and carries a generator to twice a generator if n is odd.

Proof. The map $p' : (E^{n+1}_+,S^n) \to (P^{n+1},P^n)$ is a characteristic map for the open $n+1$ cell of P^{n+1}; therefore, it induces a homology isomorphism. Consider the commutative diagram

$$\begin{array}{ccccc} H_{n+1}(E^{n+1}_+,S^n) & \xrightarrow[\simeq]{\partial_*} & H_n(S^n) & & \\ \simeq \downarrow p'_* & & \downarrow p_* & & \\ H_{n+1}(P^{n+1},P^n) & \xrightarrow{\partial_*} & H_n(P^n) & \xrightarrow{j_*} & H_n(P^n,P^{n-1}) \end{array}$$

The map ∂_* at the top of the diagram is an isomorphism, by the long exact reduced homology sequence of (E^{n+1}_+,S^n). By the preceding lemma, $(j \circ p)_*$ is zero if n is even, and multiplication by 2 if n is odd. The theorem follows. □

Theorem 40.6. *The homology of projective space is as follows:*

$$\tilde{H}_i(P^{2n+1}) \simeq \begin{cases} \mathbf{Z}/2 & \text{if } i \text{ is odd and } 0 < i < 2n+1, \\ \mathbf{Z} & \text{if } i = 2n+1, \\ 0 & \text{otherwise.} \end{cases}$$

$$\tilde{H}_i(P^{2n}) \simeq \begin{cases} \mathbf{Z}/2 & \text{if } i \text{ is odd and } 0 < i < 2n, \\ 0 & \text{otherwise.} \end{cases}$$

$$\tilde{H}_i(P^{\infty}) \simeq \begin{cases} \mathbf{Z}/2 & \text{if } i \text{ is odd and } 0 < i, \\ 0 & \text{otherwise.} \end{cases}$$

Proof. The cellular chain group $D_i(P^\infty)$ is infinite cyclic for $i \geq 0$, and the augmented chain complex has the form

$$\cdots \to D_{2i}(P^\infty) \xrightarrow{2} D_{2i-1}(P^\infty) \xrightarrow{0} \cdots \xrightarrow{0} D_0(P^\infty) \xrightarrow{1} \mathbf{Z}.$$

There are no cycles in even dimensions; while in odd dimensions, every element is a cycle and even multiples of the generator bound. Thus $\tilde{H}_i(P^\infty)$ is of order 2 if i is positive and odd, and vanishes otherwise.

The computations for P^{2n} and P^{2n+1} are similar. □

Now we prove the lemma we used in the preceding chain-level proof.

Lemma 40.7. *The spaces S^n and P^n may be triangulated so that the antipodal map $a : S^n \to S^n$ and the projection map $p : S^n \to P^n$ are simplicial.*

Proof. Step 1. We show first there is a complex L in $\mathbf{R}^{n+1}$ such that each reflection map

$$\rho_i(x_1,\ldots,x_i,\ldots,x_{n+1}) = (x_1,\ldots,-x_i,\ldots,x_{n+1})$$

induces a linear isomorphism of L with itself; and we show there is a triangulation $k : |L| \to S^n$ that commutes with each map ρ_i.

The result is trivial for $n = 0$. Assume K is a complex in $\mathbf{R}^n$ and $h : |K| \to S^{n-1}$ is a triangulation satisfying our hypotheses. Let $w_0 = (0, \dots, 0, 1)$ and $w_1 = (0, \dots, 0, -1)$ in $\mathbf{R}^{n+1}$. Let $L = (w_0 * K) \cup (w_1 * K)$. Then ρ_i induces a linear isomorphism of L with itself, for $i = 1, \dots, n+1$. The triangulation defined by

$$k(y) = (\sqrt{1 - t^2}\, h(x), t)$$

if $y = (1 - t)x + tw_0$, and

$$k(y) = (\sqrt{1 - t^2}\, h(x), -t)$$

if $y = (1 - t)x + tw_1$, commutes with each ρ_i. (This is the same triangulation we used in the proof of Theorem 21.3.)

Step 2. Let $k : |L| \to S^n$ be the triangulation of Step 1. The antipodal map commutes with k, and induces a linear isomorphism of L with itself. It also induces a linear isomorphism of $\operatorname{sd}^N L$ with itself, for any fixed N.

Let us choose N large enough that for any vertex v of $\operatorname{sd}^N L$, the closed stars of v and $a(v)$ are disjoint. Then we can use the "vertex-labelling" device of §3 to construct a complex whose underlying space is homeomorphic to P^n: Let us label the vertices of $\operatorname{sd}^N L$, giving v and $a(v)$ the same label for each vertex v. Let $g : |\operatorname{sd}^N L| \to |M|$ be the quotient map obtained from this labelling. Then the map g will identify x with $a(x)$ for each $x \in |L|$, and will identify x with no other point of $|L|$. Because the homeomorphism k commutes with a, it induces a homeomorphism of $|M|$ with P^n that is our desired triangulation of P^n. □

As a further application of these techniques, we now define a certain class of 3-dimensional manifolds called the *lens spaces* and compute their homology. It is of interest that they form one of the few classes of spaces that have been completely classified up to homeomorphism and up to homotopy type. We discuss this classification later.

Definition. Let n and k be relatively prime positive integers. We construct the lens space $L(n,k)$ as a quotient space of the ball B^3 as follows: Write the general point of B^3 in the form (z,t), where z is complex, t is real, and $|z|^2 + t^2 \le 1$. Let $\lambda = \exp(2\pi i/n)$. Define $f : S^2 \to S^2$ by the equation

$$f(x) = (\lambda^k z, -t).$$

Let us identify each point $x = (z,t)$ of the lower hemisphere E^2_- of $S^2 = \operatorname{Bd} B^3$ with the point $f(x)$ of the upper hemisphere E^2_+. The resulting quotient space is called the **lens space** $L(n,k)$.

Note that the map $z \to \lambda z$ of $\mathbf{C}$ to itself is just rotation through the angle $2\pi/n$. Thus f equals a rotation of S^2 about the z-axis through the angle $\alpha = 2\pi k/n$, followed by reflection in the xy-plane. See Figure 40.1.

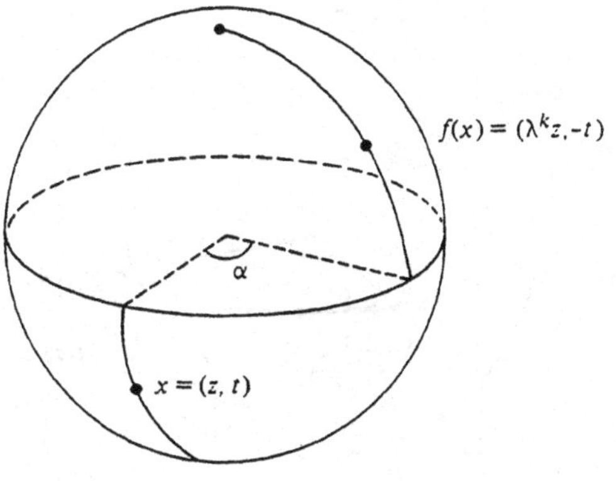

Figure 40.1

Let $p : B^3 \to L(n,k)$ be the quotient map. Each point of Int B^3 is identified only with itself under p, and each point of Int E^2_- is identified with a point of Int E^2_+. However, since a point $(z,0)$ of the equator in S^2 belongs to both E^2_- and E^2_+, its equivalence class contains n points—namely, the points

$$(\lambda^k z,0), (\lambda^{2k} z,0), \ldots, (\lambda^{nk} z,0) = (z,0).$$

Because k and n are relatively prime, these points are distinct and constitute a permutation of the points

$$(\lambda z,0), (\lambda^2 z,0), \ldots, (\lambda^n z,0) = (z,0),$$

which are evenly spaced about the equator S^1.

Theorem 40.8. *The space $L(n,k)$ is a CW complex with one cell in each dimension* 0, 1, 2, 3.

Proof. We first show that the quotient map p is closed, so that $L(n,k)$ is Hausdorff (in fact, normal). Let A be closed in B^3. The saturation $p^{-1}p(A)$ of A is the union of the set A, the following subsets of S^2:

$$f(E^2_- \cap A) \qquad \text{and} \qquad f^{-1}(E^2_+ \cap A),$$

and the following subsets of S^1:

$$f(A \cap S^1), f^2(A \cap S^1), \ldots, f^{n-1}(A \cap S^1).$$

All these sets are compact, so they are closed in B^3, and so is their union. Since $p^{-1}p(A)$ is closed, so is $p(A)$. Thus p is a closed map.

We give $L(n,k)$ the structure of CW complex as follows: First, choose a particular point a on the equator, say $a = (1,0)$; let $p(a)$ be the 0-cell e_0 of $L(n,k)$.

Let A denote the smaller arc of S^1 running from a to $b = (\lambda,0)$. Now $p|A$ is a quotient map, since A is compact and $L(n,k)$ is Hausdorff; it identifies a and b, but is one-to-one on Int A. Thus $p(\text{Int}\, A)$ is an open 1-cell; we take it to

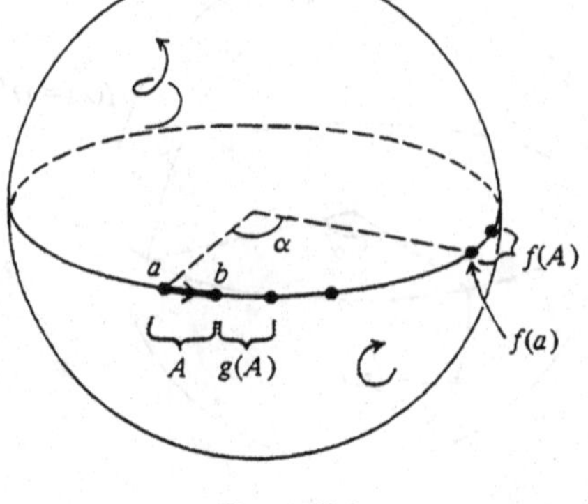

Figure 40.2

be the open 1-cell e_1 of $L(n,k)$. The map $p|A$ is a characteristic map for e_1. Note that the points of $p^{-1}p(a)$ break the circle S^1 up into n open arcs, each of which is mapped homeomorphically by p onto e_1. See Figure 40.2.

For similar reasons, the map $p|E^2_-$ is a quotient map; the set $p(\text{Int } E^2_-)$ is an open 2-cell that we take to be the open 2-cell e_2 of $L(n,k)$; and $p|E^2_-$ is a characteristic map for e_2. Finally, the set $p(\text{Int } B^3)$ is the open 3-cell e_3; and p is its characteristic map.

Note that $p(S^1)$ equals the 1-skeleton of $L(n,k)$, and $p(S^2)$ equals the 2-skeleton. □

Now we compute the homology of this lens space.

Theorem 40.9. *If $X = L(n,k)$, then the cellular chain complex of X has the form*

$$D_3(X) \xrightarrow{0} D_2(X) \xrightarrow{n} D_1(X) \xrightarrow{0} D_0(X),$$

where each group $D_i(X)$ is infinite cyclic. Therefore,

$$H_3(X) \simeq \mathbf{Z}, \qquad H_2(X) = 0, \qquad H_1(X) \simeq \mathbf{Z}/n, \qquad H_0(X) \simeq \mathbf{Z}.$$

Thus the lens spaces $L(n,k)$ and $L(m,l)$ cannot be homeomorphic, or even have the same homotopy type, unless $n = m$.

Proof. Now B^3 and $L(n,k)$ may be triangulated so that the rotation-reflection map f and the quotient map p are simplicial. We leave the proof as an exercise.

Let A be the arc of S^1 having end points $a = (1,0)$ and $b = (\lambda,0)$, as before. Let c_1 be a cycle generating $H_1(A, \text{Bd } A)$; let c_2 be a cycle generating $H_2(E^2_-, S^1)$; and let c_3 be a cycle generating $H_3(B^3, S^2)$; their signs will be chosen shortly. The chains $p_\#(c_1)$, $p_\#(c_2)$, and $p_\#(c_3)$ generate the chain groups $D_1(X)$, $D_2(X)$, and $D_3(X)$, respectively.

Choose the sign of c_1 so that $\partial c_1 = b - a$. Then we shall show that the chain

$$(*) \qquad z_1 = c_1 + f_\#(c_1) + f_\#^2(c_1) + \cdots + f_\#^{n-1}(c_1)$$

generates $H_1(S^1)$. Once this fact is proved, we note that since ∂c_2 also generates $H_1(S^1)$, we can choose the sign of c_2 so that $\partial c_2 = z_1$. We then show that the chain

$$(**) \qquad z_2 = c_2 - f_\#(c_2)$$

generates $H_2(S^2)$. Given this fact, we choose the sign of c_3 so that $\partial c_3 = z_2$.

First, we consider z_1. Let $g: S^1 \to S^1$ be the map $g(z,0) = (\lambda z,0)$; it equals rotation through angle $2\pi/n$. Rearranging terms in the expression for z_1, we have

$$z_1 = c_1 + g_\#(c_1) + \cdots + g_\#^{n-1}(c_1).$$

Now $\partial c_1 = g_\#(a) - a$. Therefore,

$$\begin{aligned}\partial z_1 &= [g_\#(a) - a] + [g_\#^2(a) - g_\#(a)] + \cdots + [g_\#^n(a) - g_\#^{n-1}(a)]\\ &= 0.\end{aligned}$$

Thus z_1 is a cycle. Because its restriction to A equals c_1, which is a fundamental cycle for $H_1(A,\operatorname{Bd} A)$, it is a fundamental cycle for S^1. Thus z_1 generates $H_1(S^1)$.

Now let us consider z_2. To show z_2 is a cycle, we compute

$$\partial z_2 = \partial c_2 - f_\#(\partial c_2) = z_1 - f_\#(z_1) = 0,$$

for by direct computation with formula (*), we have $f_\#(z_1) = z_1$. Because the restriction of z_2 to E^2_- is a fundamental cycle for (E^2_-,S^1), the chain z_2 is a fundamental cycle for S^2, so it generates $H_2(S^2)$.

Now we are ready to compute the boundary operators in the cellular chain complex of X. First,

$$\partial p_\#(c_1) = p_\#(b) - p_\#(a) = 0,$$

so the boundary operator $D_1(X) \to D_0(X)$ is trivial. Second,

$$\begin{aligned}\partial p_\#(c_2) &= p_\#(z_1) = p_\#(c_1 + f_\#(c_1) + \cdots + f_\#^{n-1}(c_1))\\ &= np_\#(c_1),\end{aligned}$$

because $p(f^j(x)) = p(x)$ for all x in S^1 and all j. Thus the map $D_2(X) \to D_1(X)$ is multiplication by n. Third,

$$\partial p_\#(c_3) = p_\#(z_2) = p_\#(c_2 - f_\#(c_2)) = 0,$$

because $p(f(x)) = p(x)$ for x in E^2_-. Thus $D_3(X) \to D_2(X)$ is trivial. □

EXERCISES

1. Show that P^n is homeomorphic to the quotient space of B^n obtained by identifying x with $-x$ for each $x \in S^{n-1}$.

2. (a) Show P^n is an n-manifold and CP^n is a $2n$-manifold.
   (b) Show more generally that if a finite-dimensional CW complex is homogeneous, it is a manifold.

3. Let A be a regular n-sided polygonal region in the plane; let B be the suspension of A. Describe $L(n,k)$ as a quotient space of B; conclude that B^3 and $L(n,k)$ can be triangulated so the quotient map is simplicial.

4. *Theorem.* *$L(n,k)$ is homeomorphic to $L(n,l)$ if either*

$$k \equiv \pm l \pmod{n} \qquad \textit{or} \qquad kl \equiv \pm 1 \pmod{n}.$$

Proof. (a) Prove the case $k \equiv -l \pmod{n}$ by considering the reflection map $(z,t) \to (z,-t)$ in B^3.

(b) Let $1 \leq k < n$, for convenience. Consider n disjoint 3-simplices

$$a_1b_1c_1d_1,\ a_2b_2c_2d_2,\ \ldots,\ a_nb_nc_nd_n,$$

indexed with the elements of $\mathbf{Z}/n$. See Figure 40.3. Show that $L(n,k)$ can be obtained from these simplices by first pasting

$$a_ib_id_i \quad \text{to} \quad a_{i+1}b_{i+1}c_{i+1},$$

for each i in $\mathbf{Z}/n$, by a linear homeomorphism that preserves the order of vertices, and then pasting

$$c_id_ib_i \quad \text{to} \quad c_{i+k}d_{i+k}a_{i+k},$$

again by a linear homeomorphism that preserves the order of the vertices.

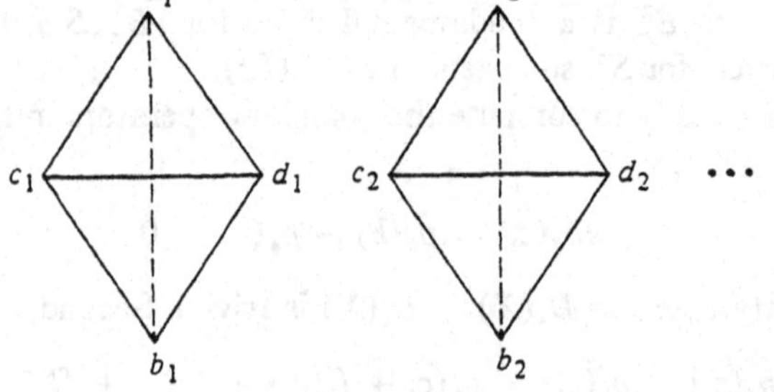

Figure 40.3

(c) Rewrite the simplices of (b) in the order

$$a_kb_kc_kd_k,\ a_{2k}b_{2k}c_{2k}d_{2k},\ \ldots,\ a_{nk}b_{nk}c_{nk}d_{nk}.$$

See Figure 40.4. Carry out the second pasting operating of (b), and then carry out the first pasting operation of (b). Show this gives a description of $L(n,l)$, where l is the integer between 1 and n such that $kl \equiv 1 \pmod{n}$.

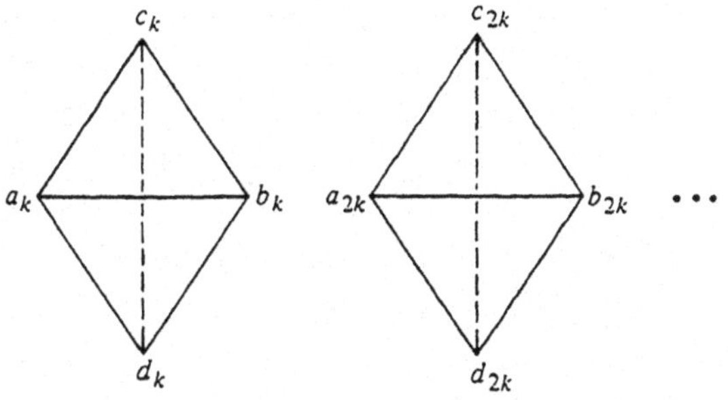

Figure 40.4

(d) Prove the theorem.

5. Show that $L(n,k)$ is a compact 3-manifold.

We note that the converse of the theorem stated in Exercise 4 holds, but the proof is very difficult.

In the 1930's R. Reidemeister defined a number associated with a simplicial complex, called its *torsion*. This number is a *combinatorial invariant*, which means that two complexes that have isomorphic subdivisions necessarily have the same torsion. By computing the torsion for lens spaces, Reidemeister showed that if the simplicial complexes $L(n,k)$ and $L(n,l)$ have isomorphic subdivisions, then either $k \equiv \pm l$ or $kl \equiv \pm 1 \pmod{n}$.

To complete the proof of the converse, it remained for E. E. Moise (in the 1950's) to prove that two triangulated 3-manifolds that are homeomorphic necessarily have subdivisions that are isomorphic.

The homeomorphism classification of lens spaces is thus known. We shall discuss their homotopy-type classification in a later chapter.

5

Cohomology

With each topological space X, we have associated a sequence of abelian groups called its *homology groups*. Now we associate with X another sequence of abelian groups, called its *cohomology groups*. These groups were not defined until long after the homology groups. The reason is not hard to understand, for they are geometrically much less natural than the homology groups. Their origins lie in algebra rather than geometry; in a certain algebraic sense (to be made precise), they are "dual" to the homology groups. In the past, topologists have used such terms as "pseudo-cycle" for representatives of these group elements, implying a certain skepticism as to their legitimacy as objects of study. However, it eventually became clear that these groups are both important in theory and useful in practice.

The duality theorems for manifolds, the connections between topology and differential geometry (de Rham's theorem) and between topology and analysis (cohomology with sheaf coefficients)—all these results are formulated in terms of cohomology. Even such purely topological problems as classifying spaces up to homeomorphism, or maps up to homotopy, are problems about which cohomology has a good deal to say. We will return to some of these problems later.

Throughout, we shall assume familiarity with the language of categories and functors (§28).

§41. THE HOM FUNCTOR

Associated with any pair of abelian groups A, G is a third abelian group, the group $\text{Hom}(A,G)$ of all homomorphisms of A into G. This group will be involved in an essential way in the definition of the cohomology groups. In this section we study some of its properties.

Definition. If A and G are abelian groups, then the set $\text{Hom}(A,G)$ of all homomorphisms of A into G becomes an abelian group if we add two homomorphisms by adding their values in G.

That is, for $a \in A$ we define $(\phi + \psi)(a) = \phi(a) + \psi(a)$. The map $\phi + \psi$ is a homomorphism, because $(\phi + \psi)(0) = 0$ and

$$\begin{aligned}(\phi + \psi)(a + b) &= \phi(a + b) + \psi(a + b)\\ &= \phi(a) + \psi(a) + \phi(b) + \psi(b)\\ &= (\phi + \psi)(a) + (\phi + \psi)(b).\end{aligned}$$

The identity element of $\text{Hom}(A,G)$ is the function mapping A to the identity element of G. The inverse of the homomorphism ϕ is the homomorphism that maps a to $-\phi(a)$, for each $a \in A$.

Example 1. $\text{Hom}(\mathbf{Z},G)$ is isomorphic with the group G itself; the isomorphism assigns to the homomorphism $\phi : \mathbf{Z} \to G$, the element $\phi(1)$.

More generally, if A is a free abelian group of finite rank with basis $e_1, \ldots, e_n$, then $\text{Hom}(A,G)$ is isomorphic with the direct sum $G \oplus \cdots \oplus G$ of n copies of G. The isomorphism assigns to the homomorphism $\phi : A \to G$, the n-tuple $(\phi(e_1), \ldots, \phi(e_n))$. Note that this isomorphism is not "natural," but depends on the choice of a basis for A. Note also that it depends on the finiteness of the rank of A. If A is free abelian with non-finite basis $\{e_\alpha\}_{\alpha \in J}$, then the correspondence $\phi \to (\phi(e_\alpha))_{\alpha \in J}$ carries ϕ not to an element of the direct *sum* $\bigoplus_{\alpha \in J} G_\alpha$ of copies of G, but rather to an element of the direct product $\prod_{\alpha \in J} G_\alpha$. (See §4 for definitions.)

We will state these facts formally as a theorem later on.

Definition. A homomorphism $f : A \to B$ gives rise to a **dual homomorphism**

$$\text{Hom}(A,G) \xleftarrow{\tilde{f}} \text{Hom}(B,G)$$

going in the reverse direction. The map $\tilde{f}$ assigns to the homomorphism $\phi : B \to G$, the composite

$$A \xrightarrow{f} B \xrightarrow{\phi} G.$$

That is, $\tilde{f}(\phi) = \phi \circ f$.

The map $\tilde{f}$ is a homomorphism, since $\tilde{f}(0) = 0$ and

$$\begin{aligned}[\tilde{f}(\phi + \psi)](a) &= (\phi + \psi)(f(a)) = \phi(f(a)) + \psi(f(a))\\ &= [\tilde{f}(\phi)](a) + [\tilde{f}(\psi)](a).\end{aligned}$$

Note that for fixed G, the assignment

$$A \to \text{Hom}(A,G) \quad \text{and} \quad f \to \tilde{f}$$

defines a *contravariant functor* from the category of abelian groups and homomorphisms to itself. For if $i_A : A \to A$ is the identity homomorphism, then

$\tilde{i}_A(\phi) = \phi \circ i_A = \phi$, so $\tilde{i}_A$ is the identity map of $\mathrm{Hom}(A,G)$. Furthermore, if the left diagram following commutes, so does the right diagram:

$$\begin{array}{ccc} A & \xrightarrow{h} & C \\ & {\scriptstyle f}\searrow \quad \nearrow{\scriptstyle g} & \\ & B & \end{array} \qquad \begin{array}{ccc} \mathrm{Hom}(A,G) & \xleftarrow{\tilde{h}} & \mathrm{Hom}(C,G) \\ & {\scriptstyle \tilde{f}}\nwarrow \quad \swarrow{\scriptstyle \tilde{g}} & \\ & \mathrm{Hom}(B,G) & \end{array}$$

For $\tilde{h}(\phi) = \phi \circ h = \phi \circ (g \circ f)$; while $\tilde{f}(\tilde{g}(\phi)) = \tilde{f}(\phi \circ g) = (\phi \circ g) \circ f$, by definition.

We list some consequences of this fact:

Theorem 41.1. *Let f be a homomorphism; let $\tilde{f}$ be the dual homomorphism.*
(a) *If f is an isomorphism, so is $\tilde{f}$.*
(b) *If f is the zero homomorphism, so is $\tilde{f}$.*
(c) *If f is surjective, then $\tilde{f}$ is injective. That is, exactness of*

$$B \xrightarrow{f} C \to 0$$

implies exactness of

$$\mathrm{Hom}(B,G) \xleftarrow{\tilde{f}} \mathrm{Hom}(C,G) \leftarrow 0.$$

Proof. (a) and (b) are immediate. To prove (c), suppose f is surjective. Let $\psi \in \mathrm{Hom}(C,G)$ and suppose $\tilde{f}(\psi) = 0 = \psi \circ f$. Then $\psi(f(b)) = 0$ for every $b \in B$. As b ranges over B, the element $f(b)$ ranges over all elements of C. Thus $\psi(c) = 0$ for every $c \in C$. □

More generally, we have the following result concerning the dual of an exact sequence.

Theorem 41.2. *If the sequence*

$$A \xrightarrow{f} B \xrightarrow{g} C \to 0$$

is exact, then the dual sequence

$$\mathrm{Hom}(A,G) \xleftarrow{\tilde{f}} \mathrm{Hom}(B,G) \xleftarrow{\tilde{g}} \mathrm{Hom}(C,G) \leftarrow 0$$

is exact. Furthermore, if f is injective and the first sequence splits, then $\tilde{f}$ is surjective and the second sequence splits.

Proof. Injectivity of $\tilde{g}$ follows from the preceding theorem. We check exactness at $\mathrm{Hom}(B,G)$. Because $h = g \circ f$ is the zero homomorphism, so is $\tilde{h} = \tilde{f} \circ \tilde{g}$. On the other hand, supposing $\tilde{f}(\psi) = 0$, we show $\psi = \tilde{g}(\phi)$ for some $\phi \in \mathrm{Hom}(C,G)$. Since $\tilde{f}(\psi) = \psi \circ f$ is the zero homomorphism, ψ vanishes on the group $f(A)$. Thus ψ induces a homomorphism $\psi' : B/f(A) \to G$.

Exactness of the original sequence implies that g induces an *isomorphism* $g' : B/f(A) \to C$, as in the following diagram:

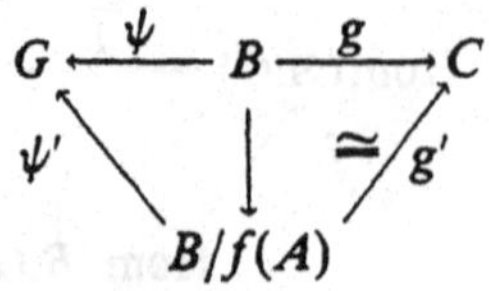

The map $\phi = \psi' \circ (g')^{-1}$ is a homomorphism of C into G, and as desired,

$$\tilde{g}(\phi) = \phi \circ g = \psi' \circ (g')^{-1} \circ g = \psi.$$

Suppose now that f maps A injectively onto a direct summand in B. Let $\pi : B \to A$ be a homomorphism such that $\pi \circ f = i_A$. Then $\tilde{f} \circ \tilde{\pi}$ is the identity of $\mathrm{Hom}(A,G)$, so $\tilde{f}$ is surjective and $\tilde{\pi} : \mathrm{Hom}(A,G) \to \mathrm{Hom}(B,G)$ splits the dual sequence. □

We remark that, in general, exactness of a short exact sequence does not imply exactness of the dual sequence. For instance, if $f : \mathbf{Z} \to \mathbf{Z}$ equals multiplication by 2, then the sequence

$$0 \to \mathbf{Z} \xrightarrow{f} \mathbf{Z} \to \mathbf{Z}/2 \to 0$$

is exact. But $\tilde{f}$ is not surjective. Indeed, if $\phi \in \mathrm{Hom}(\mathbf{Z},\mathbf{Z})$, then $\tilde{f}(\phi) = \phi \circ f$ is a homomorphism that maps $\mathbf{Z}$ into the set of even integers. Thus the image of $\tilde{f}$ is not all of $\mathrm{Hom}(\mathbf{Z},\mathbf{Z})$.

We have considered Hom as a functor of the first variable alone. But it may also be considered as a functor of both variables. In this case, it has a mixed variance; it is contravariant in the first variable and covariant in the second. We formalize this statement as follows:

Definition. Given homomorphisms $\alpha : A \to A'$ and $\beta : G' \to G$, we define a map

$$\mathrm{Hom}(\alpha,\beta) : \mathrm{Hom}(A',G') \to \mathrm{Hom}(A,G)$$

by letting it map the homomorphism $\phi' : A' \to G'$ to the homomorphism $\beta \circ \phi' \circ \alpha : A \to G$.

You can check that $\mathrm{Hom}(\alpha,\beta)$ is indeed a homomorphism. Functoriality follows: The map $\mathrm{Hom}(i_A,i_G)$ is the identity. And if $\alpha' : A' \to A''$ and $\beta' : G'' \to G'$, then

$$\mathrm{Hom}(\alpha' \circ \alpha, \beta \circ \beta') = \mathrm{Hom}(\alpha,\beta) \circ \mathrm{Hom}(\alpha',\beta')$$

by definition. (Both sides carry $\phi'' : A'' \to G''$ to $\beta \circ \beta' \circ \phi'' \circ \alpha' \circ \alpha$.)

In this notation, the "dual homomorphism" $\tilde{\alpha}$ obtained when we consider Hom as a functor of the first variable alone is just the map $\mathrm{Hom}(\alpha,i_G)$.

One can also consider Hom as a functor of the second variable alone; this case we leave to the exercises.

Now we prove some properties of the Hom functor.

Theorem 41.3. (a) *One has the following isomorphisms:*

$$\operatorname{Hom}(\oplus_{\alpha\in J} A_\alpha, G) \simeq \Pi_{\alpha\in J} \operatorname{Hom}(A_\alpha, G),$$
$$\operatorname{Hom}(A, \Pi_{\alpha\in J} G_\alpha) \simeq \Pi_{\alpha\in J} \operatorname{Hom}(A, G_\alpha).$$

(b) *There is a natural isomorphism of* $\operatorname{Hom}(\mathbf{Z},G)$ *with* G. *If* $f:\mathbf{Z}\to\mathbf{Z}$ *equals multiplication by* m, *then so does* $\tilde{f}$.

(c) $\operatorname{Hom}(\mathbf{Z}/m,G) \simeq \ker(G \xrightarrow{m} G)$.

Proof. Property (a) follows immediately from standard facts of algebra concerning homomorphisms of products. The proof of (b) is also direct; the homomorphism $\lambda: \operatorname{Hom}(\mathbf{Z},G) \to G$ assigns to the homomorphism $\phi: \mathbf{Z} \to G$ the value of ϕ at 1. That is, $\lambda(\phi) = \phi(1)$. Since ϕ is entirely determined by its value at 1, and since this value can be chosen arbitrarily, λ is an isomorphism of $\operatorname{Hom}(\mathbf{Z},G)$ with G.

Let $f:\mathbf{Z}\to\mathbf{Z}$ be multiplication by m. Then

$$\tilde{f}(\phi)(x) = \phi(f(x)) = \phi(mx) = m\phi(x),$$

so $\tilde{f}(\phi) = m\phi$. Thus $\tilde{f}$ equals multiplication by m in $\operatorname{Hom}(\mathbf{Z},G)$. Under the isomorphism λ of $\operatorname{Hom}(\mathbf{Z},G)$ with G, the map $\tilde{f}$ in turn corresponds to multiplication by m in G.

Now we prove (c). Begin with the exact sequence

$$0 \to \mathbf{Z} \xrightarrow{m} \mathbf{Z} \to \mathbf{Z}/m \to 0.$$

Then the sequence

$$\operatorname{Hom}(\mathbf{Z},G) \xleftarrow{m} \operatorname{Hom}(\mathbf{Z},G) \leftarrow \operatorname{Hom}(\mathbf{Z}/m,G) \leftarrow 0$$

is exact, and (c) follows. □

We remark that the isomorphisms given in (a) of this theorem are "natural." Specifically, suppose one is given homomorphisms $\phi_\alpha : A_\alpha \to B_\alpha$ and $\psi: H \to G$. Then it follows immediately from the definition of the isomorphism that the diagram

$$\begin{array}{ccc} \operatorname{Hom}(\oplus A_\alpha,G) & \simeq & \Pi \operatorname{Hom}(A_\alpha,G) \\ {\scriptstyle \operatorname{Hom}(\oplus\phi_\alpha,\psi)} \uparrow & & \uparrow {\scriptstyle \Pi \operatorname{Hom}(\phi_\alpha,\psi)} \\ \operatorname{Hom}(\oplus B_\alpha,H) & \simeq & \Pi \operatorname{Hom}(B_\alpha,H) \end{array}$$

commutes. A similar comment applies to the other isomorphism in (a).

The preceding theorem enables us to compute $\operatorname{Hom}(A,G)$ whenever A is finitely generated, for $\operatorname{Hom}(A,G)$ equals a direct sum of terms of the form $\operatorname{Hom}(\mathbf{Z},G)$ and $\operatorname{Hom}(\mathbf{Z}/m,G)$, which we compute by applying the rules

$$\operatorname{Hom}(\mathbf{Z},G) \simeq G, \qquad \operatorname{Hom}(\mathbf{Z}/m,G) = \ker(G \xrightarrow{m} G).$$

When G is also finitely generated, these groups can be written as direct sums of cyclic groups. One needs the following lemma, whose proof is left to the exercises.

Lemma 41.4. *There is an exact sequence*

$$0 \to \mathbf{Z}/d \to \mathbf{Z}/n \xrightarrow{m} \mathbf{Z}/n \to \mathbf{Z}/d \to 0,$$

where $d = gcd(m,n)$. □

EXERCISES

1. Show that if T is the torsion subgroup of G, then $\text{Hom}(G,\mathbf{Z}) \simeq \text{Hom}(G/T,\mathbf{Z})$.
2. Let G be fixed. Consider the following functor from the category of abelian groups to itself:

$$A \to \text{Hom}(G,A) \quad \text{and} \quad f \to \text{Hom}(i_G,f).$$

   (a) Show that this functor preserves exactness of

$$0 \to A \to B \to C.$$

   (b) Show that this functor preserves split short exact sequences.

3. (a) Show that the kernel of $\mathbf{Z}/n \xrightarrow{m} \mathbf{Z}/n$ is generated by $\{n/d\}$, where $d = gcd(m,n)$.
   (b) Show that a quotient of a cyclic group is cyclic.
   (c) Prove Lemma 41.4.
4. The abelian group G is said to be **divisible** if for each $x \in G$ and each positive integer n, there exists $y \in G$ such that $ny = x$. For instance, the rationals form a divisible group under addition.

   Theorem. Let G be divisible. Then if

$$0 \to A \to B \to C \to 0$$

   is exact, so is

$$0 \leftarrow \text{Hom}(A,G) \leftarrow \text{Hom}(B,G) \leftarrow \text{Hom}(C,G) \leftarrow 0.$$

   Proof. It suffices to show that if $A \subset B$ and $\phi : A \to G$ is a homomorphism, then ϕ extends to a homomorphism $\psi : B \to G$.
   (a) Prove this fact when B is generated by the elements of A and a single additional element b.
   (b) Let $\mathcal{B}$ be a collection of subgroups of B that is simply ordered by inclusion. Let $\{\psi_H \mid H \in \mathcal{B}\}$ be a collection of homomorphisms, where ψ_H maps H into G for each H, such that any two agree on the common part of their domains. Show that the union of the elements of $\mathcal{B}$ is a subgroup of B, and these homomorphisms extend to a homomorphism of this union into G.
   (c) Use a Zorn's lemma argument to complete the proof.
5. Let R be a commutative ring with unity element 1. Let A and B be R-modules. (See §48 if you've forgotten the definitions.)

(a) Let $\text{Hom}_R(A,B)$ denote the set of all R-module homomorphisms of A into B. Show it has the structure of R-module in a natural way. Show that if f,g are R-module homomorphisms, so is $\text{Hom}(f,g)$.
(b) State and prove the analogues of Theorems 41.2 and 41.3 for R-modules.
(c) Consider the special case where R is a field F. Then A and B are vector spaces over F, and so is $\text{Hom}_F(A,B)$. Show that in this case every exact sequence splits, so the functor Hom_F preserves exact sequences.

§42. SIMPLICIAL COHOMOLOGY GROUPS

In this section, we define the cohomology groups of a simplicial complex, and we compute some elementary examples.

Definition. Let K be a simplicial complex; let G be an abelian group. The group of ***p*-dimensional cochains** of K, with coefficients in G, is the group

$$C^p(K;G) = \text{Hom}(C_p(K),G).$$

The **coboundary operator** δ is defined to be the dual of the boundary operator $\partial : C_{p+1}(K) \rightarrow C_p(K)$. Thus

$$C^{p+1}(K;G) \xleftarrow{\delta} C^p(K;G),$$

so that δ *raises* dimension by one. We define $Z^p(K;G)$ to be the kernel of this homomorphism, $B^{p+1}(K;G)$ to be its image, and (noting that $\delta^2 = 0$ because $\partial^2 = 0$),

$$H^p(K;G) = Z^p(K;G)/B^p(K;G).$$

These groups are called the group of **cocycles,** the group of **coboundaries,** and the **cohomology group,** respectively, of K with coefficients in G. We omit G from the notation when G equals the group of integers.

If c^p is a p-dimensional cochain, and c_p is a p-dimensional chain, we commonly use the notation $\langle c^p,c_p\rangle$ to denote the value of c^p on c_p, rather than the more familiar functional notation $c^p(c_p)$. In this notation, the definition of the coboundary operator becomes

$$\langle \delta c^p,d_{p+1}\rangle = \langle c^p,\partial d_{p+1}\rangle.$$

The definition of cohomology is, as promised, highly algebraic in nature. Is it at all possible to picture the groups involved geometrically? The answer is a qualified "yes," as we now observe.

Recall that the group $C_p(K)$ of p-chains is free abelian; it has a standard basis obtained by orienting the p-simplices of K arbitrarily and using the corresponding elementary chains as a basis. Let $\{\sigma_\alpha\}_{\alpha\in J}$ be this collection of oriented simplices. Then the elements of $C_p(K)$ are represented as finite linear combinations $\Sigma\, n_\alpha\sigma_\alpha$ of the elementary chains σ_α. Now an element c^p of $\text{Hom}(C_p(K),G)$

is determined by its value g_α on each basis element σ_α, and these values may be assigned arbitrarily. There is, however, *no* requirement that c^p vanish on all but finitely many σ_α.

Suppose we let σ_α^* denote the elementary cochain, with **Z** coefficients, whose value is 1 on the basis element σ_α and 0 on all other basis elements. Then if $g \in G$, we let $g\sigma_\alpha^*$ denote the cochain whose value is g on σ_α and 0 on all other basis elements. Using this notation, we often represent c^p by the (possibly infinite) formal sum

$$c^p = \Sigma\, g_\alpha \sigma_\alpha^*.$$

Why is this representation of c^p reasonable? We justify it as follows.

Suppose we let C_α denote the infinite cyclic subgroup of $C_p(K)$ generated by σ_α. Then $C_p(K) = \oplus_\alpha C_\alpha$, and as noted earlier,

$$(*) \qquad C^p(K; G) = \operatorname{Hom}(\oplus_\alpha C_\alpha, G) \simeq \Pi_\alpha \operatorname{Hom}(C_\alpha, G);$$

the latter group is a direct product of copies of G. Under the isomorphism (*), the cochain c^p corresponds to the element $(g_\alpha \sigma_\alpha^*)_{\alpha \in J}$ of the direct product. Instead of using "tuple" notation to represent this element of the direct product, we shall use formal sum notation.

This notation is especially convenient when it comes to computing the coboundary operator δ. We claim that if $c^p = \Sigma\, g_\alpha \sigma_\alpha^*$, then

$$(**) \qquad \delta c^p = \Sigma\, g_\alpha (\delta \sigma_\alpha^*),$$

just as if we had an honest sum rather than formal one. To verify this equation, let us orient each $p + 1$ simplex τ and show that the right side makes sense and that both sides agree, when evaluated on τ. Suppose

$$\partial \tau = \sum_{i=0}^{p+1} \epsilon_i \sigma_{\alpha_i},$$

where $\epsilon_i = \pm 1$ for each i. Then

$$\langle \delta c^p, \tau \rangle = \langle c^p, \partial \tau \rangle = \sum_{i=0}^{p+1} \epsilon_i \langle c^p, \sigma_{\alpha_i} \rangle$$

$$= \sum_{i=0}^{p+1} \epsilon_i g_{\alpha_i}.$$

Furthermore,

$$\langle g_\alpha (\delta \sigma_\alpha^*), \tau \rangle = g_\alpha \langle \delta \sigma_\alpha^*, \tau \rangle = g_\alpha \langle \sigma_\alpha^*, \partial \tau \rangle$$

$$= \begin{cases} \epsilon_i g_{\alpha_i} & \text{if } \alpha = \alpha_i \text{ for } i = 0, \dots, p+1, \\ 0 & \text{otherwise.} \end{cases}$$

Thus δc^p and $\Sigma\, g_\alpha (\delta \sigma_\alpha^*)$ have the same value on τ, so (**) holds.

By (**), in order to compute δc^p, it suffices to compute $\delta\sigma^*$ for each oriented p-simplex σ. That we can compute by using the formula

$$\delta\sigma^* = \Sigma\, \epsilon_j \tau_j^*,$$

where the summation extends over all $p + 1$ simplices τ_j having σ as a face, and $\epsilon_j = \pm 1$ is the sign with which σ appears in the expression for $\partial\tau_j$. One verifies this formula by simply evaluating both sides on the general $p + 1$ simplex τ.

Now let us apply these facts to some examples. We make only a few computations here, reserving the general problem of computing cohomology groups until a later section.

Example 1. Consider the complex K pictured in Figure 42.1. Let us compute the coboundaries of a few cochains. Let $\{v_i\}$ denote the set of vertices; let $\{e_i\}$ denote the edges, oriented as indicated; let $\{\sigma_i\}$ denote the 2-simplices, oriented as indicated. We compute δe_5^*; it has value 1 on σ_1 and value -1 on σ_2, because e_5 appears in $\partial\sigma_1$ and $\partial\sigma_2$ with signs $+1$ and -1, respectively. Thus

$$\delta e_5^* = \sigma_1^* - \sigma_2^*.$$

A similar remark shows that

$$\delta v_1^* = e_1^* + e_5^* - e_2^*.$$

Are there any cocycles in this complex? Yes, both σ_1^* and σ_2^* are cocycles, for the trivial reason that K has no 3-simplices. Each of them is a coboundary, since

$$\delta e_1^* = -\sigma_1^* \qquad \text{and} \qquad \delta e_3^* = -\sigma_2^*.$$

The 1-dimensional cochain

$$c^1 = e_1^* + e_5^* - e_3^*$$

is also a cocycle, as you can check; it happens to be a coboundary, since it equals $\delta(v_1^* + v_2^*)$. Similarly, you can check that the 0-cochain

$$c^0 = v_0^* + v_1^* + v_2^* + v_3^*$$

is a cocycle; it cannot be a coboundary because there are no cochains of dimension -1.

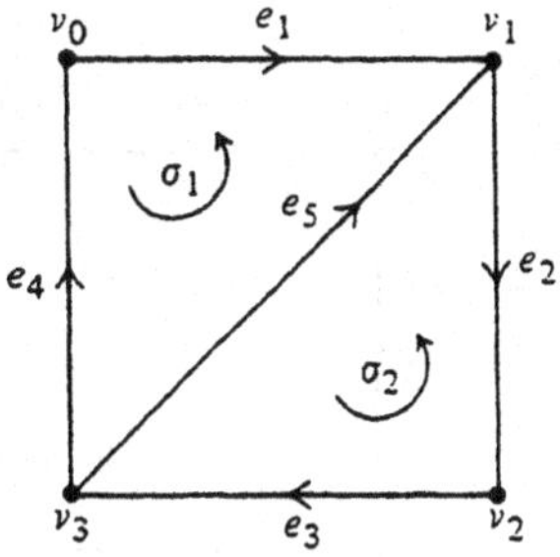

Figure 42.1

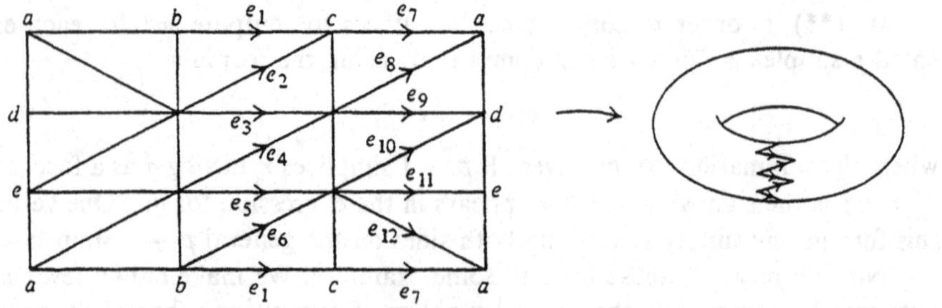

Figure 42.2

Example 2. Consider the torus, in a slightly simplified version of the triangulation considered earlier. See Figure 42.2. Consider the 1-cochain $z^1 = e_1^* + \cdots + e_6^*$ pictured in the figure. It is a cocycle, as you can check, and so is the cochain $d^1 = e_7^* + \cdots + e_{12}^*$. They happen to be cohomologous, since $\delta(c^* + h^* + j^*) = z^1 - d^1$. (Here h and j lie on the vertical line through c.)

This example illustrates the fact that while we can think of a 1-*cycle* as being a closed curve, the best way to think of a 1-*cocycle* is as a picket fence!

Later we shall compute the cohomology of the torus T and show that the cocycle z^1 represents one of the generators of $H^1(T)$. (See §47.)

Example 3. Consider the complex K pictured in Figure 42.3. We compute its cohomology groups. The general 0-cochain is a sum of the form $c^0 = \Sigma\, n_i v_i^*$. Since $\langle \delta c^0, e_i \rangle = \langle c^0, \partial e_i \rangle$, we see that δc^0 has value $n_2 - n_1$ on e_1, value $n_3 - n_2$ on e_2, and so on. If c^0 is a cocycle, then necessarily $n_1 = n_2 = n_3 = n_4$, so c^0 is of the form $n\Sigma\, v_i^*$. We conclude that $H^0(K) \simeq \mathbf{Z}$, and is generated by $\Sigma\, v_i^*$.

Now let c^1 be a 1-cochain; it is a cocycle, trivially. We show that c^1 is cohomologous to some multiple of e_1^*. It suffices to show that e_i^* is cohomologous to e_1^* for each i, and this can be done directly. For instance, e_3^* is cohomologous to e_1^* because $\delta(v_4^* + v_1^*) = e_3^* - e_1^*$. A similar remark applies to the other e_i.

Furthermore, no multiple of e_1^* is a coboundary: For let z be the cycle $e_1 + e_2 + e_3 + e_4$. Then for any 0-cochain c^0, we have $\langle \delta c^0, z \rangle = \langle c^0, \partial z \rangle = 0$. But $\langle n e_1^*, z \rangle = n$; thus $n e_1^*$ is not a coboundary unless $n = 0$.

We conclude that $H^1(K) \simeq \mathbf{Z}$ and is generated by e_1^*. It is also generated by e_2^*, by e_3^*, and by e_4^*.

Note that this same argument applies if K is a general n-gon instead of a 4-gon.

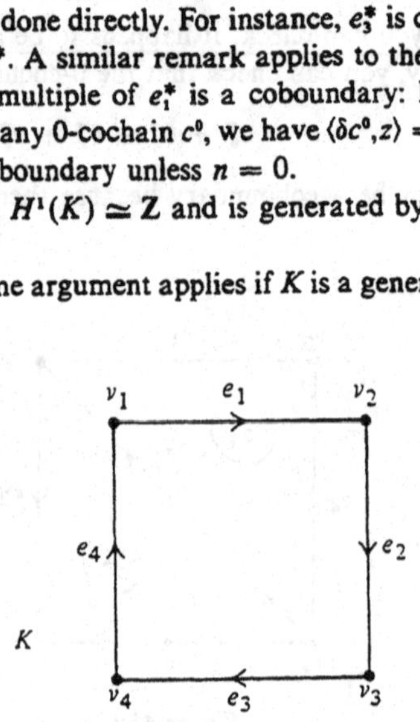

Figure 42.3

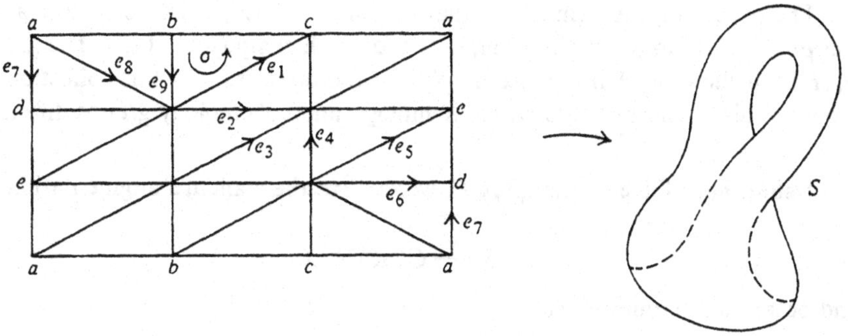

Figure 42.4

Example 4. In the preceding example, the homology and cohomology groups of K are equal. Lest you think this always occurs, consider the following example.

Let S denote the Klein bottle, represented by the labelled rectangle of Figure 42.4. We show that $H^2(S)$ is nontrivial, whereas we know that $H_2(S) = 0$. Orient the 2-simplices of L counterclockwise. Use the induced orientation of the 2-simplices of S, and let γ denote their sum. Now γ is not a cycle, because $\partial\gamma = 2z_1$, where $z_1 = [a,d] + [d,e] + [e,a]$.

Let σ denote a single oriented 2-simplex of S, as pictured. Then σ^* is a cocycle of S, because S has no 3-simplices. Furthermore, σ^* is not a coboundary. For if c^1 is an arbitrary 1-cochain, then $\langle\delta c^1,\gamma\rangle = \langle c^1,\partial\gamma\rangle = 2\langle c^1,z_1\rangle$, which is an even integer, while $\langle\sigma^*,\gamma\rangle = 1$. Thus σ^* represents a non-trivial element of $H^2(S)$.

Now in fact σ^* represents an element of order 2 in $H^2(S)$. You can check that the coboundary of the 1-cochain $(e_1^* + \cdots + e_9^*)$ pictured in the figure equals $2\sigma^*$.

Now we consider the zero-dimensional cohomology groups, and compute them.

Theorem 42.1. *Let K be a complex. Then $H^0(K; G)$ equals the group of all 0-cochains c^0 such that $\langle c^0,v\rangle = \langle c^0,w\rangle$ whenever v and w belong to the same component of $|K|$.*

In particular, if $|K|$ is connected, then $H^0(K) \simeq \mathbf{Z}$ and is generated by the cochain whose value is 1 on each vertex of K.

Proof. Note that $H^0(K; G)$ equals the group of 0-cocycles, because there are no coboundaries in dimension 0. If v and w belong to the same component of $|K|$, then there is a 1-chain c_1 of K such that $\partial c_1 = v - w$. Then for any cocycle c^0, we must have

$$0 = \langle\delta c^0,c_1\rangle = \langle c^0,\partial c_1\rangle = \langle c^0,v\rangle - \langle c^0,w\rangle.$$

Conversely, let c^0 be a cochain such that $\langle c^0,v\rangle - \langle c^0,w\rangle = 0$ whenever v and w lie in the same component of $|K|$. Then for each oriented 1-simplex σ of K,

$$\langle\delta c^0,\sigma\rangle = \langle c^0,\partial\sigma\rangle = 0.$$

We conclude that $\delta c^0 = 0$. The theorem follows. □

The preceding theorem shows that in general $H^0(K)$ is isomorphic to a direct *product* of infinite cyclic groups, one for each component of $|K|$. The group $H_0(K)$, on the other hand, is isomorphic to the direct *sum* of this collection of groups. This is another case where homology and cohomology groups differ.

Definition. Given a complex K, we dualize the standard augmentation

$$C_1(K) \xrightarrow{\partial_1} C_0(K) \xrightarrow{\epsilon} \mathbf{Z},$$

and obtain a homomorphism $\tilde{\epsilon}$,

$$C^1(K; G) \xleftarrow{\delta_1} C^0(K; G) \xleftarrow{\tilde{\epsilon}} G,$$

called a **coaugmentation.** It is injective, and $\delta_1 \circ \tilde{\epsilon} = 0$. We define the **reduced cohomology** of K by setting $\tilde{H}^q(K; G) = H^q(K; G)$ if $q > 0$, and

$$\tilde{H}^0(K; G) = \ker \delta_1 / \operatorname{im} \tilde{\epsilon}.$$

Theorem 42.2. *If $|K|$ is connected, then $\tilde{H}^0(K; G) = 0$. More generally, for any complex K,*

$$H^0(K; G) \simeq \tilde{H}^0(K; G) \oplus G.$$

Proof. If $|K|$ is connected, then $\tilde{H}_0(K)$ vanishes, so $C_1(K) \to C_0(K) \to \mathbf{Z} \to 0$ is exact. It follows that

$$C^1(K; G) \leftarrow C^0(K; G) \leftarrow G \leftarrow 0$$

is exact, so $\tilde{H}^0(K; G)$ vanishes. The rest of the theorem we leave as an exercise. □

EXERCISES

1. Consider the complex K of Example 1. Find a basis for the cocycle group in each dimension. Show that the cohomology of K vanishes in positive dimensions.
2. Check the computations of Examples 2, 3, and 4.
3. (a) Suppose one is given homomorphisms

$$C_1 \xrightarrow{\phi} C_0 \xrightarrow{\epsilon} \mathbf{Z},$$

where $\epsilon \circ \phi = 0$ and ϵ is an epimorphism. Consider the dual sequence

$$\operatorname{Hom}(C_1, G) \xleftarrow{\tilde{\phi}} \operatorname{Hom}(C_0, G) \xleftarrow{\tilde{\epsilon}} \operatorname{Hom}(\mathbf{Z}, G).$$

Show that $\operatorname{im} \tilde{\epsilon}$ is a direct summand in $\operatorname{Hom}(C_0, G)$ and hence in $\ker \tilde{\phi}$. [*Hint:* See the exercises of §7.]

(b) Conclude that

$$\ker \tilde{\phi} \simeq \frac{\ker \tilde{\phi}}{\operatorname{im} \tilde{\epsilon}} \oplus G,$$

so that in particular,

$$H^0(K; G) \simeq \tilde{H}^0(K; G) \oplus G.$$

4. Let K be the complex whose space is the real line and whose vertices are the integers. Let $\sigma_n = [n,n+1]$. Show that $H^1(K) = 0$ by finding a specific cochain whose coboundary is

$$\sum_{i=-\infty}^{\infty} a_i \sigma_i^*.$$

*5. Let K be a finite complex.
   (a) Use the theorem on standard bases for chain complexes (§11) to express $H^p(K)$ in terms of the betti numbers and torsion coefficients of K.
   (b) Express $H^p(K;G)$ in terms of the betti numbers and torsion coefficients of K; the answer will involve the groups G, G/mG, and $\ker(G \xrightarrow{m} G)$.
   (c) Compute $H^p(X; G)$ if X is the torus, Klein bottle, or projective plane.

§43. RELATIVE COHOMOLOGY

Continuing our discussion of simplicial cohomology, we define the relative cohomology groups. We also consider the homomorphism induced by a simplicial map, and the long exact sequence in cohomology. In some respects, relative cohomology is similar to relative homology; in other respects it is rather different, as we shall see.

Definition. Let K be a complex; let K_0 be a subcomplex of K; let G be an abelian group. We define the group of **relative cochains** in dimension p by the equation

$$C^p(K,K_0; G) = \operatorname{Hom}(C_p(K,K_0),G).$$

The relative coboundary operator δ is defined as the dual of the relative boundary operator. We define $Z^p(K,K_0; G)$ to be the kernel of the homomorphism

$$\delta : C^p(K,K_0; G) \longrightarrow C^{p+1}(K,K_0; G),$$

$B^{p+1}(K,K_0; G)$ to be its image, and

$$H^p(K,K_0; G) = Z^p(K,K_0; G)/B^p(K,K_0; G).$$

These groups are called the group of **relative cocycles**, the group of **relative coboundaries**, and the **relative cohomology group**, respectively.

We have an idea of how to picture cochains and cocycles. How shall we picture a *relative* cochain or cocycle? Here the situation is rather different from the situation in homology. We explain the difference as follows:

For chains, we had an exact sequence

$$0 \to C_p(K_0) \xrightarrow{i} C_p(K) \xrightarrow{j} C_p(K,K_0) \to 0,$$

where $C_p(K_0)$ is a subgroup of $C_p(K)$, and $C_p(K,K_0)$ is their quotient. The sequence splits because the relative chain group is free. Therefore, the sequence

$$0 \leftarrow C^p(K_0; G) \xleftarrow{\tilde{i}} C^p(K; G) \xleftarrow{\tilde{j}} C^p(K,K_0; G) \leftarrow 0$$

is exact and splits. This leads to the following surprising fact:

It is natural to consider $C^p(K,K_0; G)$ as a *subgroup* of $C^p(K; G)$, and $C^p(K_0; G)$ as a *quotient group* of $C^p(K; G)$.

Let us examine this situation more closely. A relative cochain is a homomorphism mapping $C_p(K,K_0)$ into G. The group of such homomorphisms corresponds precisely to the group of all homomorphisms of $C_p(K)$ into G that vanish on the subgroup $C_p(K_0)$. This is just a subgroup of the group of *all* homomorphisms of $C_p(K)$ into G. Thus $C^p(K,K_0; G)$ can be naturally considered to be the subgroup of $C^p(K; G)$ consisting of those cochains that vanish on every oriented simplex of K_0. In some sense, $C^p(K,K_0; G)$ is the group of those cochains of K that are "carried by" $K - K_0$. The coboundary operator maps this subgroup of $C^p(K; G)$ into itself: Suppose c^p vanishes on every simplex of K_0. If τ is a $p + 1$ simplex of K_0, then $\partial\tau$ is carried by K_0, so

$$\langle \delta c^p, \tau \rangle = \langle c^p, \partial\tau \rangle = 0.$$

Thus the map $\tilde{j}$ can be interpreted as an inclusion map. To interpret $\tilde{i}$, we note that it carries the cochain c^p of K to the cochain $c^p \circ i$, which is just the restriction of c^p to $C_p(K_0)$. We summarize these results as follows:

If we begin with the sequence

$$0 \to C_p(K_0) \xrightarrow{i} C_p(K) \xrightarrow{j} C_p(K,K_0) \to 0,$$

the dual of the projection map j is an inclusion map $\tilde{j}$, and the dual of the inclusion map i is a restriction map $\tilde{i}$.

Let us now consider the homomorphism of cohomology induced by a simplicial map.

Recall that if $f : (K,K_0) \to (L,L_0)$ is a simplicial map, then one has a corresponding chain map

$$f_\# : C_p(K,K_0) \to C_p(L,L_0).$$

The dual of $f_\#$ maps cochains to cochains; we usually denote it by $f^\#$. Because $f_\#$ commutes with ∂, the map $f^\#$ commutes with δ, since the dual of the equation $f_\# \circ \partial = \partial \circ f_\#$ is the equation $\delta \circ f^\# = f^\# \circ \delta$. Hence $f^\#$ carries cocycles to cocycles and coboundaries to coboundaries. It is called a **cochain map**; it induces a homomorphism of cohomology groups,

$$H^p(K,K_0;G) \xleftarrow{f^*} H^p(L,L_0;G).$$

Functoriality holds, even on the cochain level. For if i is the identity, then $i_\#$ is the identity and so is $i^\#$. Similarly, the equation $(g \circ f)_\# = g_\# \circ f_\#$ gives, when dualized, the equation $(g \circ f)^\# = f^\# \circ g^\#$.

Just as in the case of homology, one has a long exact sequence in cohomology involving the relative groups. But again, there are a few differences.

Theorem 43.1. *Let K be a complex; let K_0 be a subcomplex. There exists an exact sequence*

$$\cdots \leftarrow H^p(K_0;G) \leftarrow H^p(K;G) \leftarrow H^p(K,K_0;G) \xleftarrow{\delta^*} H^{p-1}(K_0;G) \leftarrow \cdots.$$

A similar sequence exists in reduced cohomology if K_0 is not empty. A simplicial map $f:(K,K_0) \to (L,L_0)$ induces a homomorphism of long exact cohomology sequences.

Proof. This theorem follows from applying the zig-zag lemma to the diagram

$$\begin{array}{ccccccccc}
0 & \longleftarrow & C^{p+1}(K_0;G) & \xleftarrow{\tilde{i}} & C^{p+1}(K;G) & \xleftarrow{\tilde{j}} & C^{p+1}(K,K_0;G) & \longleftarrow & 0 \\
 & & \uparrow \delta & & \uparrow \delta & & \uparrow \delta & & \\
0 & \longleftarrow & C^p(K_0;G) & \xleftarrow{\tilde{i}} & C^p(K;G) & \xleftarrow{\tilde{j}} & C^p(K,K_0;G) & \longleftarrow & 0.
\end{array}$$

Since i and j commute with ∂, the dual maps $\tilde{i}$ and $\tilde{j}$ commute with δ. The fact that δ^* *raises* dimension by 1 follows from the proof of the zig-zag lemma. (If you turn the page upside down, the arrows look like those in the proof of the lemma.)

The sequence in reduced cohomology is derived similarly, once one adjoins the sequence

$$\begin{array}{ccccccccc}
0 & \longleftarrow & C^0(K_0;G) & \longleftarrow & C^0(K;G) & \longleftarrow & C^0(K,K_0;G) & \longleftarrow & 0 \\
 & & \uparrow \tilde{\epsilon} & & \uparrow \tilde{\epsilon} & & \uparrow & & \\
0 & \longleftarrow & G & \longleftarrow & G & \longleftarrow & 0 & \longleftarrow & 0
\end{array}$$

at the bottom of the diagram. Since $\tilde{\epsilon}$ is injective, no nontrivial cohomology groups appear in dimension -1. □

Now let us compute cohomology groups in a few examples, using this sequence.

Example 1. Consider the case of a square K modulo its boundary K_0, as pictured in Figure 43.1. We treat the group of relative cochains as those cochains of K that

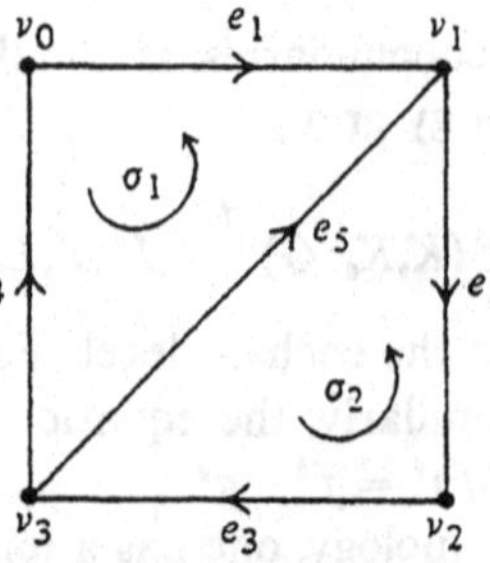

Figure 43.1

are carried by $K - K_0$. Both σ_1^* and σ_2^* are such cochains, and each is a cocycle (trivially). There is only one cochain e_5^* in dimension 1 carried by $K - K_0$; its coboundary is

$$\delta e_5^* = \sigma_1^* - \sigma_2^*.$$

Thus the group $H^2(K,K_0)$ is infinite cyclic, and is generated by the cohomology class $\{\sigma_1^*\} = \{\sigma_2^*\}$.

The group $H^1(K,K_0)$ vanishes because the only 1-cochain carried by $K - K_0$ is not a cocycle. The group $H^0(K,K_0)$ vanishes because there are no 0-cochains carried by $K - K_0$.

Now consider the exact sequence

$$\begin{array}{ccccccccccc} H^2(K_0) & \leftarrow & H^2(K) & \leftarrow & H^2(K,K_0) & \leftarrow & H^1(K_0) & \leftarrow & H^1(K) & \leftarrow & H^1(K,K_0) \\ 0 & \leftarrow & (?) & \leftarrow & \mathbf{Z} & \overset{\delta^*}{\leftarrow} & \mathbf{Z} & \leftarrow & (?) & \leftarrow & 0 \end{array}$$

We just computed the cohomology of (K,K_0), and we found the cohomology of K_0 in Example 3 of §42. What is δ^*? The group $H^1(K_0)$ is generated by the cocycle e_1^*. To compute $\delta^*\{e_1^*\}$, we first "pull e_1^* back" to K (considering it as a cochain of K) taking its coboundary δe_1^* in K, and then considering the result as a cocycle of K modulo K_0. By direct computation, $\delta e_1^* = -\sigma_1^*$ as cochains of K. Since σ_1^* *generates* $H^2(K,K_0)$, as just proved, it follows that δ^* is an isomorphism.

Therefore, both the unknown groups in this exact cohomology sequence must vanish.

Example 2. Consider the Möbius band M modulo its edge E, as pictured in Figure 43.2. We calculate the cohomology of (M,E) and of M.

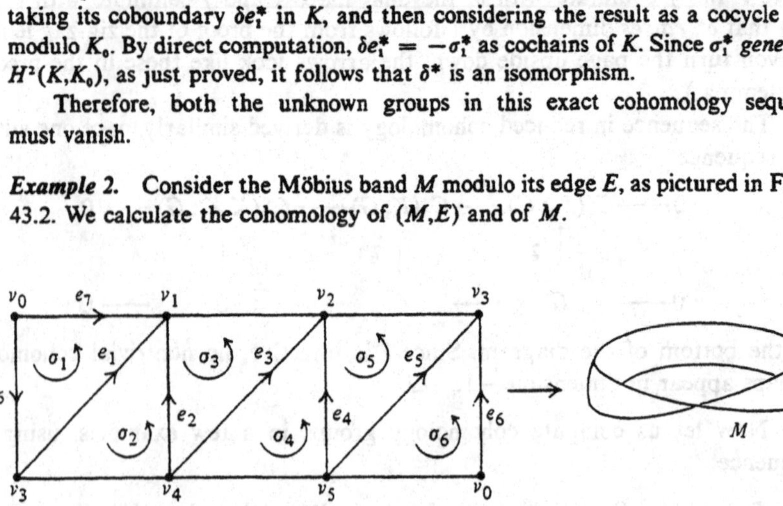

Figure 43.2

Each of the cochains σ_i^* is a cocycle (trivially), so they form a basis for the group $Z^2(M,E)$ of relative 2-cocycles. Similarly, $e_1^*, \ldots, e_6^*$ form a basis for the group $C^1(M,E)$ of relative 1-cochains. (The other 1-simplices of M lie in the edge E.) It is convenient to replace these bases for $Z^2(M,E)$ and $C^1(M,E)$ by different bases. Let us take

$$\sigma_1^*, \quad \sigma_1^* - \sigma_2^*, \quad \sigma_2^* - \sigma_3^*, \quad \sigma_3^* - \sigma_4^*, \quad \sigma_4^* - \sigma_5^*, \quad \sigma_5^* - \sigma_6^*$$

as a basis for $Z^2(M,E)$, and

$$e_1^*, \quad e_2^*, \quad e_3^*, \quad e_4^*, \quad e_5^*, \quad e_1^* + \cdots + e_6^*$$

as a basis for $C^1(M,E)$. Then we can calculate $H^2(M,E)$ readily. We see that $\delta e_i^* = \sigma_i^* - \sigma_{i+1}^*$ for $i = 1,\ldots,5$; and $\delta(e_1^* + \cdots + e_6^*) = 2\sigma_1^*$. Thus $H^2(M,E) \simeq \mathbf{Z}/2$ and is generated by σ_1^*.

The group $H^1(M,E)$ vanishes, for there are no relative 1-cocycles: This follows from the fact that δ carries our chosen basis for $C^1(M,E)$ to a basis for a subgroup of $C^2(M,E)$. (You can also prove directly that $n_1e_1^* + \cdots + n_6e_6^*$ is a cocycle only if $n_i = 0$ for all i.)

Consider now the exact sequence

$$\begin{array}{ccccccccccc} H^2(E) & \leftarrow & H^2(M) & \leftarrow & H^2(M,E) & \leftarrow & H^1(E) & \leftarrow & H^1(M) & \leftarrow & H^1(M,E) \\ 0 & \leftarrow & (?) & \leftarrow & \mathbf{Z}/2 & \overset{\delta^*}{\leftarrow} & \mathbf{Z} & \leftarrow & (?) & \leftarrow & 0 \end{array}$$

We have just computed $H^2(M,E)$; the proof that $H^1(E)$ is infinite cyclic is easy, since E is a 6-gon. (See Example 3 of §42.) Let us compute δ^*. The group $H^1(E)$ is generated by e_7^*, where e_7 is any oriented 1-simplex of E. Choose e_7 as indicated in Figure 43.2. Then by direct computation $\delta^* e_7^* = -\sigma_1^*$; so δ^* is surjective. It follows at once that $H^2(M) = 0$ and $H^1(M) \simeq \mathbf{Z}$.

EXERCISES

1. Let M denote the Möbius band, as in Example 2. Find a cocycle generating $H^1(M)$.

2. Consider the cylinder C modulo one edge E.
   (a) Compute $H^i(C,E)$.
   (b) Use the long exact cohomology sequence of the pair (C,E) to compute $H^i(C)$.

Relative pseudo n-manifolds

A simplicial pair (K,K_0) is called a **relative pseudo n-manifold** if:

(i) The closure of $|K| - |K_0|$ equals a union of n-simplices.

(ii) Each $n-1$ simplex of K not in K_0 is a face of exactly two n-simplices of K.

(iii) Given two n-simplices σ, σ' of K not in K_0, there is a sequence of n-simplices of K not in K_0

$$\sigma = \sigma_0, \sigma_1, \ldots, \sigma_k = \sigma'$$

such that $\sigma_i \cap \sigma_{i+1}$ is an $n-1$ simplex not in K_0, for each i.

If $K_0 = \varnothing$, we call K simply a **pseudo n-manifold**.

3. Which of the following spaces are pseudo manifolds in their familiar triangulations?
   (a) S^1
   (b) S^2
   (c) The letter θ.
   (d) The union of S^2 and a circle that intersects S^2 in one point.
   (e) The union of two copies of S^2 with a point in common.
   (f) S^2 with the north and south poles identified.

4. Show that the compact 2-manifolds, in their usual triangulations, are pseudo manifolds. (We will see later that *any* connected, triangulated n-manifold is a pseudo n-manifold.)

5. Let (K,K_0) be a relative pseudo n-manifold.
   (a) Given $\sigma \neq \sigma'$, neither in K_0, show there exists a sequence

$$\sigma = \sigma_0, \sigma_1, \ldots, \sigma_k = \sigma'$$

   as in (iii) *with no repetitions.*

   In this situation, once σ is oriented, there are unique orientations of each σ_i such that $\partial\sigma_{i-1} + \partial\sigma_i$ has coefficient 0 on $\sigma_{i-1} \cap \sigma_i$ for each i. The resulting orientation of $\sigma_k = \sigma'$ is said to be *induced by* the given orientation of σ, relative to the given sequence.

   (b) Let σ be fixed and oriented. If for every $\sigma' \neq \sigma$, the orientation of σ' induced by that of σ is independent of the sequence joining them, then (K,K_0) is said to be **orientable.** Otherwise, (K,K_0) is said to be **non-orientable.** Show that if K is finite,

$$H_n(K,K_0) \simeq \mathbf{Z} \quad \text{and} \quad H^n(K,K_0) \simeq \mathbf{Z} \qquad \text{if} \quad (K,K_0) \text{ is orientable,}$$

$$H_n(K,K_0) = 0 \quad \text{and} \quad H^n(K,K_0) \simeq \mathbf{Z}/2 \quad \text{if} \quad (K,K_0) \text{ is non-orientable.}$$

   [*Hint:* If γ is the sum of all the n-simplices of K not in K_0, oriented arbitrarily, then for each relative $n-1$ cochain c^{n-1}, the number $\langle\delta c^{n-1},\gamma\rangle$ is even. Therefore, σ^* does not cobound.]

   (c) Conclude that in the finite case at least, orientability is independent of the choice of σ, and in fact depends only on the topological pair $(|K|,|K_0|)$, not on the particular triangulation involved.

   (d) Show that if K is finite,

$$H_n(K,K_0;\mathbf{Z}/2) \simeq \mathbf{Z}/2 \simeq H^n(K,K_0;\mathbf{Z}/2).$$

§44. COHOMOLOGY THEORY

Now that we have some feeling for what the simplicial cohomology groups look like, let us deal with cohomology theory more generally. We construct both the simplicial and the singular cohomology theories, we show they are naturally isomorphic for triangulable spaces, and we verify the cohomology versions of the Eilenberg-Steenrod axioms.

First, let us work on the level of chain complexes.

The cohomology of a chain complex

Let $\mathcal{C} = \{C_p,\partial\}$ be a chain complex; let G be an abelian group. We define the **p-dimensional cochain group** of $\mathcal{C}$, with coefficients in G, by the equation

$$C^p(\mathcal{C}; G) = \text{Hom}(C_p,G).$$

We define the **coboundary operator** δ to be the dual of the boundary operator; it follows that $\delta^2 = 0$. The family of groups and homomorphisms $\{C^p(\mathcal{C}; G),\delta\}$ is called the **cochain complex** of $\mathcal{C}$ with coefficients in G. As usual, the kernel of the homomorphism

$$\delta : C^p(\mathcal{C}; G) \to C^{p+1}(\mathcal{C}; G)$$

is denoted $Z^p(\mathcal{C}; G)$, its image is denoted $B^{p+1}(\mathcal{C}; G)$, and the **cohomology group** of $\mathcal{C}$ in dimension p, with coefficients in G, is defined by the equation

$$H^p(\mathcal{C}; G) = Z^p(\mathcal{C}; G)/B^p(\mathcal{C}; G).$$

If $\{\mathcal{C},\epsilon\}$ is an augmented chain complex, then one has a corresponding cochain complex

$$\cdots \leftarrow C^1(\mathcal{C}; G) \overset{\delta_1}{\leftarrow} C^0(\mathcal{C}; G) \overset{\tilde{\epsilon}}{\leftarrow} \text{Hom}(\mathbf{Z},G),$$

where $\tilde{\epsilon}$ is injective. We define the **reduced cohomology groups** of $\mathcal{C}$ by setting $\tilde{H}^q(\mathcal{C}; G) = H^q(\mathcal{C}; G)$ if $q > 0$, and

$$\tilde{H}^0(\mathcal{C}; G) = \ker \delta_1/\text{im}\,\tilde{\epsilon}.$$

It is easy to see that if $\tilde{H}_0(\mathcal{C})$ vanishes, then $\tilde{H}^0(\mathcal{C}; G)$ vanishes as well, for exactness of $C_1 \to C_0 \to \mathbf{Z} \to 0$ implies exactness of the dual sequence. In general (see the exercises of §42), we have the equation

$$H^0(\mathcal{C}; G) \simeq \tilde{H}^0(\mathcal{C}; G) \oplus G.$$

Definition. Suppose $\mathcal{C} = \{C_p,\partial\}$ and $\mathcal{C}' = \{C'_p,\partial'\}$ are chain complexes. Suppose $\phi : \mathcal{C} \to \mathcal{C}'$ is a chain map, so that $\partial' \circ \phi = \phi \circ \partial$. Then the dual homomorphism

$$C^p(\mathcal{C}; G) \overset{\tilde{\phi}}{\leftarrow} C^p(\mathcal{C}'; G)$$

commutes with δ; such a homomorphism is called a **cochain map.** It carries cocycles to cocycles, and coboundaries to coboundaries, so it induces a homomorphism of cohomology groups,

$$H^p(\mathcal{C}; G) \overset{\phi^*}{\leftarrow} H^p(\mathcal{C}'; G)$$

The assignment

$$\mathcal{C} \to H^p(\mathcal{C}; G) \qquad \text{and} \qquad \phi \to \phi^*$$

satisfies the usual functorial properties; in fact, they hold already on the chain level.

If $\{\mathcal{C},\epsilon\}$ and $\{\mathcal{C}',\epsilon'\}$ are augmented chain complexes and if $\phi : \mathcal{C} \to \mathcal{C}'$ is an augmentation-preserving chain map, then $\epsilon' \circ \phi = \epsilon$, so that $\tilde{\phi} \circ \tilde{\epsilon}' = \tilde{\epsilon}$. In this case, $\tilde{\phi}$ induces a homomorphism of reduced as well as ordinary cohomology.

Suppose now that $\phi, \psi : \mathcal{C} \to \mathcal{C}'$ are chain maps, and that D is a chain homotopy between them, so that

$$D\partial + \partial' D = \phi - \psi.$$

Then $\tilde{D} : C^{p+1}(\mathcal{C}'; G) \to C^p(\mathcal{C}; G)$ is a homomorphism satisfying the equation

$$\delta\tilde{D} + \tilde{D}\delta' = \tilde{\phi} - \tilde{\psi}.$$

It is said to be a **cochain homotopy** between $\tilde{\phi}$ and $\tilde{\psi}$. If such a $\tilde{D}$ exists, it follows at once that the induced cohomology homomorphisms ϕ^* and ψ^* are equal. For given any cocycle z^p, we have

$$\tilde{\phi}(z^p) - \tilde{\psi}(z^p) = \delta\tilde{D}z^p + 0.$$

This observation has the following consequence:

Theorem 44.1. *Let $\mathcal{C}$ and $\mathcal{C}'$ be chain complexes; let $\phi : \mathcal{C} \to \mathcal{C}'$ be a chain equivalence. Then ϕ_* and ϕ^* are isomorphisms of homology and cohomology, respectively. If $\mathcal{C}$ and $\mathcal{C}'$ are augmented and ϕ is augmentation-preserving, then ϕ_* and ϕ^* are isomorphisms of reduced homology and cohomology, respectively.*

Proof. Since ϕ is a chain equivalence, there is a chain map $\psi : \mathcal{C}' \to \mathcal{C}$ such that $\phi \circ \psi$ and $\psi \circ \phi$ are chain homotopic to identity maps. Then $\tilde{\psi} \circ \tilde{\phi}$ and $\tilde{\phi} \circ \tilde{\psi}$ are cochain homotopic to identity maps, so $\psi^* \circ \phi^*$ and $\phi^* \circ \psi^*$ equal the identity maps of $H^p(\mathcal{C}')$ and $H^p(\mathcal{C})$, respectively.

The same argument holds for reduced cohomology. □

Finally, suppose

$$0 \to \mathcal{C} \xrightarrow{\phi} \mathcal{D} \xrightarrow{\psi} \mathcal{E} \to 0$$

is a short exact sequence of chain complexes *that splits in each dimension.* (This occurs, for example, when $\mathcal{E}$ is free.) Then the dual sequence

$$0 \leftarrow C^p(\mathcal{C}; G) \xleftarrow{\tilde{\phi}} C^p(\mathcal{D}; G) \xleftarrow{\tilde{\psi}} C^p(\mathcal{E}; G) \leftarrow 0$$

is exact. Applying the zig-zag lemma, we have a long exact sequence in cohomology,

$$\cdots \leftarrow H^p(\mathcal{C}; G) \xleftarrow{\phi^*} H^p(\mathcal{D}; G) \xleftarrow{\psi^*} H^p(\mathcal{E}; G) \xleftarrow{\delta^*} H^{p-1}(\mathcal{C}; G) \leftarrow \cdots,$$

where δ^* is induced in the usual manner by the coboundary operator. This sequence is natural in the sense that if f is a homomorphism of short exact sequences of chain complexes, then $\tilde{f}$ is a homomorphism of the dual sequences, and f^* is a homomorphism of the corresponding long exact cohomology sequences.

The axioms for cohomology

Now we state the cohomology versions of the Eilenberg-Steenrod axioms.

Given an admissible class $\mathcal{A}$ of pairs of spaces (X,A) and an abelian group G, a **cohomology theory** on $\mathcal{A}$ with coefficients in G consists of the following:

(1) A function defined for each integer p and each pair (X,A) in $\mathcal{A}$, whose value is an abelian group $H^p(X,A;G)$.

(2) A function that assigns to each continuous map $h:(X,A) \to (Y,B)$ and each integer p, a homomorphism

$$H^p(X,A;G) \xleftarrow{h^*} H^p(Y,B;G).$$

(3) A function that assigns to each pair (X,A) in $\mathcal{A}$ and each integer p, a homomorphism

$$H^p(X,A;G) \xleftarrow{\delta^*} H^{p-1}(A;G).$$

The following axioms are to be satisfied:

Axiom 1. If i is the identity, then i^* is the identity.

Axiom 2. $(k \circ h)^* = h^* \circ k^*$.

Axiom 3. δ^* is a natural transformation of functors.

Axiom 4. The following sequence is exact, where i and j are inclusions:

$$\cdots \leftarrow H^p(A;G) \xleftarrow{i^*} H^p(X;G) \xleftarrow{j^*} H^p(X,A;G) \xleftarrow{\delta^*} H^{p-1}(A;G) \leftarrow \cdots.$$

Axiom 5. If $h \simeq k$, then $h^* = k^*$.

Axiom 6. Given (X,A), let U be an open set in X such that $\bar{U} \subset \text{Int}\, A$. If $(X - U, A - U)$ is admissible, then inclusion j induces a cohomology isomorphism

$$H^p(X-U, A-U;G) \xleftarrow{j^*} H^p(X,A;G).$$

Axiom 7. If P is a one-point space, then $H^p(P;G) = 0$ for $p \neq 0$ and

$$H^0(P;G) \simeq G.$$

The axiom of compact support has no counterpart in cohomology theory.

Singular cohomology theory

Now we consider singular theory and show it satisfies the axioms.

The **singular cohomology groups** of a topological pair (X,A), with coefficients in the abelian group G, are defined by the equation

$$H^p(X,A;G) = H^p(\mathcal{S}(X,A);G),$$

where $\mathcal{S}(X,A)$ is the singular chain complex of (X,A). As usual, we delete A from the notation if $A = \varnothing$, and we delete G if it equals the group of integers.

The reduced cohomology groups are defined by the equation

$$\tilde{H}^p(X; G) = \tilde{H}^p(\mathcal{S}(X); G),$$

relative to the standard augmentation for $\mathcal{S}(X)$.

Given a continuous map $h : (X,A) \to (Y,B)$, there is a chain map

$$h_\# : S_p(X,A) \to S_p(Y,B),$$

defined by $h_\#(T) = h \circ T$. We customarily denote the dual cochain map by $h^\#$. It induces a homomorphism

$$H^p(X,A; G) \xleftarrow{h^*} H^p(Y,B; G).$$

(The same holds in reduced cohomology if A and B are empty, since $h_\#$ is augmentation-preserving.) The functorial properties (Axioms 1 and 2) hold even on the cochain level. The short exact sequence of chain complexes

$$0 \to S_p(A) \to S_p(X) \to S_p(X,A) \to 0$$

splits because $S_p(X,A)$ is free. Therefore, the zig-zag lemma gives us an exact sequence

$$\cdots \leftarrow H^p(A; G) \leftarrow H^p(X; G) \leftarrow H^p(X,A; G) \xleftarrow{\delta^*} H^{p-1}(A; G) \leftarrow \cdots.$$

A continuous map $h : (X,A) \to (Y,B)$ induces a homomorphism of long exact cohomology sequences, by the naturality property of the zig-zag lemma. (A similar result holds for reduced cohomology if A is non-empty.) Thus Axioms 3 and 4 hold.

If $h, k : (X,A) \to (Y,B)$ are homotopic, then $h_\#$ and $k_\#$ are chain homotopic, as we proved in Theorem 30.7. Then $h^\#$ and $k^\#$ are cochain homotopic, so that $h^* = k^*$.

To compute the cohomology of a 1-point space P, we recall (see Theorem 30.3) that the singular chain complex of P has the form

$$\cdots \xrightarrow{\cong} \mathbf{Z} \xrightarrow{0} \mathbf{Z} \xrightarrow{\cong} \mathbf{Z} \xrightarrow{0} \mathbf{Z} \to 0.$$

The cochain complex of P then has the form

$$\cdots \xleftarrow{\cong} G \xleftarrow{0} G \xleftarrow{\cong} G \xleftarrow{0} G \leftarrow 0.$$

It follows that $H^i(P; G)$ is isomorphic to G if $i = 0$, and vanishes otherwise.

Finally, we come to the excision property of singular cohomology. Here is a point where the arguments we gave for homology do not go through automatically for cohomology. Suppose

$$j : (X - U, A - U) \to (X,A)$$

is an excision map. If we had showed that $j_\#$ is a chain equivalence, then there would be no difficulty, for $j^\#$ would be a cochain equivalence, and it would follow that j^* is an isomorphism. But we proved only the weaker result that j_* is an isomorphism. (See Theorem 31.7.) So we have some work to do to carry this result over to cohomology.

What we need is the following fact, which will be proved in the next section: *Let $\mathcal{C}$ and $\mathcal{D}$ be free chain complexes; let $\phi : \mathcal{C} \to \mathcal{D}$ be a chain map that induces a homology isomorphism in all dimensions. Then ϕ induces a cohomology isomorphism in all dimensions, for all coefficient groups G.*

The excision property of singular cohomology is an immediate consequence: Given $U \subset A \subset X$, with $\bar{U} \subset \text{Int } A$, consider the inclusion map

$$j : (X - U, A - U) \to (X,A).$$

Since the chain complexes involved are free, and since $j_\#$ induces an isomorphism in homology, it induces an isomorphism in cohomology as well.

Note that singular cohomology, like singular homology, satisfies an excision property slightly stronger than that stated in the axiom. One needs to have $\bar{U} \subset \text{Int } A$, but one does not need U to be open, in order for excision to hold.

Simplicial cohomology theory

We have already dealt with some aspects of simplicial cohomology. We have defined the cohomology groups of a simplicial pair (K,K_0), and have showed how a simplicial map f induces a homomorphism of these groups. Just as with homology, showing that an arbitrary continuous map induces a homomorphism requires a bit of work.

The construction follows the pattern of §14–§18. First, we recall that if f and g are two simplicial approximations to the same continuous map, then they are contiguous, so the corresponding chain maps $f_\#$ and $g_\#$ are chain homotopic. It follows that $f^\#$ and $g^\#$ are cochain homotopic, so the cohomology homomorphisms f^* and g^* are equal. Furthermore, if K' is a subdivision of K, and if $g : (K',K_0') \to (K,K_0)$ is a simplicial approximation to the identity, then $g_\#$ is a chain equivalence, so that $g^\#$ is a cochain equivalence, and g^* is an isomorphism.

One then defines the homomorphism induced by the continuous map $h : (|K|,|K_0|) \to (|L|,|L_0|)$ as follows: Choose a subdivision K' of K so that h has a simplicial approximation $f : (K',K_0') \to (L,L_0)$. Choose $g : (K',K_0') \to (K,K_0)$ to be a simplicial approximation to the identity. Finally, define

$$h^* = (g^*)^{-1} \circ f^*.$$

To verify that h^* is independent of the choices involved, and to check its functorial properties, involves arguments very similar to those given in §18, when we verified the corresponding properties for the homomorphism h_*. In fact, one can use exactly the same diagrams as appear in that section, simply reversing the arrows for all the induced homomorphisms! We leave the details to you.

Verifying the Eilenberg-Steenrod axioms is now straightforward. The existence of the long exact cohomology sequence we have already noted. Naturality of the sequence reduces to proving naturality in the case of a simplicial map, which we have already done. The homotopy axiom, the simplicial version of the excision axiom, and the dimension axiom follow just as they did in the case of singular cohomology. Nothing new of interest occurs.

The topological invariance, indeed the homotopy-type invariance, of the simplicial cohomology groups follows at once.

The isomorphism between simplicial and singular cohomology

Let K be a simplicial complex. We defined in §34 a chain map

$$\eta : \mathcal{C}(K) \to \mathcal{S}(|K|)$$

that induces an isomorphism in homology. Although it depends on a choice of a partial ordering for the vertices of K, the induced homomorphism η_* does not. Because η commutes with inclusions of subcomplexes, it induces a homomorphism on the relative groups, which is an isomorphism as well. We now show the same result holds for cohomology.

Theorem 44.2. *Let (K,K_0) be a simplicial pair. Then η induces a cohomology isomorphism*

$$H^p(\mathcal{C}(K,K_0); G) \xleftarrow{\eta^*} H^p(\mathcal{S}(|K|,|K_0|); G)$$

that is independent of the choice of partial ordering of the vertices of K. It commutes with δ^* and with homomorphisms induced by continuous maps.

Proof. The chain map η carries the oriented simplex $[v_0, \dots ,v_p]$ of K to the linear singular simplex $l(v_0, \dots ,v_p)$ of K, provided $v_0 < \cdots < v_p$ in the chosen ordering. Because the chain complexes involved are free and η induces a homology isomorphism, it induces a cohomology isomorphism as well. Furthermore, because η commutes with inclusions, it induces a homomorphism of relative cohomology. This homomorphism commutes with the coboundary operator δ^*, by the naturality of the zig-zag lemma; therefore, it is an isomorphism in relative cohomology, by the Five-lemma.

To prove the rest of the theorem, we must examine the definition of η more closely. The map η was defined as the composite

$$C_p(K) \xrightarrow{\phi} C_p'(K) \xrightarrow{\theta} S_p(|K|),$$

where $\phi([v_0, \dots ,v_p])$ equals the ordered simplex $(v_0, \dots ,v_p)$ if $v_0 < \cdots < v_p$ in the chosen ordering, and

$$\theta((w_0, \dots ,w_p)) = l(w_0, \dots ,w_p).$$

In §13, we defined a chain-homotopy inverse to ϕ by the equation

$$\psi((w_0, \dots ,w_p)) = \begin{cases} [w_0, \dots ,w_p] & \text{if the } w_i \text{ are distinct} \\ 0 & \text{otherwise.} \end{cases}$$

The maps ψ and θ do not depend on the ordering, although ϕ does. The fact that η^* is independent of the ordering follows from the equation $\eta^* = (\psi^*)^{-1} \circ \theta^*$.

To show η^* commutes with induced homomorphisms, we first consider the case of a simplicial map f. We showed in proving Theorem 13.7 that $f_\#$ commutes with ψ, and we showed in proving Theorem 34.4 that $f_\#$ commutes with θ. It follows that $f^\#$ commutes with the duals of ψ and θ, so that f^* commutes with η^*. To extend this result to arbitrary continuous maps, one follows the pattern of Theorem 34.5. □

EXERCISES

1. If X is a path-connected space, show $H^0(X) \simeq \mathbf{Z}$. Find a generating cocycle.
2. State and prove a Mayer-Vietoris theorem in simplicial cohomology. Can you prove it for arbitrary coefficients?
3. (a) Let $A_1, A_2 \subset X$. Show that if $\{A_1, A_2\}$ is an excisive couple, then inclusion

$$\mathcal{S}(A_1) + \mathcal{S}(A_2) \to \mathcal{S}(A_1 \cup A_2)$$

induces an isomorphism in cohomology as well as homology.

(b) State and prove a Mayer-Vietoris theorem in singular cohomology.

Simplicial cohomology with compact support.

Let K be a complex. Let $C^p_c(K; G)$ denote the group of homomorphisms of $C_p(K)$ into G that vanish on all but finitely many oriented simplices of K. These homomorphisms are called **cochains with compact support.**

4. (a) Show that if K is locally finite, then δ maps C^p_c to C^{p+1}_c. The resulting cohomology groups are denoted $H^p_c(K; G)$ and called the **cohomology groups with compact support.**

(b) If K is the complex whose space is $\mathbf{R}$ and whose vertices are the integers, show that $H^1_c(K) \simeq \mathbf{Z}$ and $H^0_c(K) = 0$.

(c) Show that if $|K|$ is connected and non-compact, then $H^0_c(K) = 0$.

5. A map $h : X \to Y$ is said to be **proper** if for each compact subset D of Y, the set $h^{-1}(D)$ is a compact subset of X. A **proper homotopy** is a homotopy that is itself a proper map.

(a) Show that the assignment

$$K \to H^p_c(K; G) \qquad \text{and} \qquad h \to h^*$$

defines a functor from the category of locally finite simplicial complexes and proper continuous maps of their polytopes to abelian groups and homomorphisms. [*Hint:* If h is proper, so is any simplicial approximation to h. If f and g are contiguous and proper, and D is the chain homotopy between them, show that $\tilde{D}$ carries finite cochains to finite cochains. If $\lambda : \mathcal{C}(K) \to \mathcal{C}(K')$ is the subdivision operator and if $g : K' \to K$ is a simplicial approximation to the identity, show that the duals of λ and $g_\#$ carry finite cochains to finite cochains. Prove the same for the dual of the chain homotopy between $\lambda \circ g_{\#}$ and the identity.]

(b) Show that if

$$h, k : |K| \to |L|$$

are properly homotopic, then $h^* = k^*$ as homomorphisms of cohomology with compact support. [*Hint:* Show that if D is the chain homotopy constructed in proving Theorem 19.2, then $\tilde{D}$ carries finite cochains to finite cochains.]

(c) Extend the results of (a) and (b) to relative cohomology. Does there exist a long exact sequence of cohomology with compact support? If so, is it natural with respect to induced homomorphisms?

(d) Is there an excision theorem for cohomology with compact support?

*6. Repeat Exercise 5 for the homology groups based on infinite chains, which were introduced in the Exercises of §5.

§45. THE COHOMOLOGY OF FREE CHAIN COMPLEXES

Until now, we have computed the cohomology groups only for a few simple spaces. We wish to compute them more generally. We shall prove that for a CW complex X, the cellular chain complex $\mathcal{D}(X)$, which we know can be used to compute the homology of X, can be used to compute the cohomology as well. This we shall prove in §47. The proof depends on two theorems about free chain complexes, which we prove in this section.

The first theorem states that for free chain complexes $\mathcal{C}$ and $\mathcal{D}$, *any* homomorphism $H_p(\mathcal{C}) \to H_p(\mathcal{D})$ of homology groups is induced by a chain map ϕ. And the second states that if a chain map $\phi : \mathcal{C} \to \mathcal{D}$ of free chain complexes induces an isomorphism in homology, it induces an isomorphism in cohomology as well. An independent proof of this second theorem will be given in Chapter 7, so you can simply assume this theorem for the time being if you wish.

Definition. A short exact sequence of abelian groups,

$$0 \to A \to B \to C \to 0,$$

where A and B are free, is called a **free resolution** of C. Any abelian group C has a free resolution; one can take B to be the free abelian group generated by the elements of C, and take A to be the kernel of the natural projection $B \to C$. This gives what is called the **canonical free resolution** of C.

Free resolutions have the following useful property: Suppose one is given the diagram

$$\begin{array}{ccccccccc} 0 & \longrightarrow & A & \xrightarrow{\phi} & B & \xrightarrow{\psi} & C & \longrightarrow & 0 \\ & & & & & & \downarrow{\scriptstyle\gamma} & & \\ 0 & \longrightarrow & A' & \xrightarrow{\phi'} & B' & \xrightarrow{\psi'} & C' & \longrightarrow & 0 \end{array}$$

where the horizontal sequences are exact and A and B are free. In this situation, there exist homomorphisms $\alpha : A \to A'$ and $\beta : B \to B'$ that make this diagram commute.

The proof is easy. Choose a basis for B; if b is a basis element, let β map b into any element of the set $(\psi')^{-1}(\gamma(\psi(b)))$; this set is non-empty because ψ' is surjective. A bit of diagram-chasing shows that β carries $\operatorname{im}\phi$ into $\operatorname{im}\phi'$; because ϕ' is a monomorphism, β induces a homomorphism $\alpha : A \to A'$.

Now we prove the first of our basic theorems.

Theorem 45.1. *Let $\mathcal{C}$ and $\mathcal{C}'$ be free chain complexes. If $\gamma : H_p(\mathcal{C}) \to H_p(\mathcal{C}')$ is a homomorphism defined for all p, then there is a chain map $\phi : \mathcal{C} \to \mathcal{C}'$ that induces γ.*

Indeed, if $\beta : Z_p \to Z_p'$ is any homomorphism of cycle groups inducing γ, then β extends to a chain map ϕ.

Proof. Let Z_p denote the p-cycles, and B_p, the p-boundaries, in the chain complex $\mathcal{C}$. Similarly, let Z_p' and B_p' be the p-cycles and p-boundaries of $\mathcal{C}'$. Since these groups are free abelian, there exist, for all p, homomorphisms α, β making the following diagram commute:

$$\begin{array}{ccccccccc} 0 & \longrightarrow & B_p & \longrightarrow & Z_p & \longrightarrow & H_p(\mathcal{C}) & \longrightarrow & 0 \\ & & \downarrow{\scriptstyle\alpha} & & \downarrow{\scriptstyle\beta} & & \downarrow{\scriptstyle\gamma} & & \\ 0 & \longrightarrow & B_p' & \longrightarrow & Z_p' & \longrightarrow & H_p(\mathcal{C}') & \longrightarrow & 0. \end{array}$$

We seek to extend β to a chain map ϕ of C_p into C_p'. For that purpose, consider the short exact sequences

$$\begin{array}{ccccccccc} 0 & \longrightarrow & Z_p & \longrightarrow & C_p & \xrightarrow{\partial_0} & B_{p-1} & \longrightarrow & 0 \\ & & {\scriptstyle\beta}\downarrow & & {\scriptstyle\phi}\,\vdots & & \downarrow{\scriptstyle\alpha} & & \\ 0 & \longrightarrow & Z_p' & \longrightarrow & C_p' & \xrightarrow{\partial_0'} & B_{p-1}' & \longrightarrow & 0. \end{array}$$

Because B_{p-1} and B_{p-1}' are free, the sequences split. Choose subgroups U_p and U_p' so that

$$C_p = Z_p \oplus U_p \qquad \text{and} \qquad C_p' = Z_p' \oplus U_p'.$$

Then $\partial_0 : U_p \to B_{p-1}$ and $\partial_0' : U_p' \to B_{p-1}'$ are isomorphisms. We define $\phi : C_p \to C_p'$ by letting it equal $\beta : Z_p \to Z_p'$ on the summand Z_p, and the map $U_p \to U_p'$

induced by α on the summand U_p. Then the first square commutes, automatically. The second square commutes for any element of U_p, by definition. And it commutes for any element of Z_p by exactness, for

$$\alpha(\partial_0 z_p) = \alpha(0) = 0 \qquad \text{and} \qquad \partial_0'\phi(z_p) = \partial_0'(\beta(z_p)) = 0.$$

Thus it commutes for any element of C_p.

Now we show that ϕ is a chain map. Consider the diagram

$$\begin{array}{ccccccc} C_p & \xrightarrow{\partial_0} & B_{p-1} & \longrightarrow & Z_{p-1} & \longrightarrow & C_{p-1} \\ \downarrow \phi & & \downarrow \alpha & & \downarrow \beta & & \downarrow \phi \\ C_p' & \xrightarrow{\partial_0'} & B_{p-1}' & \longrightarrow & Z_{p-1}' & \longrightarrow & C_{p-1}' \end{array}$$

where the unlabelled maps are inclusions. The middle square commutes by definition of α and β; the two end squares commute as we have just proved. Thus ϕ is a chain map.

The fact that ϕ induces the original homology homomorphism γ follows from the definition of β. □

Corollary 45.2. *Suppose $\{\mathcal{C}, \epsilon\}$ and $\{\mathcal{C}', \epsilon'\}$ are free augmented chain complexes. If $\gamma : \tilde{H}_p(\mathcal{C}) \to \tilde{H}_p(\mathcal{C}')$ is a homomorphism defined for all p, then γ is induced by an augmentation-preserving chain map $\phi : \mathcal{C} \to \mathcal{C}'$.*

Proof. Consider the augmented chain complexes obtained from $\mathcal{C}$ and $\mathcal{C}'$; they have $\mathbf{Z}$ as their (-1)-dimensional groups and ϵ, ϵ', respectively, as the boundary operators from dimension 0 to dimension -1. We define β to equal the identity in dimension -1 (where the homology vanishes), and to be any homomorphism of cycle groups inducing γ in other dimensions. The preceding theorem applies; the resulting chain map ϕ will automatically preserve augmentation. □

Now we prove the second of our basic theorems. We begin by considering a special case.

Lemma 45.3. *Let*

$$0 \to \mathcal{C} \xrightarrow{\phi} \mathcal{D} \to \mathcal{E} \to 0$$

be an exact sequence of free chain complexes. If ϕ induces homology isomorphisms in all dimensions, it induces cohomology isomorphisms as well.

Proof. The existence of the long exact sequence in homology and the fact that ϕ_* is an isomorphism imply that $H_p(\mathcal{E}) = 0$ for all p. To prove that ϕ^* is an isomorphism, it suffices to show that $H^p(\mathcal{E}; G) = 0$ for all p.

Let $B_p \subset Z_p \subset E_p$ denote boundaries, cycles, and chains of $\mathcal{E}$, respectively, in dimension p. The short exact sequence

$$0 \to Z_p \to E_p \xrightarrow{\partial_0} B_{p-1} \to 0$$

splits because B_{p-1} is free. Furthermore, $Z_p = B_p$ because the homology of $\mathcal{E}$ vanishes. Therefore, we can write $E_p = B_p \oplus U_p$, where ∂ maps B_p to zero and carries U_p isomorphically onto B_{p-1}. Now

$$\operatorname{Hom}(E_p,G) \simeq \operatorname{Hom}(B_p,G) \oplus \operatorname{Hom}(U_p,G).$$

You can check that δ induces a homomorphism that carries $\operatorname{Hom}(B_p,G)$ isomorphically onto $\operatorname{Hom}(U_{p+1},G)$ and carries $\operatorname{Hom}(U_p,G)$ to zero. Then $\operatorname{Hom}(U_p,G)$ represents the cocycle group of $\mathcal{E}$ and it equals the image of δ. Thus $H^p(\mathcal{E};G) = 0$. □

This special case of our theorem is all we actually used in the preceding section. Later, however, we shall need the general version. The general case is reduced to the special case by means of the following lemma.

Lemma 45.4. *Let $\mathcal{C}$ and $\mathcal{D}$ be free chain complexes; let $\phi : \mathcal{C} \to \mathcal{D}$ be a chain map. There is a free chain complex $\mathcal{D}'$ and injective chain maps $i : \mathcal{C} \to \mathcal{D}'$ and $j : \mathcal{D} \to \mathcal{D}'$ such that j induces homology isomorphisms in all dimensions, and the diagram*

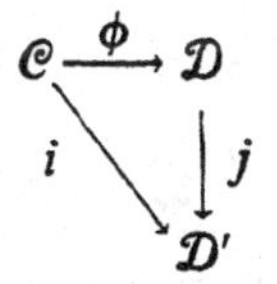

commutes up to chain homotopy. Furthermore, the quotients $\mathcal{D}'/\operatorname{im} i$ and $\mathcal{D}'/\operatorname{im} j$ are free.

Proof. The definition of $\mathcal{D}'$ is one we shall simply "pull out of a hat." Later, we shall explain its geometric motivation. It is sometimes called the "algebraic mapping cylinder" of ϕ.

Define $\mathcal{D}'$ to be the chain complex whose chain group in dimension p is given by

$$D'_p = C_p \oplus D_p \oplus C_{p-1}.$$

Let the boundary operator in $\mathcal{D}'$ be defined by the equations

$$\begin{aligned} \partial'(c_p,0,0) &= (\partial c_p,0,0), \\ \partial'(0,d_p,0) &= (0,\partial d_p,0), \\ \partial'(0,0,c_{p-1}) &= (-c_{p-1},\phi(c_{p-1}),-\partial c_{p-1}). \end{aligned}$$

You can check without difficulty that $\partial' \circ \partial' = 0$. It is clear from the definitions that the natural inclusions $i : C_p \to D'_p$ and $j : D_p \to D'_p$ are chain maps. It is also clear that D'_p and the quotients $D'_p/\operatorname{im} i$ and $D'_p/\operatorname{im} j$ are free.

We define a chain homotopy $D : C_p \to D'_{p+1}$ by the equation

$$D(c_p) = (0,0,c_p).$$

You can check that it satisfies the equation

$$\partial' D + D\partial = j \circ \phi - i.$$

Finally, we show that j induces an isomorphism in homology. For that purpose, it suffices to show that the homology of the chain complex $\mathcal{D}'/\mathcal{D}$ vanishes. The pth chain group of $\mathcal{D}'/\mathcal{D}$ is isomorphic to $C_p \oplus C_{p-1}$, and the induced boundary operator ∂'' is given by the equation

$$\partial''(c_p, c_{p-1}) = (\partial c_p - c_{p-1}, -\partial c_{p-1}).$$

If (c_p, c_{p-1}) is a cycle of this chain complex, then it follows in particular that $\partial c_p - c_{p-1} = 0$. We compute directly

$$(c_p, c_{p-1}) = (c_p, \partial c_p) = -\partial''(0, c_p).$$

Thus (c_p, c_{p-1}) is a boundary. □

Theorem 45.5. *Let $\mathcal{C}$ and $\mathcal{D}$ be free chain complexes; let $\phi : \mathcal{C} \to \mathcal{D}$ be a chain map. If ϕ induces homology isomorphisms in all dimensions, then ϕ induces cohomology isomorphisms in all dimensions.*

Proof. Given ϕ, let $i : \mathcal{C} \to \mathcal{D}'$ and $j : \mathcal{D} \to \mathcal{D}'$ be as in the preceding lemma. One has exact sequences of free chain complexes

$$0 \to \mathcal{C} \xrightarrow{i} \mathcal{D}' \to \mathcal{D}'/\mathcal{C} \to 0,$$

$$0 \to \mathcal{D} \xrightarrow{j} \mathcal{D}' \to \mathcal{D}'/\mathcal{D} \to 0.$$

The map j induces a homology isomorphism by the preceding lemma, while i induces a homology isomorphism because both j and ϕ do and i is chain homotopic to $j \circ \phi$. Therefore, by Lemma 45.3, i and j induce cohomology isomorphisms i^* and j^*, respectively, in all dimensions. Since i is chain homotopic to $j \circ \phi$, we have $i^* = \phi^* \circ j^*$; therefore, ϕ^* is a cohomology isomorphism as well. □

Corollary 45.6. *Let $\mathcal{C}$ and $\mathcal{D}$ be free chain complexes. If $H_p(\mathcal{C}) \simeq H_p(\mathcal{D})$ for all p, then $H^p(\mathcal{C}; G) \simeq H^p(\mathcal{D}; G)$ for all p and G.* □

Remark. Let us explain the geometric motivation underlying the definition of the chain complex $\mathcal{D}'$.

In homotopy theory, there is a standard construction for, roughly speaking, replacing an arbitrary continous map $h : X \to Y$ by an imbedding of X in a space that is homotopy equivalent to Y. More precisely, there is a space Y' and imbeddings i and j such that the diagram

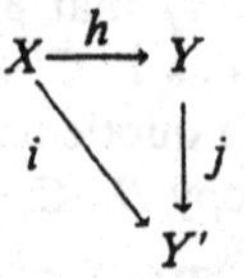

commutes up to homotopy, and such that j is a homotopy equivalence. It follows at once that h is a homotopy equivalence if and only if i is. In this way problems concerning the map h are reduced to problems concerning the imbedding i.

This construction is called the *mapping cylinder* construction. We describe it here, and explain how the chain complex $\mathcal{D}'$ is an algebraic analogue.

Given $h : X \to Y$, let us form a quotient space from the disjoint union of $X \times I$ and Y by identifying each point $(x,0)$ of $X \times 0$ with the point $h(x)$ of Y. The resulting adjunction space Y' is called the **mapping cylinder** of h. We picture it as looking something like a "top hat." See Figure 45.1.

Let $\pi : (X \times I) \cup Y \to Y'$ be the quotient map. The restriction of π to Y defines an imbedding j of Y in Y', and the map $i(x) = \pi(x,1)$ defines an imbedding of X in Y'. Clearly $j(Y)$ is a deformation retract of Y'; one just "pushes down the top hat" onto $j(Y)$. Just as clearly, the map $i : X \to Y'$ is homotopic to the map $j \circ h$; again, one just "pushes $i(X)$ down."

We seek to imitate this construction algebraically. So let us suppose for convenience that Y' is triangulated in such a way that $i(X)$ and $j(Y)$ are subcomplexes, and so is each set $\pi(\sigma \times I)$, for $\sigma \in X$. Let us identify X with $i(X)$, and Y, with $j(Y)$, for simplicity of notation. Now $\mathcal{C}(X)$ plays the role of the chain complex $\mathcal{C}$, and $\mathcal{C}(Y)$ plays the role of $\mathcal{D}$, and $\mathcal{C}(Y')$ plays the role of $\mathcal{D}'$. The map $h_\# : \mathcal{C}(X) \to \mathcal{C}(Y)$ plays the role of the chain map ϕ.

What does the chain complex $\mathcal{D}'$ look like algebraically? Suppose we break Y' up into the cells of a CW complex. The open p-cells will consist of the open p-simplices Int σ_p of X, the open p-simplices Int τ_p of Y, and the open cells of the form $\pi(\text{Int}\,\sigma_{p-1} \times \text{Int}\,I)$ that lie "between" X and Y. Then the pth chain group of Y' is essentially just

$$C_p(X) \oplus C_p(Y) \oplus C_{p-1}(X),$$

since the group of "in between" cells is isomorphic to $C_{p-1}(X)$. How does the boundary operator of Y' act on these chains? Clearly it acts just like ∂_X in X, and like ∂_Y in Y. What does it do to a cell of the third kind? In the space $X \times I$, it is easy to see that

$$\partial(\sigma \times I) = \sigma \times 0 - \sigma \times 1 \pm (\partial\sigma) \times I.$$

(One has to be careful with signs.) When $X \times 0$ is pasted onto Y via h, this formula becomes

$$\partial'(\text{Int}\,\sigma \times \text{Int}\,I) = h_\#(\sigma) - \sigma \pm (\partial\sigma) \times I.$$

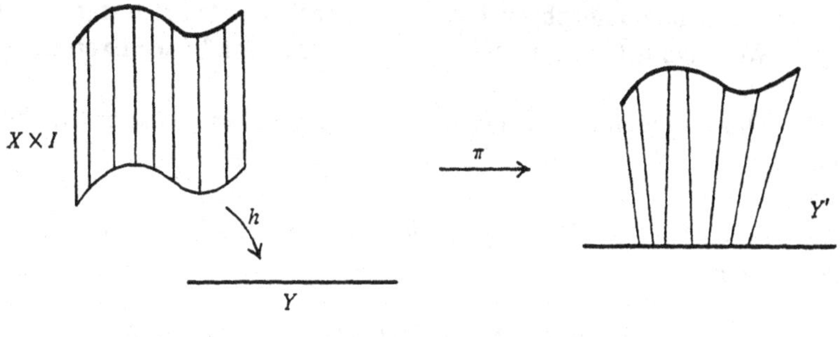

Figure 45.1

Finally, when we identify the p-chains of Y' with the group $C_p(X) \oplus C_p(Y) \oplus C_{p-1}(X)$, this formula becomes

$$\partial'(0,0,\sigma_{p-1}) = (-\sigma_{p-1}, h_\#(\sigma_{p-1}), \pm \partial\sigma_{p-1}).$$

(The last sign must in fact be $-$, in order that $\partial' \circ \partial' = 0$.) By now it should be quite clear what the connection is between the algebraic and topological mapping cylinders!

Let us give one application of this theorem. It is a formula that relates cohomology with homology. It will be generalized in Chapter 7.

Definition. If $\mathcal{C} = \{C_i,\partial\}$ is a chain complex, there is a map

$$\operatorname{Hom}(C_p,G) \times C_p \to G$$

which carries the pair (c^p,c_p) to the element $\langle c^p,c_p\rangle$ of G. It is bilinear, and is called the "evaluation map." It induces a bilinear map

$$H^p(\mathcal{C};G) \times H_p(\mathcal{C}) \to G,$$

which we call the **Kronecker index.** We denote the image of α^p and β_p under the map also by $\langle\alpha^p,\beta_p\rangle$.

It is easy to see that the Kronecker index is well-defined, since

$$\langle z^p + \delta d^{p-1}, z_p\rangle = \langle z^p,z_p\rangle + \langle d^{p-1},\partial z_p\rangle$$

and

$$\langle z^p, z_p + \partial d_{p+1}\rangle = \langle z^p,z_p\rangle + \langle \delta z^p, d_{p+1}\rangle.$$

The final terms vanish if z^p is a cocycle and z_p is a cycle.

Definition. It is convenient to define the **Kronecker map**

$$\kappa : H^p(\mathcal{C};G) \to \operatorname{Hom}(H_p(\mathcal{C}),G)$$

as the map that sends α to the homomorphism $\langle\alpha, \quad\rangle$. Formally, we define

$$(\kappa\alpha^p)(\beta_p) = \langle\alpha^p,\beta_p\rangle.$$

The map κ is a homomorphism because Kronecker index is linear in the first variable. We leave it to you to check that κ is "natural." (See Exercise 2.)

The following lemma is elementary; its proof makes no use of the theorems of this section.

Lemma 45.7. *Let $\mathcal{C}$ be a free chain complex. Then there is a natural exact sequence*

$$0 \leftarrow \operatorname{Hom}(H_p(\mathcal{C}),G) \xleftarrow{\kappa} H^p(\mathcal{C};G) \leftarrow \ker\kappa \leftarrow 0.$$

It splits, but not naturally.

Proof. We shall construct a homomorphism

$$\lambda^* : \text{Hom}(H_p(\mathcal{C}),G) \to H^p(\mathcal{C}; G)$$

such that $\kappa \circ \lambda^*$ is the identity. It follows that κ is surjective and that the sequence splits.

Let B_p, Z_p, and C_p denote boundaries, cycles, and chains, respectively, in $\mathcal{C}$. We begin with the projection homomorphism

$$\pi : Z_p \to Z_p/B_p = H_p(\mathcal{C}).$$

The exact sequence $0 \to Z_p \to C_p \to B_{p-1} \to 0$ shows that Z_p is a direct summand in C_p. Therefore, π extends to a homomorphism $\lambda : C_p \to H_p(\mathcal{C})$. Define a chain complex $\mathcal{E}$ by letting $E_p = H_p(\mathcal{C})$ and letting all boundary operators in $\mathcal{E}$ vanish. Then

$$H_p(\mathcal{E}) = E_p = H_p(\mathcal{C}),$$
$$H^p(\mathcal{E}; G) = \text{Hom}(E_p,G) = \text{Hom}(H_p(\mathcal{C}),G).$$

Because the boundary operators in $\mathcal{E}$ vanish, the map $\lambda : \mathcal{C} \to \mathcal{E}$ is a chain map. For $\lambda(\partial c_{p+1}) = \{\partial c_{p+1}\} = 0$. The induced homomorphism in homology

$$\lambda_* : H_p(\mathcal{C}) \to H_p(\mathcal{E}) = H_p(\mathcal{C})$$

is the identity map (and hence an isomorphism). For if z_p is a p-cycle,

$$\lambda_*(\{z_p\}) = \lambda(z_p) = \pi(z_p) = \{z_p\}.$$

The induced homomorphism in cohomology

$$H^p(\mathcal{C}; G) \xleftarrow{\lambda^*} H^p(\mathcal{E}; G) = \text{Hom}(H_p(\mathcal{C}),G)$$

is not in general an isomorphism.

Now the composite $\kappa \circ \lambda^*$ is the identity map of $\text{Hom}(H_p(\mathcal{C}),G)$, for if $\{z_p\} \in H_p(\mathcal{C})$ and $\gamma \in \text{Hom}(H_p(\mathcal{C}),G)$, then

$$\begin{aligned}(\kappa\lambda^*(\gamma))(\{z_p\}) &= \langle\lambda^*(\gamma),\{z_p\}\rangle = \langle\tilde{\lambda}(\gamma),z_p\rangle \\ &= \langle\gamma,\lambda(z_p)\rangle = \gamma(\{z_p\}). \quad \square\end{aligned}$$

Theorem 45.8. *Let $\mathcal{C}$ be a free chain complex. If $H_p(\mathcal{C})$ is free for all p, then κ is an isomorphism for all p.*

Proof. Let $\lambda : \mathcal{C} \to \mathcal{E}$ be as in the preceding lemma. Since the homology of $\mathcal{C}$ is free, $\mathcal{E}$ is a free chain complex and Theorem 45.5 applies. Since the chain map $\lambda : \mathcal{C} \to \mathcal{E}$ induces homology isomorphisms λ_* in all dimensions, λ^* is an isomorphism. The fact that $\kappa \circ \lambda^*$ is the identity implies that κ is an isomorphism as well. $\square$

This theorem says that if the homology of $\mathcal{C}$ is free in all dimensions, then the cohomology group $H^p(\mathcal{C}; G)$ can be considered in a natural way to be the dual group $\text{Hom}(H_p(\mathcal{C}),G)$ of the homology group $H_p(\mathcal{C})$. We will generalize this theorem in Chapter 7; all one in fact needs is for the single group $H_{p-1}(\mathcal{C})$ to be free.

EXERCISES

1. Check that the operator ∂' defined in the proof of Lemma 45.4 satisfies $\partial' \circ \partial' = 0$.

2. (a) Show that the Kronecker map κ is natural with respect to homomorphisms induced by chain maps. That is, show that if $\phi : \mathcal{C} \to \mathcal{D}$ is a chain map, the following diagram commutes:

$$\begin{array}{ccc} \operatorname{Hom}(H_p(\mathcal{C}),G) & \xleftarrow{\kappa} & H^p(\mathcal{C};G) \\ \uparrow \tilde{\phi}_* & & \uparrow \phi^* \\ \operatorname{Hom}(H_p(\mathcal{D}),G) & \xleftarrow{\kappa} & H^p(\mathcal{D};G). \end{array}$$

(b) Naturality of the Kronecker index itself is a bit awkward to formulate, since it is covariant in one variable and contravariant in the other. Let $\phi : \mathcal{C} \to \mathcal{D}$ be a chain map; show that if $\alpha \in H^p(\mathcal{D}; G)$ and $\beta \in H_p(\mathcal{C})$, then

$$\langle \phi^*(\alpha),\beta\rangle = \langle \alpha,\phi_*(\beta)\rangle.$$

3. Let X and Y be spaces such that $H_n(X)$, $H^n(X)$, $H_n(Y)$, and $H^n(Y)$ are infinite cyclic. Let $f : X \to Y$ be a continuous map. Show that if

$$f_* : H_n(X) \to H_n(Y)$$

equals multiplication by d, then (up to sign) so does

$$f^* : H^n(Y) \to H^n(X).$$

[*Hint:* Show κ is an isomorphism.]

4. Since arbitrary choices are made in the definition of the homomorphism $\lambda : C_p \to H_p(\mathcal{C})$ in the proof of Lemma 45.7, one would not expect the splitting homomorphism

$$\lambda^* : \operatorname{Hom}(H_p(\mathcal{C}),G) \to H^p(\mathcal{C}; G)$$

to be natural. Show that it is in fact not natural, as follows:

(a) Find free chain complexes $\mathcal{C}$, $\mathcal{D}$ and a chain map $\phi : \mathcal{C} \to \mathcal{D}$ such that for no choices of λ does the following diagram commute:

$$\begin{array}{ccc} \operatorname{Hom}(H_p(\mathcal{C}),G) & \xrightarrow{\lambda^*} & H^p(\mathcal{C};G) \\ \uparrow \tilde{\phi}_* & & \uparrow \phi^* \\ \operatorname{Hom}(H_p(\mathcal{D}),G) & \xrightarrow{\lambda^*} & H^p(\mathcal{D};G). \end{array}$$

*(b) Find spaces X and Y and a continuous map $f : X \to Y$, such that setting $\mathcal{C} = \mathcal{S}(X)$, $\mathcal{D} = \mathcal{S}(Y)$, and $\phi = f_\#$ realizes the situation of (a).

*§46. CHAIN EQUIVALENCES IN FREE CHAIN COMPLEXES[†]

We now prove a second version of Theorem 45.5. Again we assume we are given a chain map $\phi : \mathcal{C} \to \mathcal{D}$ of free chain complexes that induces homology isomorphisms in all dimensions. We show that if $\mathcal{C}$ and $\mathcal{D}$ satisfy the (fairly mild) additional condition that both vanish below a certain dimension, then it follows not just that ϕ^* is an isomorphism, but that the chain map ϕ is itself a chain equivalence. The proof involves the "algebraic mapping cylinder" we constructed in the last section.

First we need an elementary lemma, which for later use is stated in slightly greater generality than we presently need.

Lemma 46.1. *Let $\mathcal{E}$ and $\mathcal{F}$ be non-negative chain complexes. Suppose E_p is free for $p > 0$ and $H_p(\mathcal{F}) = 0$ for $p > 0$. Then any two chain maps $f, g : \mathcal{E} \to \mathcal{F}$ that agree in dimension 0 are chain homotopic.*

Proof. Define $D : E_0 \to F_1$ to be zero. Then the equation $\partial D + D\partial = g - f$ holds in dimension 0, because $g = f$ in dimension 0. Suppose D is defined in dimension $p - 1$, where $p > 0$. Choose a basis for E_p. If e is a basis element, then $g(e) - f(e) - D(\partial e)$ is well-defined, and it is a cycle (by the usual computation). Define $D(e)$ to be an element of F_{p+1} whose boundary equals this cycle. □

As you might suspect, one can with care derive this result from the acyclic carrier theorem. It isn't worth the effort.

Theorem 46.2. *Let $\mathcal{C}$ and $\mathcal{D}$ be free chain complexes that vanish below a certain dimension; let $\phi : \mathcal{C} \to \mathcal{D}$ be a chain map. If ϕ induces homology isomorphisms in all dimensions, then ϕ is a chain equivalence.*

Proof. We return to the chain complex $\mathcal{D}'$ defined in the proof of Lemma 45.4. The inclusion mapping

$$i : C_p \to D'_p = C_p \oplus D_p \oplus C_{p-1}$$

is chain homotopic to $j \circ \phi$, where $j : D_p \to D'_p$ is inclusion. Since j and ϕ induce homology isomorphisms, so does i. It follows that the chain complex $\mathcal{E} = \mathcal{D}'/\mathcal{C}$ has vanishing homology in all dimensions. Now

$$E_p \simeq D_p \oplus C_{p-1}.$$

The induced boundary operator satisfies the formula

$$\partial(d_p, c_{p-1}) = (\partial d_p + \phi(c_{p-1}), -\partial c_{p-1}).$$

[†]We shall use the results of this section in §56 and §60 in proving the naturality of certain exact sequences.

We apply the preceding lemma to the chain complex $\mathcal{E}$ and any two chain maps f, g from $\mathcal{E}$ to itself. The chain complex $\mathcal{E}$ vanishes below a certain dimension (since $\mathcal{C}$ and $\mathcal{D}$ do); we may as well take this dimension to be dimension 1. We know that E_p is free and $H_p(\mathcal{E}) = 0$, for all p. Since any two chain maps $f, g : \mathcal{E} \to \mathcal{E}$ agree (trivially) in dimension 0, they are chain homotopic. In particular, there is a chain homotopy between the identity map and the zero chain map. That is, there is a homomorphism $D : E_p \to E_{p+1}$ satisfying the equation

$$\partial D + D\partial = \text{identity}.$$

Hidden in this formula are all the chain maps and chain homotopies we need.

Recalling that $E_p \simeq D_p \oplus C_{p-1}$, we define homomorphisms θ, ψ, λ, μ by the equations

$$D(d_p,0) = (\theta(d_p),\psi(d_p)) \qquad \in D_{p+1} \oplus C_p,$$
$$D(0,c_{p-1}) = (\lambda(c_{p-1}),\mu(c_{p-1})) \in D_{p+1} \oplus C_p.$$

Then we compute like mad! First we compute

$$D\partial(d_p,0) = (\theta(\partial d_p),\psi(\partial d_p)),$$
$$\partial D(d_p,0) = (\partial\theta(d_p) + \phi\psi(d_p),-\partial\psi(d_p)).$$

Adding these two equations, we obtain the equations

$$d_p = \theta(\partial d_p) + \partial\theta(d_p) + \phi\psi(d_p),$$
$$0 = \psi(\partial d_p) - \partial\psi(d_p).$$

The second equation says that $\psi : D_p \to C_p$ is a chain map, and the first says that $\theta : D_p \to D_{p+1}$ is a chain homotopy between $\phi \circ \psi$ and the identity. Second, we compute

$$\begin{aligned} D\partial(0,c_{p-1}) &= D(\phi(c_{p-1}),0) - D(0,\partial c_{p-1}) \\ &= (\theta\phi(c_{p-1}),\psi\phi(c_{p-1})) - (\lambda(\partial c_{p-1}),\mu(\partial c_{p-1})), \\ \partial D(0,c_{p-1}) &= \partial(\lambda(c_{p-1}),\mu(c_{p-1})) \\ &= (\partial\lambda(c_{p-1}) + \phi\mu(c_{p-1}),-\partial\mu(c_{p-1})). \end{aligned}$$

Adding the second coordinates of these equations, we obtain the equation

$$c_{p-1} = \psi\phi(c_{p-1}) - \mu(\partial c_{p-1}) - \partial\mu(c_{p-1}),$$

which says that μ is a chain homotopy between $\psi \circ \phi$ and the identity.

Thus our theorem is proved. □

EXERCISES

1. Let X be a space; let $\mathcal{A}$ be a collection of subsets of X whose interiors cover X. Show that inclusion

$$i : \mathcal{S}^{\mathcal{A}}(X) \to \mathcal{S}(X)$$

is a chain equivalence.

2. Let $\eta : \mathcal{C}(K) \to \mathcal{S}(|K|)$ be the chain map of §34, which induces an isomorphism of simplicial with singular homology. Show η is a chain equivalence.

§47. THE COHOMOLOGY OF CW COMPLEXES

Now we can compute the cohomology groups of some familiar spaces, and can find specific cocycles that generate these groups. The basic theorem we shall use in carrying out these computations is the following:

Theorem 47.1. *Let X be a CW complex; let $\mathcal{D}(X)$ be its cellular chain complex. Then*

$$H^p(\mathcal{D}(X);G) \simeq H^p(X;G)$$

for all p and G. If X is a triangulable CW complex, triangulated by a complex K, then the isomorphism is induced by inclusion $\mathcal{D}(X) \to \mathcal{C}(K)$.

Proof. Both $\mathcal{D}(X)$ and $\mathcal{S}(X)$ are free chain complexes. Since their homology groups are isomorphic, so are their cohomology groups, by Corollary 45.6. In the case where X is triangulable, the inclusion map $i : \mathcal{D}(X) \to \mathcal{C}(K)$ induces the homology isomorphism in question. (See Theorem 39.5.) Then i induces a cohomology isomorphism as well. □

Corollary 47.2. *Let $n > 0$. Then*

$$H^i(S^n;G) \simeq G \text{ for } i = 0 \text{ and } i = n,$$
$$H^i(B^n,S^{n-1};G) \simeq G \text{ for } i = n.$$

These cohomology groups vanish for other values of i.

Proof. The first statement follows from the fact that the cellular chain complex of S^n is infinite cyclic in dimensions 0 and n and vanishes otherwise, and all the boundary operators vanish. The second then follows from the long exact sequence in reduced cohomology, using the fact that the reduced cohomology of B^n vanishes, since it is contractible. □

Example 1. Let X denote either the torus T or the Klein bottle S, expressed as a CW complex having one open cell in dimension 2, two in dimension 1, and one in dimension 0. We computed the cellular chain complex of X, in Example 2 of §39, to be of the form

$$\cdots \to 0 \to \mathbf{Z} \xrightarrow{\partial_2} \mathbf{Z} \oplus \mathbf{Z} \xrightarrow{\partial_1} \mathbf{Z} \to 0.$$

Let γ generate $D_2(X)$; let w_1 and z_1 be a basis for $D_1(X)$. We know ∂_2 and ∂_1 vanish in the case of the torus. Passing to the dual sequence, we compute

$$H^2(T;G) \simeq G, \qquad H^1(T;G) \simeq G \oplus G, \qquad H^0(T;G) \simeq G.$$

In the case of the Klein bottle, we know that ∂_1 vanishes, and that we can choose w_1 and z_1 so that $\partial_2\gamma = 2z_1$. The dual sequence is of the form

$$\cdots \leftarrow 0 \leftarrow G \xleftarrow{\delta_2} G \oplus G \xleftarrow{\delta_1} G \leftarrow 0.$$

Here $\operatorname{Hom}(D_1(S),G) \simeq G \oplus G$, where the first summand represents those homomorphisms $\phi : D_1(S) \to G$ that vanish on z_1, and the second represents those homomorphisms $\psi : D_1(S) \to G$ that vanish on w_1. Because ∂_1 is trivial, so is its dual δ_1. We compute δ_2 as follows:

$$\langle \delta_2\phi, \gamma\rangle = \langle \phi, \partial_2\gamma\rangle = 2\langle \phi, z_1\rangle = 0,$$
$$\langle \delta_2\psi, \gamma\rangle = \langle \psi, \partial_2\gamma\rangle = 2\langle \psi, z_1\rangle.$$

Thus δ_2 carries the first summand to zero, and equals multiplication by 2 on the second summand. We conclude that

$$H^2(S;G) \simeq G/2G, \qquad H^1(S;G) \simeq G \oplus \ker(G \xrightarrow{2} G), \qquad H^0(S;G) \simeq G.$$

In particular,

$$H^2(S) \simeq \mathbf{Z}/2, \qquad H^1(S) \simeq \mathbf{Z}, \qquad H^0(S) \simeq \mathbf{Z}.$$

The computations given in the preceding example are typical. Once one has the cellular chain complex $\mathcal{D}(X)$, computing the cohomology groups is not hard.

However, there is an associated problem that is more difficult—namely, the problem of finding specific simplicial cocycles that generate these groups. In the next section, when we study cup products, we shall need to have such cocycles at hand. How can one find them?

In the case of *homology*, finding simplicial *cycles* of X that generate the homology is not difficult. Let us represent T and S as quotient spaces of the rectangle L, as pictured in Figure 47.1. The chain d of L that is the sum of all the 2-simplices of L, oriented counterclockwise, is a fundamental cycle for $(L, \operatorname{Bd} L)$, so its image $\gamma = g_\#(d)$ generates the cellular chain group $D_2(X)$. Similarly, the chains

$$w_1 = [a,b] + [b,c] + [c,a],$$
$$z_1 = [a,d] + [d,e] + [e,a],$$

are a basis for the cellular chain group $D_1(X)$. As we know, these chains are cycles representing certain elements of the homology of $\mathcal{D}(X)$. Now *because* the isomorphism $H_i(\mathcal{D}(X)) \simeq H_i(X)$ is induced by inclusion $i : \mathcal{D}(X) \to \mathcal{C}(X)$, *these same chains*, considered now as chains in $\mathcal{C}(X)$, represent elements of the simplicial homology of X.

However, cohomology is not so easy. What happens in this case? We can of course find generators for the cohomology of the cellular chain complex $\mathcal{D}(X)$: The homomorphism $\lambda : D_2(X) \to \mathbf{Z}$ that maps the fundamental cycle γ to 1 generates $\operatorname{Hom}(D_2(X),\mathbf{Z}) \simeq \mathbf{Z}$. And the homomorphisms $\phi, \psi : D_1(X) \to \mathbf{Z}$ defined by

$$\langle \phi, w_1\rangle = 1 \quad \text{and} \quad \langle \phi, z_1\rangle = 0,$$
$$\langle \psi, w_1\rangle = 0 \quad \text{and} \quad \langle \psi, z_1\rangle = 1,$$

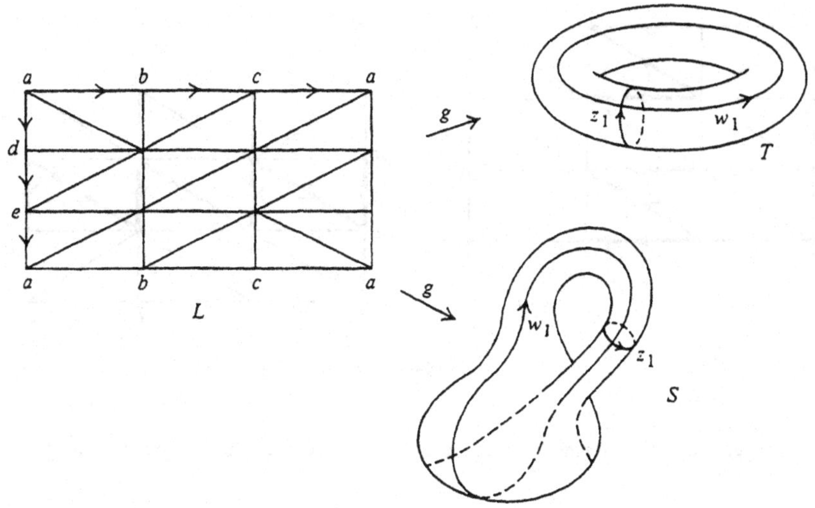

Figure 47.1

are a basis for the group $\operatorname{Hom}(D_1(X),\mathbf{Z}) \simeq \mathbf{Z} \oplus \mathbf{Z}$. For the torus, δ vanishes, so λ generates $H^2(\mathcal{D}(T))$ and ϕ and ψ represent a basis for $H^1(\mathcal{D}(T))$. For the Klein bottle, we have $\delta\phi = 0$ and $\delta\psi = 2\lambda$, so λ represents the non-zero element of $H^2(\mathcal{D}(S)) \simeq \mathbf{Z}/2$ and ϕ represents a generator of $H^1(\mathcal{D}(S)) \simeq \mathbf{Z}$.

However, unlike the situation in homology, the homomorphisms λ and ϕ and ψ *cannot* be "considered" as cocycles of the simplicial complex X. For the inclusion map $i: D_p(X) \rightarrow C_p(X)$ induces a homomorphism in the *opposite* direction

$$\operatorname{Hom}(D_p(X),\mathbf{Z}) \xleftarrow{\tilde{i}} \operatorname{Hom}(C_p(X),\mathbf{Z}).$$

This homomorphism is a *restriction* map. To find cocycles of the simplicial complex X that generate the simplicial cohomology of X, we must *pull* λ, ϕ, *and* ψ *back* to cocycles

$$z^2 : C_2(X) \rightarrow \mathbf{Z},$$
$$w^1, z^1 : C_1(X) \rightarrow \mathbf{Z},$$

whose *restrictions* to the subgroups $D_2(X)$ and $D_1(X)$, equal λ, ϕ, and ψ, respectively.

There is no general procedure for finding such cocycles. But in the present case, since we know that a cocycle is supposed to look like a "picket fence," we can find the desired cocycles without too much difficulty, as we now show.

Example 2. *Generators for the cohomology of the torus.* We represent T as a quotient space of the rectangle, as in the preceding example. Let w_1 and z_1 be as in that example. The cochains w^1 and z^1 pictured in Figure 47.2 are cocycles of T, *by direct*

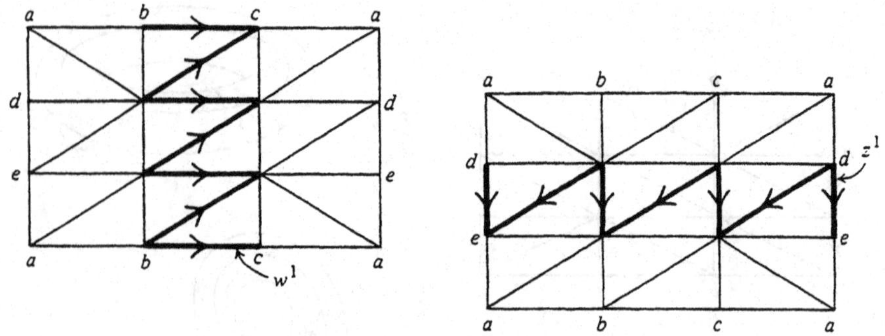

Figure 47.2

computation. Furthermore, when evaluated on the cycles w_1 and z_1 that generate $D_1(X)$, we have

$$\langle w^1, w_1 \rangle = 1 \quad \text{and} \quad \langle w^1, z_1 \rangle = 0,$$
$$\langle z^1, w_1 \rangle = 0 \quad \text{and} \quad \langle z^1, z_1 \rangle = 1.$$

Thus w^1 is a "pull-back" of ϕ, and z^1 is a pull-back of ψ. *Therefore, they represent a basis for* $H^1(T)$.

Similarly, if σ is any 2-simplex of T, oriented counterclockwise, then σ^* is a cocycle of T. Because $\langle \sigma^*, \gamma \rangle = 1$, the restriction of σ^* to the subgroup $D_2(X)$ of $C_2(X)$ equals λ, so σ^* is a pull-back of λ. Thus σ^* represents a generator of $H^2(T)$. (More generally, if all 2-simplices of T are oriented counterclockwise, then the cochain $\Sigma n_i \sigma_i^*$ of T is a pull-back of λ if and only if $\Sigma n_i = 1$.)

Example 3. *Generators for the cohomology of the Klein bottle, with integer coefficients.* We follow the pattern of the preceding example. Switch the labels d and e on the right side of the rectangles in Figure 47.2 so they represent the Klein bottle. Then w^1 still represents a cocycle; it generates $H^1(S)$. And the cochain σ^* represents the non-zero element of $H^2(S)$. (More generally, the cochain $\Sigma n_i \sigma_i^*$ represents the non-zero element of $H^2(S)$ if and only if Σn_i is odd.)

Example 4. *Generators for the cohomology of the Klein bottle, with* **Z**/2 *coefficients.* The cohomology groups are given by

$$H^2(S; \mathbf{Z}/2) \simeq \mathbf{Z}/2, \qquad H^1(S; \mathbf{Z}/2) \simeq \mathbf{Z}/2 \oplus \mathbf{Z}/2, \qquad H^0(S; \mathbf{Z}/2) \simeq \mathbf{Z}/2.$$

The pattern of the preceding argument applies to show that the cochains w^1 and z^1 of Figure 47.2 generate the 1-dimensional cohomology. (You can erase the arrows if you like, since $1 = -1$ in the group **Z**/2. Thus there is no problem in making z^1 a cocycle.) The cochain σ^* (or more generally $\sigma_1^* + \cdots + \sigma_k^*$, where k is odd), generates $H^2(S; \mathbf{Z}/2)$.

Example 5. *The cohomology of* P^2 *with* **Z**/2 *coefficients.* One has

$$H^i(P^2; \mathbf{Z}/2) \simeq \mathbf{Z}/2 \qquad \text{for} \qquad i = 0,1,2.$$

If σ is a 2-simplex, the cochain σ^* generates the 2-dimensional group. And the co-

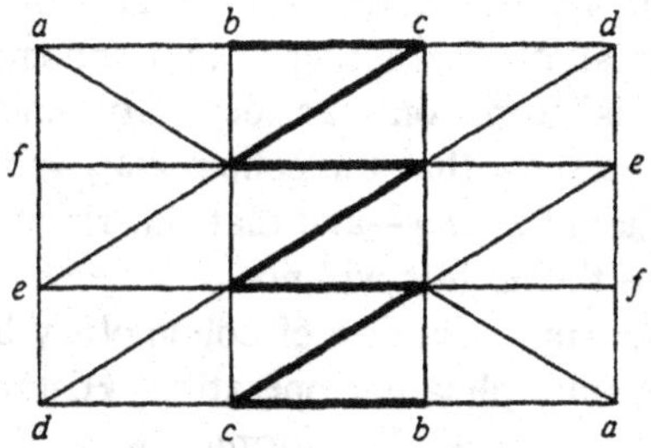

Figure 47.3

cycle pictured in Figure 47.3 generates $H^1(P^2; \mathbf{Z}/2)$, for its value is 1 on the cycle

$$[a,b] + [b,c] + [c,d] + [d,e] + [e,f] + [f,a]$$

that generates $D_1(X)$.

EXERCISES

1. Compute the cohomology of P^n and P^∞ with integer coefficients, with $\mathbf{Z}/2$ coefficients, and with rational coefficients.
2. Compute the cohomology of CP^n and CP^∞.
3. Compute $H^i(T \# T)$. Find simplicial cocycles that generate the cohomology, if $T \# T$ is triangulated as indicated in the exercises of §6.
4. Compute the cohomology of $P^2 \# P^2 \# P^2 \# P^2$, with $\mathbf{Z}/2$ coefficients. Find simplicial cocycles that generate the cohomology.
5. Compute the cohomology of the Klein bottle S with $\mathbf{Z}/6$ coefficients; find representative cocycles as in Examples 3 and 4.
6. Compute the cohomology of the 5-fold dunce cap X with $\mathbf{Z}$ and $\mathbf{Z}/5$ coefficients. (See Exercise 6 of §6.) Triangulate X and find cocycles that generate the cohomology.
7. Compute the cohomology of the lens space $L(n,k)$ with $\mathbf{Z}$ and $\mathbf{Z}/n$ coefficients.
8. Triangulate S^2 and the torus T; let $f : T \to S^2$ be a simplicial map. Show that if $f_* : H_2(T) \to H_2(S^2)$ equals multiplication by d, so does $f^* : H^2(S^2) \to H^2(T)$, by comparing the values of $f_\#$ and $f^\#$ on generators. Compare with Exercise 3 of §45.

§48. CUP PRODUCTS

The results of §45 tell us that, if the homology groups fail to distinguish between two spaces, then the cohomology groups will fail as well. One might be tempted to ask, "Why bother with cohomology? What possible use can it be?"

There are several answers to this question. One answer is that cohomology appears naturally when one studies the problem of classifying, up to homotopy, maps of one space into another. Another is that cohomology is involved when one integrates differential forms on manifolds. Still another answer is the one given in this section. We show that the cohomology groups have an additional algebraic structure—that of a *ring*—and that this ring will distinguish between spaces when the groups themselves will not.

We shall define the ring structure of cohomology by giving a specific cochain formula for the multiplication operation. Historically, this is how the ring structure was first obtained. The formula was discovered about 1936 by Alexander, Čech, and Whitney. At the time, it seemed very mysterious; its geometric meaning was not at all clear. Furthermore, it was very puzzling why there was a multiplication operation in cohomology but not in homology.

In the case of a compact orientable *manifold*, it had been known for some time that its homology had a ring structure. The multiplication operation in this ring was called the intersection product, and had a clear geometric meaning. But all attempts to generalize this multiplication to more general spaces failed. It was not until 1942, when Lefschetz gave a new definition of the multiplication operation in cohomology, did it become clear why, for general spaces, there exists a cohomology ring but not a homology ring. It also became clear about the same time what the relation was between the homology and cohomology rings when both were defined; the Poincaré duality theorem showed that the two rings were isomorphic. We shall return to these matters later. (See §61 and §69.)

Review of rings, modules, and fields

We begin by reviewing some basic facts from algebra concerning rings and modules.

A **ring** R is an abelian group, written additively, with a multiplication operation satisfying two axioms:

(1) (Associativity) $\alpha \cdot (\beta \cdot \gamma) = (\alpha \cdot \beta) \cdot \gamma$.

(2) (Distributivity) $\alpha \cdot (\beta + \gamma) = \alpha \cdot \beta + \alpha \cdot \gamma$;
$(\alpha + \beta) \cdot \gamma = \alpha \cdot \gamma + \beta \cdot \gamma$.

If $\alpha \cdot \beta = \beta \cdot \alpha$ for all α,β, then R is said to be **commutative.** If there is an element 1 in R such that $\alpha \cdot 1 = 1 \cdot \alpha = \alpha$ for all α, then 1 is called a **unity element** in R. If R has a unity element, this element is easily seen to be unique; furthermore, one has $(-1) \cdot \alpha = -\alpha$ and $0 \cdot \alpha = 0$ for all α.

If R is a commutative ring with unity, and if R satisfies the additional condition that for every $\alpha \neq 0$, there is a β such that $\alpha \cdot \beta = 1$, then R is called a **field.**

Example 1. Examples of rings include, among many others, the following:

(i) The integers $\mathbf{Z}$.

(ii) The set $\mathbf{Z}/n$ of integers modulo n.

(iii) The set of $n \times n$ matrices with integer entries.

(iv) The set of polynomials with integer coefficients.

Examples of fields include:

(v) $\mathbf{Z}/p$, where p is prime.

(vi) The rationals $\mathbf{Q}$.

(vii) The real numbers $\mathbf{R}$.

(viii) The complex numbers $\mathbf{C}$.

In each case, the multiplication operation is the usual one.

Now suppose A is an additive group and R is a commutative ring with unity. We say A has the structure of **module** over R if there is a binary operation $R \times A \to A$ (written as scalar multiplication) such that for $\alpha,\beta \in R$ and $a,b \in A$, we have:

(1) $\alpha(a + b) = \alpha a + \alpha b$.

(2) $(\alpha + \beta)a = \alpha a + \beta a$.

(3) $\alpha(\beta a) = (\alpha \cdot \beta)a$.

(4) $1a = a$.

If A and B are R-modules, a **module homomorphism** is a homomorphism $\phi : A \to B$ such that $\phi(\alpha a) = \alpha\phi(a)$ for $\alpha \in R$ and $a \in A$. The kernel and the cokernel of such a homomorphism have natural R-module structures. In the special case where R is a field F, we call A a **vector space** over F, and we call the homomorphism ϕ a **linear transformation.**

We shall not have much occasion to deal with modules in this book. Our primary concern will be with rings and vector spaces.

Example 2. Given R, it can always be considered as an R-module over itself. More generally, the cartesian product R^n becomes an R-module if we define

$$\alpha(\beta_1, \ldots, \beta_n) = (\alpha\beta_1, \ldots, \alpha\beta_n).$$

Example 3. If G is an abelian group, then G has a natural structure of $\mathbf{Z}$-module, obtained by defining ng to be the n-fold sum $g + \cdots + g$, as usual.

Cup products

Throughout this section and the next, *we shall let R denote a commutative ring with unity element* 1.

Definition. Let X be a topological space. Let $S^p(X; R) = \operatorname{Hom}(S_p(X), R)$ denote the group of singular p-cochains of X, with coefficients in R. We define a map

$$S^p(X; R) \times S^q(X; R) \xrightarrow{\cup} S^{p+q}(X; R)$$

by the following equation: If $T:\Delta_{p+q}\rightarrow X$ is a singular $p+q$ simplex, let

$$\langle c^p \cup c^q, T\rangle = \langle c^p, T\circ l(\epsilon_0,\ldots,\epsilon_p)\rangle \cdot \langle c^q, T\circ l(\epsilon_p,\ldots,\epsilon_{p+q})\rangle.$$

The cochain $c^p \cup c^q$ is called the **cup product** of the cochains c^p and c^q.

Recall that $l(w_0,\ldots,w_p)$ is the linear singular simplex mapping ϵ_i into w_i for $i=0,\ldots,p$. The map $T\circ l(\epsilon_0,\ldots,\epsilon_p)$ is just the restriction of T to the "front p-face" Δ_p of Δ_{p+q}; it is a singular p-simplex on X. Similarly, $T\circ l(\epsilon_p,\ldots,\epsilon_{p+q})$ is, roughly speaking, the restriction of T to the "back q-face" of Δ_{p+q}; it is a singular q-simplex on X. The multiplication indicated on the right side of this equation is multiplication in the ring R, of course.

What, if anything, this mysterious formula means remains to be seen.

Theorem 48.1. *Cup product of cochains is bilinear and associative. The cochain z^0 whose value is 1 on each singular 0-simplex acts as a unity element. Furthermore, the following coboundary formula holds:*

$$(*)\qquad \delta(c^p \cup c^q) = (\delta c^p)\cup c^q + (-1)^p c^p \cup (\delta c^q).$$

Proof. Bilinearity is immediate, since two cochains are added by adding their values, and multiplication in R is distributive. Associativity is also immediate; the value of $(c^p \cup c^q)\cup c^r$ on $T:\Delta_{p+q+r}\rightarrow X$ equals the product of

$$\langle c^p, T\circ l(\epsilon_0,\ldots,\epsilon_p)\rangle,$$
$$\langle c^q, T\circ l(\epsilon_p,\ldots,\epsilon_{p+q})\rangle, \quad\text{and}$$
$$\langle c^r, T\circ l(\epsilon_{p+q},\ldots,\epsilon_{p+q+r})\rangle.$$

The value of $c^p \cup (c^q \cup c^r)$ on T equals the same. The fact that $c^p \cup z^0 = z^0 \cup c^p = c^p$ follows directly from the definition.

To check the coboundary formula, we compute the value of both sides of (*) on $T:\Delta_r\rightarrow X$, where we let $r=p+q+1$ for convenience. The cochains $(\delta c^p)\cup c^q$ and $(-1)^p c^p \cup (\delta c^q)$, evaluated on T, equal the two following expressions, respectively:

$$\sum_{i=0}^{p+1}(-1)^i\langle c^p, T\circ l(\epsilon_0,\ldots,\hat{\epsilon}_i,\ldots,\epsilon_{p+1})\rangle \cdot \langle c^q, T\circ l(\epsilon_{p+1},\ldots,\epsilon_r)\rangle$$

and

$$(-1)^p\sum_{i=p}^{r}(-1)^{i-p}\langle c^p, T\circ l(\epsilon_0,\ldots,\epsilon_p)\rangle \cdot \langle c^q, T\circ l(\epsilon_p,\ldots,\hat{\epsilon}_i,\ldots,\epsilon_r)\rangle.$$

If we add these expressions, the last term of the first expression cancels the first term of the second expression; what remains is precisely the formula for $\langle c^p \cup c^q, \partial T\rangle$. □

Theorem 48.2. *The cochain cup product induces an operation*

$$H^p(X;R) \times H^q(X;R) \xrightarrow{\cup} H^{p+q}(X;R)$$

that is bilinear and associative. The cohomology class $\{z^0\}$ acts as a unity element.

Proof. If z^p and z^q are cocycles, then their cup product is a cocycle as well, since

$$\delta(z^p \cup z^q) = \delta z^p \cup z^q + (-1)^p z^p \cup \delta z^q = 0.$$

The cohomology class of this product depends only on the cohomology classes of z^p and z^q, since

$$(z^p + \delta d^{p-1}) \cup z^q = z^p \cup z^q + \delta(d^{p-1} \cup z^q)$$

and

$$z^p \cup (z^q + \delta d^{q-1}) = z^p \cup z^q + (-1)^p \delta(z^p \cup d^{q-1}). \quad \square$$

Theorem 48.3. *If $h : X \to Y$ is a continuous map, then h^* preserves cup products.*

Proof. In fact, the cochain map $h^\#$ preserves cup products of cochains. For by definition, the value of $h^\#(c^p \cup c^q)$ on T equals the value of $c^p \cup c^q$ on $h \circ T$, which is

$$\langle c^p, h \circ T \circ l(\epsilon_0, \ldots, \epsilon_p)\rangle \cdot \langle c^q, h \circ T \circ l(\epsilon_p, \ldots, \epsilon_{p+q})\rangle,$$

and the value of $h^\#(c^p) \cup h^\#(c^q)$ on T equals the same. $\square$

Definition. Let $H^*(X;R)$ denote the external direct sum $\oplus H^i(X;R)$. The cup product operation makes this group into a ring with a unity element. It is called the **cohomology ring** of X with coefficients in R.

If $h : X \to Y$ is a continuous map, then h^* is a ring homomorphism. Therefore, a homotopy equivalence induces a ring isomorphism. It follows that the cohomology ring is a topological invariant, in fact, a homotopy-type invariant.

Commutativity

We have not yet discussed whether or not the cohomology ring is commutative. In fact it is not, in general. Instead, it has a property commonly called **anti-commutativity**. Specifically, if $\alpha^p \in H^p(X;R)$ and $\beta^q \in H^q(X;R)$, then

$$\alpha^p \cup \beta^q = (-1)^{pq} \beta^q \cup \alpha^p.$$

We shall not prove this formula now, for the proof will become much easier when we give an alternate definition of the cup product operation later on.

In the next section, we shall compute the cohomology ring in several specific cases. But first, let us introduce several generalized versions of our cup product operation.

Cup products with general coefficients

Let G be an arbitrary abelian group. Then we note that the cup product formula makes sense if we interpret it as a function

$$S^p(X) \times S^q(X; G) \xrightarrow{\cup} S^{p+q}(X; G).$$

In this case, c^p is a cochain with integral values, c^q is a cochain with values in G, and the multiplication on the right side of the formula is the usual product operation sending (n, g) to ng. Bilinearity is immediate. Associativity holds when it makes sense—that is, when it involves a map

$$S^p(X) \times S^q(X) \times S^r(X; G) \to S^{p+q+r}(X; G).$$

The cochain z^0 whose value is 1 on each 0-simplex T acts as a left unity element. The proof of the coboundary formula is unchanged, as is the proof that the homomorphism h^* induced by a continuous map preserves cup products.

Therefore, we have a well-defined cup product operation

$$H^p(X) \times H^q(X; G) \xrightarrow{\cup} H^{p+q}(X; G).$$

The most general cup product operation usually considered begins with a bilinear map $\alpha : G \times G' \to G''$, called a "coefficient pairing"; using this map to replace the multiplication operation in the cochain formula, one has a well-defined cup product

$$H^p(X; G) \times H^q(X; G') \xrightarrow{\cup} H^{p+q}(X; G'').$$

We shall not need this degree of generality.

Relative cup products

Sometimes, one wishes to define a cup product operation on relative cohomology groups. One can use the same cup product formula as before. The relative cup products we shall need are the following:

$$H^p(X,A; R) \times H^q(X; R) \to H^{p+q}(X,A; R),$$
$$H^p(X,A; R) \times H^q(X,A; R) \to H^{p+q}(X,A; R).$$

(The second of these is in fact just a restriction of the first.) The existence of these cup products is easy to demonstrate. For if $c^p : S_p(X) \to R$ vanishes on all singular p-simplices carried by A, then $c^p \cup c^q$ automatically vanishes on all

$p + q$ simplices carried by A. The coboundary formula holds just as before, so one has an induced operation on the cohomology level. Bilinearity and associativity are immediate, as is the fact that the homomorphism induced by a continuous map preserves cup products. The class $\{z^0\} \in H^0(X;R)$ acts as a right unity element for the first of these operations.

The most general relative cup product operation is a bilinear map

$$H^p(X,A;R) \times H^q(X,B;R) \to H^{p+q}(X,A \cup B;R).$$

It is defined whenever $\{A,B\}$ is an excisive couple. See the exercises.

EXERCISES

1. Let A be a path component of X; let B be the union of the remaining path components of X; assume $B \neq \varnothing$. Let c^0 be the cochain whose value is 1 on each $T : \Delta_0 \to A$ and 0 on each $T : \Delta_0 \to B$. Show c^0 is a cocycle, and describe $\{c^0\} \cup \beta^p$ for a general cohomology class β^p.

2. Let $A \subset X$; let $i : A \to X$ be inclusion. Let $\eta \in H^q(X;R)$; let $\eta | A$ denote $i^*(\eta) \in H^q(A;R)$. Show that the following diagram commutes:

$$\begin{array}{ccccccc}
H^{p+1}(X,A;R) & \longleftarrow & H^p(A;R) & \longleftarrow & H^p(X;R) & \longleftarrow & H^p(X,A;R) \\
\downarrow \cup \eta & & \downarrow \cup (\eta|A) & & \downarrow \cup \eta & & \downarrow \cup \eta \\
H^{p+q+1}(X,A;R) & \longleftarrow & H^{p+q}(A;R) & \longleftarrow & H^{p+q}(X;R) & \longleftarrow & H^{p+q}(X,A;R)
\end{array}$$

What happens if you replace $\cup\, \eta$ with $\eta\, \cup$ throughout?

3. Show that if $\{A,B\}$ is an excisive couple, then the cup product formula induces a bilinear map

$$H^p(X,A;R) \times H^q(X,B;R) \to H^{p+q}(X,A \cup B;R).$$

Interpret associativity in this case.

4. (a) If G is an abelian group, show that the group $\operatorname{Hom}(G,R)$ can be given the structure of R-module by defining $\langle \alpha\phi, g \rangle = \alpha \cdot \langle \phi, g \rangle$ if $\phi \in \operatorname{Hom}(G,R)$ and $\alpha \in R$ and $g \in G$. Show that if $f : G \to G'$ is a homomorphism, then $\tilde{f}$ is an R-module homomorphism.
   (b) Give $S^p(X;R) = \operatorname{Hom}(S_p(X),R)$ the structure of R-module as in (a). Show that δ is an R-module homomorphism, so that $H^p(X;R)$ has the structure of R-module.
   (c) Show that if $h : X \to Y$ is a continuous map, h^* is an R-module homomorphism.
   (d) Show that cup product, as a function of each variable separately, is an R-module homomorphism. (This means that $H^*(X;R)$ is what is sometimes called a **ring with operators** R. In the special case where R is a field F, it is called an **algebra** over F.)

§49. COHOMOLOGY RINGS OF SURFACES

There is no general method for computing the singular cohomology ring of X, even if X is a CW complex. The reason is not hard to find: It turns out that the cellular chain complex of X does not *determine* the cohomology ring of X. That is, two CW complexes can have isomorphic cellular chain complexes without having isomorphic cohomology rings!

Therefore, to compute cup products, we turn to simplicial cohomology. In this section, we define a simplicial cup product formula that corresponds to the previous formula for singular cochains under the standard isomorphism of simplicial with singular theory. Then we use this formula to compute a number of examples.

Definition. Given a complex K, choose a partial ordering of the vertices of K that linearly orders the vertices of each simplex of K. Define

$$C^p(K;R) \times C^q(K;R) \xrightarrow{\cup} C^{p+q}(K;R)$$

by the formula

$$\langle c^p \cup c^q, [v_0, \ldots, v_{p+q}]\rangle = \langle c^p, [v_0, \ldots, v_p]\rangle \cdot \langle c^q, [v_p, \ldots, v_{p+q}]\rangle$$

if $v_0 < \cdots < v_{p+q}$ in the given ordering.

The similarity of this formula to the corresponding formula in singular theory is striking.

Theorem 49.1. *Given an ordering of the vertices of K, the corresponding simplicial cup product is bilinear and associative. The cochain z^0 whose value is 1 on each vertex of K acts as a unity element. The coboundary formula (*) of Theorem 48.1 holds. If $\eta : C_p(K) \to S_p(|K|)$ is the chain map of §34, determined by the given ordering, then its dual $\tilde{\eta}$ carries singular cup product to simplicial cup product.*

Proof. The proofs are straightforward. Only the coboundary formula requires comment. One can prove it by the same computations we used in proving Theorem 48.1; only slight changes of notation are needed. Alternatively, one can use the fact that since η carries basis elements to basis elements, η is injective and its image is a direct summand in $S_p(|K|)$. Hence its dual $\tilde{\eta}$ is surjective. Given simplicial cochains c^p and c^q, they can thus be pulled back to singular cochains of $|K|$, say $\bar{c}^p$ and $\bar{c}^q$. We know the coboundary formula holds in singular theory for $\bar{c}^p \cup \bar{c}^q$. Since $\tilde{\eta}$ preserves both cup products and coboundaries, the same coboundary formula must hold in simplicial theory for $c^p \cup c^q$. □

Theorem 49.2. *The simplicial cup product induces an operation*

$$H^p(K;R) \times H^q(K;R) \xrightarrow{\cup} H^{p+q}(K;R)$$

that is bilinear and associative. It is independent of the ordering of vertices of K. The cohomology class $\{z^0\}$ acts as unity element. If $h : |K| \to |L|$ is a continuous map, then h^* preserves cup products.

Proof. The existence of $\cup$ follows from the coboundary formula as before. The chain map η induces an isomorphism η^* of singular with simplicial cohomology that preserves cup products. Since η^* is independent of the chosen ordering in K, so is the cup product in simplicial cohomology.

Because h^* preserves cup products in singular cohomology, and η^* commutes with h^*, the homomorphism h^* in simplicial theory necessarily preserves cup products. □

Note that in general, if $f : K \to L$ is a simplicial map, the cochain map $f^\#$ need not preserve cup products on the cochain level. For the simplicial cup product is defined using a particular ordering of the vertices, and the simplicial map f need not preserve the ordering of vertices in K and L. This would be a serious problem if one wanted to work entirely within oriented simplicial theory; it would be difficult to prove naturality of the cohomology cup product.

We remark that the more general cup products

$$H^p(K) \times H^q(K; G) \xrightarrow{\cup} H^{p+q}(K; G)$$

and

$$H^p(K,A; R) \times H^q(K,B; R) \xrightarrow{\cup} H^{p+q}(K,A \cup B; R)$$

exist in simplicial theory just as they do in singular theory. The relative cup product is in fact easier to define in simplicial than in singular theory, for if c^p vanishes on $C_p(A)$ and c^q vanishes on $C_q(B)$, then $c^p \cup c^q$ automatically vanishes on $C_r(A \cup B)$, because simplex of $A \cup B$ must lie in either A or B. (Of course, $\{A,B\}$ is excisive in this case, so that singular cup product is defined as well. See the exercises of §34.)

Now let us compute some examples. First, we need some terminology.

Definition. Since the cohomology ring has a unity element in dimension 0, multiplication by this element is never trivial. It can happen, however, that every product of positive-dimensional cohomology classes vanishes. In that case, we say that the cohomology ring is the **trivial ring.**

Here are some cohomology rings that are not trivial.

Example 1. Consider the torus T. Let w^1 and z^1 denote the cocycles pictured in Figure 49.1. We know that $\alpha = \{w^1\}$ and $\beta = \{z^1\}$ generate $H^1(T)$. If we orient each 2-simplex counterclockwise, then $\Lambda = \{\sigma^*\}$ generates $H^2(T) \simeq \mathbf{Z}$, where σ is any oriented 2-simplex of T. In general, if $\sigma_1, \dots, \sigma_k$ are oriented 2-simplices of T, the cochain $\Sigma n_i \sigma_i^*$ is cohomologous to $(\Sigma n_i)\sigma^*$.

Order the vertices of T alphabetically. Using this ordering, we compute the value of $w^1 \cup z^1$ on each oriented 2-simplex σ. Note that $\langle w^1 \cup z^1, \sigma \rangle = 0$ unless a

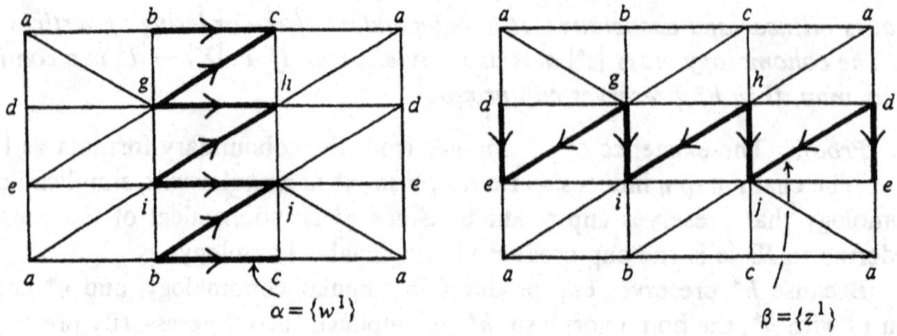

Figure 49.1

face of σ is in the carrier of w^1 and a face is in the carrier of z^1. Thus the only possible non-zero values occur when σ is one of the simplices ghi or hij. We compute

$$\langle w^1 \cup z^1, [g,h,i]\rangle = \langle w^1, [g,h]\rangle \cdot \langle z^1, [h,i]\rangle = 1 \cdot 1 = 1,$$
$$\langle w^1 \cup z^1, [h,i,j]\rangle = \langle w^1, [h,i]\rangle \cdot \langle z^1, [i,j]\rangle = (-1) \cdot 0 = 0.$$

Thus $w^1 \cup z^1 = [g,h,i]^*$. Now the orientation of $[g,h,i]$ is clockwise. Therefore, in terms of our standard generators, $\alpha \cup \beta = -\Lambda$.

A similar computation shows that

$$\langle z^1 \cup w^1, [g,h,i]\rangle = 0 \cdot (-1) = 0,$$
$$\langle z^1 \cup w^1, [h,i,j]\rangle = 1 \cdot 1 = 1,$$

so that $z^1 \cup w^1 = [h,i,j]^*$. Thus $\beta \cup \alpha = \Lambda$. (This is exactly what anticommutativity would have led us to expect.)

A similar direct computation shows that $\alpha \cup \alpha = 0$ and $\beta \cup \beta = 0$. Alternatively, we note that w^1 is cohomologous to the cochain y^1 pictured in Figure 49.2. Sincc no 2-simplex has one face in the carrier of w^1 and another face in the carrier of y^1, necessarily $w^1 \cup y^1 = 0$. Hence $\alpha \cup \alpha = 0$. A similar argument shows that $\beta \cup \beta = 0$.

Another alternative computation comes by noting that anticommutativity implies that $\alpha \cup \alpha = -(\alpha \cup \alpha)$. Since $H^2(T)$ has no elements of order 2, $\alpha \cup \alpha$ must vanish. A similar remark applies to $\beta \cup \beta$.

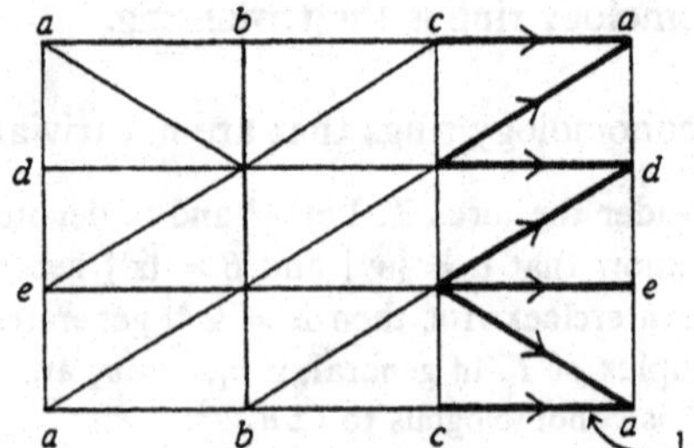

Figure 49.2

We can specify the ring structure of $H^*(T)$ by writing its multiplication table in terms of generators for $H^*(T)$. This table becomes (omitting the unity element)

$\cup$	α	β	Λ
α	0	$-\Lambda$	0
β	Λ	0	0
Λ	0	0	0

where the last row and column vanish for dimensional reasons.

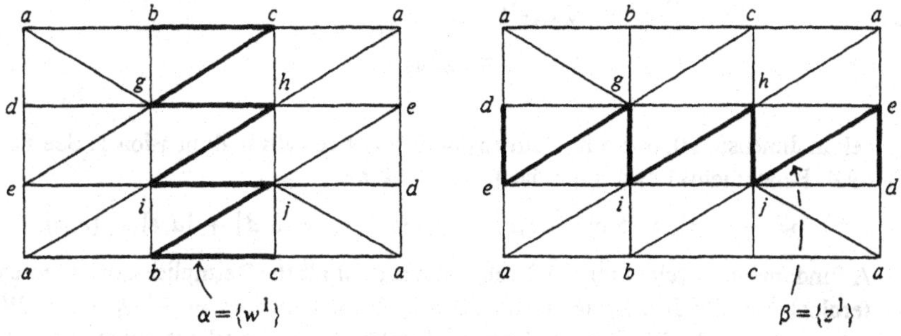

Figure 49.3

***Example* 2.** Consider now the Klein bottle S. Let us compute the cohomology ring with $\mathbf{Z}/2$ coefficients. We know that $H^1(S;\mathbf{Z}/2)$ is generated by the cocycles w^1 and z^1 pictured in Figure 49.3. Furthermore, $H^2(S;\mathbf{Z}/2)$ is generated by σ^*, for any 2-simplex σ. (We omit orientations, since $1 = -1$ in $\mathbf{Z}/2$.) Let $\alpha = \{w^1\}$ and $\beta = \{z^1\}$ and $\Lambda = \{\sigma^*\}$. Some of the computations we carried out in Example 1 apply without change, provided we reduce the coefficients modulo 2. In particular,

$$w^1 \cup z^1 = [g,h,i]^*, \quad \text{and}$$

$$w^1 \cup y^1 = 0,$$

where y^1 is the cochain pictured in Figure 49.2 (without the arrows). We conclude that $\alpha \cup \beta = \Lambda$ and $\alpha \cup \alpha = 0$.

Computing $z^1 \cup z^1$ must be done directly, since we cannot "pull it off itself" as we did with w^1. (Why?) The cochain $z^1 \cup z^1$ has value 1 on $[d,e,g]$, on $[e,g,i]$, and on $[d,e,j]$; and it vanishes on all other 2-simplices. Thus it is cohomologous to $3\sigma^* = \sigma^*$. We conclude that $\beta \cup \beta = \Lambda$.

The multiplication table for $H^*(S;\mathbf{Z}/2)$ thus has the form

$\cup$	α	β	Λ
α	0	Λ	0
β	Λ	Λ	0
Λ	0	0	0

***Example* 3.** Consider the connected sum $P^2 \# P^2$. We compute its cohomology ring with $\mathbf{Z}/2$ coefficients. Let us express $P^2 \# P^2$ as a CW complex X having one

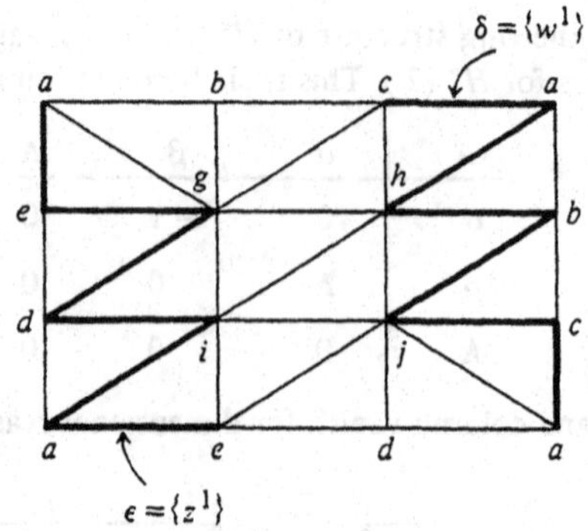

Figure 49.4

cell in dimension 0, one cell in dimension 2, and two cells in dimension 1. See Figure 49.4. Fundamental cycles for the 1-cells of X are

$$w^1 = [a,b] + [b,c] + [c,a] \qquad \text{and} \qquad z_1 = [a,d] + [d,e] + [e,a].$$

A fundamental cycle γ for the 2-cell is the sum of all the 2-simplices, oriented counterclockwise. By direct computation, $\partial w_1 = \partial z_1 = 0$, and $\partial \gamma = -2w_1 - 2z_1$. When we pass to the dual (cochain) complex $\text{Hom}(\mathcal{D}(X), \mathbf{Z}/2)$, all coboundary operators vanish (since we are using $\mathbf{Z}/2$ coefficients). Thus in singular cohomology, we have

$$H^1(P^2 \# P^2; \mathbf{Z}/2) \simeq \mathbf{Z}/2 \oplus \mathbf{Z}/2,$$
$$H^2(P^2 \# P^2; \mathbf{Z}/2) \simeq \mathbf{Z}/2.$$

Passing now to simplicial cohomology, we see that the cocycles w^1 and z^1 pictured in the figure, when restricted to $D_1(X)$, serve as a basis for $\text{Hom}(D_1(X), \mathbf{Z}/2)$, since

$$\langle w^1, w_1 \rangle = 1 \qquad \text{and} \qquad \langle w^1, z_1 \rangle = 0,$$
$$\langle z^1, w_1 \rangle = 0 \qquad \text{and} \qquad \langle z^1, z_1 \rangle = 1.$$

The classes $\delta = \{w^1\}$ and $\epsilon = \{z^1\}$ thus generate H^1, and $\Lambda = \{\sigma^*\}$ generates H^2 (where σ is any 2-simplex). Direct computation shows that

$$w^1 \cup z^1 = 0$$
$$w^1 \cup w^1 = [a,c,j]^*,$$
$$z^1 \cup z^1 = [a,e,g]^*.$$

Thus $H^*(P^2 \# P^2; \mathbf{Z}/2)$ has the multiplication table

$\cup$	δ	ϵ	Λ
δ	Λ	0	0
ϵ	0	Λ	0
Λ	0	0	0

But we know that $P^2 \# P^2$ is homeomorphic to the Klein bottle S. (See Figure 6.9.) Thus their cohomology rings are isomorphic, even though this multiplication table is quite unlike the one we computed in Example 2. We leave it to you to construct an isomorphism between these two rings.

This example illustrates the following important fact: *One cannot, in general, by examining the multiplication tables of two rings, determine at once whether or not the rings are isomorphic.*

Example 4. Consider the space X pictured in Figure 49.5; it is the union of two topological circles and a topological 2-sphere with a point in common. It is called the **wedge product** $S^1 \vee S^1 \vee S^2$. The space X can be expressed as a CW complex with one cell in dimension 0, one cell in dimension 2, and two cells in dimension 1. When we write fundamental cycles

$$w_1 = [a,b] + [b,c] + [c,a],$$

$$z_1 = [a,d] + [d,e] + [e,a],$$

$$z_2 = \partial[a,f,g,h],$$

for these cells, we see that the boundary operators in the cellular chain complex all vanish. *Therefore, the cellular chain complex $\mathcal{D}(X)$ of X is isomorphic to the cellular chain complex $\mathcal{D}(T)$ of the torus T.*

It follows that the homology groups and cohomology groups of X are isomorphic to those of T. However, their cohomology *rings* are not isomorphic. For it is easy to see that the cohomology ring of X is trivial. Consider the cocycles

$$w^1 = [b,c]^* \quad \text{and} \quad z^1 = [d,e]^*.$$

Since the cycles w_1 and z_1 are a basis for the chain group $D_1(X)$, the cocycles w^1 and z^1 give the dual basis for the cochain group $\text{Hom}(D_1(X), \mathbf{Z})$. All the cup products $w^1 \cup z^1$ and $w^1 \cup w^1$ and $z^1 \cup z^1$ vanish, because no 2-simplex has a face in the carriers of either w^1 or z^1.

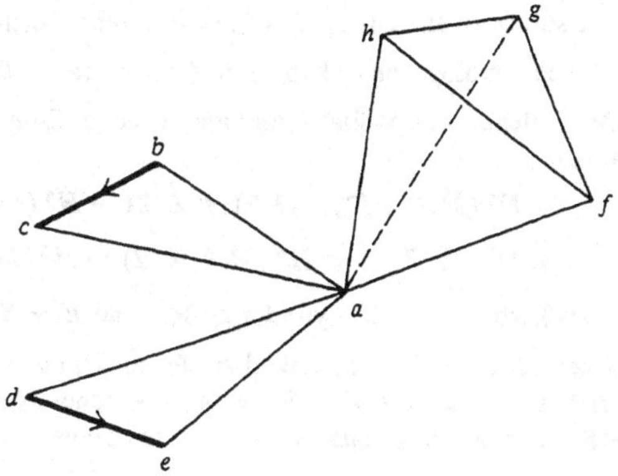

Figure 49.5

The preceding examples will, I hope, convince you that cup products are not as a rule easy to compute. The difficulty is that one must go down to the simplicial level and find specific representative cocycles, in order to use the cup product formula.

As a consequence, any theorem that tells us something about cup products in general is likely to be a useful theorem. We shall prove two such theorems in later chapters. One will tell us something about the cohomology ring of a product space $X \times Y$. The other will give us information about the cohomology ring of a manifold, which will in particular enable us to compute the cohomology rings of the projective spaces.

EXERCISES

Throughout, let T denote the torus, and let S denote the Klein bottle.

1. Let $f : S^2 \to T$ be continuous. Show that $f^* : H^2(T) \to H^2(S^2)$ is trivial, and conclude that $f_* : H_2(S^2) \to H_2(T)$ is trivial. What can you say about a continuous map $g : T \to S^2$?
2. If $f : X \to Y$, show that
$$f^* : H^2(Y; \mathbf{Z}/2) \to H^2(X; \mathbf{Z}/2)$$
is trivial in the following cases:
(a) $X = S^2$ and $Y = S$.
(b) $X = S$ and $Y = T$.
(c) $X = T$ and $Y = S$.
3. Give multiplication tables for the following cohomology rings:
(a) $H^*(T \# \cdots \# T)$.
(b) $H^*(P^2; \mathbf{Z}/2)$.
(c) $H^*(P^2 \# \cdots \# P^2; \mathbf{Z}/2)$.
4. Define an isomorphism between the rings of Examples 2 and 3.
5. Consider the cohomology rings of the Klein bottle S and the space $P^2 \vee S^1$.
(a) Show that these rings are isomorphic with integer coefficients.
(b) Show that these rings are not isomorphic with $\mathbf{Z}/2$ coefficients. [*Note:* It does not suffice to show that they have different multiplication tables!]
6. Compute the cohomology ring of the 3-fold dunce cap with $\mathbf{Z}/3$ coefficients.
7. (a) Let (M,E) denote the Möbius band and its edge. Compute the cup product operations
$$H^*(M,E; \mathbf{Z}/2) \times H^*(M; \mathbf{Z}/2) \to H^*(M,E; \mathbf{Z}/2),$$
$$H^*(M,E; \mathbf{Z}/2) \times H^*(M,E; \mathbf{Z}/2) \to H^*(M,E; \mathbf{Z}/2).$$
(b) Repeat (a) when M is the cylinder $S^1 \times I$ and $E = S^1 \times \text{Bd } I$.
8. Let A be the union of two once-linked circles in S^3; let B be the union of two unlinked circles, as in Figure 49.6. Show that the cohomology groups of $S^3 - A$ and $S^3 - B$ are isomorphic, but the cohomology rings are not.

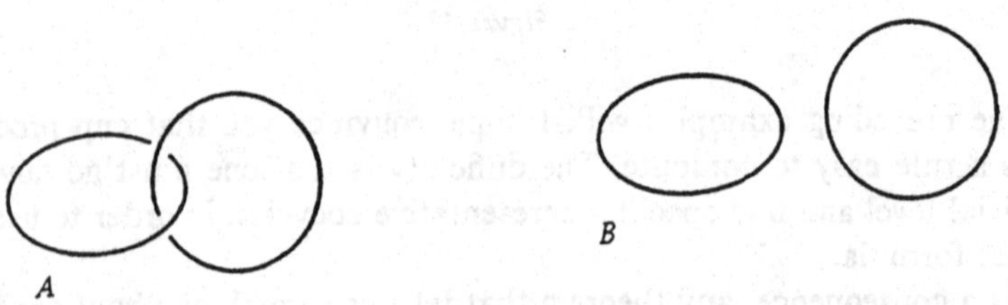

Figure 49.6

6

Homology with Coefficients

Having studied cohomology with arbitrary coefficients, we now return to a subject introduced briefly for simplicial theory in Chapter 1—homology with arbitrary coefficients.

First, we introduce an algebraic functor called the tensor product, and study its properties. It plays a role in homology theory similar to that played by the Hom functor for cohomology. Then we study homology with arbitrary coefficients in general.

§50. TENSOR PRODUCTS

If A and B are abelian groups, then their Cartesian product $A \times B$ is of course an abelian group, and one often considers homomorphisms of the group $A \times B$ into an abelian group C. However, one sometimes wishes rather to consider functions from $A \times B$ to C that are **bilinear,** that is, functions that are homomorphisms when considered as functions of each variable separately. In this section, we define an abelian group called the *tensor product* of A and B and denoted by $A \otimes B$. It has the property that *bilinear* functions from $A \times B$ to C can naturally be considered as *homomorphisms* from $A \otimes B$ to C, and conversely. By this means the study of bilinear functions is reduced to something familiar, the study of homomorphisms.

Definition. Let A and B be abelian groups. Let $F(A,B)$ be the free abelian group generated by the set $A \times B$. Let $R(A,B)$ be the subgroup generated by all elements of the form

$$(a + a',b) - (a,b) - (a',b),$$
$$(a,b + b') - (a,b) - (a,b'),$$

for $a, a' \in A$ and $b, b' \in B$. We define

$$A \otimes B = F(A,B)/R(A,B),$$

and call it the **tensor product** of A and B. The coset of the pair (a,b) is denoted by $a \otimes b$.

Now any function f from the set $A \times B$ to the abelian group C determines a unique homomorphism of $F(A,B)$ into C, since the elements of $A \times B$ are the basis elements for $F(A,B)$. This function f is bilinear if and only if it maps the subgroup $R(A,B)$ to zero. Thus every homomorphism of $A \otimes B$ into C gives rise to a bilinear function from $A \times B$ into C, and conversely.

Note that any element of $F(A,B)$ is a finite linear combination of pairs (a,b), so any element of $A \otimes B$ is a finite linear combination of elements of the form $a \otimes b$.

NOTE WELL: The element $a \otimes b$ is not the typical *element* of $A \otimes B$, but rather a typical *generator*.

We have the following relations in $A \otimes B$:

$$(a + a') \otimes b = a \otimes b + a' \otimes b,$$
$$a \otimes (b + b') = a \otimes b + a \otimes b',$$

by definition. It is immediate that $0 \otimes b = 0$, since

$$a \otimes b = (0 + a) \otimes b = 0 \otimes b + a \otimes b.$$

Similarly, $a \otimes 0 = 0$ for all a. It follows that

$$(-a) \otimes b = -(a \otimes b) = a \otimes (-b),$$

since adding $a \otimes b$ to each expression gives zero. An immediate consequence is the relation

$$(na) \otimes b = n(a \otimes b) = a \otimes (nb),$$

when n is an arbitrary integer.

Definition. Let $f : A \to A'$ and $g : B \to B'$ be homomorphisms. There is a unique homomorphism

$$f \otimes g : A \otimes B \to A' \otimes B'$$

such that $(f \otimes g)(a \otimes b) = f(a) \otimes g(b)$ for all a, b; it is called the **tensor product** of f and g.

This fact is an immediate consequence of the fact that the function from $A \times B$ into $A' \otimes B'$ carrying (a,b) to $f(a) \otimes g(b)$ is bilinear, as you can check. We also leave it to you to check the following:

Lemma 50.1. *The function mapping (A,B) to $A \otimes B$ and (f,g) to $f \otimes g$ is a covariant functor from the category of pairs of abelian groups and homomorphisms to the category of abelian groups and homomorphisms.* □

Later we will show how to compute tensor products. For the present, we merely note the following:

Theorem 50.2. *There is an isomorphism*

$$\mathbf{Z} \otimes G \simeq G$$

that maps $n \otimes g$ to ng; it is natural with respect to homomorphisms of G.

Proof. The function mapping $\mathbf{Z} \times G$ to G that sends (n,g) to ng is bilinear, so it induces a homomorphism $\phi : \mathbf{Z} \otimes G \to G$ sending $n \otimes g$ to ng.

Let $\psi : G \to \mathbf{Z} \otimes G$ be defined by the equation $\psi(g) = 1 \otimes g$; then ψ is a homomorphism. For $g \in G$, we have

$$\phi\psi(g) = \phi(1 \otimes g) = g;$$

while on a typical generator $n \otimes g$ of $\mathbf{Z} \otimes G$, we have

$$\psi\phi(n \otimes g) = \psi(ng) = 1 \otimes (ng) = n \otimes g.$$

Thus ψ is an inverse for ϕ.

Naturality is a consequence of the commutativity of the diagram

$$\begin{array}{ccc} \mathbf{Z} \otimes G & \xrightarrow{\simeq} & G \\ {\scriptstyle i_{\mathbf{Z}} \otimes f} \downarrow & & \downarrow {\scriptstyle f} \\ \mathbf{Z} \otimes H & \xrightarrow{\simeq} & H. \end{array}$$ □

We now derive some general properties of tensor products.

First, let us note a common fallacy. Suppose A' is a subgroup of A, and B' is a subgroup of B. Then it is tempting to assume that $A' \otimes B'$ can be considered as a subgroup of $A \otimes B$. *But this is not in general correct.* The inclusion mappings $i : A' \to A$ and $j : B' \to B$ do give rise to a homomorphism

$$i \otimes j : A' \otimes B' \to A \otimes B,$$

but this homomorphism is not in general injective. For example, the integers $\mathbf{Z}$ are a subgroup of the additive group of rationals $\mathbf{Q}$. But $\mathbf{Z} \otimes \mathbf{Z}/2$ is a non-trivial group, and $\mathbf{Q} \otimes \mathbf{Z}/2$ is trivial, for in $\mathbf{Q} \otimes \mathbf{Z}/2$,

$$a \otimes b = (a/2) \otimes 2b = (a/2) \otimes 0 = 0.$$

Although tensor products of injective maps are not in general injective, tensor product of *surjective* maps are always surjective. This is the substance of the following lemma:

Lemma 50.3. *Suppose the homomorphisms $\phi : B \to C$ and $\phi' : B' \to C'$ are surjective. Then*

$$\phi \otimes \phi' : B \otimes B' \to C \otimes C'$$

is surjective, and its kernel is the subgroup of $B \otimes B'$ generated by all elements of the form $b \otimes b'$ for which $b \in \ker \phi$ or $b' \in \ker \phi'$.

Proof. Let G denote the subgroup of $B \otimes B'$ generated by these elements $b \otimes b'$. Clearly $\phi \otimes \phi'$ maps G to zero, so it induces a homomorphism

$$\Phi : (B \otimes B')/G \to C \otimes C'.$$

We show that Φ is an isomorphism by defining an inverse Ψ for Φ.

We begin by defining a function

$$\psi : C \times C' \to (B \otimes B')/G$$

by the rule $\psi(c,c') = b \otimes b' + G$, where b is chosen so that $\phi(b) = c$ and b' is chosen so that $\phi'(b') = c'$. We show that ψ is well-defined. Suppose $\phi(b_0) = c$ and $\phi'(b_0') = c'$. Then

$$b \otimes b' - b_0 \otimes b_0' = ((b - b_0) \otimes b') + (b_0 \otimes (b' - b_0')).$$

This element lies in G because $b - b_0 \in \ker \phi$ and $b' - b_0' \in \ker \phi'$. Thus ψ is well-defined. It follows from its definition that ψ is bilinear, so it induces a homomorphism

$$\Psi : C \otimes C' \to (B \otimes B')/G.$$

It is straightforward to check that $\Phi \circ \Psi$ and $\Psi \circ \Phi$ are identity maps. □

Just as we did with the case of the Hom functor, we consider what "tensoring" does to an exact sequence.

Theorem 50.4. *Suppose the sequence*

$$A \xrightarrow{\phi} B \xrightarrow{\psi} C \to 0$$

is exact. Then the sequence

$$A \otimes G \xrightarrow{\phi \otimes i_G} B \otimes G \xrightarrow{\psi \otimes i_G} C \otimes G \to 0$$

is exact. If ϕ is injective and the first sequence splits, then $\phi \otimes i_G$ is injective and the second sequence splits.

Proof. The preceding lemma implies that $\psi \otimes i_G$ is surjective, and that its kernel is the subgroup D of $B \otimes G$ generated by all elements of the form $b \otimes g$

for $b \in \ker \psi$. The image of $\phi \otimes i_G$ is the subgroup E generated by all elements of the form $\phi(a) \otimes g$. Since image ϕ = kernel ψ, we have $D = E$.

Suppose ϕ is injective and the sequence splits. Let $p : B \to A$ be a homomorphism such that $p \circ \phi = i_A$. Then

$$(p \otimes i_G) \circ (\phi \otimes i_G) = i_A \otimes i_G = i_{A \otimes G},$$

so $\phi \otimes i_G$ is injective and $p \otimes i_G$ splits the tensored sequence. □

Corollary 50.5. *There is a natural isomorphism*

$$\mathbf{Z}/m \otimes G \simeq G/mG.$$

Proof. We take the exact sequence

$$0 \to \mathbf{Z} \xrightarrow{m} \mathbf{Z} \to \mathbf{Z}/m \to 0$$

and tensor it with G, obtaining the exact sequence

$$\mathbf{Z} \otimes G \xrightarrow{m \otimes i_G} \mathbf{Z} \otimes G \to \mathbf{Z}/m \otimes G \to 0.$$

Applying Theorem 50.2, we have the exact sequence

$$G \xrightarrow{m} G \to \mathbf{Z}/m \otimes G \to 0.$$

The corollary follows. □

Now we prove some additional properties of tensor products.

Theorem 50.6. *One has the following natural isomorphisms:*

(a) $A \otimes B \simeq B \otimes A$.

(b) $(\bigoplus A_\alpha) \otimes B \simeq \bigoplus (A_\alpha \otimes B)$,
$A \otimes (\bigoplus B_\alpha) \simeq \bigoplus (A \otimes B_\alpha)$.

(c) $A \otimes (B \otimes C) \simeq (A \otimes B) \otimes C$.

Proof. We construct the isomorphisms, and leave naturality for you to check.

(a) The map $A \times B \to B \times A$ sending (a,b) to (b,a) induces an isomorphism of $F(A,B)$ with $F(B,A)$ that carries $R(A,B)$ onto $R(B,A)$.

(b) We apply Lemma 4.1. By hypothesis, there are homomorphisms

$$j_\beta : A_\beta \to \bigoplus A_\alpha \quad \text{and} \quad \pi_\beta : \bigoplus A_\alpha \to A_\beta$$

such that $\pi_\beta \circ j_\alpha$ is trivial if $\alpha \neq \beta$ and equals the identity if $\alpha = \beta$. Let

$$f_\beta = j_\beta \otimes i_B : A_\beta \otimes B \to (\bigoplus A_\alpha) \otimes B,$$
$$g_\beta = \pi_\beta \otimes i_B : (\bigoplus A_\alpha) \otimes B \to A_\beta \otimes B.$$

Then $g_\beta \circ f_\alpha$ is trivial if $\alpha \neq \beta$ and equals the identity if $\alpha = \beta$. Now $(\bigoplus A_\alpha) \otimes B$ is generated by elements of the form $a \otimes b$, where $a \in \bigoplus A_\alpha$ and $b \in B$. Since a

is, in turn, equal to a finite sum of elements of the form $j_\alpha(a_\alpha)$, we see that $(\oplus A_\alpha) \otimes B$ is generated by the groups $f_\alpha(A_\alpha \otimes B)$. The existence of the first isomorphism of (b) follows from Lemma 4.1.

The second isomorphism of (b) follows by commutativity.

(c) To define a homomorphism mapping the tensor product $A \otimes (B \otimes C)$ into an abelian group G, we must define a bilinear function f on the *set* $A \times (B \otimes C)$. In order that f be linear in the second variable, it must for fixed $a \in A$, come from a bilinear map of the set $a \times B \times C$ into G. We conclude that a map f of the set $A \times B \times C$ into G defines a homomorphism of $A \otimes (B \otimes C)$ into G precisely if it is linear in each of the three variables separately. Such a function is called a **multilinear function.** A similar argument shows that homomorphisms of $(A \otimes B) \otimes C$ into G are obtained in exactly the same way.

Now consider the functions

$$f(a,b,c) = a \otimes (b \otimes c),$$
$$g(a,b,c) = (a \otimes b) \otimes c.$$

These are multilinear functions from $A \times B \times C$ to $A \otimes (B \otimes C)$ and $(A \otimes B) \otimes C$, respectively. They induce homomorphisms

$$(A \otimes B) \otimes C \xrightarrow{F} A \otimes (B \otimes C) \xrightarrow{G} (A \otimes B) \otimes C,$$

respectively. Both $F \circ G$ and $G \circ F$ act as identity maps on generators of these groups, so they are identity maps. □

Corollary 50.7. *If*

$$0 \to A \to B \to C \to 0$$

is exact and G is torsion-free, then

$$0 \to A \otimes G \to B \otimes G \to C \otimes G \to 0$$

is exact.

Proof. Step 1. We show first that the theorem holds if G is free. The sequence

$$(*) \qquad 0 \to A \otimes \mathbf{Z} \to B \otimes \mathbf{Z} \to C \otimes \mathbf{Z} \to 0$$

is exact because $D \otimes \mathbf{Z}$ is naturally isomorphic to D for all D. Therefore,

$$0 \to A \otimes G \to B \otimes G \to C \otimes G \to 0$$

is exact; for by the preceding theorem this sequence is isomorphic to a direct sum of sequences of type (*), and direct sums of exact sequences are exact.

Step 2. We prove the following fact: Let $a_1, \ldots, a_k \in A$ and $b_1, \ldots, b_k \in B$. Suppose the element $\sum a_i \otimes b_i$ of $A \otimes B$ vanishes. Then there are finitely generated subgroups A_0 and B_0 of A and B containing $\{a_1, \ldots, a_k\}$ and $\{b_1, \ldots, b_k\}$, respectively, such that the sum $\sum a_i \otimes b_i$ vanishes when considered as an element of $A_0 \otimes B_0$.

Recall that $A \otimes B$ equals the quotient of $F(A,B)$ by a certain relations subgroup $R(A,B)$. The equation $\Sigma\, a_i \otimes b_i = 0$ means that the element $\Sigma(a_i,b_i)$ of $F(A,B)$ lies in $R(A,B)$. That is, it can be written as a finite linear combination of terms of the form

$$(a + a',b) - (a,b) - (a',b)$$

and

$$(a,b + b') - (a,b) - (a,b').$$

Let A_0 denote the subgroup of A generated by the first components of these finitely many terms, along with $a_1, \ldots, a_k$. Let B_0 denote the subgroup of B generated by the second components of these terms, along with $b_1, \ldots, b_k$. Then when we consider the formal sum $\Sigma(a_i,b_i)$ as an element of $F(A_0,B_0)$, it lies in the relations subgroup used in defining $A_0 \otimes B_0$. Thus the sum $\Sigma\, a_i \otimes b_i$ vanishes when considered as an element of $A_0 \otimes B_0$.

Step 3. Now we complete the proof. Suppose

$$0 \to A \xrightarrow{\phi} B \xrightarrow{\psi} C \to 0$$

is exact and G is torsion-free. We wish to show that $\phi \otimes i_G$ is injective. The typical element of $A \otimes G$ is a finite sum $\Sigma\, a_i \otimes g_i$. Suppose it lies in the kernel of $\phi \otimes i_G$. Then $\Sigma\, \phi(a_i) \otimes g_i$ vanishes in $B \otimes G$. Choose finitely generated subgroups B_0, G_0 of B, G, respectively, such that this sum vanishes when considered as an element of $B_0 \otimes G_0$. Applying the map $B_0 \otimes G_0 \to B \otimes G_0$ induced by inclusion, we see that it vanishes when considered as an element of $B \otimes G_0$.

Now G_0 is torsion-free, being a subgroup of G; therefore, since G_0 is finitely generated, *it is free*. As a consequence, the sequence

$$0 \to A \otimes G_0 \to B \otimes G_0 \to C \otimes G_0 \to 0$$

is exact. We conclude that $\Sigma\, a_i \otimes g_i$ must vanish when it is considered as an element of $A \otimes G_0$. Applying the map $A \otimes G_0 \to A \otimes G$ induced by inclusion, we see that it also vanishes when considered as an element of $A \otimes G$. □

The theorems we have proved enable us to compute the group $A \otimes B$ when A is finitely generated. For $\otimes$ commutes with direct sums, and we have the rules

$$\mathbf{Z} \otimes G \simeq G, \qquad \mathbf{Z}/m \otimes G \simeq G/mG.$$

In particular, tensor products of free abelian groups are free abelian. For later use, we state this fact formally as follows:

Theorem 50.8. *If A is free abelian with basis $\{a_i\}$ and B is free abelian with basis $\{b_j\}$, then $A \otimes B$ is free abelian with basis $\{a_i \otimes b_j\}$.*

Proof. Let $\langle a_i \rangle$ and $\langle b_j \rangle$ denote the infinite cyclic subgroups of A and B generated by a_i and b_j, respectively. Then

$$A = \oplus \langle a_i \rangle \qquad \text{and} \qquad B = \oplus \langle b_j \rangle.$$

It follows that $A \otimes B \simeq \oplus(\langle a_i\rangle \otimes \langle b_j\rangle)$. Now $\mathbf{Z} \otimes \mathbf{Z}$ is infinite cyclic and is generated by $1 \otimes 1$; likewise, $\langle a_i\rangle \otimes \langle b_j\rangle$ is infinite cyclic and is generated by $a_i \otimes b_j$. The theorem follows. □

Tensor products of modules

Let R be a commutative ring with unity element, as usual. If A and B are modules over R, then (as we mentioned in the exercises of §41) the group $\text{Hom}_R(A,B)$ of all module homomorphisms of A into B has the structure of R-module. An analogous situation obtains for the tensor product functor. We shall in fact be interested only in the special case where R is a field. But we may as well consider the general case for the time being.

Let A and B be modules over the ring R. As before, let $F(A,B)$ be the free abelian group generated by the set $A \times B$. But now let $R(A,B)$ be the subgroup generated by elements of the form

$$\begin{aligned}&(a + a',b) - (a,b) - (a',b),\\&(a,b + b') - (a,b) - (a,b'),\\&(\alpha a,b) - (a,\alpha b), \qquad \text{for} \qquad \alpha \in R.\end{aligned}$$

Then $F(A,B)/R(A,B)$ has the structure of module over R: Given α, we define a map of $F(A,B)$ to itself by the rule $\alpha(a,b) = (\alpha a,b)$. This map carries $R(A,B)$ into itself. For instance, when we apply α to the first of the listed generators for $R(A,B)$, we have

$$\begin{aligned}(\alpha(a + a'),b) &- (\alpha a,b) - (\alpha a',b)\\&= (\alpha a + \alpha a',b) - (\alpha a,b) - (\alpha a',b).\end{aligned}$$

The latter element is in $R(A,B)$ by definition. A similar remark applies to the second of the listed generators. For the third, we have

$$\begin{aligned}\beta(\alpha a,b) - \beta(a,\alpha b) &= (\beta(\alpha a),b) - (\beta a,\alpha b)\\&= (\alpha(\beta a),b) - (\beta a,\alpha b),\end{aligned}$$

which is in $R(A,B)$ by definition.

Thus we have an induced operation on the quotient F/R. The module properties are easy to verify. We shall denote the resulting module by $A \otimes_R B$, and call it the **tensor product** of A and B **over the ring** R. The coset of (a,b) will be denoted $a \otimes b$, as before. Besides the usual relations in $A \otimes B$, one also has in $A \otimes_R B$ the relation

$$\alpha(a \otimes b) = (\alpha a) \otimes b = a \otimes (\alpha b).$$

Now let us consider a set map $f : A \times B \to C$ that is a module homomorphism in each variable separately. Since it maps $R(A,B)$ to zero, it induces a homomorphism

$$g : A \otimes_R B \to C,$$

which is actually a module homomorphism, as you can check.

Note that $A \otimes_R B$ is "smaller" than $A \otimes B$; in fact, it is isomorphic to the quotient group of $A \otimes B$ by the subgroup generated by all terms of the form $(\alpha a) \otimes b - a \otimes (\alpha b)$. This is analogous to the situation for the Hom functor, where $\text{Hom}_R(A,B)$ is a subgroup of $\text{Hom}(A,B)$.

If $f: A \to A'$ and $g: B \to B'$ are module homomorphisms, then there is a module homomorphism

$$f \otimes g: A \otimes_R B \to A' \otimes_R B'$$

mapping $a \otimes b$ to $f(a) \otimes g(b)$ for all a, b, for the map sending (a,b) to $f(a) \otimes g(b)$ is a module homomorphism in each variable separately.

The theorems of this section generalize to the tensor product of modules. Proofs are left as exercises.

EXERCISES

1. Let G be an abelian group with torsion subgroup T. Let H be a divisible group. Show that $G \otimes H \simeq (G/T) \otimes H$. [*Note:* T is not necessarily a direct summand in G!]
2. Show that the additive group $\mathbf{Q}$ is torsion-free but not free. [*Hint:* Compute $\mathbf{Q} \otimes \mathbf{Z}/2$.]
3. Show that if A and B are $\mathbf{Z}$-modules, then
$$A \otimes_{\mathbf{Z}} B = A \otimes B.$$
4. Let A be an R-module. Show that there is an R-module isomorphism
$$R \otimes_R A \simeq A.$$
5. Prove the analogues of Theorems 50.4 and 50.6 for $\otimes_R$.
6. Let A, B, and C be vector spaces over a field F.
   (a) Show that $\otimes_F$ preserves exact sequences of vector spaces. [*Hint:* Every such sequence splits.]
   (b) If A and B have vector space bases $\{a_i\}$ and $\{b_j\}$, respectively, show that $\{a_i \otimes b_j\}$ is a vector space basis for $A \otimes_F B$.
7. If A and B are vector spaces over $\mathbf{Q}$, show that
$$A \otimes_{\mathbf{Q}} B = A \otimes B.$$

§51. HOMOLOGY WITH ARBITRARY COEFFICIENTS

We now use the tensor product functor to define homology groups with arbitrary coefficients in general. The treatment will follow the pattern of §44, where we dealt with cohomology theory with arbitrary coefficients.

First, we work on the level of chain complexes. Then we specialize to singular theory, and finally to simplicial theory, at which point we show that the definition of homology with arbitrary coefficients we use here is equivalent to the one given in §10.

Homology of a chain complex

Let G be an abelian group. Let $\mathcal{C} = \{C_p,\partial\}$ be a chain complex. We denote the pth homology group of the chain complex $\mathcal{C} \otimes G = \{C_p \otimes G, \partial \otimes i_G\}$ by $H_p(\mathcal{C}; G)$, and call it the pth **homology group of $\mathcal{C}$ with coefficients in G.**

If $\{\mathcal{C},\epsilon\}$ is an augmented chain complex, then one has the corresponding chain complex obtained from $\mathcal{C} \otimes G$ by adjoining the group $\mathbf{Z} \otimes G \simeq G$ in dimension -1, and using $\epsilon \otimes i_G$ as the boundary operator from dimension 0 to dimension -1. Its homology groups are denoted $\tilde{H}_p(\mathcal{C}; G)$ and are called the **reduced homology groups of $\mathcal{C}$ with coefficients in** G. If $\tilde{H}_0(\mathcal{C})$ vanishes, so does $\tilde{H}_0(\mathcal{C}; G)$, since exactness of $C_1 \to C_0 \to \mathbf{Z} \to 0$ implies exactness of

$$C_1 \otimes G \to C_0 \otimes G \to \mathbf{Z} \otimes G \to 0.$$

In general, we have the equation

$$H_0(\mathcal{C}; G) \simeq \tilde{H}_0(\mathcal{C}; G) \oplus G.$$

Note that if G is the group of integers, then $\mathcal{C} \otimes G$ is naturally isomorphic with $\mathcal{C}$. Thus the usual homology of $\mathcal{C}$ can be thought of as "homology with coefficients in $\mathbf{Z}$."

If $\phi : \mathcal{C} \to \mathcal{D}$ is a chain map, then so is $\phi \otimes i_G : \mathcal{C} \otimes G \to \mathcal{D} \otimes G$. The induced homology homomorphism is for convenience denoted by

$$\phi_* : H_p(\mathcal{C}; G) \to H_p(\mathcal{D}; G),$$

rather than by $(\phi \otimes i_G)_*$. If $\mathcal{C}$ and $\mathcal{D}$ are augmented and ϕ is augmentation-preserving, then $\phi \otimes i_G$ induces a homomorphism of reduced as well as ordinary homology.

If $\phi, \psi : \mathcal{C} \to \mathcal{C}'$ are chain maps, and if D is a chain homotopy between them, then $D \otimes i_G$ is a chain homotopy between $\phi \otimes i_G$ and $\psi \otimes i_G$. It follows that if ϕ and ψ are chain homotopic, then ϕ_* and ψ_* are equal as homomorphisms of homology with arbitrary coefficients. It also follows that if ϕ is a chain equivalence, so is $\phi \otimes i_G$.

Finally, suppose one has a short exact sequence

$$0 \to \mathcal{C} \to \mathcal{D} \to \mathcal{E} \to 0$$

of chain complexes *that splits in each dimension*. Then the tensored sequence

$$0 \to C_p \otimes G \to D_p \otimes G \to E_p \otimes G \to 0$$

is exact. Applying the zig-zag lemma, one has a long exact sequence

$$\cdots \to H_p(\mathcal{C}; G) \to H_p(\mathcal{D}; G) \to H_p(\mathcal{E}; G) \xrightarrow{\partial_*} H_{p-1}(\mathcal{C}; G) \to \cdots$$

where ∂_* is induced by $\partial \otimes i_G$. This sequence is natural with respect to homomorphisms induced by chain maps.

Singular homology

We define the singular homology groups of a topological pair (X,A), with coefficients in G, by the equation

$$H_p(X,A; G) = H_p(\mathcal{S}(X,A); G).$$

As usual, we delete A from the notation if A is empty. We define reduced homology groups by the equation

$$\tilde{H}_p(X; G) = H_p(\{\mathcal{S}(X),\epsilon\}; G),$$

where ϵ is the usual augmentation for $\mathcal{S}(X)$.

Since $S_p(X,A)$ is free abelian, the group $S_p(X,A) \otimes G$ is a direct sum of copies of G. Indeed, if $\{T_\alpha\}$ is the family consisting of those singular p-simplices of X whose image sets do not lie in A, then the cosets modulo $S_p(A)$ of the singular simplices T_α form a basis for $S_p(X,A)$. Therefore each element of $S_p(X,A) \otimes G$ can be represented uniquely by a finite sum $\Sigma\, T_\alpha \otimes g_\alpha$. This is our usual way of representing a singular p-chain with coefficients in G. The **minimal carrier** of such a chain is the union of the sets $T_\alpha(\Delta_p)$, where the union is taken over those α for which $g_\alpha \neq 0$. It is of course a compact set.

A continuous map $h : (X,A) \to (Y,B)$ gives rise to a chain map

$$h_\# \otimes i_G : \mathcal{S}(X,A) \otimes G \to \mathcal{S}(Y,B) \otimes G.$$

We sometimes denote this map simply by $h_\#$, and the induced homology homomorphism by h_*. Functoriality is immediate; it holds in fact on the chain level.

The short exact sequence of chain complexes

$$0 \to S_p(A) \to S_p(X) \to S_p(X,A) \to 0$$

splits because $S_p(X,A)$ is free. Therefore, one has a long exact homology sequence with coefficients in G; it is natural with respect to homomorphisms induced by continuous maps.

If the maps $h,k : (X,A) \to (Y,B)$ are homotopic, then by the proof of Theorem 30.7 there is a chain homotopy between $h_\#$ and $k_\#$. Then $h_\# \otimes i_G$ and $k_\# \otimes i_G$ are chain homotopic, so h_* and k_* are equal as maps of homology with coefficients in G.

Direct consideration of the singular chain complex of a one-point space P leads to the result that $H_i(P; G) = 0$ for $i \neq 0$ and $H_0(P; G) \simeq G$.

The "compact support property" of singular homology carries over at once to the same property for singular homology with arbitrary coefficients.

The only property of singular homology that requires some care is the excision property.

Let (X,A) be a topological pair, and let U be a subset of X such that $\bar{U} \subset \operatorname{Int} A$. We know that inclusion

$$j : (X - U, A - U) \to (X,A)$$

induces an isomorphism in ordinary homology. We wish to show it induces an isomorphism in homology with arbitrary coefficients as well.

One way of doing this is to use Theorem 46.2, which implies that $j_{\#}$ is a chain equivalence. (Note that the chain complexes involved are free and vanish below a certain dimension.) Alternatively, one can prove the following, which is an analogue of Theorem 45.5.

Theorem 51.1. *Let $\mathcal{C}$ and $\mathcal{D}$ be free chain complexes. If the chain map $\phi : \mathcal{C} \to \mathcal{D}$ induces homology isomorphisms in all dimensions, so does the chain map*

$$\phi \otimes i_G : \mathcal{C} \otimes G \to \mathcal{D} \otimes G.$$

Proof. *Step 1.* We first consider the case where we have a short exact sequence

$$0 \to \mathcal{C} \to \mathcal{D} \to \mathcal{E} \to 0$$

of free chain complexes. We know that $H_p(\mathcal{E}) = 0$ for all p, and we wish to prove that $H_p(\mathcal{E}; G) = 0$ for all p. As in the proof of Lemma 45.3, we can write $E_p = B_p \oplus U_p$, where ∂ maps B_p to zero and carries U_p isomorphically onto B_{p-1}. Then

$$E_p \otimes G \simeq (B_p \otimes G) \oplus (U_p \otimes G),$$

and $\partial \otimes i_G$ maps $B_p \otimes G$ to zero and carries $U_p \otimes G$ isomorphically onto $B_{p-1} \otimes G$. It follows that $H_p(\mathcal{E}; G) = 0$ for all p.

Step 2. The general case now follows from Lemma 45.4. Given $\phi : \mathcal{C} \to \mathcal{D}$, there is a chain complex $\mathcal{D}'$ and injective chain maps $i : \mathcal{C} \to \mathcal{D}'$ and $j : \mathcal{D} \to \mathcal{D}'$ such that j induces homology isomorphisms in all dimensions and $j \circ \phi$ is chain homotopic to i. Furthermore, $\mathcal{D}'$ and $\mathcal{D}'/\text{im } i$ and $\mathcal{D}'/\text{im } j$ are free. If ϕ_* is an isomorphism in ordinary homology, so is $i_* = j_* \circ \phi_*$. Then by Step 1, both i and j induce isomorphisms of homology with arbitrary coefficients. Hence ϕ does the same. □

Simplicial homology

Let (K,K_0) be a simplicial pair; let G be an abelian group. We define the simplicial homology of (K,K_0) with coefficients in G by the equation

$$H_p(K,K_0; G) = H_p(\mathcal{C}(K,K_0); G).$$

We define the reduced groups by the equation

$$\tilde{H}_p(K; G) = H_p(\{\mathcal{C}(K),\epsilon\}; G),$$

where ϵ is the usual augmentation for $\mathcal{C}(K)$.

Now the group $C_p(K,K_0) \otimes G$ is the direct sum of copies of G, one for each p-simplex of K not in K_0. Indeed, if we orient the p-simplices σ_α of K not in K_0 arbitrarily, then each element of $C_p(K,K_0) \otimes G$ can be represented uniquely by a finite sum $\Sigma\, \sigma_\alpha \otimes g_\alpha$. Its boundary is then represented by $\Sigma\, (\partial\sigma_\alpha) \otimes g_\alpha$.

The connection with the definition we gave in §10 of homology with arbitrary coefficients is now clear. In that section, we represented a simplicial p-chain c_p with coefficients in G by a finite formal sum $c_p = \Sigma g_\alpha \sigma_\alpha$, and its boundary was defined by the formula

$$\partial c_p = \Sigma g_\alpha (\partial \sigma_\alpha).$$

It follows that the chain complex $\mathcal{C}(K; G)$ defined in §10 is isomorphic to the chain complex $\mathcal{C}(K) \otimes G$. Henceforth, we shall use the latter chain complex in dealing with simplicial homology with arbitrary coefficients.

The existence of induced homomorphisms and the long exact homology sequence, and the verification of the Eilenberg-Steenrod axioms, is so straightforward that we omit the details. The argument follows the pattern given in §44 for simplicial cohomology.

The isomorphism between simplicial and singular theory

We have the chain map of §34,

$$\eta : C_p(K,K_0) \to S_p(|K|,|K_0|),$$

which induces an isomorphism in ordinary homology. It follows from Theorem 51.1 that it induces an isomorphism in homology with arbitrary coefficients as well. The fact that it is independent of the ordering of vertices used in defining η, and the fact that it commutes with the boundary homomorphism ∂_* and with homomorphisms induced by continuous maps, follow as in the proof of Theorem 44.2.

The homology of CW complexes

If X is a CW complex with cellular chain complex $\mathcal{D}(X)$, we know that $H_p(\mathcal{D}(X)) \simeq H_p(X)$ for all p. It follows from Theorem 45.1 that there is a chain map inducing this isomorphism; it then follows from Theorem 51.1 that

$$H_p(\mathcal{D}(X) \otimes G) \simeq H_p(X; G).$$

Thus the cellular chain complex of X can be used to compute homology with arbitrary coefficients. If X is a triangulable CW complex, then the chain map inducing this isomorphism is the inclusion map

$$j : \mathcal{D}(X) \to \mathcal{C}(X).$$

EXERCISES

1. If $\{\mathcal{C},\epsilon\}$ is an augmented chain complex, show that

$$H_0(\mathcal{C}; G) \simeq \tilde{H}_0(\mathcal{C}; G) \oplus G.$$

[*Hint:* The sequence $0 \to \ker \epsilon \to C_0 \to \mathbf{Z} \to 0$ splits.]

2. Use the cellular chain complexes to compute the homology, with general coefficients G, of $T \# T$ and $P^2 \# P^2 \# P^2$ and P^N and the k-fold dunce cap.

3. *Theorem. Let $\mathcal{C}$ be a free chain complex. Then there is a natural exact sequence*

$$0 \rightarrow H_p(\mathcal{C}) \otimes G \xrightarrow{\phi} H_p(\mathcal{C} \otimes G) \rightarrow \operatorname{cok} \phi \rightarrow 0,$$

where ϕ is induced by inclusion. The sequence splits, but not naturally. If $H_i(\mathcal{C})$ is free for all i, then ϕ is an isomorphism.

[*Note:* This lemma is the analogue for homology of Lemma 45.7 and Theorem 45.8. It will be generalized in the next chapter.]

4. Let R be a commutative ring with unity; let $\mathcal{C}$ be a chain complex. Show that $C_p \otimes R$ can be given the structure of R-module by defining

$$\alpha(c_p \otimes \beta) = c_p \otimes (\alpha\beta)$$

for $\alpha, \beta \in R$. Show that $H_p(\mathcal{C}; R)$ has the structure of R-module, and that chain maps induce R-module homomorphisms. Show that ∂_* is an R-module homomorphism.

7

Homological Algebra

We have already seen, in Chapter 5, that two spaces with isomorphic homology groups have isomorphic cohomology groups as well. This fact leads one to suspect that the cohomology groups of a space are in some way determined by the homology groups. In this chapter this suspicion is confirmed. We show precisely *how* the cohomology groups (with arbitrary coefficients) are determined by the homology groups (with integer coefficients). The theorem involved is called the Universal Coefficient Theorem for Cohomology. Its statement involves not only the "Hom" functor, which we have already studied, but also a new functor, called "Ext," which we shall introduce.

Similarly, we know that if two spaces have isomorphic integral homology groups, then the same is true for homology with arbitrary coefficients. Just as with the cohomology groups, it turns out that the homology groups with arbitrary coefficients are determined by the homology groups with integer coefficients. The theorem involved is called the Universal Coefficient Theorem for Homology. It involves not only the tensor product functor, but also a new functor, called the "torsion product," which we shall introduce.

These functors form part of a general subject called Homological Algebra. Although its origins are topological in nature, it has come to have an independent existence nowadays within the field of algebra, having applications to many problems that are purely algebraic. Our interest in it is confined to its connections with topology.

These functors also play a role when one comes to study the homology of a product space $X \times Y$. It turns out that the homology of $X \times Y$ is determined by the homology of X and of Y. The relationship is expressed in the form of an exact sequence called the Künneth sequence. It involves the tensor and torsion product functors.

There is a similar sequence for cohomology that holds if the homology is finitely generated. It has applications to cup products in general, and to computing the cohomology ring of $X \times Y$ in particular.

§52. THE EXT FUNCTOR

Associated with the functor $\mathrm{Hom}(A,B)$ is another functor of two variables, called $\mathrm{Ext}(A,B)$. Like the Hom functor, it is contravariant in the first variable and covariant in the second. This means that, given homomorphisms $\gamma : A \to A'$ and $\delta : B' \to B$, there is a homomorphism

$$\mathrm{Ext}(\gamma,\delta) : \mathrm{Ext}(A',B') \to \mathrm{Ext}(A,B),$$

and the usual functorial properties hold.

Defining this functor involves some preliminary work. But its crucial property is easy to remember; we express it in the form of a theorem.

Theorem 52.1. *There is a function that assigns, to each free resolution*

$$0 \to R \xrightarrow{\phi} F \xrightarrow{\psi} A \to 0$$

of the abelian group A, and to each abelian group B, an exact sequence

$$0 \leftarrow \mathrm{Ext}(A,B) \xleftarrow{\pi} \mathrm{Hom}(R,B) \xleftarrow{\tilde{\phi}} \mathrm{Hom}(F,B) \xleftarrow{\tilde{\psi}} \mathrm{Hom}(A,B) \leftarrow 0.$$

This function is natural, in the sense that a homomorphism

$$\begin{array}{ccccccccc} 0 & \longrightarrow & R & \longrightarrow & F & \longrightarrow & A & \longrightarrow & 0 \\ & & \downarrow{\scriptstyle\alpha} & & \downarrow{\scriptstyle\beta} & & \downarrow{\scriptstyle\gamma} & & \\ 0 & \longrightarrow & R' & \longrightarrow & F' & \longrightarrow & A' & \longrightarrow & 0 \end{array}$$

of free resolutions and a homomorphism $\delta : B' \to B$ of abelian groups gives rise to a homomorphism of exact sequences:

$$\begin{array}{ccccccccc} 0 & \longleftarrow & \mathrm{Ext}(A,B) & \longleftarrow & \mathrm{Hom}(R,B) & \longleftarrow & \mathrm{Hom}(F,B) & \longleftarrow & \mathrm{Hom}(A,B) & \longleftarrow & 0 \\ & & \uparrow{\scriptstyle \mathrm{Ext}(\gamma,\delta)} & & \uparrow{\scriptstyle \mathrm{Hom}(\alpha,\delta)} & & \uparrow{\scriptstyle \mathrm{Hom}(\beta,\delta)} & & \uparrow{\scriptstyle \mathrm{Hom}(\gamma,\delta)} & & \\ 0 & \longleftarrow & \mathrm{Ext}(A',B') & \longleftarrow & \mathrm{Hom}(R',B') & \longleftarrow & \mathrm{Hom}(F',B') & \longleftarrow & \mathrm{Hom}(A',B') & \longleftarrow & 0. \end{array}$$

We shall prove this theorem shortly. It will then be used to derive the other properties of the Ext functor, and to compute it.

Now we define the Ext functor. We begin with a lemma.

Lemma 52.2. *Suppose one is given a homomorphism*

$$\begin{array}{ccccccccc} 0 & \longrightarrow & R & \xrightarrow{\phi} & F & \longrightarrow & A & \longrightarrow & 0 \\ & & \downarrow{\scriptstyle\alpha} & & \downarrow{\scriptstyle\beta} & & \downarrow{\scriptstyle\gamma} & & \\ 0 & \longrightarrow & R' & \xrightarrow[\phi']{} & F' & \longrightarrow & A' & \longrightarrow & 0 \end{array}$$

of free resolutions of A and A', respectively, and a homomorphism $\delta : B' \to B$. Then there is a unique homomorphism ϵ making the following diagram of exact sequences commute:

$$\begin{array}{ccccccccccc} 0 & \longleftarrow & \operatorname{cok}\tilde{\phi} & \longleftarrow & \operatorname{Hom}(R,B) & \xleftarrow{\tilde{\phi}} & \operatorname{Hom}(F,B) & \longleftarrow & \operatorname{Hom}(A,B) & \longleftarrow & 0 \\ & & \uparrow{\scriptstyle\epsilon} & & \uparrow{\scriptstyle\operatorname{Hom}(\alpha,\delta)} & & \uparrow{\scriptstyle\operatorname{Hom}(\beta,\delta)} & & \uparrow{\scriptstyle\operatorname{Hom}(\gamma,\delta)} & & \\ 0 & \longleftarrow & \operatorname{cok}\tilde{\phi}' & \longleftarrow & \operatorname{Hom}(R',B') & \xleftarrow[\tilde{\phi}']{} & \operatorname{Hom}(F',B') & \longleftarrow & \operatorname{Hom}(A',B') & \longleftarrow & 0. \end{array}$$

The homomorphism ϵ is independent of the choice of α and β.

Proof. Functoriality of Hom shows that the two right squares of the preceding diagram commute. Therefore, $\operatorname{Hom}(\alpha,\delta)$ induces a homomorphism ϵ of cokernels.

We show ϵ is independent of the choice of α and β. Suppose $\{\alpha', \beta', \gamma\}$ is another homomorphism of the two given free resolutions. Consider the free resolution of A as a chain complex $\mathcal{A}$, indexed so that A is the 0-dimensional group. Do the same for A', obtaining a chain complex $\mathcal{A}'$. Then $\{\alpha, \beta, \gamma\}$ and $\{\alpha', \beta', \gamma\}$ are chain maps of $\mathcal{A}$ to $\mathcal{A}'$. By exactness, the homology groups of $\mathcal{A}$ and $\mathcal{A}'$ vanish. The cohomology groups need not vanish; indeed, the group $\operatorname{cok}\tilde{\phi}$ is just the 2-dimensional cohomology group $H^2(\mathcal{A}; B)$, and $\operatorname{cok}\tilde{\phi}' = H^2(\mathcal{A}'; B')$. The map ϵ is just the cohomology homomorphism induced by the chain map $\{\alpha, \beta, \gamma\}$.

Now the hypotheses of Lemma 46.1 are satisfied by the chain complexes $\mathcal{A}$ and $\mathcal{A}'$. Therefore there is a chain homotopy D between the chain map $\{\alpha, \beta, \gamma\}$ and the chain map $\{\alpha', \beta', \gamma\}$. Then $\operatorname{Hom}(D,\delta)$ is a cochain homotopy between the corresponding cochain maps; it follows that they induce the same homomorphism ϵ of 2-dimensional cohomology groups. □

Definition. The homomorphism ϵ constructed in the preceding lemma is said to be **induced** by γ and δ, since it depends only on the homomorphisms γ and δ and the free resolutions involved.

We show that a version of functoriality holds. First, we show that *composites* behave correctly. Let $\alpha, \beta, \gamma, \delta, \epsilon$ be as in the lemma; and suppose

$$\begin{array}{ccccccccc} 0 & \longrightarrow & R' & \longrightarrow & F' & \longrightarrow & A' & \longrightarrow & 0 \\ & & \downarrow{\scriptstyle\alpha'} & & \downarrow{\scriptstyle\beta'} & & \downarrow{\scriptstyle\gamma'} & & \\ 0 & \longrightarrow & R'' & \longrightarrow & F'' & \longrightarrow & A'' & \longrightarrow & 0 \end{array}$$

is another homomorphism of free resolutions, with $\delta' : B'' \to B'$ another homomorphism of abelian groups. Let ϵ' be the homomorphism induced by γ' and δ'. The fact that

$$\text{Hom}(\gamma,\delta) \circ \text{Hom}(\gamma',\delta') = \text{Hom}(\gamma' \circ \gamma, \delta \circ \delta')$$

and the similar equations for α, α' and β, β', imply that $\epsilon \circ \epsilon'$ is the homomorphism induced by $\gamma' \circ \gamma$ and $\delta \circ \delta'$.

Second, we show that if i_A and i_B are identity maps, then the homomorphism ϵ they induce is an *isomorphism*. Certainly this is true in the situation where the two free resolutions are the same, since α and β can then be chosen to be identity maps, so that ϵ is the identity. But it is also true in the situation

$$\begin{array}{ccccccccc} 0 & \longrightarrow & R & \xrightarrow{\phi} & F & \longrightarrow & A & \longrightarrow & 0 \\ & & \downarrow{\scriptstyle\alpha} & & \downarrow{\scriptstyle\beta} & & \downarrow{\scriptstyle i_A} & & \\ 0 & \longrightarrow & R' & \xrightarrow{\phi'} & F' & \longrightarrow & A & \longrightarrow & 0, \end{array}$$

as we now prove. Let $\epsilon : \text{cok}\,\tilde{\phi}' \to \text{cok}\,\tilde{\phi}$ be the map induced by (i_A,i_B), relative to these free resolutions. Choose α' and β' making the following diagram commute:

$$\begin{array}{ccccccccc} 0 & \longrightarrow & R' & \xrightarrow{\phi'} & F' & \longrightarrow & A & \longrightarrow & 0 \\ & & \downarrow{\scriptstyle\alpha'} & & \downarrow{\scriptstyle\beta'} & & \downarrow{\scriptstyle i_A} & & \\ 0 & \longrightarrow & R & \xrightarrow{\phi} & F & \longrightarrow & A & \longrightarrow & 0. \end{array}$$

(Here we use the fact that F' is free.) Let $\epsilon' : \text{cok}\,\tilde{\phi} \to \text{cok}\,\tilde{\phi}'$ be the map induced by (i_A,i_B) relative to these free resolutions. By the previous remarks, both composites $\epsilon \circ \epsilon'$ and $\epsilon' \circ \epsilon$ equal identity maps. Thus ϵ is an isomorphism.

It follows from these comments that, given A and B, if one chooses *any* free resolution of A, the group $\text{cok}\,\tilde{\phi}$ will be independent (up to isomorphism) of the choice. This fact leads us to the following definition of a "canonical version" of $\text{cok}\,\tilde{\phi}$, which we shall call $\text{Ext}(A,B)$.

Definition. If A is an abelian group, let $F(A)$ denote the free abelian group generated by the elements of A, and let $R(A)$ be the kernel of the natural projection $F(A) \to A$. The sequence

$$0 \to R(A) \xrightarrow{\phi} F(A) \to A \to 0$$

is called the canonical free resolution of A. (See §45.) The group

$$\text{cok}\,\tilde{\phi} = \text{Hom}(R(A),B)/\tilde{\phi}(\text{Hom}(F(A),B))$$

is denoted $\text{Ext}(A,B)$. If $\gamma : A \to A'$ and $\delta : B' \to B$ are homomorphisms, we extend γ to a homomorphism of the canonical free resolution of A with that of A', and define

$$\operatorname{Ext}(\gamma,\delta) : \operatorname{Ext}(A',B') \to \operatorname{Ext}(A,B)$$

to be the homomorphism induced by γ and δ relative to these free resolutions.

The previous remark shows that Ext is a functor of two variables, contravariant in the first variable and covariant in the second.

The group $\operatorname{Ext}(A,B)$ is sometimes called the *group of extensions of B by A*; for an explanation of this terminology, see [MacL].

Now we prove our basic theorem.

Proof of Theorem 52.1. The exact sequence

$$0 \to R \xrightarrow{\phi} F \to A \to 0$$

gives rise to the exact sequence

$$0 \leftarrow \operatorname{cok} \tilde{\phi} \leftarrow \operatorname{Hom}(R,B) \leftarrow \operatorname{Hom}(F,B) \leftarrow \operatorname{Hom}(A,B) \leftarrow 0.$$

In view of the preceding remarks, there is an isomorphism of $\operatorname{cok} \tilde{\phi}$ with $\operatorname{Ext}(A,B)$ that is induced by (i_A,i_B). We use this isomorphism to replace $\operatorname{cok} \tilde{\phi}$ by $\operatorname{Ext}(A,B)$ in this exact sequence.

It remains to check naturality. Let α, β, γ define a homomorphism of free resolutions of A and A', respectively, as in the statement of the theorem; let $\delta : B' \to B$. Consider the following diagram:

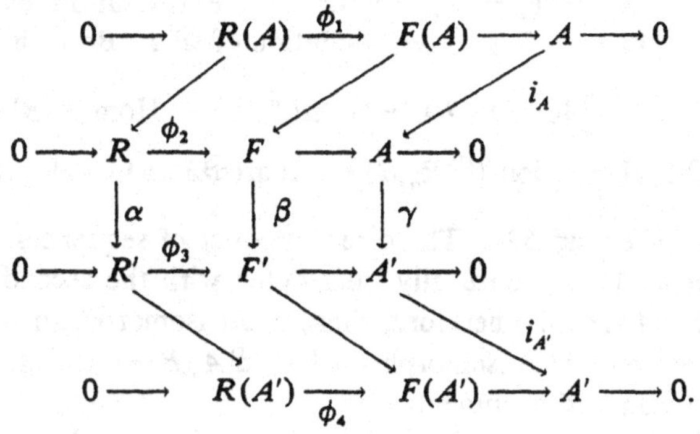

There are homomorphisms

$$\operatorname{cok} \tilde{\phi}_4 \xrightarrow{\epsilon_3} \operatorname{cok} \tilde{\phi}_3 \xrightarrow{\epsilon_2} \operatorname{cok} \tilde{\phi}_2 \xrightarrow{\epsilon_1} \operatorname{cok} \tilde{\phi}_1$$

induced by $(i_{A'},i_{B'})$ and (γ,δ) and (i_A,i_B), respectively. Both ϵ_3 and ϵ_1 are isomorphisms. The composite $\epsilon_1 \circ \epsilon_2 \circ \epsilon_3$ is, by functoriality, the unique homomorphism induced by

$$(i_{A'} \circ \gamma \circ i_A, i_B \circ \delta \circ i_{B'}) = (\gamma,\delta),$$

relative to the canonical free resolutions, so it must equal $\operatorname{Ext}(\gamma,\delta)$. Therefore, the diagram

$$\begin{array}{ccccc}
\operatorname{Ext}(A,B) & \xleftarrow[\simeq]{\epsilon_1} & \operatorname{cok}\tilde{\phi}_2 & \longleftarrow & \operatorname{Hom}(R,B) \\
{\scriptstyle \operatorname{Ext}(\gamma,\delta)}\uparrow & & \uparrow{\scriptstyle \epsilon_2} & & \uparrow{\scriptstyle \operatorname{Hom}(\alpha,\delta)} \\
\operatorname{Ext}(A',B') & \xrightarrow[\simeq]{\epsilon_3} & \operatorname{cok}\tilde{\phi}_3 & \longleftarrow & \operatorname{Hom}(R',B')
\end{array}$$

commutes, and the proof is complete. □

We now prove further properties of the Ext functor.

Theorem 52.3. (a) *There are natural isomorphisms*

$$\operatorname{Ext}(\oplus A_\alpha,B) \simeq \Pi \operatorname{Ext}(A_\alpha,B),$$
$$\operatorname{Ext}(A,\Pi B_\alpha) \simeq \Pi \operatorname{Ext}(A,B_\alpha).$$

(b) $\operatorname{Ext}(A,B) = 0$ *if A is free.*
(c) *Given B, there is an exact sequence*

$$0 \leftarrow \operatorname{Ext}(\mathbf{Z}/m,B) \leftarrow B \xleftarrow{m} B \leftarrow \operatorname{Hom}(\mathbf{Z}/m,B) \leftarrow 0.$$

Proof. In the proof, we shall use the fact that direct sums and direct products of exact sequences are exact, and the fact that a direct sum (but not a direct product) of free abelian groups is free abelian.

(a) Let $0 \to R_\alpha \to F_\alpha \to A_\alpha \to 0$ be a free resolution of A_α; then $0 \to \oplus R_\alpha \to \oplus F_\alpha \to \oplus A_\alpha \to 0$ is a free resolution of $\oplus A_\alpha$. Both the sequences

$$0 \leftarrow \operatorname{Ext}(A_\alpha,B) \leftarrow \operatorname{Hom}(R_\alpha,B) \leftarrow \operatorname{Hom}(F_\alpha,B) \leftarrow \operatorname{Hom}(A,B) \leftarrow 0,$$
$$0 \leftarrow \operatorname{Ext}(\oplus A_\alpha,B) \leftarrow \operatorname{Hom}(\oplus R_\alpha,B) \leftarrow \operatorname{Hom}(\oplus F_\alpha,B) \leftarrow \operatorname{Hom}(\oplus A_\alpha,B) \leftarrow 0,$$

are exact, by Theorem 52.1. The direct product of sequences of the first type gives a sequence that is naturally isomorphic with the second sequence in its three right-hand terms. Therefore, there is an isomorphism between the left-hand terms as well. This isomorphism $\operatorname{Ext}(\oplus A_\alpha,B) \to \Pi \operatorname{Ext}(A_\alpha,B)$ is in fact natural, by a standard argument.

Similarly, if $0 \to R \to F \to A \to 0$ is a free resolution of A, then the sequences

$$0 \leftarrow \operatorname{Ext}(A,B_\alpha) \leftarrow \operatorname{Hom}(R,B_\alpha) \leftarrow \operatorname{Hom}(F,B_\alpha) \leftarrow \operatorname{Hom}(A,B_\alpha) \leftarrow 0,$$
$$0 \leftarrow \operatorname{Ext}(A,\Pi B_\alpha) \leftarrow \operatorname{Hom}(R,\Pi B_\alpha) \leftarrow \operatorname{Hom}(F,\Pi B_\alpha) \leftarrow \operatorname{Hom}(A,\Pi B_\alpha) \leftarrow 0,$$

are exact. Since the direct product of sequences of the first type agrees with the second in the three right-hand terms (the isomorphisms in question being natural), these sequences agree in the left-hand group as well. Again, the isomorphism is natural.

To check (b), we recall that a free resolution of A splits if A is free. Then the dual sequence is exact, so $\operatorname{Ext}(A,B)$ vanishes.

To prove (c), one begins with the free resolution

$$0 \to \mathbf{Z} \xrightarrow{m} \mathbf{Z} \to \mathbf{Z}/m \to 0.$$

Applying Theorem 52.1, one obtains the exact sequence

$$0 \leftarrow \operatorname{Ext}(\mathbf{Z}/m,B) \leftarrow \operatorname{Hom}(\mathbf{Z},B) \xleftarrow{m} \operatorname{Hom}(\mathbf{Z},B) \leftarrow \operatorname{Hom}(\mathbf{Z}/m,B) \leftarrow 0,$$

from which (c) follows. □

This theorem enables us to compute $\operatorname{Ext}(A,B)$ when A is finitely generated. For Ext commutes with finite direct sums, and one has the rules

$$\operatorname{Ext}(\mathbf{Z},G) = 0, \qquad \operatorname{Ext}(\mathbf{Z}/m,G) \simeq G/mG.$$

EXERCISES

1. Show that if B is divisible, then $\operatorname{Ext}(A,B) = 0$.
2. Compute $\operatorname{Hom}(A,B)$ and $\operatorname{Ext}(A,B)$ if

$$A = \mathbf{Z} \oplus \mathbf{Z}/2 \oplus \mathbf{Z}/4 \oplus \mathbf{Z}/6, \qquad B = \mathbf{Z} \oplus \mathbf{Z} \oplus \mathbf{Z}/9 \oplus \mathbf{Z}/12.$$

3. Let S^1 denote the additive group $\mathbf{R}/\mathbf{Z}$. It is isomorphic to the multiplicative group of complex numbers of unit modulus and is often called the **circle group.**
   (a) If A is finitely generated, compute $\operatorname{Hom}(A,G)$ and $\operatorname{Ext}(A,G)$ in terms of the betti number and torsion coefficients of A, if $G = S^1$.
   (b) Repeat (a) with $G = \mathbf{Q}$.
4. If we "Hom" a short exact sequence with a group G (either on the left or the right), the resulting sequence may fail to be exact. The Ext functor measures in some sense the extent to which exactness fails. One has the following theorem:

   Theorem. There are functors assigning to each short exact sequence of abelian groups

$$0 \to A \to B \to C \to 0,$$

   and each abelian group G, two exact sequences:

$$0 \leftarrow \operatorname{Ext}(A,G) \leftarrow \operatorname{Ext}(B,G) \leftarrow \operatorname{Ext}(C,G) \leftarrow \operatorname{Hom}(A,G) \leftarrow \operatorname{Hom}(B,G) \leftarrow \operatorname{Hom}(C,G) \leftarrow 0,$$

$$0 \to \operatorname{Hom}(G,A) \to \operatorname{Hom}(G,B) \to \operatorname{Hom}(G,C) \to \operatorname{Ext}(G,A) \to \operatorname{Ext}(G,B) \to \operatorname{Ext}(G,C) \to 0.$$

   Proof. (a) To obtain the second sequence let $0 \to R \to F \to G \to 0$ be a free resolution of G, and apply the serpent lemma to the diagram

$$\begin{array}{ccccccccc}
0 & \longrightarrow & \operatorname{Hom}(F,A) & \longrightarrow & \operatorname{Hom}(F,B) & \longrightarrow & \operatorname{Hom}(F,C) & \longrightarrow & 0 \\
 & & \downarrow & & \downarrow & & \downarrow & & \\
0 & \longrightarrow & \operatorname{Hom}(R,A) & \longrightarrow & \operatorname{Hom}(R,B) & \longrightarrow & \operatorname{Hom}(R,C) & \longrightarrow & 0.
\end{array}$$

(b) Given an abelian group G, show there is an exact sequence

$$0 \to G \to H \to K \to 0$$

where H and K are divisible. Such an exact sequence is sometimes called an **injective resolution** of G. [*Hint:* It suffices to find H, since a quotient of a divisible group is divisible. Write $G = F/R$, where F is free, and construct a monomorphism of F into a direct sum of copies of $\mathbf{Q}$.]

(c) Show there is a functor assigning to each injective resolution $0 \to G \to H \to K \to 0$ of G, and each abelian group A, an exact sequence

$$0 \to \operatorname{Hom}(A,G) \to \operatorname{Hom}(A,H) \to \operatorname{Hom}(A,K) \to \operatorname{Ext}(A,G) \to 0.$$

(d) Use the serpent lemma to derive the first sequence of the theorem.

§53. THE UNIVERSAL COEFFICIENT THEOREM FOR COHOMOLOGY

We now show how the homology groups of a space determine the cohomology groups. The answer is expressed in the form of a short exact sequence involving the groups H^p and H_p and H_{p-1}, and the functors Hom and Ext.

We already know that if $\mathcal{C}$ is a free chain complex, there is a natural exact sequence

$$0 \leftarrow \operatorname{Hom}(H_p(\mathcal{C}),G) \xleftarrow{\kappa} H^p(\mathcal{C};G) \leftarrow \ker \kappa \leftarrow 0.$$

(See Lemma 45.7.) We now identify the group $\ker \kappa$, and show it depends only on the groups $H_{p-1}(\mathcal{C})$ and G.

Theorem 53.1 (The universal coefficient theorem for cohomology). *Let $\mathcal{C}$ be a free chain complex; let G be an abelian group. There is an exact sequence*

$$0 \leftarrow \operatorname{Hom}(H_p(\mathcal{C}),G) \xleftarrow{\kappa} H^p(\mathcal{C};G) \leftarrow \operatorname{Ext}(H_{p-1}(\mathcal{C}),G) \leftarrow 0$$

that is natural with respect to homomorphisms induced by chain maps. It splits, but not naturally.

Proof. Step 1. Let C_p, Z_p, and B_p denote the groups of p-chains, p-cycles, and p-boundaries of $\mathcal{C}$, respectively. Consider the exact sequence

$$0 \to Z_p \xrightarrow{i} C_p \xrightarrow{\partial_0} B_{p-1} \to 0.$$

It splits because B_{p-1} is free. Define a chain complex $\mathcal{Z}$ by letting its p-dimensional group be Z_p and by letting its boundary operator be the restriction of ∂. Then all boundary operators in $\mathcal{Z}$ vanish. Similarly, define a chain complex $\mathcal{D}$ by letting $D_p = B_{p-1}$ and by letting its boundary operator be the restriction of ∂; all boundary operators in $\mathcal{D}$ vanish.

We then have a (split) short exact sequence of chain complexes

$$0 \to Z_p \xrightarrow{i} C_p \xrightarrow{\partial_0} D_p \to 0.$$

The map i is a chain map by definition, and the map ∂_0 is a chain map trivially (since $\partial \circ \partial = 0$).

Applying the Hom functor, we obtain a short exact sequence of cochain complexes and hence a long exact cohomology sequence, of which we consider five terms:

$$(*) \quad H^{p+1}(\mathcal{D}; G) \xleftarrow{\beta} H^p(\mathcal{Z}; G) \xleftarrow{i^*} H^p(\mathcal{C}; G) \xleftarrow{\partial_0^*} H^p(\mathcal{D}; G) \xleftarrow{\beta} H^{p-1}(\mathcal{Z}; G).$$

To avoid confusion, we use β instead of δ^* to denote the zig-zag homomorphism.

Step 2. We now identify the terms of this sequence and the homomorphism β. Since the chain complex $\mathcal{D}$ has trivial boundary operators, so does the corresponding cochain complex. Therefore, the group $H^{p+1}(\mathcal{D}; G)$ equals the cochain group $\operatorname{Hom}(D_{p+1},G) = \operatorname{Hom}(B_p,G)$. For similar reasons, $H^p(\mathcal{Z}; G) = \operatorname{Hom}(Z_p,G)$.

Thus β is a map

$$\operatorname{Hom}(B_p,G) \xleftarrow{\beta} \operatorname{Hom}(Z_p,G).$$

We show that it is just the dual $\tilde{j}_p$ of the inclusion map $j_p : B_p \to Z_p$.

One obtains the zig-zag homomorphism β by following through the diagram

$$\begin{array}{ccccc} & & \operatorname{Hom}(C_{p+1},G) & \xleftarrow{\tilde{\partial}_0} & \operatorname{Hom}(B_p,G) \\ & & \uparrow \delta & & \\ \operatorname{Hom}(Z_p,G) & \xleftarrow{\tilde{i}} & \operatorname{Hom}(C_p,G) & & \end{array}$$

from left to right, as follows: Let f be an element of $\operatorname{Hom}(Z_p,G)$. Note that $\delta f = 0$, because all coboundary operations in $\mathcal{Z}$ vanish. Pull f back via $\tilde{i}$ to an element g of $\operatorname{Hom}(C_p,G)$. (That is, extend f to a homomorphism $g : C_p \to G$.) Form δg, and pull it back via $\tilde{\partial}_0$ to an element of $\operatorname{Hom}(B_p,G)$. We show that $\tilde{j}_p(f)$ is such a pull-back of δg; that is, we show that $\tilde{\partial}_0 \tilde{j}_p(f) = \delta g$. Then our result is proved.

We compute

$$\langle \delta g, c_{p+1} \rangle = \langle g, \partial c_{p+1} \rangle = \langle f, \partial c_{p+1} \rangle,$$

the last equation holding because ∂c_{p+1} is in Z_p. Similarly,

$$\langle \tilde{\partial}_0 \tilde{j}_p(f), c_{p+1} \rangle = \langle f, j_p(\partial_0 c_{p+1}) \rangle = \langle f, \partial c_{p+1} \rangle.$$

Step 3. We now prove the theorem. The five-term exact sequence (*) gives rise to the short exact sequence

$$(**) \qquad 0 \leftarrow \ker \tilde{j}_p \leftarrow H^p(\mathcal{C};G) \leftarrow \operatorname{cok} \tilde{j}_{p-1} \leftarrow 0.$$

We need only to identify the kernel and the cokernel of the map

$$\mathrm{Hom}(B_p,G) \xleftarrow{\tilde{j}_p} \mathrm{Hom}(Z_p,G).$$

Now the sequence

$$0 \to B_p \xrightarrow{j_p} Z_p \xrightarrow{\pi} H_p(\mathcal{C}) \to 0$$

is a free resolution of $H_p(\mathcal{C})$. Therefore, the sequence

$$(***) \quad 0 \leftarrow \mathrm{Ext}(H_p(\mathcal{C}),G) \leftarrow \mathrm{Hom}(B_p,G) \xleftarrow{\tilde{j}_p} \mathrm{Hom}(Z_p,G) \xleftarrow{\tilde{\pi}} \mathrm{Hom}(H_p(\mathcal{C}),G) \leftarrow 0$$

is exact. The kernel and cokernel of $\tilde{j}_p$ are now apparent. The existence of the exact sequence of our theorem is thus established.

Step 4. We check naturality of the sequence. Let $\phi : \mathcal{C} \to \mathcal{C}'$ be a chain map; let Z'_p, B'_p, and C'_p be the cycles, boundaries, and chains of $\mathcal{C}'$, respectively. The chain maps

$$\begin{array}{ccccccccc} 0 & \longrightarrow & Z_p & \xrightarrow{i} & C_p & \xrightarrow{\partial_0} & B_{p-1} & \longrightarrow & 0 \\ & & \downarrow{\scriptstyle\phi} & & \downarrow{\scriptstyle\phi} & & \downarrow{\scriptstyle\phi} & & \\ 0 & \longrightarrow & Z'_p & \xrightarrow{i'} & C'_p & \xrightarrow{\partial'_0} & B'_{p-1} & \longrightarrow & 0 \end{array}$$

define a homomorphism of short exact sequences, for ϕ commutes with i because i is inclusion, and ϕ commutes with ∂_0 because ϕ is a chain map. It follows that ϕ induces a homomorphism to the five-term exact sequence (*) from the corresponding sequence for $\mathcal{C}'$, and hence induces a homomorphism to the short exact sequence (**) from the corresponding sequence for $\mathcal{C}'$. It remains only to comment that the exact sequence (***) is natural with respect to homomorphisms induced by chain maps, so the isomorphisms of ker $\tilde{j}_p$ and cok $\tilde{j}_{p-1}$ with the appropriate Hom and Ext groups are natural as well.

Step 5. To complete the proof, we show that the homomorphism

$$\mathrm{Hom}(H_p(\mathcal{C}),G) \leftarrow H^p(\mathcal{C};G)$$

of our exact sequence equals the Kronecker map κ. Then the splitting of the sequence is a consequence of Lemma 45.7.

The homomorphism in our exact sequence equals the composite $(\tilde{\pi})^{-1} \circ i^*$ of the following diagram:

$$\begin{array}{ccc} \mathrm{Hom}(Z_p,G) \supset \ker \tilde{j}_p & \xleftarrow{i^*} & H^p(\mathcal{C};G) \\ \tilde{\pi}\uparrow\cong & \swarrow{\scriptstyle\kappa} & \\ \mathrm{Hom}(H_p(\mathcal{C}),G) & & \end{array}$$

where $\tilde{j}_p$ is the dual of the inclusion map $j_p : B_p \to Z_p$, where i^* is induced by inclusion $i : Z_p \to C_p$, and where $\tilde{\pi}$ is the dual of projection $\pi : Z_p \to H_p(\mathcal{C})$. We need to show that the diagram commutes.

Let $\{z^p\}$ be an element of $H^p(\mathcal{C}; G)$; we show that

$$\tilde{\pi}\kappa(\{z^p\}) = i^*(\{z^p\}) = \tilde{i}(z^p),$$

and the proof is complete. Let $z_p \in Z_p$; we compute

$$[\tilde{\pi}\kappa(\{z^p\})](z_p) = [\kappa(\{z^p\})](\pi(z_p)) = \langle\{z^p\},\{z_p\}\rangle$$
$$= \langle z^p,z_p\rangle = \langle z^p,i(z_p)\rangle = \langle\tilde{i}(z^p),z_p\rangle. \quad \square$$

Corollary 53.2. *Let (X,A) be a topological pair. There is an exact sequence*

$$0 \leftarrow \operatorname{Hom}(H_p(X,A),G) \leftarrow H^p(X,A;G) \leftarrow \operatorname{Ext}(H_{p-1}(X,A),G) \leftarrow 0,$$

which is natural with respect to homomorphisms induced by continuous maps. It splits, but not naturally. $\square$

Corollary 53.3. *Let $\mathcal{C}$ and $\mathcal{D}$ be free chain complexes; let $\phi : \mathcal{C} \to \mathcal{D}$ be a chain map. If $\phi_* : H_i(\mathcal{C}) \to H_i(\mathcal{D})$ is an isomorphism for $i = p$ and $i = p - 1$, then*

$$H^p(\mathcal{C}; G) \xleftarrow{\phi^*} H^p(\mathcal{D}; G)$$

is an isomorphism.

Proof. Apply naturality of the universal coefficient sequence and the Five-lemma. $\square$

This corollary provides an alternate proof of Theorem 45.5. It in fact proves something more than stated in that theorem, since we do not need to assume that ϕ_* is an isomorphism in *all* dimensions.

The universal coefficient theorem with field coefficients

There is a second version of the universal coefficient theorem that is often useful. It concerns the case where the coefficient group is a field F. It relates cohomology with homology if *both* are taken with coefficients in F.

We have already remarked that if $\mathcal{C}$ is a chain complex and F is a field, then both

$$\operatorname{Hom}(C_p,F) \quad \text{and} \quad C_p \otimes F$$

have, in a natural way, the structure of vector space over F. Given $\alpha \in F$, and $c^p \in \operatorname{Hom}(C_p,F)$, and $c_p \in C_p$, one defines

$$\langle \alpha c^p, c_p\rangle = \alpha \cdot \langle c^p,c_p\rangle,$$
$$\alpha(c_p \otimes \beta) = c_p \otimes (\alpha\beta).$$

(See the exercises of §48 and §51.) Both δ and ∂ are vector space homomorphisms (linear transformations), so that

$$H^p(\mathcal{C}; F) \qquad \text{and} \qquad H_p(\mathcal{C}; F)$$

also have the structure of vector space over F. We show that they are in fact dual as vector spaces.

We recall that if A and B are vector spaces over F, then $\operatorname{Hom}_F(A,B)$ is also a vector space over F; it is the set of linear transformations of A into B. If $f: A \to A'$ and $g: B' \to B$ are linear transformations, then $\operatorname{Hom}(f,g)$ is also a linear transformation. (See the exercises of §41.)

We wish to reprove the universal coefficient theorem for cohomology in this context. One difficulty is that we have several possible contenders for the title of the space of cochains—namely,

$$\operatorname{Hom}(C_p,F), \qquad \operatorname{Hom}(C_p \otimes F, F), \qquad \operatorname{Hom}_F(C_p \otimes F, F).$$

By definition, the first of these gives the cohomology $H^p(\mathcal{C}; F)$ of $\mathcal{C}$ with coefficients in F. We shall show that the third does, as well.

Lemma 53.4. *Let $\mathcal{C}$ be a chain complex. Let*

$$\omega: \operatorname{Hom}(C_p,F) \to \operatorname{Hom}_F(C_p \otimes F, F)$$

be defined by the equation

$$\langle \omega(f), c_p \otimes \alpha \rangle = \langle f, c_p \rangle \cdot \alpha,$$

where $f \in \operatorname{Hom}(C_p,F)$, and $c_p \in C_p$, and $\alpha \in F$. Then ω is a vector space isomorphism that commutes with δ.

Proof. Strictly speaking, we use the preceding formula to define $\omega(f)$ as a function on the cartesian product $C_p \times F$, and note that it is bilinear. To check that $\omega(f)$ is a linear transformation, we compute

$$\begin{aligned}\langle \omega(f), \alpha(c_p \otimes \beta)\rangle &= \langle \omega(f), c_p \otimes \alpha\beta\rangle = \langle f, c_p\rangle \cdot (\alpha\beta) \\ &= \alpha \cdot (\langle f, c_p\rangle \cdot \beta) = \alpha \cdot \langle \omega(f), c_p \otimes \beta\rangle.\end{aligned}$$

The map ω is injective. Suppose $\omega(f)$ is the zero linear transformation. Then in particular

$$\langle \omega(f), c_p \otimes 1\rangle = 0 = \langle f, c_p\rangle \cdot 1$$

for all $c_p \in C_p$. This implies that f is the zero homomorphism.

The map ω is surjective. Let $\phi: C_p \otimes F \to F$ be a linear transformation. Let us define $f: C_p \to F$ by the equation $f(c_p) = \phi(c_p \otimes 1)$. It follows that f is a homomorphism of abelian groups, because $f(0) = 0$ and

$$\begin{aligned}f(c_p + d_p) &= \phi((c_p + d_p) \otimes 1) \\ &= \phi(c_p \otimes 1 + d_p \otimes 1) = f(c_p) + f(d_p).\end{aligned}$$

Furthermore, $\omega(f) = \phi$, since

$$\langle \omega(f), c_p \otimes \alpha \rangle = \langle f, c_p \rangle \cdot \alpha = \alpha \cdot \phi(c_p \otimes 1),$$

$$\phi(c_p \otimes \alpha) = \phi(\alpha \cdot (c_p \otimes 1)) = \alpha \cdot \phi(c_p \otimes 1).$$

This last equation holds *because* ϕ is a linear transformation, not just a homomorphism of abelian groups.

The map ω *commutes with* δ. We compute

$$\begin{aligned}\langle \delta\omega(f), c_{p+1} \otimes \alpha \rangle &= \langle \omega(f), (\partial \otimes i_F)(c_{p+1} \otimes \alpha) \rangle \\ &= \langle f, \partial c_{p+1} \rangle \cdot \alpha = \langle \delta f, c_{p+1} \rangle \cdot \alpha \\ &= \langle \omega(\delta f), c_{p+1} \otimes \alpha \rangle. \quad \square\end{aligned}$$

Theorem 53.5. *Let $\mathcal{C}$ be a free chain complex; let F be a field. Then there is a natural vector space isomorphism*

$$\operatorname{Hom}_F(H_p(\mathcal{C}; F), F) \leftarrow H^p(\mathcal{C}; F).$$

Proof. We imitate the proof of the universal coefficient theorem. First, we note that if

$$0 \to A \to B \to C \to 0$$

is a short exact sequence of vector spaces over F and linear transformations, then for any vector space V over F, the dual sequence

$$0 \leftarrow \operatorname{Hom}_F(A,V) \leftarrow \operatorname{Hom}_F(B,V) \leftarrow \operatorname{Hom}_F(C,V) \leftarrow 0$$

is exact. The proof is easy, for since any vector space has a basis, the first sequence splits.

Let $\mathcal{E}$ denote the chain complex $\mathcal{C} \otimes F$. Then $E_p = C_p \otimes F$ is a vector space over F. Let B_p and Z_p denote boundaries and cycles, respectively, in the chain complex $\mathcal{E}$; they are also vector spaces over F. Consider the short exact sequence of vector spaces

$$0 \to Z_p \to E_p \to B_{p-1} \to 0.$$

It gives rise to the dual sequence

$$0 \leftarrow \operatorname{Hom}_F(Z_p,F) \leftarrow \operatorname{Hom}_F(E_p,F) \leftarrow \operatorname{Hom}_F(B_{p-1},F) \leftarrow 0.$$

We apply the zig-zag lemma, as before. The cochain complex in the middle is isomorphic to the cochain complex $\operatorname{Hom}(C_p,F)$, by the preceding lemma. Therefore, by the same argument as used before, we obtain an exact sequence

$$0 \leftarrow \ker \tilde{j}_p \leftarrow H^p(\mathcal{C}; F) \leftarrow \operatorname{cok} \tilde{j}_{p-1} \leftarrow 0,$$

where $j_p : B_p \to Z_p$ is inclusion.

Now the proof takes a different tack. Consider the exact sequence

$$0 \to B_p \xrightarrow{j_p} Z_p \to H_p(\mathcal{E}) \to 0.$$

Because it is a sequence of vector spaces and linear transformations, the dual sequence is exact:

$$0 \leftarrow \mathrm{Hom}_F(B_p,F) \xleftarrow{\tilde{j}_p} \mathrm{Hom}_F(Z_p,F) \leftarrow \mathrm{Hom}_F(H_p(\mathscr{E}),F) \leftarrow 0.$$

Therefore cok $\tilde{j}_p = 0$ and

$$\ker \tilde{j}_p \simeq \mathrm{Hom}_F(H_p(\mathscr{C}; F), F).$$

The theorem is now proved. □

Corollary 53.6. *If (X,A) is a topological pair, there is a natural vector space isomorphism*

$$\mathrm{Hom}_F(H_p(X,A; F), F) \leftarrow H^p(X,A; F). \quad \square$$

This theorem shows that if F is a field, then the vector space $H^p(X,A; F)$ can be identified in a natural way with the *dual* vector space of the vector space $H_p(X,A; F)$. In the case where the dimension of $H_p(X,A; F)$ is finite, this means that $H^p(X,A; F)$ and $H_p(X,A; F)$ are isomorphic as vector spaces (though not naturally).

In differential geometry, it is common to deal with compact manifolds and to use the field of reals as coefficients. Because the cohomology and homology vector spaces are dual in this case, differential geometers sometimes treat homology and cohomology as if they were the same object. Needless to say, this can lead to confusion.

EXERCISES

1. Let $\mathscr{C}$ be a free chain complex. Show that if $H_{p-1}(\mathscr{C})$ is free or if G is divisible, then the Kronecker map κ is an isomorphism.
2. Use the universal coefficient theorem to compute the cohomology with general coefficients G of $T \# T$ and $P^2 \# P^2 \# P^2$ and P^N and the k-fold dunce cap.
3. Assume $H_i(X)$ is finitely generated for all i.
   (a) Compute $H^p(X; G)$ in terms of the betti numbers and torsion coefficients of X. Compare with Exercise 5 of §42.
   (b) Repeat (a) when G is the circle group $S^1 = \mathbf{R}/\mathbf{Z}$.
   (c) Repeat (a) when $G = \mathbf{Q}$.
4. A homomorphism $\alpha : G \to G'$ gives rise to homomorphisms

$$\alpha_* : H_i(\mathscr{C}; G) \to H_i(\mathscr{C}; G'),$$
$$\alpha^* : H^i(\mathscr{C}; G) \to H^i(\mathscr{C}; G').$$

   They are called **coefficient homomorphisms.** Show the universal coefficient sequence for cohomology is natural with respect to coefficient homomorphisms.
5. Let (K,K_0) be a relative pseudo n-manifold. (See the exercises of §43.) Suppose K is finite. Show that the torsion subgroup of $H_{n-1}(K,K_0)$ vanishes if (K,K_0) is orientable, and has order 2 otherwise.

§54. TORSION PRODUCTS

Associated with the functor Hom and derived from it is another functor, called Ext. Both were involved in the statement of the Universal Coefficient Theorem for Cohomology. Similarly, associated with the tensor product functor and derived from it is a second functor, which we call the torsion product. Both of *these* functors will be involved in the statement of the Universal Coefficient Theorem for Homology. Construction of the torsion product is so similar to construction of the Ext functor that we abbreviate some of the details.

The torsion product is a functor that assigns to an ordered pair A, B of abelian groups, an abelian group $A * B$, and to an ordered pair of homomorphisms $\gamma : A \to A'$ and $\delta : B \to B'$, a homomorphism

$$\gamma * \delta : A * B \to A' * B'.$$

Like the tensor product, it is covariant in both variables.

The crucial property of the torsion product is expressed in the following theorem, whose proof we shall give later:

Theorem 54.1. *There is a function that assigns to each free resolution*

$$0 \to R \to F \to A \to 0$$

of the abelian group A, and to each abelian group B, an exact sequence

$$0 \to A * B \to R \otimes B \to F \otimes B \to A \otimes B \to 0.$$

This function is natural, in the sense that a homomorphism of a free resolution of A to a free resolution of A', and a homomorphism of B to B', induce a homomorphism of the corresponding exact sequences.

First, we prove a lemma.

Lemma 54.2. *Given a homomorphism of free resolutions*

$$\begin{array}{ccccccccc} 0 & \longrightarrow & R & \xrightarrow{\phi} & F & \xrightarrow{\psi} & A & \longrightarrow & 0 \\ & & \downarrow{\scriptstyle\alpha} & & \downarrow{\scriptstyle\beta} & & \downarrow{\scriptstyle\gamma} & & \\ 0 & \longrightarrow & R' & \xrightarrow{\phi'} & F' & \xrightarrow{\psi'} & A' & \longrightarrow & 0 \end{array}$$

and a homomorphism $\delta : B \to B'$, there exists a unique homomorphism ϵ making the following diagram commute:

$$\begin{array}{ccccccccccc} 0 & \longrightarrow & \ker(\phi \otimes i_B) & \longrightarrow & R \otimes B & \xrightarrow{\phi \otimes i_B} & F \otimes B & \xrightarrow{\psi \otimes i_B} & A \otimes B & \longrightarrow & 0 \\ & & \downarrow{\scriptstyle\epsilon} & & \downarrow{\scriptstyle\alpha \otimes \delta} & & \downarrow{\scriptstyle\beta \otimes \delta} & & \downarrow{\scriptstyle\gamma \otimes \delta} & & \\ 0 & \longrightarrow & \ker(\phi' \otimes i_{B'}) & \longrightarrow & R' \otimes B' & \xrightarrow{\phi' \otimes i_{B'}} & F' \otimes B' & \xrightarrow{\psi' \otimes i_{B'}} & A' \otimes B' & \longrightarrow & 0. \end{array}$$

The homomorphism ϵ is independent of the choice of α and β.

Proof. Functoriality of $\otimes$ shows the two right-hand squares of the preceding diagram commute. Therefore, $\alpha \otimes \delta$ induces a homomorphism ϵ of kernels.

The proof that ϵ' is independent of the choice of α and β proceeds as in the proof of Lemma 52.2. Let $\{\alpha', \beta', \gamma\}$ be another choice. We treat $\{\alpha, \beta, \gamma\}$ and $\{\alpha', \beta', \gamma\}$ as chain maps of one chain complex $\mathcal{A}$ to another $\mathcal{A}'$. There is a chain homotopy D between them; then $D \otimes \delta$ is a chain homotopy between the corresponding chain maps of $\mathcal{A} \otimes B$ to $\mathcal{A}' \otimes B'$. Thus they induce the same homomorphism

$$\begin{array}{ccc} \epsilon : H_2(\mathcal{A}; B) & \longrightarrow & H_2(\mathcal{A}'; B') \\ \| & & \| \\ \ker(\phi \otimes i_B) & & \ker(\phi' \otimes i_{B'}). \quad \square \end{array}$$

The homomorphism ϵ is said to be **induced** by γ and δ, relative to the free resolutions involved.

Just as in §52, we see that ϵ depends functorially on γ and δ. That is, the composite of the homomorphisms induced by (γ,δ) and (γ',δ') is the homomorphism induced by $(\gamma \circ \gamma', \delta \circ \delta')$. And the homomorphism induced by (i_A, i_B) is an isomorphism.

Definition. Given A, let

$$0 \rightarrow R(A) \xrightarrow{\phi} F(A) \rightarrow A \rightarrow 0$$

be the canonical free resolution of A. The group $\ker(\phi \otimes i_B)$ is denoted $A * B$, and called the **torsion product** of A and B. If $\gamma : A \rightarrow A'$ and $\delta : B \rightarrow B'$ are homomorphisms, we extend γ to a homomorphism of canonical free resolutions, and define

$$\gamma * \delta : A * B \rightarrow A' * B'$$

to be the homomorphism induced by γ and δ relative to these free resolutions.

The preceding remarks show that torsion product is a functor of two variables, covariant in both.

The proof of Theorem 54.1 is now straightforward. Given any free resolution

$$0 \rightarrow R \xrightarrow{\phi} F \rightarrow A \rightarrow 0,$$

the preceding remarks show that (i_A, i_B) induces an isomorphism of $\ker(\phi \otimes i_B)$ with $A * B$. Thus the exact sequence of the theorem exists. Naturality is proved as in the proof of Theorem 52.1.

One property possessed by the torsion product that is not possessed by Ext is commutativity. We prove it now. First, we need a lemma.

Lemma 54.3. *There is a function assigning to each short exact sequence of abelian groups*

$$0 \rightarrow A \rightarrow B \rightarrow C \rightarrow 0$$

and each abelian group D, an exact sequence

$$0 \to D * A \to D * B \to D * C \to D \otimes A \to D \otimes B \to D \otimes C \to 0.$$

This function is natural with respect to homomorphisms of short exact sequences and abelian groups.

Proof. This result is analogous to the theorem stated in Exercise 4 of §52. Let

$$0 \to R \xrightarrow{\phi} F \to D \to 0$$

be a free resolution of D. Because R and F are free, we have horizontal exactness in the diagram

$$\begin{array}{ccccccccc} 0 & \longrightarrow & R \otimes A & \longrightarrow & R \otimes B & \longrightarrow & R \otimes C & \longrightarrow & 0 \\ & & \downarrow \phi \otimes i_A & & \downarrow \phi \otimes i_B & & \downarrow \phi \otimes i_C & & \\ 0 & \longrightarrow & F \otimes A & \longrightarrow & F \otimes B & \longrightarrow & F \otimes C & \longrightarrow & 0. \end{array}$$

(See Corollary 50.7.) Treating this diagram as a short exact sequence of chain complexes, we apply the zig-zag lemma to obtain an exact sequence in homology, which has the form

$$0 \to \ker(\phi \otimes i_A) \to \ker(\phi \otimes i_B) \to \ker(\phi \otimes i_C) \to$$
$$\operatorname{cok}(\phi \otimes i_A) \to \operatorname{cok}(\phi \otimes i_B) \to \operatorname{cok}(\phi \otimes i_C) \to 0.$$

(This result is called the "serpent lemma." See Exercise 2 of §24.) Theorem 54.1 enables us to identify these terms;

$$\ker(\phi \otimes i_A) \simeq D * A \quad \text{and} \quad \operatorname{cok}(\phi \otimes i_A) \simeq D \otimes A.$$

Similar results hold for B and C.

The naturality of the zig-zag lemma and of the sequence in Theorem 54.1 gives us naturality of this exact sequence. □

Theorem 54.4. (a) *There is a natural isomorphism*

$$A * B \simeq B * A.$$

(b) *There are natural isomorphisms*

$$(\oplus A_\alpha) * B \simeq \oplus (A_\alpha * B),$$
$$A * (\oplus B_\alpha) \simeq \oplus (A * B_\alpha).$$

(c) *$A * B = 0$ if A or B is torsion-free.*
(d) *Given B, there is an exact sequence*

$$0 \to (Z/m) * B \to B \xrightarrow{m} B \to (Z/m) \otimes B \to 0.$$

Proof. Note first that if B is torsion-free, then when one tensors a free resolution of A with B, exactness is preserved, by Corollary 50.7. It follows that $A * B = 0$.

(a) Apply the preceding lemma to the free resolution $0 \to R \to F \to A \to 0$ of A. One obtains a six-term exact sequence. The first terms

$$0 \to B * R \to B * F$$

vanish because R and F are torsion-free. What remains is the exact sequence

$$0 \to B * A \to B \otimes R \to B \otimes F \to B \otimes A \to 0.$$

In its last three terms, this sequence is naturally isomorphic to the sequence of Theorem 54.1,

$$0 \to A * B \to R \otimes B \to F \otimes B \to A \otimes B \to 0.$$

Therefore, the first terms are isomorphic as well. Naturality is straightforward.

(b) Let

$$0 \to R_\alpha \to F_\alpha \to A_\alpha \to 0$$

be a free resolution of A_α. Then

$$0 \to \oplus R_\alpha \to \oplus F_\alpha \to \oplus A_\alpha \to 0$$

is a free resolution of $\oplus A_\alpha$. We tensor the first sequence with B and sum; and we tensor the second sequence with B. We obtain the two sequences

$$0 \to \oplus(A_\alpha * B) \to \oplus(R_\alpha \otimes B) \to \oplus(F_\alpha \otimes B) \to \oplus(A_\alpha \otimes B) \to 0,$$
$$0 \to (\oplus A_\alpha) * B \to (\oplus R_\alpha) \otimes B \to (\oplus F_\alpha) \otimes B \to (\oplus A_\alpha) \otimes B \to 0.$$

Since the last three terms of these two sequences are naturally isomorphic, the first terms are isomorphic as well. Naturality is straightforward. This proves the first isomorphism of (b). The second follows from commutativity.

(c) This follows from the remark made at the beginning of the proof and commutativity.

(d) We begin with the free resolution

$$0 \to \mathbf{Z} \xrightarrow{m} \mathbf{Z} \to \mathbf{Z}/m \to 0.$$

Tensoring with B, we obtain the sequence

$$0 \to \mathbf{Z}/m * B \to \mathbf{Z} \otimes B \xrightarrow{m \otimes i_B} \mathbf{Z} \otimes B \to \mathbf{Z}/m \otimes B \to 0.$$

The sequence in (d) follows. □

Remark. We have studied the Ext and torsion product functors only for abelian groups; they are "derived" from the functors Hom and ⊗, respectively. If one deals instead with modules over a ring R, one has the functors Hom_R and $\otimes_R$, as already indicated. A question arises: Can one also introduce Ext and torsion product for modules? The answers to this question constitute, basically, the subject matter of homological algebra. (See [MacL].)

One begins, much as in the case of abelian groups, with the notion of a free R-module. Then, given an R-module A, one lets F_0 denote the free module generated by the elements of A. There is a natural epimorphism $\phi : F_0 \to A$. However, in the case of a general ring R, a submodule of a free module is not necessarily free. Thus there need be no short exact sequence $0 \to R \to F \to A \to 0$, where R and F are free. However, one *can* let F_1 denote the free R-module generated by the elements of $\ker \phi$ and obtain an exact sequence $F_1 \to F_0 \to A \to 0$, where F_1 and F_0 are free. Continuing similarly, one obtains an exact sequence

$$\cdots \to F_k \xrightarrow{\phi_k} F_{k-1} \to \cdots \xrightarrow{\phi_1} F_0 \to A \to 0,$$

where each F_i is free. Such a sequence is called a **free resolution** of the R-module A. Applying the functor Hom_R, one obtains a sequence

$$\cdots \xleftarrow{\tilde{\phi}_2} \mathrm{Hom}_R(F_1,G) \xleftarrow{\tilde{\phi}_1} \mathrm{Hom}_R(F_0,G) \leftarrow \mathrm{Hom}_R(A,G) \leftarrow 0.$$

Exactness holds at the two right-hand terms. For $n \geq 1$, we define

$$\mathrm{Ext}_R^n(A,G) = \ker \tilde{\phi}_{n+1}/\mathrm{im}\, \tilde{\phi}_n.$$

Thus we obtain an entire sequence of Ext groups! Of course, if $R = \mathbf{Z}$, then Ext_R^n vanishes for $n > 1$ and equals Ext for $n = 1$.

A similar construction, using $\otimes_R$, gives one a sequence of groups $\mathrm{Tor}_n^R(A,G)$. Many of the theorems we have proved about Ext and torsion product generalize to statements about these new groups. The applications, both to topology and algebra, are numerous and fruitful.

The theorems of this section enable us to compute tensor and torsion products when the groups are finitely generated. We summarize the rules for these products, as well as those for Hom and Ext, as follows:

$\mathbf{Z} \otimes G \simeq G$	$\mathbf{Z} * G = 0$
$\mathrm{Hom}(\mathbf{Z},G) \simeq G$	$\mathrm{Ext}(\mathbf{Z},G) = 0$
$\mathbf{Z}/m \otimes G \simeq G/mG$	$\mathbf{Z}/m * G \simeq \ker(G \xrightarrow{m} G)$
$\mathrm{Hom}(\mathbf{Z}/m,G) \simeq \ker(G \xrightarrow{m} G)$	$\mathrm{Ext}(\mathbf{Z}/m,G) \simeq G/mG$

In particular, these rules imply the following, where $d = \gcd(m,n)$:

$\mathbf{Z}/m \otimes \mathbf{Z} \simeq \mathbf{Z}/m$	$\mathbf{Z}/m * \mathbf{Z} = 0$
$\mathrm{Hom}(\mathbf{Z}/m,\mathbf{Z}) = 0$	$\mathrm{Ext}(\mathbf{Z}/m,\mathbf{Z}) \simeq \mathbf{Z}/m$

$\mathbf{Z}/m \otimes \mathbf{Z}/n \simeq \mathbf{Z}/m * \mathbf{Z}/n \simeq \mathrm{Hom}(\mathbf{Z}/m,\mathbf{Z}/n) \simeq \mathrm{Ext}(\mathbf{Z}/m,\mathbf{Z}/n) \simeq \mathbf{Z}/d$

EXERCISES

1. Compute $A \otimes B$ and $A * B$ if

$$A = \mathbf{Z} \oplus \mathbf{Z}/2 \oplus \mathbf{Z}/4 \oplus \mathbf{Z}/6, \qquad B = \mathbf{Z} \oplus \mathbf{Z} \oplus \mathbf{Z}/9 \oplus \mathbf{Z}/12.$$

2. Let A be finitely generated.
   (a) If S^1 is the circle group, compute $A \otimes S^1$ and $A * S^1$ in terms of the betti numbers and torsion coefficients of A.
   (b) Repeat (a) with S^1 replaced by Q.
3. Suppose $\phi : A \rightarrow B$ is a monomorphism. Show that

$$A * C \xrightarrow{\phi * i_C} B * C$$

is a monomorphism. Conclude that if $A' \subset A$ and $B' \subset B$, then $A' * B'$ can be naturally considered as a subgroup of $A * B$.

4. Let T_A and T_B be the torsion subgroups of A and B, respectively. Use Lemma 54.3 to show that inclusion induces an isomorphism

$$A * T_B \rightarrow A * B.$$

Conclude that $A * B \simeq T_A * T_B$.

5. Show that in general, $A * B$ is a torsion group. [*Hint:* Consider the sequence

$$0 \rightarrow A * T_B \rightarrow R \otimes T_B \rightarrow F \otimes T_B \rightarrow A \otimes T_B \rightarrow 0.]$$

§55. THE UNIVERSAL COEFFICIENT THEOREM FOR HOMOLOGY

Just as there is with the cohomology group $H^p(X; G)$, there is a theorem expressing the homology group $H_p(X; G)$ in terms of the homology groups $H_p(X)$ and $H_{p-1}(X)$. The statements of the two theorems are similar, except that here the tensor product and torsion products replace the Hom and Ext functors, respectively, and the arrows are reversed.

Theorem 55.1 (The universal coefficient theorem for homology). *Let $\mathcal{C}$ be a free chain complex; let G be an abelian group. There is an exact sequence*

$$0 \rightarrow H_p(\mathcal{C}) \otimes G \rightarrow H_p(\mathcal{C}; G) \rightarrow H_{p-1}(\mathcal{C}) * G \rightarrow 0,$$

which is natural with respect to homomorphisms induced by chain maps. It splits, but not naturally.

One can give a direct proof of this theorem that is very similar to the proof of the universal coefficient theorem for cohomology. Instead, we shall postpone the proof, and derive it from a more general theorem called the Künneth theorem, which we shall prove in §58.

Corollary 55.2. *Let (X,A) be a topological pair. There is an exact sequence*

$$0 \rightarrow H_p(X,A) \otimes G \rightarrow H_p(X,A; G) \rightarrow H_{p-1}(X,A) * G \rightarrow 0,$$

which is natural with respect to homomorphisms induced by continuous maps. It splits, but not naturally. □

Corollary 55.3. *Let $\mathcal{C}$ and $\mathcal{D}$ be free chain complexes; let $\phi : \mathcal{C} \to \mathcal{D}$ be a chain map. If $\phi_* : H_i(\mathcal{C}) \to H_i(\mathcal{D})$ is an isomorphism for $i = p$ and $i = p - 1$, then*

$$\phi_* : H_p(\mathcal{C}; G) \to H_p(\mathcal{D}; G)$$

is an isomorphism for arbitrary G.

Proof. This result follows from naturality of the universal coefficient sequence and the Five-lemma. □

This corollary provides an alternate proof of Theorem 51.1.

EXERCISES

1. Let $\mathcal{C}$ be a free chain complex. Show that if $H_{p-1}(\mathcal{C})$ or G is torsion-free then $H_p(\mathcal{C}) \otimes G \simeq H_p(\mathcal{C}; G)$.
2. Use the universal coefficient theorem to compute the homology, with general coefficients G, of $T \# T$ and $P^2 \# P^2 \# P^2$ and P^N and the k-fold dunce cap. Compare with Exercise 2 of §51.
3. Assume $H_i(X)$ is finitely generated for all i.
   (a) Compute $H_p(X; G)$ in terms of the betti numbers and torsion coefficients of X.
   (b) Repeat (a) when G is the circle group S^1.
   (c) Repeat (a) when $G = \mathbf{Q}$.
4. Let $\mathcal{C}$ be a free chain complex. Show that if $H_i(\mathcal{C})$ is finitely generated for all i, then
$$H_p(\mathcal{C}; \mathbf{Z}/n) \simeq H^p(\mathcal{C}; \mathbf{Z}/n).$$
   (The isomorphism is not natural.)
5. Show that if K is a finite complex, the homology group $H_n(K)$ is uniquely determined by the groups $H_i(K; \mathbf{Z}/p^k)$, as p^k ranges over all prime powers and i ranges from 0 to n.
6. Show that if K is a finite complex, the homology group $H_p(K)$ is uniquely determined by the groups $H_p(K; S^1)$ and $H_{p+1}(K; S^1)$.
7. Show that if T is the torus and S is the Klein bottle, and if $f : T \to S$, then
$$f_* : H_2(T; \mathbf{Z}/2) \to H_2(S; \mathbf{Z}/2)$$
   is trivial.
8. Show by example that the splitting of the sequence in Corollary 55.2 is not natural.

***9.** Prove Theorem 55.1. [*Hint:* Let $\mathcal{Z}$, $\mathcal{C}$, and $\mathcal{D}$ be as in the proof of Theorem 53.1. One has an exact sequence

$$\cdots \to H_{p+1}(\mathcal{D}; G) \xrightarrow{\beta} H_p(\mathcal{Z}; G) \to H_p(\mathcal{C}; G) \to H_p(\mathcal{D}; G) \to \cdots.$$

Show that β is induced by inclusion $j_p: B_p \to Z_p$.]

*§56. OTHER UNIVERSAL COEFFICIENT THEOREMS[†]

We have stated two universal coefficient theorems, which express the homology and cohomology groups with arbitrary coefficients in terms of homology with integer coefficients. One might ask whether cohomology and homology are symmetric in some sense—that is, whether these same groups can be expressed in terms of *cohomology* with integer coefficients. There are in fact two short exact sequences that do this. However, they do not hold in complete generality, but only if the homology groups $H_i(\mathcal{C})$ are finitely generated. We consider them now.

These sequences are derived by re-indexing the groups of the cochain complex $\{\text{Hom}(C_i,\mathbf{Z}),\delta\}$ to make it into a chain complex, and then applying the universal coefficient theorems we have already considered. The problem with this approach is that the group $\text{Hom}(C_i,\mathbf{Z})$ is not in general free, so some restrictive hypotheses are needed.

Theorem 56.1. *Let $\mathcal{C}$ be a free chain complex; let G be an abelian group. If $\mathcal{C}$ is finitely generated in each dimension, then there are exact sequences*

$$0 \leftarrow \text{Hom}(H^p(\mathcal{C}),G) \leftarrow H_p(\mathcal{C}; G) \leftarrow \text{Ext}(H^{p+1}(\mathcal{C}),G) \leftarrow 0,$$
$$0 \to H^p(\mathcal{C}) \otimes G \to H^p(\mathcal{C}; G) \to H^{p+1}(\mathcal{C}) * G \to 0,$$

which are natural with respect to homomorphisms induced by chain maps. They split, but not naturally.

Proof. *Step 1.* Since C_p is free and finitely generated, the cochain complex $\{\text{Hom}(C_p,\mathbf{Z}),\delta\}$ is free. We define a chain complex $\mathcal{E}$ by letting $E_p = \text{Hom}(C_{-p},\mathbf{Z})$ and taking as the boundary operator ∂_E in $\mathcal{E}$ the usual coboundary operator δ of $\mathcal{C}$. One then has the universal coefficient sequences

$$0 \leftarrow \text{Hom}(H_{-p}(\mathcal{E}),G) \leftarrow H^{-p}(\mathcal{E}; G) \leftarrow \text{Ext}(H_{-p-1}(\mathcal{E}),G) \leftarrow 0,$$
$$0 \to H_{-p}(\mathcal{E}) \otimes G \to H_{-p}(\mathcal{E}; G) \to H_{-p-1}(\mathcal{E}) * G \to 0,$$

where we have for convenience replaced p by $-p$ throughout. It is immediate

[†]This section uses the results of §46.

from the definition that $H_{-p}(\mathcal{E}) = H^p(\mathcal{C})$. We shall show that there are natural isomorphisms

$$(*) \qquad H_{-p}(\mathcal{E}; G) \simeq H^p(\mathcal{C}; G),$$

$$(**) \qquad H^{-p}(\mathcal{E}; G) \simeq H_p(\mathcal{C}; G).$$

The theorem follows.

Step 2. We show first that if A is free of finite rank, then there is a natural isomorphism

$$\Phi : \text{Hom}(A,\mathbf{Z}) \otimes G \to \text{Hom}(A,G),$$

defined by the equation

$$[\Phi(\phi \otimes g)](a) = \phi(a) \cdot g.$$

(Strictly speaking, we define Φ on the pair (ϕ,g) by this formula. We note that $\Phi(\phi,g)$ is indeed a homomorphism of A into G. Then we note that Φ is bilinear.)

Let $a_1, \ldots, a_k$ be a basis for A; let $\tilde{a}_1, \ldots, \tilde{a}_k$ be the dual basis for $\text{Hom}(A,\mathbf{Z})$, defined by $\tilde{a}_i(a_j) = \delta_{ij}$, where $\delta_{ij} = 0$ if $i \neq j$ and $\delta_{ij} = 1$ if $i = j$. The group $\text{Hom}(A,\mathbf{Z}) \otimes G$ is a direct sum of k copies of G, the ith copy consisting of all elements of the form $\tilde{a}_i \otimes g$, for $g \in G$. The group $\text{Hom}(A,G)$ is also a direct sum of k copies of G, the ith copy consisting of all homomorphisms that vanish on a_j for $j \neq i$. These two copies of G are isomorphic under Φ, since $\Phi(\tilde{a}_i \otimes g)$ is the homomorphism of A into G that maps a_i to g and maps a_j to 0 for $j \neq i$.

Step 3. To prove (*), we recall that $E_{-p} = \text{Hom}(C_p,\mathbf{Z})$ and apply Step 2 to obtain an isomorphism

$$E_{-p} \otimes G \simeq \text{Hom}(C_p,G).$$

The naturality of the isomorphism means that $\partial_E \otimes i_G$ corresponds to δ. Thus $H_{-p}(\mathcal{E}; G) \simeq H^p(\mathcal{C}; G)$, as desired.

Step 4. We show that if A is free of finite rank, then there is a natural isomorphism

$$e : A \to \text{Hom}(\text{Hom}(A,\mathbf{Z}),\mathbf{Z}),$$

defined by the equation

$$[e(a)](\phi) = \phi(a),$$

where $a \in A$ and $\phi \in \text{Hom}(A,\mathbf{Z})$.

First, one checks that $e(a)$ is a homomorphism of $\text{Hom}(A,\mathbf{Z})$ into $\mathbf{Z}$; then one checks that e itself is a homomorphism. To prove e is an isomorphism, choose a basis $a_1, \ldots, a_k$ for A. Let $\tilde{a}_1, \ldots, \tilde{a}_k$ be the dual basis for $\text{Hom}(A,\mathbf{Z})$; then let $\tilde{\tilde{a}}_1, \ldots, \tilde{\tilde{a}}_k$ be the corresponding dual basis for $\text{Hom}(\text{Hom}(A,\mathbf{Z}),\mathbf{Z})$. All the groups involved are free abelian of rank k; the homomorphism e carries a_i to $\tilde{\tilde{a}}_i$, since for all j,

$$[e(a_i)](\tilde{a}_j) = \tilde{a}_j(a_i) = \delta_{ij} = \tilde{\tilde{a}}_i(\tilde{a}_j).$$

Step 5. To prove (**), we apply first Step 4 and then Step 2 to define natural isomorphisms

$$A \otimes G \to \text{Hom}(\text{Hom}(A,\mathbf{Z}),\mathbf{Z}) \otimes G \to \text{Hom}(\text{Hom}(A,\mathbf{Z}),G).$$

Setting $A = C_p$, we have a natural isomorphism

$$C_p \otimes G \to \text{Hom}(\text{Hom}(C_p,\mathbf{Z}),G) = \text{Hom}(E_{-p},G).$$

By naturality, the boundary operator $\partial \otimes i_G$ corresponds to the coboundary operator $\tilde{\partial}_E$, so one has a natural isomorphism

$$H_p(\mathcal{C}; G) \simeq H^{-p}(\mathcal{E}; G),$$

as desired. □

The preceding theorem is of limited interest topologically. The only interesting case to which it applies is the homology of a finite simplicial complex or a finite CW complex. However, one can obtain a much more general theorem with very little effort:

Theorem 56.2. *The preceding theorem holds if the hypothesis that $\mathcal{C}$ is finitely generated in each dimension is replaced by the hypothesis that $\mathcal{C}$ vanishes below a certain dimension, and the homology of $\mathcal{C}$ is finitely generated in each dimension.*

To prove this theorem, we need the following lemma:

Lemma 56.3. *Let $\mathcal{C}$ be a free chain complex such that $H_i(\mathcal{C})$ is finitely generated for each i. Then there is a free chain complex $\mathcal{C}'$ that is finitely generated in each dimension, whose homology is isomorphic to that of $\mathcal{C}$. If $\mathcal{C}$ vanishes below a certain dimension, so does $\mathcal{C}'$.*

Proof. Let β_p be the betti number of $H_p(\mathcal{C})$, and let $t_1^{(p)}, \ldots, t_{k_p}^{(p)}$ be its torsion coefficients. Let U_p, V_p, and W_p be free abelian groups, where U_p has rank k_{p-1}, and V_p has rank β_p, and W_p has rank k_p. Let

$$C'_p = U_p \oplus V_p \oplus W_p.$$

Let $\partial' : C'_{p+1} \to C'_p$ vanish on $V_{p+1} \oplus W_{p+1}$. Choose bases for U_{p+1} and W_p, and define $\partial' : U_{p+1} \to W_p$ to be the homomorphism whose matrix relative to these bases is

$$\begin{bmatrix} t_1^{(p)} & & 0 \\ & \ddots & \\ 0 & & t_{k_p}^{(p)} \end{bmatrix}.$$

Then ∂' satisfies the requirements of the lemma. □

Proof of Theorem 56.2. Choose a chain complex $\mathcal{C}'$ as in the preceding lemma. In view of Theorem 45.1, we can choose a chain map $\phi : \mathcal{C}' \to \mathcal{C}$ inducing the given isomorphism $H_i(\mathcal{C}') \to H_i(\mathcal{C})$. It follows from the universal coefficient theorems already stated (or from Theorems 45.5 and 51.1) that ϕ also induces isomorphisms

$$H_i(\mathcal{C}'; G) \simeq H_i(\mathcal{C}; G) \qquad \text{and} \qquad H^i(\mathcal{C}'; G) \simeq H^i(\mathcal{C}; G)$$

for all i and G. Now the universal coefficient sequences in question hold for $\mathcal{C}'$. Using the isomorphisms induced by ϕ, we can replace $\mathcal{C}'$ by $\mathcal{C}$ in these exact sequences.

The resulting exact sequences for $\mathcal{C}$ are in fact independent of the choices of $\mathcal{C}'$ and ϕ. This is a consequence of naturality, which we now prove.

Let $\theta : \mathcal{C} \to \mathcal{D}$ be a chain map, where $\mathcal{C}$ and $\mathcal{D}$ are free chain complexes vanishing below a certain dimension, and $H_i(\mathcal{C})$ and $H_i(\mathcal{D})$ are finitely generated for each i. We wish to show θ induces homomorphisms of the universal coefficient sequences of our theorem.

Here we need (for the first time) the fact that the chain map ϕ is actually a chain equivalence (Theorem 46.2).

Let $\phi : \mathcal{C}' \to \mathcal{C}$ and $\psi : \mathcal{D}' \to \mathcal{D}$ be chain maps that induce homology isomorphisms in all dimensions, as in the preceding lemma; here $\mathcal{C}'$ and $\mathcal{D}'$ are free, finitely generated in each dimension, and vanishing below a certain dimension. In view of Theorem 46.2, there are chain-homotopy inverses $\lambda : \mathcal{C} \to \mathcal{C}'$ and $\mu : \mathcal{D} \to \mathcal{D}'$ to ϕ and ψ, respectively.

Now ϕ induces an isomorphism of the exact sequences for $\mathcal{C}$ with the exact sequences for $\mathcal{C}'$; indeed, this is how we *obtained* the exact sequences for $\mathcal{C}$. A similar remark applies to ψ. Consider the diagram

$$\begin{array}{ccc} \mathcal{C}' & \underset{\lambda}{\overset{\phi}{\rightleftarrows}} & \mathcal{C} \\ & & \big\downarrow \theta \\ \mathcal{D}' & \underset{\mu}{\overset{\psi}{\rightleftarrows}} & \mathcal{D}. \end{array}$$

The composite $\mu \circ \theta \circ \phi$ induces homomorphisms of the exact sequences for $\mathcal{C}'$ with those for $\mathcal{D}'$, by the naturality property of Theorem 56.1. It follows that $\psi \circ (\mu \circ \theta \circ \phi) \circ \lambda$ induces homomorphisms of the exact sequences for $\mathcal{C}$ with those for $\mathcal{D}$. But this map is chain homotopic to θ.

Applying this result to the special case in which $\mathcal{C} = \mathcal{D}$ and θ is the identity, we see that the exact sequences are independent of the choices of $\mathcal{C}'$ and ϕ. $\square$

Corollary 56.4. *Let (X,A) be a topological pair such that $H_i(X,A)$ is finitely generated for each i. Then there are natural exact sequences*

$$0 \leftarrow \operatorname{Hom}(H^p(X,A),G) \leftarrow H_p(X,A;G) \leftarrow \operatorname{Ext}(H^{p+1}(X,A),G) \leftarrow 0,$$
$$0 \to H^p(X,A) \otimes G \to H^p(X,A;G) \to H^{p+1}(X,A) * G \to 0.$$

They split, but not naturally. $\square$

EXERCISE

1. The argument used in proving Lemma 56.3 has a flavor that makes one suspect there is a functorial version of it. Here is that formulation:

   Let **C** be the category whose objects are free chain complexes $\mathcal{C}$, each of which vanishes below some dimension, such that $H_i(\mathcal{C})$ is finitely generated for each i. Let **C′** be the subcategory whose objects consist of those objects of **C** whose chain groups are finitely generated in each dimension. In both cases, the morphisms are chain-homotopy classes of chain maps.

   Let $J : \mathbf{C}' \rightarrow \mathbf{C}$ be the inclusion functor. Show there is a functor $G : \mathbf{C} \rightarrow \mathbf{C}'$ such that $G \circ J$ is the identity functor of **C′** and $J \circ G$ is naturally equivalent to the identity functor of **C**.

§57. TENSOR PRODUCTS OF CHAIN COMPLEXES

In this section, we introduce an algebraic device that will be used for studying the homology of a product space. We define the tensor product of two chain complexes, and show how to make it into a chain complex.

Definition. Let $\mathcal{C} = \{C_p, \partial\}$ and $\mathcal{C}' = \{C'_p, \partial'\}$ be chain complexes. We define their **tensor product** $\mathcal{C} \otimes \mathcal{C}'$ to be the chain complex whose chain group in dimension m is defined by

$$(\mathcal{C} \otimes \mathcal{C}')_m = \bigoplus_{p+q=m} C_p \otimes C'_q,$$

and whose boundary operator $\bar{\partial}$ is given by the equation

$$\bar{\partial}(c_p \otimes c'_q) = \partial c_p \otimes c'_q + (-1)^p c_p \otimes \partial' c'_q.$$

Formally, we define $\bar{\partial}$ as a function on the set $C_p \times C'_q$; since it is bilinear, it induces a homomorphism of the tensor product.

It is straightforward to check that $\bar{\partial}^2 = 0$. It is also straightforward to check that, if $\phi : \mathcal{C} \rightarrow \mathcal{D}$ and $\phi' : \mathcal{C}' \rightarrow \mathcal{D}'$ are chain maps, so is $\phi \otimes \phi'$.

Note that if $\mathcal{C}$ and $\mathcal{C}'$ are free chain complexes, then so is $\mathcal{C} \otimes \mathcal{C}'$. Indeed, if $\{a^p_\alpha\}$ is a basis for C_p and $\{b^q_\beta\}$ is a basis for C'_q, then the collection

$$\{a^p_\alpha \otimes b^q_\beta \mid p + q = m\}$$

is a basis for the group $(\mathcal{C} \otimes \mathcal{C}')_m$.

Definition. Let $\{\mathcal{C}, \epsilon\}$ and $\{\mathcal{C}', \epsilon'\}$ be augmented chain complexes. We augment $\mathcal{C} \otimes \mathcal{C}'$ by the homomorphism $\bar{\epsilon}$, which is the composite

$$C_0 \otimes C'_0 \xrightarrow{\epsilon \otimes \epsilon'} \mathbf{Z} \otimes \mathbf{Z} \xrightarrow{\simeq} \mathbf{Z}.$$

You can check that $\bar{\epsilon}$ is surjective, and that $\bar{\epsilon} \circ \bar{\partial} = 0$.

We now give the geometric motivation underlying our definition of $\mathcal{C} \otimes \mathcal{C}'$.

Let K and L be simplicial complexes, with K locally finite. Then $|K| \times |L|$ is a CW complex with cells $\sigma \times \tau$, for $\sigma \in K$ and $\tau \in L$. In fact, it is a special kind of CW complex called a "regular cell complex"; therefore, it can be triangulated so that each cell is the polytope of a subcomplex. (See Example 4 and Exercise 5 of §38.)

Let $\mathcal{D}(K \times L)$ denote the cellular chain complex of the CW complex $X = |K| \times |L|$. Let M denote the underlying simplicial complex of X. Because X is triangulated, we can use simplicial homology in the definition of the chain group $D_m(K \times L)$. This chain group is a free abelian group, with one basis element for each cell $\sigma^p \times \tau^q$ for which $p + q = m$. This basis element is a fundamental cycle for $\sigma^p \times \tau^q$.

If we orient the simplices of K and L, once and for all, then they form bases for $\mathcal{C}(K)$ and $\mathcal{C}(L)$, respectively. A basis for the m-dimensional chain group of $\mathcal{C}(K) \otimes \mathcal{C}(L)$ consists of all the elements $\sigma^p \otimes \tau^q$, for $p + q = m$.

It follows that the *groups*

$$(\mathcal{C}(K) \otimes \mathcal{C}(L))_m \qquad \text{and} \qquad D_m(K \times L)$$

are isomorphic, for they are both free abelian and there is a one-to-one correspondence ϕ between their bases, assigning to the basis element $\sigma^p \otimes \tau^q$ for $(\mathcal{C}(K) \otimes \mathcal{C}(L))_m$, a fundamental cycle for the $p + q$ cell $\sigma \times \tau$. We shall prove that this isomorphism can be chosen so that it preserves boundary operators, so that it is thus an isomorphism of chain complexes as well.

Theorem 57.1. *If K and L are simplicial complexes, and if $|K| \times |L|$ is triangulated so each cell $\sigma \times \tau$ is the polytope of a subcomplex, then*

$$\mathcal{C}(K) \otimes \mathcal{C}(L) \simeq \mathcal{D}(K \times L).$$

Thus $\mathcal{C}(K) \otimes \mathcal{C}(L)$ can be used to compute the homology of $|K| \times |L|$.

Proof. We proceed by induction. A basis element for $(\mathcal{C}(K) \otimes \mathcal{C}(L))_0$ is of the form $v \otimes w$, where v is a vertex of K and w is a vertex of L. Define $\phi(v \otimes w)$ to be a fundamental 0-cycle for $v \times w$. (That is, $\phi(v \otimes w) = v \times w$.)

The augmentation $\bar{\epsilon}$ for $\mathcal{C}(K) \otimes \mathcal{C}(L)$ satisfies the equation $\bar{\epsilon}(v \otimes w) = 1$ whenever v is a vertex of K and w is a vertex of L. The chain complex $\mathcal{D}(K \times L)$ is augmented as a subchain complex of $\mathcal{C}(M)$, so $\epsilon(v \times w) = 1$. The map ϕ is thus augmentation-preserving, since $\epsilon \circ \phi = \bar{\epsilon}$ for 0-chains.

For the induction step, we assume that

$$\phi : (\mathcal{C}(K) \otimes \mathcal{C}(L))_i \to D_i(K \times L)$$

is defined in dimensions i less than m, where $m \geq 1$, and that $\phi \circ \bar{\partial} = \partial \circ \phi$ in those dimensions. We suppose further that for $p + q < m$, the chain $\phi(\sigma^p \otimes \tau^q)$ is a fundamental cycle for the cell $\sigma^p \times \tau^q$. Then we define ϕ in dimension m.

Consider the basis element $\sigma^p \otimes \tau^q$, where $p + q = m$. Let

$$c_{m-1} = \phi(\bar{\partial}(\sigma^p \otimes \tau^q)).$$

Then c_{m-1} is an $m-1$ chain of the simplicial complex M underlying X.

First we note that if $m = 1$, then $c_0 \in \ker \epsilon$, for

$$\epsilon(c_0) = \epsilon\phi(\bar{\partial} c_1) = \bar{\epsilon}(\bar{\partial} c_1) = 0.$$

Second, we note that if $m > 1$, then c_{m-1} is a cycle. For by the induction hypothesis,

$$\partial c_{m-1} = \partial\phi(\bar{\partial}(\sigma^p \otimes \tau^q)) = \phi\bar{\partial}(\bar{\partial}(\sigma^p \otimes \tau^q)) = 0.$$

Now we show that c_{m-1} is carried by $\operatorname{Bd}(\sigma \times \tau)$. Let us write $\partial\sigma = \Sigma s_i$ if $\dim \sigma > 0$, where the s_i are oriented simplices of dimension $p-1$. Similarly, we write $\partial\tau = \Sigma t_j$ if $\dim \tau > 0$. Then we have the formula

$$c_{m-1} = \sum_i \phi(s_i \otimes \tau) + (-1)^p \sum_j \phi(\sigma \otimes t_j)$$

if $\dim \sigma > 0$ and $\dim \tau > 0$. If $\dim \sigma = 0$, the first summation is missing from this formula; while if $\dim \tau = 0$, the second is missing. Since the chain $\phi(s_i \otimes \tau)$ is carried by $s_i \times \tau$, and $\phi(\sigma \otimes t_j)$ is carried by $\sigma \times t_j$, the cycle c_{m-1} is carried by $\operatorname{Bd}(\sigma \times \tau)$.

Now $\operatorname{Bd}(\sigma \times \tau)$ is, topologically, an $m-1$ sphere; we show that c_{m-1} is a fundamental cycle for this $m-1$ sphere. That is, it generates the group $\tilde{H}_{m-1}(\operatorname{Bd}(\sigma \times \tau))$, which is infinite cyclic and equals the group of $m-1$ cycles if $m > 1$, and the group $\ker \epsilon$ if $m = 1$.

Either $\dim \sigma > 0$ or $\dim \tau > 0$; assume the former. Then $\partial\sigma = \Sigma s_i$, and the restriction of c_{m-1} to the cell $s_i \times \tau$ equals $\phi(s_i \otimes \tau)$, which is by hypothesis, a fundamental cycle for this cell. It follows that c_{m-1} is not zero, and that it is not a multiple of another cycle. Thus c_{m-1} is a generator of the infinite cyclic group $\tilde{H}_{m-1}(\operatorname{Bd}(\sigma \times \tau))$.

Consider now the exact sequence

$$0 \to H_m(\sigma \times \tau, \operatorname{Bd}(\sigma \times \tau)) \xrightarrow{\partial_*} \tilde{H}_{m-1}(\operatorname{Bd}(\sigma \times \tau)) \to 0.$$

Both end groups vanish because $\sigma \times \tau$ is acyclic. Because homology groups equal cycle groups in this instance, this sequence is actually the sequence

$$0 \to Z_m(\sigma \times \tau, \operatorname{Bd}(\sigma \times \tau)) \to \tilde{Z}_{m-1}(\operatorname{Bd}(\sigma \times \tau)) \to 0,$$

where $\tilde{Z}_{m-1}$ denotes $\ker \epsilon$ if $m = 1$. We define $\phi(\sigma^p \otimes \tau^q)$ to be the unique fundamental cycle for the cell $\sigma \times \tau$ such that

$$\partial\phi(\sigma^p \otimes \tau^q) = c_{m-1}.$$

Then $\partial\phi = \phi\bar{\partial}$, and our result is proved. □

EXERCISES

1. (a) Show that $\bar{\partial}^2 = 0$.
   (b) Show that if $\mathcal{C}$ and $\mathcal{C}'$ are augmented by ϵ, ϵ', respectively, then $\mathcal{C} \otimes \mathcal{C}'$ is augmented by

$$C_0 \otimes C_0' \xrightarrow{\epsilon \otimes \epsilon'} Z \otimes Z \simeq Z.$$

2. Let $\phi, \psi : \mathcal{C} \to \mathcal{D}$ and $\phi', \psi' : \mathcal{C}' \to \mathcal{D}'$ be chain maps.
   (a) Show that $\phi \otimes \phi' : \mathcal{C} \otimes \mathcal{C}' \to \mathcal{D} \otimes \mathcal{D}'$ is a chain map.
   (b) Show that if D is a chain homotopy of ϕ to ψ, then $D \otimes \phi'$ is a chain homotopy of $\phi \otimes \phi'$ to $\psi \otimes \phi'$.
   (c) If D is a chain homotopy of ϕ' to ψ', find a chain homotopy of $\psi \otimes \phi'$ to $\psi \otimes \psi'$.
   (d) Show that if ϕ and ϕ' are chain equivalences, so is $\phi \otimes \phi'$.
3. Our formula for a boundary operator in $\mathcal{C} \otimes \mathcal{C}'$ seems rather arbitrary. It is; there are other formulas that would do as well.

   Let K and L be complexes; let $\hat{\partial}$ be a boundary operator for $\mathcal{C}(K) \otimes \mathcal{C}(L)$ such that:
   (i) If K_0 and L_0 are subcomplexes of K and L, respectively, then $\hat{\partial}$ maps $\mathcal{C}(K_0) \otimes \mathcal{C}(L_0)$ to itself.
   (ii) If v is a vertex of K and w is a vertex of L, then
$$\hat{\partial} = \pm \partial_K \otimes i \quad \text{on} \quad C_1(K) \otimes C_0(w),$$
   and
$$\hat{\partial} = \pm i \otimes \partial_L \quad \text{on} \quad C_0(v) \otimes C_1(L).$$
   Assuming $X = |K| \times |L|$ triangulated as in Theorem 57.1, show that the chain complex $\{\mathcal{C}(K) \otimes \mathcal{C}(L), \hat{\partial}\}$ is isomorphic to the cellular chain complex $\mathcal{D}(X)$.
4. Show that the formula
$$\hat{\partial}(c_p \otimes c_q') = (-1)^q \partial c_p \otimes c_q' + c_p \otimes \partial c_q'$$
   defines a boundary operator for $\mathcal{C} \otimes \mathcal{C}'$ that satisfies the requirements of Exercise 3.

§58. THE KÜNNETH THEOREM

Now we prove a basic theorem that enables us to compute the homology of the tensor product of two chain complexes. The proof is, as you would expect, highly algebraic in nature.

The first thing we show is that, under suitable hypotheses, there is a natural monomorphism of the tensor product $H_p(\mathcal{C}) \otimes H_q(\mathcal{C}')$ into the group $H_{p+q}(\mathcal{C} \otimes \mathcal{C}')$.

Definition. We define a homomorphism
$$\Theta : H_p(\mathcal{C}) \otimes H_q(\mathcal{C}') \to H_{p+q}(\mathcal{C} \otimes \mathcal{C}')$$
as follows: If z_p is a p-cycle of $\mathcal{C}$ and z_q' is a q-cycle of $\mathcal{C}'$, let
$$\Theta(\{z_p\} \otimes \{z_q'\}) = \{z_p \otimes z_q'\}.$$

Using the formula for the boundary operator $\bar{\partial}$, we see that $z_p \otimes z'_q$ is a cycle of $\mathcal{C} \otimes \mathcal{C}'$. To show Θ well-defined, one computes

$$(z_p + \partial d_{p+1}) \otimes z'_q = z_p \otimes z'_q + \bar{\partial}(d_{p+1} \otimes z'_q),$$
$$z_p \otimes (z'_q + \partial' d'_{q+1}) = z_p \otimes z'_q + (-1)^p \bar{\partial}(z_p \otimes d'_{q+1}).$$

Because Θ is induced by the inclusion map $Z_p \otimes Z'_q \to (\mathcal{C} \otimes \mathcal{C}')_{p+q}$, it is natural with respect to chain maps.

Lemma 58.1. *Let $\mathcal{C}$ and $\mathcal{C}'$ be chain complexes such that in each dimension the cycles form a direct summand in the chains. (This occurs, for instance, when $\mathcal{C}$ and $\mathcal{C}'$ are free.) Then*

$$\Theta : \oplus_{p+q=m} H_p(\mathcal{C}) \otimes H_q(\mathcal{C}') \to H_m(\mathcal{C} \otimes \mathcal{C}')$$

is a monomorphism, and its image is a direct summand.

Proof. We define a homomorphism λ in the opposite direction to Θ, such that $\lambda \circ \Theta$ equals the identity. This suffices.

Let Z_p denote the group of p-cycles in $\mathcal{C}$; let Z'_q denote the q-cycles in $\mathcal{C}'$. We begin with the natural projections $Z_p \to H_p(\mathcal{C})$ and $Z'_q \to H_q(\mathcal{C}')$. Because the cycles form a direct summand in the chains, these maps extend to homomorphisms

$$\phi : C_p \to H_p(\mathcal{C}) \qquad \text{and} \qquad \phi' : C'_q \to H_q(\mathcal{C}'),$$

respectively. Let $\mathcal{E}$ be the chain complex whose pth group is $H_p(\mathcal{C})$, and whose boundary operators vanish. Similarly, let $\mathcal{E}'$ be the chain complex whose qth group is $H_q(\mathcal{C}')$ and whose boundary operators vanish. Then $\phi : \mathcal{C} \to \mathcal{E}$ and $\phi' : \mathcal{C}' \to \mathcal{E}'$ are chain maps, as you can check. Now

$$\phi \otimes \phi' : \mathcal{C} \otimes \mathcal{C}' \to \mathcal{E} \otimes \mathcal{E}'$$

is a chain map, so it induces a homomorphism

$$\lambda : H_m(\mathcal{C} \otimes \mathcal{C}') \to H_m(\mathcal{E} \otimes \mathcal{E}').$$

Since the boundary operator in $\mathcal{E} \otimes \mathcal{E}'$ vanishes, $H_m(\mathcal{E} \otimes \mathcal{E}')$ equals the m-dimensional chain group

$$(\mathcal{E} \otimes \mathcal{E}')_m = \oplus_{p+q=m} H_p(\mathcal{C}) \otimes H_q(\mathcal{C}').$$

It is trivial to check that $\lambda \circ \Theta$ is the identity, for

$$\begin{aligned} \lambda\Theta(\{z\} \otimes \{z'\}) &= \lambda(\{z \otimes z'\}) \\ &= \phi(z) \otimes \phi'(z') = \{z\} \otimes \{z'\}. \quad \square \end{aligned}$$

Now we prove our main theorem, which identifies the group $\operatorname{cok} \Theta$.

Theorem 58.2 (The Künneth theorem for chain complexes). *Let $\mathcal{C}$ be a free chain complex; let $\mathcal{C}'$ be a chain complex. There is an exact sequence*

$$0 \to \oplus_{p+q=m} H_p(\mathcal{C}) \otimes H_q(\mathcal{C}') \xrightarrow{\Theta} H_m(\mathcal{C} \otimes \mathcal{C}') \to$$
$$\oplus_{p+q=m} H_{p-1}(\mathcal{C}) * H_q(\mathcal{C}') \to 0$$

which is natural with respect to homomorphisms induced by chain maps. If the cycles of $\mathcal{C}'$ are a direct summand in the chains, the sequence splits, but not naturally.

Proof. Let B_p, Z_p, and C_p be the p-boundaries, p-cycles, and p-chains, respectively, of $\mathcal{C}$. Choose B'_p, Z'_p, and C'_p similarly for $\mathcal{C}'$.

Step 1. We begin with the exact sequence

$$0 \to Z_p \xrightarrow{j} C_p \xrightarrow{\partial_0} B_{p-1} \to 0,$$

which splits because B_{p-1} is free. Let $\mathcal{Z}$ be the chain complex whose p-dimensional group is Z_p and whose boundary operators vanish. Let $\mathcal{D}$ be the chain complex whose p-dimensional group D_p equals B_{p-1}, and whose boundary operators vanish. Then the preceding sequence becomes a short exact sequence of chain complexes.

Now tensor with C'_q and sum over $p + q = m$. Exactness is retained, because the original sequence splits and direct sums of exact sequences are exact. One obtains the exact sequence

$$0 \longrightarrow \oplus(Z_p \otimes C'_q) \xrightarrow{j \otimes i'} \oplus(C_p \otimes C'_q) \xrightarrow{\partial_0 \otimes i'} \oplus(D_p \otimes C'_q) \longrightarrow 0,$$

where i' denotes the identity map of $\mathcal{C}'$. Since the maps involved are chain maps, this is a short exact sequence of chain complexes

$$0 \to (\mathcal{Z} \otimes \mathcal{C}')_m \to (\mathcal{C} \otimes \mathcal{C}')_m \to (\mathcal{D} \otimes \mathcal{C}')_m \to 0.$$

The boundary operator in $\mathcal{C} \otimes \mathcal{C}'$ is denoted as usual by $\bar{\partial}$. The boundary operators in the end groups have the form $\pm i \otimes \partial'$, where i is the identity map in $\mathcal{C}$. For instance, in $\mathcal{Z} \otimes \mathcal{C}'$, one computes

$$\bar{\partial}(z_p \otimes c'_q) = (-1)^p z_p \otimes \partial' c'_q,$$

since $\partial z_p = 0$. A similar remark applies to $\mathcal{D} \otimes \mathcal{C}'$.

We obtain from this short exact sequence of chain complexes, a long exact homology sequence, of which we consider five terms:

$$H_{m+1}(\mathcal{D} \otimes \mathcal{C}') \xrightarrow{\beta_{m+1}} H_m(\mathcal{Z} \otimes \mathcal{C}') \longrightarrow H_m(\mathcal{C} \otimes \mathcal{C}') \longrightarrow$$
$$H_m(\mathcal{D} \otimes \mathcal{C}') \xrightarrow{\beta_m} H_{m-1}(\mathcal{Z} \otimes \mathcal{C}'),$$

where β_m denotes the zig-zag homomorphism. This sequence is natural with respect to homomorphisms induced by chain maps. Consider the induced sequence

(*) $$0 \to \operatorname{cok} \beta_{m+1} \to H_m(\mathcal{C} \otimes \mathcal{C}') \to \ker \beta_m \to 0.$$

It is also natural. We identify the terms of this short exact sequence.

Before proceeding, let us recall that if B is a subgroup of C, and if A is free, then the map

$$A \otimes B \to A \otimes C$$

induced by inclusion is injective. Hence in this case, we can consider $A \otimes B$ as a subgroup of $A \otimes C$. We shall use this fact freely in what follows.

Step 2. We compute the group $H_m(Z \otimes \mathcal{C}')$. We show that there is an isomorphism, induced by inclusion,

$$\oplus_{p+q=m} Z_p \otimes H_q(\mathcal{C}') \simeq H_m(Z \otimes \mathcal{C}').$$

The chain group of $Z \otimes \mathcal{C}'$ in dimension m is

$$(Z \otimes \mathcal{C}')_m = \oplus_{p+q=m} Z_p \otimes C'_q.$$

To compute the cycles and boundaries of $Z \otimes \mathcal{C}'$, we begin with the exact sequence

$$0 \to Z'_q \to C'_q \xrightarrow{\partial'_0} B'_{q-1} \to 0.$$

Since Z_p is free, when we tensor with Z_p we have horizontal and vertical exactness in the diagram

$$\begin{array}{ccccccccc} & & & & & & 0 & & \\ & & & & & & \downarrow & & \\ 0 & \longrightarrow & Z_p \otimes Z'_q & \longrightarrow & Z_p \otimes C'_q & \xrightarrow{i \otimes \partial'_0} & Z_p \otimes B'_{q-1} & \longrightarrow & 0. \\ & & & & & \searrow^{i \otimes \partial'} & \downarrow & & \\ & & & & & & Z_p \otimes C'_{q-1} & & \end{array}$$

It follows that the kernel of $i \otimes \partial'$ is the subgroup $Z_p \otimes Z'_q$ of $C_p \otimes C'_q$; and the image is the subgroup $Z_p \otimes B'_{q-1}$ of $Z_p \otimes C'_{q-1}$. Hence the groups of cycles and boundaries of $Z \otimes \mathcal{C}'$ in dimension m are the respective groups

$$\oplus_{p+q=m} Z_p \otimes Z'_q \quad \text{and} \quad \oplus_{p+q=m} Z_p \otimes B'_q.$$

Now consider the exact sequence

$$0 \to B'_q \to Z'_q \to H_q(\mathcal{C}') \to 0.$$

Tensoring and summing, we obtain the exact sequence

$$0 \to \oplus Z_p \otimes B'_q \to \oplus Z_p \otimes Z'_q \to \oplus Z_p \otimes H_q(\mathcal{C}') \to 0,$$

where the summations extend over $p + q = m$. Since the first two groups of this sequence are the m-boundaries and m-cycles of $Z \otimes \mathcal{C}'$, the last is isomorphic to $H_m(Z \otimes \mathcal{C}')$, as claimed.

Step 3. An entirely similar argument shows that the m-cycle group of $\mathcal{D} \otimes \mathcal{C}'$ is the group

$$\oplus_{p+q=m} D_p \otimes Z'_q,$$

and that inclusion induces an isomorphism

$$\oplus_{p+q=m} D_p \otimes H_q(\mathcal{C}') \simeq H_m(\mathcal{D} \otimes \mathcal{C}').$$

Step 4. We now show that the zig-zag homomorphism β_m is induced by inclusion $j : B_p \to Z_p$. More precisely, we show that the following diagram commutes:

$$\begin{array}{ccc} H_{m+1}(\mathcal{D} \otimes \mathcal{C}') & \xrightarrow{\beta_{m+1}} & H_m(\mathcal{Z} \otimes \mathcal{C}') \\ \uparrow \simeq & & \uparrow \simeq \\ \oplus_{p+q=m} D_{p+1} \otimes H_q(\mathcal{C}') & \xrightarrow{j \otimes i'_*} & \oplus_{p+q=m} Z_p \otimes H_q(\mathcal{C}'). \end{array}$$

(Recall that $B_p = D_{p+1}$.)

The map β_{m+1} is defined via the following zig-zag diagram:

$$\begin{array}{ccc} & \oplus_{p+q=m} C_{p+1} \otimes C'_q & \xrightarrow{\partial_0 \otimes i'} \oplus_{p+q=m} D_{p+1} \otimes C'_q \\ & \downarrow \bar{\partial} & \\ \oplus_{r+s=m} Z_r \otimes C'_s \xrightarrow{j \otimes i'} & \oplus_{r+s=m} C_r \otimes C'_s & \end{array}$$

One begins with an element of the $m + 1$ cycle group of $\mathcal{D} \otimes \mathcal{C}'$. By the previous step, this cycle group is

$$\oplus_{p+q=m} D_{p+1} \otimes Z'_q.$$

Since $D_{p+1} = B_p$, a typical generator for this group has the form $b_p \otimes z'_q$, where $b_p \in B_p$ and $z'_q \in Z'_q$ and $p + q = m$. Pulling back to $\mathcal{C} \otimes \mathcal{C}'$, one obtains an element $c_{p+1} \otimes z'_q$, where $\partial_0 c_{p+1} = b_p$. Applying $\bar{\partial}$, we obtain the element

$$\begin{aligned} \bar{\partial}(c_{p+1} \otimes z'_q) &= \partial c_{p+1} \otimes z'_q \pm c_{p+1} \otimes \partial' z'_q \\ &= b_p \otimes z'_q. \end{aligned}$$

Finally, we pull this back to $\mathcal{Z} \otimes \mathcal{C}'$, obtaining $b_p \otimes z'_q$, which is the same element we started with!

Step 5. Now we can identify the kernel and cokernel of β_m. Begin with the free resolution

$$0 \to B_p \to Z_p \to H_p(\mathcal{C}) \to 0.$$

Tensor with $H_q(\mathcal{C}')$, obtaining the sequence

$$0 \to H_p(\mathcal{C}) * H_q(\mathcal{C}') \to B_p \otimes H_q(\mathcal{C}') \to \\ Z_p \otimes H_q(\mathcal{C}') \to H_p(\mathcal{C}) \otimes H_q(\mathcal{C}') \to 0.$$

The map in the middle is induced by inclusion. Summing over $p + q = m$ and applying Step 4, we obtain the sequence

$$0 \to \oplus_{p+q=m} H_p(\mathcal{C}) * H_q(\mathcal{C}') \to H_{m+1}(\mathcal{D} \otimes \mathcal{C}') \xrightarrow{\beta_{m+1}} \\ H_m(\mathcal{Z} \otimes \mathcal{C}') \to \oplus_{p+q=m} H_p(\mathcal{C}) \otimes H_q(\mathcal{C}') \to 0.$$

The kernel and cokernel of β_{m+1} are now apparent. From this sequence and sequence (*) of Step 1, the theorem follows. Because both sequences are natural, the Künneth sequence is also natural.

Finally, we note that the map

$$\operatorname{cok} \beta_{m+1} \to H_m(\mathcal{C} \otimes \mathcal{C}')$$

in sequence (*) is induced by inclusion $Z \otimes \mathcal{C}' \to \mathcal{C} \otimes \mathcal{C}'$. Therefore, the first map of the Künneth sequence is just the map Θ of Lemma 58.1. □

Corollary 58.3. *Let $\mathcal{C}$, $\mathcal{C}'$, $\mathcal{D}$, $\mathcal{D}'$ be chain complexes with $\mathcal{C}$ and $\mathcal{D}$ free; let $\phi : \mathcal{C} \to \mathcal{D}$ and $\phi' : \mathcal{C}' \to \mathcal{D}'$ be chain maps that induce homology isomorphisms in all dimensions. Then $\phi \otimes \phi' : \mathcal{C} \otimes \mathcal{C}' \to \mathcal{D} \otimes \mathcal{D}'$ induces a homology isomorphism in all dimensions.*

Proof. We apply naturality of the Künneth sequence, and the Five-lemma. □

Example 1. Let K and L be simplicial complexes, with K or L locally finite. The results of the preceding section show that the chain complex $\mathcal{C}(K) \otimes \mathcal{C}(L)$ can be used to compute the homology of $|K| \times |L|$. Since $\mathcal{C}(K)$ and $\mathcal{C}(L)$ are free, the Künneth theorem implies that

$$H_m(|K| \times |L|) \simeq \oplus_{p+q=m}[H_p(K) \otimes H_q(L) \oplus H_{p-1}(K) * H_q(L)].$$

Example 2. In particular, for the product $S^r \times S^s$, we have

$$H_m(S^r \times S^s) \simeq \oplus_{p+q=m} H_p(S^r) \otimes H_q(S^s).$$

Hence if $r \neq s$,

$$H_m(S^r \times S^s) \simeq \begin{cases} \mathbf{Z} & \text{if } m = 0, r, s, r+s, \\ 0 & \text{otherwise.} \end{cases}$$

$$H_m(S^r \times S^r) \simeq \begin{cases} \mathbf{Z} & \text{if } m = 0, 2r, \\ \mathbf{Z} \oplus \mathbf{Z} & \text{if } m = r, \\ 0 & \text{otherwise.} \end{cases}$$

Example 3. Given $\alpha \in H_p(K)$ and $\beta \in H_q(L)$, it is geometrically clear that there should be a corresponding element in $H_{p+q}(|K| \times |L|)$. The image of $\alpha \otimes \beta$ in $H_{p+q}(|K| \times |L|)$ under the natural monomorphism Θ is called the **homology cross product** of α and β, and denoted $\alpha \times \beta$. It behaves in some sense like a cartesian product. In the torus $T = S^1 \times S^1$, for example, the cross product of the two 1-dimensional homology classes pictured in Figure 58.1 is the generator of the 2-dimensional homology, by the Künneth theorem. This is quite plausible; if $z = \Sigma s_i$ is a fundamental cycle for a triangulation K of S^1, then $z \otimes z = \Sigma s_i \otimes s_j$ represents the sum of all the 2-cells of $K \times K$, suitably oriented.

It is geometrically less clear why, given a cycle z in K representing a torsion element in $H_p(K)$ of order k, and a cycle z' in L representing a torsion element in $H_q(L)$ of order l, there should be *two* elements in the homology of $|K| \times |L|$ of order $n = \gcd(k,l)$. One, of course, is their cross product, in dimension $p + q$. The other is an element in dimension $p + q + 1$ arising from $H_p(K) * H_q(L)$. Where does this unexpected one come from?

Let us write

$$kz = \partial d \quad \text{and} \quad lz' = \partial' d',$$

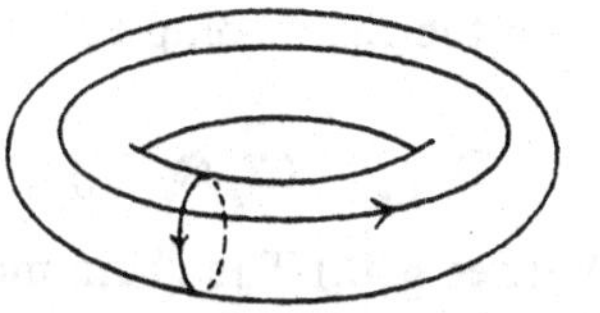

Figure 58.1

where d and d' are chains of K and L, respectively. The unexpected element in dimension $p + q + 1$ is exhibited by computing

$$\begin{aligned}\bar{\partial}(d \otimes d') &= (kz) \otimes d' + (-1)^{p+1} d \otimes (lz') \\ &= n\left[\frac{k}{n}(z \otimes d') + (-1)^{p+1}\frac{l}{n}(d \otimes z')\right].\end{aligned}$$

The expression in brackets is thus a cycle of dimension $p + q + 1$, and n times it is a boundary! This is the element that we were seeking.

The universal coefficient theorem for homology

We now derive the universal coefficient theorem for homology as a corollary of the Künneth theorem.

Proof of Theorem 55.1. Let $\mathcal{C}$ be a free chain complex; let G be an abelian group. We define a chain complex $\mathcal{C}'$ by the equation

$$C'_p = \begin{cases} G & \text{if } p = 0, \\ 0 & \text{if } p \neq 0. \end{cases}$$

All the boundary operators in $\mathcal{C}'$ vanish. The chain complex $\mathcal{C}'$ is not in general free, but the cycles do form a direct summand in the chains, in each dimension (trivially).

We compute

$$\begin{aligned}(\mathcal{C} \otimes \mathcal{C}')_m &= \oplus_{p+q=m} C_p \otimes C'_q = C_m \otimes G, \\ \bar{\partial}(c_m \otimes g) &= \partial c_m \otimes g + 0.\end{aligned}$$

Thus $\{\mathcal{C} \otimes \mathcal{C}', \bar{\partial}\}$ equals the chain complex $\{\mathcal{C} \otimes G, \partial \otimes i_G\}$. We now apply the Künneth theorem, and use the facts

$$\begin{aligned}H_0(\mathcal{C}') &= G, \\ H_p(\mathcal{C}') &= 0 \quad \text{for } p \neq 0.\end{aligned}$$

The universal coefficient theorem for homology is an immediate consequence. □

The Künneth theorem with field coefficients

Suppose $\mathcal{E}$ and $\mathcal{E}'$ are chain complexes whose chain groups are vector spaces over a field F, and whose boundary operators are linear transformations.

In this situation, we can form the chain complex $\mathscr{E} \otimes_F \mathscr{E}'$, whose chain group in dimension m is

$$\oplus_{p+q=m}(E_p \otimes_F E'_q).$$

It is a vector space over F. (See §50.) The chain map $\bar{\partial}$ is a linear transformation, so its kernel and image are vector spaces.

The Künneth theorem simplifies considerably in this case:

Theorem 58.4. *Suppose the chain complexes $\mathscr{E}$ and $\mathscr{E}'$ are vector spaces over the field F, and the boundary operators are vector space homomorphisms. Then $H_p(\mathscr{E})$ and $H_q(\mathscr{E}')$ are vector spaces over F, and there is a natural isomorphism of vector spaces*

$$\oplus_{p+q=m} H_p(\mathscr{E}) \otimes_F H_q(\mathscr{E}') \to H_m(\mathscr{E} \otimes_F \mathscr{E}').$$

Proof. The proof of the Künneth theorem proceeds unchanged through its first four steps, if $\otimes$ is replaced throughout by $\otimes_F$. A change first appears in Step 5, where we take the sequence

$$0 \to B_p \to Z_p \to H_p(\mathscr{E}) \to 0$$

and tensor it with $H_q(\mathscr{E}')$. Because we are in the context of vector spaces, we obtain the sequence

$$0 \longrightarrow B_p \otimes_F H_q(\mathscr{E}') \xrightarrow{\beta_{m+1}} Z_p \otimes_F H_q(\mathscr{E}') \longrightarrow H_p(\mathscr{E}) \otimes_F H_q(\mathscr{E}') \longrightarrow 0.$$

That is, *no torsion product term appears*. As a result, the homomorphism β_{m+1} has vanishing kernel for all m. □

The Künneth theorem with field coefficients follows:

Theorem 58.5. *Let $\mathscr{C}$ and $\mathscr{C}'$ be free chain complexes; let F be a field. There is a natural isomorphism*

$$\oplus_{p+q=m} H_p(\mathscr{C}; F) \otimes_F H_q(\mathscr{C}'; F) \to H_m(\mathscr{C} \otimes \mathscr{C}'; F).$$

Proof. We apply the preceding theorem to the chain complexes $\mathscr{E} = \mathscr{C} \otimes F$ and $\mathscr{E}' = \mathscr{C}' \otimes F$, which are vector spaces over F. We obtain a natural isomorphism

$$\oplus_{p+q=m} H_p(\mathscr{E}) \otimes_F H_q(\mathscr{E}') \to H_m(\mathscr{E} \otimes_F \mathscr{E}').$$

Now $H_p(\mathscr{E}) = H_p(\mathscr{C}; F)$ and $H_q(\mathscr{E}') = H_q(\mathscr{C}'; F)$ by definition. It remains to show there is a natural vector space isomorphism

$$f : \mathscr{E} \otimes_F \mathscr{E}' \to (\mathscr{C} \otimes \mathscr{C}') \otimes F$$

that is also a chain map. Then our theorem is proved. Let us define

$$f : C_p \times F \times C'_q \times F \to C_p \otimes C'_q \otimes F$$

by the equation

$$(*) \qquad f(x,\alpha,y,\beta) = (x \otimes y) \otimes \alpha\beta.$$

This map is multilinear, so it gives us a homomorphism

$$f' : (C_p \otimes F) \otimes (C'_q \otimes F) \to C_p \otimes C'_q \otimes F.$$

Now f' is in fact a vector space homomorphism in each variable separately. For instance, if y and β are fixed,

$$\begin{aligned} f'(\gamma(x \otimes \alpha) \otimes (y \otimes \beta)) &= f'((x \otimes \gamma\alpha) \otimes (y \otimes \beta)) \\ &= x \otimes y \otimes (\gamma\alpha)\beta = \gamma(x \otimes y \otimes \alpha\beta). \end{aligned}$$

Thus f' induces a homomorphism

$$f'' : (C_p \otimes F) \otimes_F (C'_q \otimes F) \to C_p \otimes C'_q \otimes F.$$

The fact that f'' is an isomorphism is easy to prove. If $\{x_\alpha\}$ is a basis for $\mathcal{C}$ and $\{y_\beta\}$ is a basis for $\mathcal{C}'$, then $\{x_\alpha \otimes y_\beta\}$ is a basis for $\mathcal{C} \otimes \mathcal{C}'$. The elements $\{x_\alpha \otimes 1\}$ and $\{y_\beta \otimes 1\}$ form vector space bases for $\mathcal{C} \otimes F$ and $\mathcal{C}' \otimes F$, respectively; hence the elements $\{(x_\alpha \otimes 1) \otimes_F (y_\beta \otimes 1)\}$ form a vector space basis for $(\mathcal{C} \otimes F) \otimes_F (\mathcal{C}' \otimes F)$. Their images under f'' are the elements $\{(x_\alpha \otimes y_\beta) \otimes 1\}$, which form a vector space basis for $(\mathcal{C} \otimes \mathcal{C}') \otimes F$. □

Example 4. If K and L are simplicial complexes, with K or L locally finite, there is a vector space isomorphism

$$\oplus_{p+q=m} H_p(K; F) \otimes_F H_q(L; F) \to H_m(|K| \times |L|; F).$$

This fact follows from the preceding theorem, once we note that since for all i,

$$H_i(\mathcal{C}(K) \otimes \mathcal{C}(L)) \simeq H_i(|K| \times |L|),$$

the same holds with arbitrary coefficients F, by Theorem 45.1 and Theorem 51.1.

EXERCISES

1. Show the Künneth theorem fails in reduced homology.

2. Let K_0 and L_0 be subcomplexes of K and L, respectively; assume K or L locally finite. Then $|K| \times |L|$ is triangulable, as usual. Show there is a chain map that induces homology isomorphisms

$$\mathcal{C}(K,K_0) \otimes \mathcal{C}(L,L_0) \to \mathcal{D}(K \times L, (K_0 \times L) \cup (K \times L_0)).$$

Derive a Künneth sequence in relative simplicial homology.

3. Let $\mathcal{C}$ be a free chain complex. Show there is a natural exact sequence

$$0 \to \oplus_{p+q=m} H_p(\mathcal{C}) \otimes H_q(\mathcal{C}'; G) \to H_m(\mathcal{C} \otimes \mathcal{C}'; G) \to \oplus_{p+q=m} H_{p-1}(\mathcal{C}) * H_q(\mathcal{C}'; G) \to 0.$$

Show this sequence splits if $\mathcal{C}'$ is free and either G is torsion-free or $H_i(\mathcal{C}')$ is torsion-free for all i.

§59. THE EILENBERG-ZILBER THEOREM

Now we show that for an arbitrary pair of topological spaces, the chain complex $\mathcal{S}(X) \otimes \mathcal{S}(Y)$ can be used to compute the singular homology of $X \times Y$. The crucial tool for the proof is the acyclic model theorem (§32).

Lemma 59.1. *If $\{\mathcal{C},\epsilon\}$ and $\{\mathcal{C}',\epsilon'\}$ are acyclic augmented chain complexes, and if $\mathcal{C}$ is free, then $\{\mathcal{C} \otimes \mathcal{C}',\bar{\epsilon}\}$ is acyclic.*

Proof. We show that $H_m(\mathcal{C} \otimes \mathcal{C}')$ is infinite cyclic if $m = 0$, and vanishes otherwise. This proves that $\mathcal{C} \otimes \mathcal{C}'$ is acyclic relative to any augmentation.

We can apply the Künneth theorem, since $\mathcal{C}$ is free. We have

$$H_p(\mathcal{C}) * H_q(\mathcal{C}') = 0$$

for all p and q, because each group involved either vanishes or is infinite cyclic. Therefore,

$$H_m(\mathcal{C} \otimes \mathcal{C}') \simeq \oplus_{p+q=m} H_p(\mathcal{C}) \otimes H_q(\mathcal{C}').$$

If m is positive, at least one of the groups $H_p(\mathcal{C})$ or $H_q(\mathcal{C}')$ vanishes, so $H_m(\mathcal{C} \otimes \mathcal{C}') = 0$. If $m = 0$, then

$$H_0(\mathcal{C} \otimes \mathcal{C}') \simeq H_0(\mathcal{C}) \otimes H_0(\mathcal{C}') \simeq \mathbf{Z} \otimes \mathbf{Z} \simeq \mathbf{Z}. \quad \square$$

Theorem 59.2 (The Eilenberg-Zilber theorem). *For every pair X, Y of topological spaces, there are chain maps*

$$\mathcal{S}(X) \otimes \mathcal{S}(Y) \underset{\nu}{\overset{\mu}{\rightleftarrows}} \mathcal{S}(X \times Y)$$

that are chain-homotopy inverse to each other; they are natural with respect to chain maps induced by continuous maps.

Proof. Consider the category of pairs of topological spaces and pairs of continuous maps. We define two functors from this category to the category of augmented chain complexes, as follows:

$$G(X,Y) = \mathcal{S}(X) \otimes \mathcal{S}(Y), \qquad G'(X,Y) = \mathcal{S}(X \times Y),$$
$$G(f,g) = f_\# \otimes g_\#. \qquad G'(f,g) = (f \times g)_\#.$$

It is easy to check the functorial properties.

Let $\mathcal{M}$ be the collection of all pairs (Δ_p,Δ_q), where Δ_k denotes the standard k-simplex. We show that G and G' are both acyclic *and* free relative to the collection of models $\mathcal{M}$. Then the theorem follows immediately from the acyclic model theorem.

Acyclicity is easy to check. The augmented chain complex $\{\mathcal{S}(\Delta_p \times \Delta_q),\epsilon\}$ is acyclic because $\Delta_p \times \Delta_q$ is a contractible space. The chain complex $\{\mathcal{S}(\Delta_p) \otimes \mathcal{S}(\Delta_q),\bar{\epsilon}\}$ is acyclic by the preceding lemma.

We show that G' is free. Given a non-negative integer m, let the indexed family of objects from $\mathcal{M}$ corresponding to this integer consist of the single pair

(Δ_m,Δ_m), and let the corresponding element of $S_m(\Delta_m \times \Delta_m)$ be the diagonal map $d : \Delta_m \to \Delta_m \times \Delta_m$. We show that as (f,g) ranges over all pairs of continuous maps

$$(f,g) : (\Delta_m,\Delta_m) \to (X,Y),$$

then the elements

$$[G'(f,g)](d) = (f \times g)_\#(d) = (f \times g) \circ d$$

are distinct and range over a *basis* for the group $S_m(X \times Y)$. But

$$(f \times g) \circ d : \Delta_m \to X \times Y$$

is just the map T given in coordinates by $T(z) = (f(z),g(z))$. For each pair (f,g) of maps, we have a different singular simplex T, and every singular simplex on $X \times Y$ is of this form. Thus these elements do form a basis for $S_m(X \times Y)$.

Now we show G is free. Given a non-negative integer m, let the indexed family of objects from $\mathcal{M}$ corresponding to this integer be the family

$$\{(\Delta_p,\Delta_q) \mid p + q = m\}.$$

The index set in this case consists of all pairs (p,q) of non-negative integers for which $p + q = m$. Let the element of $(\mathcal{S}(\Delta_p) \otimes \mathcal{S}(\Delta_q))_m$ corresponding to the index (p,q) be the element $i_p \otimes i_q$, where $i_k : \Delta_k \to \Delta_k$ denotes the identity singular simplex on the space Δ_k. We need to show that as (p,q) ranges over its index set and (f,g) ranges over all maps

$$(f,g) : (\Delta_p,\Delta_q) \to (X,Y),$$

then the elements

$$[G(f,g)](i_p \otimes i_q) = (f_\# \otimes g_\#)(i_p \otimes i_q) = f \otimes g$$

are distinct and form a basis for the group $(\mathcal{S}(X) \otimes \mathcal{S}(Y))_m$. But $S_p(X)$ is free with the set of all singular simplices $f : \Delta_p \to X$ as a basis. Similarly, $S_q(Y)$ has as basis the set of all singular simplices $g : \Delta_q \to Y$. By Theorem 50.8, $S_p(X) \otimes S_q(Y)$ has the elements $f \otimes g$ as a basis. The proof is now complete. □

Theorem 59.3 (The Künneth theorem for topological spaces). *Given topological spaces X, Y, there is an exact sequence*

$$0 \to \bigoplus_{p+q=m} H_p(X) \otimes H_q(Y) \to H_m(X \times Y) \to \bigoplus_{p+q=m} H_{p-1}(X) * H_q(Y) \to 0.$$

It is natural with respect to homomorphisms induced by continuous maps. It splits, but not naturally. □

The monomorphism

$$H_p(X) \otimes H_q(Y) \to H_m(X \times Y)$$

of this theorem is called the **homology cross product.** It equals the composite

$$H_p(X) \otimes H_q(Y) \xrightarrow{\Theta} H_m(\mathcal{S}(X) \otimes \mathcal{S}(Y)) \xrightarrow{\mu_*} H_m(X \times Y),$$

where Θ is induced by inclusion and μ is the Eilenberg-Zilber chain equivalence. We shall not have much occasion to use the homology cross product, but a similar cross product in cohomology will prove to be quite important, as we shall see in the next two sections.

Theorem 59.4. *Let X and Y be topological spaces; let F be a field. There is a natural isomorphism*

$$\oplus_{p+q=m} H_p(X;F) \otimes_F H_q(Y;F) \rightarrow H_m(X \times Y;F). \quad \square$$

For later applications, it will be convenient to have a specific formula for one of the Eilenberg-Zilber chain equivalences. Here is such a formula.

Theorem 59.5. *Let* $\pi_1 : X \times Y \rightarrow X$ *and* $\pi_2 : X \times Y \rightarrow Y$ *be projections. Define*

$$\nu : S_m(X \times Y) \rightarrow \oplus_{p+q=m} S_p(X) \otimes S_q(Y)$$

by the equation

$$\nu(T) = \sum_{i=0}^{m} [\pi_1 \circ T \circ l(\epsilon_0, \ldots, \epsilon_i)] \otimes [\pi_2 \circ T \circ l(\epsilon_i, \ldots, \epsilon_m)].$$

Then ν *is a natural chain equivalence that preserves augmentation.*

Proof. That ν is natural and augmentation-preserving is easy to check. The fact that ν is a chain map is a straightforward computation, about as messy as such computations usually are. In fact, when one computes directly, one finds that the expression for $\bar{\partial}\nu(T)$ equals the expression for $\nu(\partial T)$ plus some extra terms. One has

$$\begin{aligned} &\bar{\partial}\nu(T) - \nu(\partial T) \\ &= \sum_{i=1}^{m} (-1)^i [T_X \circ l(\epsilon_0, \ldots, \hat{\epsilon}_i)] \otimes [T_Y \circ l(\epsilon_i, \ldots, \epsilon_m)] \\ &+ \sum_{i=0}^{m-1} (-1)^i [T_X \circ l(\epsilon_0, \ldots, \epsilon_i)] \otimes [T_Y \circ l(\hat{\epsilon}_i, \ldots, \epsilon_m)]. \end{aligned}$$

where $T_X = \pi_1 \circ T$ and $T_Y = \pi_2 \circ T$. These terms cancel in pairs.

It then follows from the proof of the Eilenberg-Zilber theorem that ν is a chain equivalence. $\square$

The original proof of the Eilenberg-Zilber theorem did not use acyclic models. It used the chain map ν, and a chain map μ defined by a similar formula. One wrote down formulas to show that both $\nu \circ \mu$ and $\mu \circ \nu$ were chain homotopic to identity maps.

EXERCISES

1. Show there is a (non-natural) isomorphism

$$H_m(X \times Y) \simeq \oplus_{p+q=m} H_p(X; H_q(Y)).$$

2. (a) Let $A \subset X$ and $B \subset Y$. Use the Eilenberg-Zilber theorem to show that there is a natural chain equivalence

$$\mathcal{S}(X,A) \otimes \mathcal{S}(Y,B) \to \frac{\mathcal{S}(X \times Y)}{\mathcal{S}(X \times B) + \mathcal{S}(A \times Y)}.$$

(b) State and prove a Künneth theorem in relative singular homology, assuming $\{X \times B, A \times Y\}$ is excisive.

3. Let ν be the chain map defined in Theorem 59.5. Show that it is natural and preserves augmentation. Check that ν is a chain map.

4. (a) Let X and Y be CW complexes. Show that the tensor product $\mathcal{D}(X) \otimes \mathcal{D}(Y)$ of their cellular chain complexes can be used to compute the homology of $X \times Y$.
*(b) Show in fact that $\mathcal{D}(X) \otimes \mathcal{D}(Y)$ and $\mathcal{S}(X) \otimes \mathcal{S}(Y)$ are chain equivalent.
(c) Compute the homology of $P^2 \times P^3$ by computing the groups and boundary operators in the chain complex $\mathcal{D}(P^2) \otimes \mathcal{D}(P^3)$.
(d) Check the results of (c) by using the Künneth theorem.

5. Compute the homology of the following:
(a) $S^2 \times P^5$
(b) $P^3 \times P^5$
(c) $S^3 \times L(n,k)$
(d) $L(n,k) \times L(m,j)$
(e) $S^1 \times S^1 \times S^3$
(f) $CP^2 \times CP^3$

6. Suppose M is a compact, connected n-manifold, such that $H_n(M) \simeq \mathbf{Z}$. There are relations among the betti numbers and torsion coefficients of M. Use the results of the preceding exercise to formulate a conjecture concerning these relations.

*§60. THE KÜNNETH THEOREM FOR COHOMOLOGY

One can always compute the cohomology of a product space by first computing its homology from the Künneth theorem, and then applying the universal coefficient theorem for cohomology. But one might hope for a version of the Künneth theorem that applies to cohomology directly. We prove such a theorem now. It has applications to computing the cohomology ring of a product space, which we shall study in the next section.

First, we introduce the cross product operation in cohomology. We give only the definition here, postponing discussion of its properties to the next section.

Throughout, let R be a commutative ring with unity element.

Let $\mathcal{C}$ and $\mathcal{C}'$ be chain complexes. Corresponding to the chain complex $\mathcal{C}$, there is a cochain complex $\operatorname{Hom}(\mathcal{C},R)$, whose p-dimensional group is $\operatorname{Hom}(C_p,R)$. Similarly, one has the cochain complex $\operatorname{Hom}(\mathcal{C}',R)$. We make the tensor product of these cochain complexes into a cochain complex in the way one might expect: The m-dimensional cochain group of $\operatorname{Hom}(\mathcal{C},R)\otimes \operatorname{Hom}(\mathcal{C}',R)$ is the group

$$\bigoplus_{p+q=m} \operatorname{Hom}(C_p,R)\otimes \operatorname{Hom}(C'_q,R).$$

And the coboundary operator is given by the rule

$$\bar{\delta}(\phi^p\otimes\psi^q)=\delta\phi^p\otimes\psi^q+(-1)^p\phi^p\otimes\delta'\psi^q.$$

Roughly speaking, this cochain complex is obtained by Homming first, and then tensoring. Of course, one could instead tensor first and then Hom. One obtains the cochain complex $\operatorname{Hom}((\mathcal{C}\otimes\mathcal{C}'),R)$. Its group in dimension m is

$$\operatorname{Hom}((\mathcal{C}\otimes\mathcal{C}')_m,R).$$

Its coboundary operator is the dual of $\bar{\partial}$, so we denote it by $\bar{\bar{\partial}}$.

There is a natural homomorphism of the first of these cochain complexes to the second, described as follows:

Definition. Let $p+q=m$; let

$$\theta:\operatorname{Hom}(C_p,R)\otimes \operatorname{Hom}(C'_q,R)\to \operatorname{Hom}((\mathcal{C}\otimes\mathcal{C}')_m,R)$$

be the homomorphism defined by the equation

$$\langle\theta(\phi^p\otimes\psi^q),c_r\otimes c'_s\rangle=\langle\phi^p,c_r\rangle\cdot\langle\psi^q,c'_s\rangle.$$

Here we make the convention that $\langle\phi^p,c_r\rangle$ equals 0 unless $p=r$, and $\langle\psi^q,c'_s\rangle=0$ unless $q=s$. Said differently, we agree that any element ϕ^p of $\operatorname{Hom}(C_p,R)$ is automatically extended, without change of notation, to a homomorphism of $\oplus C_r$ into R, by letting it be zero on all summands other than C_p.

Lemma 60.1. *The homomorphism θ is a natural cochain map.*

Proof. Let $\phi\in\operatorname{Hom}(C_p,R)$ and $\psi\in\operatorname{Hom}(C'_q,R)$, where $p+q=m$. Let $r+s=m+1$. We compute

$$\langle\theta(\bar{\delta}(\phi\otimes\psi)),c_r\otimes c'_s\rangle=\langle\delta\phi,c_r\rangle\langle\psi,c'_s\rangle+(-1)^p\langle\phi,c_r\rangle\langle\delta'\psi,c'_s\rangle,$$

$$\begin{aligned}\langle\bar{\bar{\partial}}\theta(\phi\otimes\psi),c_r\otimes c'_s\rangle&=\langle\theta(\phi\otimes\psi),\bar{\partial}(c_r\otimes c'_s)\rangle\\&=\langle\phi,\partial c_r\rangle\langle\psi,c'_s\rangle+(-1)^r\langle\phi,c_r\rangle\langle\psi,\partial' c'_s\rangle.\end{aligned}$$

These expressions are equal except for the signs $(-1)^p$ and $(-1)^r$ on the final terms. However, if $p \neq r$, then $\langle \phi, c_r \rangle = 0$ and these final terms vanish.

It is apparent from its definition that if $f : C_p \to D_p$ and $g : C'_q \to D'_q$ are homomorphisms, then

$$\theta \circ (\tilde{f} \otimes \tilde{g}) = (\widetilde{f \otimes g}) \circ \theta.$$

Thus θ is natural. □

Now we define a homomorphism in cohomology analogous to the map Θ that appeared in the homology version of the Künneth theorem.

Definition. Define a homomorphism

$$\Theta : H^p(\mathcal{C}; R) \otimes H^q(\mathcal{C}'; R) \to H^m(\mathcal{C} \otimes \mathcal{C}'; R),$$

where $m = p + q$, as follows: Let ϕ and ψ be cocycles of $\mathcal{C}$ and $\mathcal{C}'$ of dimensions p and q, respectively. Then define

$$\Theta(\{\phi\} \otimes \{\psi\}) = \{\theta(\phi \otimes \psi)\}.$$

Note that $\theta(\phi \otimes \psi)$ is a cocycle, because

$$\bar{\bar{\delta}}\theta(\phi \otimes \psi) = \theta(\bar{\delta}(\phi \otimes \psi)) = \theta(\delta\phi \otimes \psi \pm \phi \otimes \delta'\psi),$$

which vanishes because ϕ and ψ are cocycles. A similar computation shows Θ is well-defined.

Definition. Let $m = p + q$. We define the **cohomology cross product** as the composite

$$H^p(X; R) \otimes H^q(Y; R) \xrightarrow{\Theta} H^m(\mathcal{S}(X) \otimes \mathcal{S}(Y); R) \xrightarrow{\nu^*} H^m(X \times Y; R),$$

where ν is the Eilenberg-Zilber chain equivalence. The image of $\alpha^p \otimes \beta^q$ under this homomorphism is denoted $\alpha^p \times \beta^q$.

Let us note that just as was the case with the cup product, the homomorphism Θ can be defined for other coefficients than R. For instance, our formula for θ can be used to define a homomorphism

$$\theta : \operatorname{Hom}(C_p, \mathbf{Z}) \otimes \operatorname{Hom}(C'_q, G) \to \operatorname{Hom}((\mathcal{C} \otimes \mathcal{C}')_m, G)$$

that is a natural cochain map. It is the same as the previous map when $R = G = \mathbf{Z}$.

Similarly, one notes that both $\operatorname{Hom}(C_p, R)$ and $\operatorname{Hom}(C'_q, R)$ have R-module structures. From the definition of θ, it follows that

$$\theta(\alpha\phi \otimes \psi) = \theta(\phi \otimes \alpha\psi) = \alpha\theta(\phi \otimes \psi).$$

This means that θ induces a homomorphism

$$\theta : \operatorname{Hom}(C_p, R) \otimes_R \operatorname{Hom}(C'_q, R) \to \operatorname{Hom}((\mathcal{C} \otimes \mathcal{C}')_m, R)$$

that is in fact a homomorphism of R-modules. The case of particular interest to us occurs when R is a field F; in this case θ is a linear transformation of vector spaces.

Now in both these cases, one has an induced homomorphism Θ on the cohomology level, and thus a cohomology cross product operation as well. In one case, the cross product is a homomorphism

$$H^p(X) \otimes H^q(Y; G) \rightarrow H_m(X \times Y; G).$$

In the other case, it is an R-module homomorphism

$$H^p(X; R) \otimes_R H^q(Y; R) \rightarrow H^m(X \times Y; R).$$

The most general cross product is defined relative to an arbitrary **coefficient pairing,** which is a homomorphism $A \otimes_R A' \rightarrow A''$ of R-modules. We shall not need this degree of generality.

The Künneth theorem with integer coefficients

Now we prove the Künneth theorem in cohomology. We restrict ourselves to integer coefficients for the present.

The idea of the proof is to re-index the groups $\text{Hom}(C_p, \mathbf{Z})$ so as to make $\text{Hom}(\mathcal{C}, \mathbf{Z})$ into a chain complex, and then apply the Künneth theorem we have already proved. One difficulty is that $\text{Hom}(C_p, \mathbf{Z})$ is not free in general, so some restrictive hypotheses are needed.

There is an additional difficulty as well. By definition, the cohomology of $\mathcal{C} \otimes \mathcal{C}'$ is computed from the cochain complex $\text{Hom}((\mathcal{C} \otimes \mathcal{C}')_m, \mathbf{Z})$. This is not the same as the cochain complex

$$\text{Hom}(\mathcal{C}, \mathbf{Z}) \otimes \text{Hom}(\mathcal{C}', \mathbf{Z})$$

to which we are going to apply the Künneth theorem. Here is where the homomorphism θ comes in. We prove the following lemma:

Lemma 60.2. *Let $\mathcal{C}$ and $\mathcal{C}'$ be chain complexes that vanish below a certain dimension. Assume $\mathcal{C}$ is free and finitely generated in each dimension. Then*

$$\theta : \oplus_{p+q=m} \text{Hom}(C_p, \mathbf{Z}) \otimes \text{Hom}(C'_q, \mathbf{Z}) \rightarrow \text{Hom}((\mathcal{C} \otimes \mathcal{C}')_m, \mathbf{Z})$$

is an isomorphism.

Proof. For purposes of this proof, we break θ up into a composite of two maps:

$$\text{Hom}(C_p, \mathbf{Z}) \otimes \text{Hom}(C'_q, \mathbf{Z}) \xrightarrow{M} \text{Hom}(C_p \otimes C'_q, \mathbf{Z}),$$

$$\oplus \text{Hom}(C_p \otimes C'_q, \mathbf{Z}) \xrightarrow{e} \text{Hom}(\oplus C_p \otimes C'_q, \mathbf{Z}),$$

where the summations extend over $p + q = m$. The map M satisfies the equation

$$\langle M(\phi^p \otimes \psi^q), c_p \otimes c'_q \rangle = \langle \phi^p, c_p \rangle \langle \psi^q, c'_q \rangle.$$

The map e is defined as follows: If $F : C_p \otimes C'_q \rightarrow \mathbf{Z}$, then $e(F)$ is the map of

$\oplus_{r+s=m} C_r \otimes C'_s$ into $\mathbf{Z}$ that equals F on the summand $C_p \otimes C'_q$ and vanishes on all other summands.

Step 1. Because C_p is free of finite rank, it follows that

$$M : \text{Hom}(C_p,\mathbf{Z}) \otimes \text{Hom}(C'_q,\mathbf{Z}) \to \text{Hom}(C_p \otimes C'_q,\mathbf{Z})$$

is an isomorphism: Let $a_1, \ldots, a_k$ be a basis for C_p; let $\tilde{a}_1, \ldots, \tilde{a}_k$ be the dual basis for $\text{Hom}(C_p,\mathbf{Z})$. The group on the left is the direct sum of k copies of $\text{Hom}(C'_q,\mathbf{Z})$, the ith summand being the subgroup

$$\langle \tilde{a}_i \rangle \otimes \text{Hom}(C'_q,\mathbf{Z}),$$

where $\langle \tilde{a}_i \rangle$ is the infinite cyclic group generated by $\tilde{a}_i$. Now $C_p \otimes C'_q$ is the direct sum of k copies of C'_q, the ith copy being $\langle a_i \rangle \otimes C'_q$. So $\text{Hom}(C_p \otimes C'_q, \mathbf{Z})$ is the direct sum of k copies of $\text{Hom}(C'_q,\mathbf{Z})$, the ith copy consisting of all homomorphisms that vanish on $\langle a_j \rangle \otimes C'_q$ for $j \neq i$. Now M is an isomorphism of $\langle \tilde{a}_i \rangle \otimes \text{Hom}(C'_q,\mathbf{Z})$ with the ith summand of $\text{Hom}(C_p \otimes C'_q,\mathbf{Z})$, so M is an isomorphism.

Step 2. The map e is clearly injective, and its image consists of those homomorphisms of $\oplus C_p \otimes C'_q$ into $\mathbf{Z}$ that vanish on all but finitely many summands. Thus e is not in general surjective. However, since $\mathcal{C}$ and $\mathcal{C}'$ vanish below a certain dimension, the summation

$$\oplus_{p+q=m}(C_p \otimes C'_q)$$

is finite. Hence under our present hypotheses, e is an isomorphism. □

Theorem 60.3 (The Künneth theorem for cohomology). *Let $\mathcal{C}$ and $\mathcal{C}'$ be chain complexes that vanish below a certain dimension. Suppose $\mathcal{C}$ is free and finitely generated in each dimension. Then there is a natural exact sequence*

$$0 \to \oplus_{p+q=m} H^p(\mathcal{C}) \otimes H^q(\mathcal{C}') \xrightarrow{\Theta} H^m(\mathcal{C} \otimes \mathcal{C}') \to$$
$$\oplus_{p+q=m} H^{p+1}(\mathcal{C}) * H^q(\mathcal{C}') \to 0.$$

It splits (but not naturally) if $\mathcal{C}'$ is free and finitely generated in each dimension.

Proof. Let $\mathcal{E}$ and $\mathcal{E}'$ be the chain complexes whose chain groups in dimension $-p$ are defined by the equations

$$E_{-p} = \text{Hom}(C_p,\mathbf{Z}) \quad \text{and} \quad E'_{-p} = \text{Hom}(C'_p,\mathbf{Z}),$$

and whose boundary operators are δ and δ', respectively. Since $\mathcal{E}$ is free, the Künneth theorem applies. There is an exact sequence

$$0 \to \oplus(H_{-p}(\mathcal{E}) \otimes H_{-q}(\mathcal{E}')) \to H_{-m}(\mathcal{E} \otimes \mathcal{E}') \to$$
$$\oplus(H_{-p-1}(\mathcal{E}) * H_{-q}(\mathcal{E}')) \to 0,$$

where the summations extend over $-p - q = -m$. Now $H_{-p}(\mathcal{E}) = H^p(\mathcal{C})$

and $H_{-q}(\mathcal{E}') = H^q(\mathcal{C}')$, by definition. Furthermore, by preceding lemma, there is a natural isomorphism of cochain complexes

$$(\mathcal{E} \otimes \mathcal{E}')_{-m} = (\operatorname{Hom}(\mathcal{C},\mathbf{Z}) \otimes \operatorname{Hom}(\mathcal{C}',\mathbf{Z}))_m \xrightarrow{\theta} \operatorname{Hom}((\mathcal{C} \otimes \mathcal{C}')_m,\mathbf{Z}).$$

It induces a natural isomorphism

$$H_{-m}(\mathcal{E} \otimes \mathcal{E}') \to H^m(\mathcal{C} \otimes \mathcal{C}').$$

Combining these facts, we obtain the Künneth sequence for cohomology. Because the maps

$$H_{-p}(\mathcal{E}) \otimes H_{-q}(\mathcal{E}') \to H_{-m}(\mathcal{E} \otimes \mathcal{E}') \to H^m(\mathcal{C} \otimes \mathcal{C}')$$

are induced by inclusion and θ, respectively, their composite equals Θ.

If $\mathcal{C}'$ is free and finitely generated in each dimension, then $\mathcal{E}'$ is free and the sequence splits. □

Corollary 60.4. *In the preceding theorem, the hypothesis that $\mathcal{C}$ be finitely generated in each dimension can be replaced by the hypothesis that $H_i(\mathcal{C})$ be finitely generated for each i. Similarly, the hypothesis that $\mathcal{C}'$ be finitely generated in each dimension can be replaced by the hypothesis that $H_i(\mathcal{C}')$ be finitely generated for each i.*

Proof. The proof is similar to that of Theorem 56.2. If $\mathcal{C}$ is free and vanishes below a certain dimension, and $H_i(\mathcal{C})$ is finitely generated for each i, we choose a free chain complex $\mathcal{D}$ that vanishes below a certain dimension such that $H_i(\mathcal{D}) \simeq H_i(\mathcal{C})$ and $\mathcal{D}$ is finitely generated in each dimension. By Theorem 45.1, there is a chain map $\phi : \mathcal{C} \to \mathcal{D}$ that induces homology isomorphisms in all dimensions. By Theorem 46.2, it is a chain equivalence. Then

$$\phi \otimes i' : \mathcal{C} \otimes \mathcal{C}' \to \mathcal{D} \otimes \mathcal{C}'$$

is also a chain equivalence. (See the exercises of §57.)

Since the Künneth theorem holds for $\mathcal{D} \otimes \mathcal{C}'$, it holds for $\mathcal{C} \otimes \mathcal{C}'$ as well. Naturality of the sequence follows as in the proof of Theorem 56.2.

One proceeds similarly if $H_i(\mathcal{C}')$ is finitely generated for each i. One replaces $\mathcal{C}'$ by a chain-equivalent chain complex $\mathcal{D}'$ that is finitely generated in each dimension. □

Theorem 60.5. *Let X and Y be topological spaces; suppose $H_i(X)$ is finitely generated for each i. There is a natural exact sequence*

$$0 \to \oplus_{p+q=m} H^p(X) \otimes H^q(Y) \xrightarrow{\times} H^m(X \times Y) \to$$
$$\oplus_{p+q=m} H^{p+1}(X) * H^q(Y) \to 0.$$

It splits (but not naturally) if $H_i(Y)$ is finitely generated for each i. □

The Künneth theorem with field coefficients

Just as with homology, the Künneth theorem simplifies considerably if the cohomology groups are taken with coefficients in a field.

Theorem 60.6. *Let $\mathcal{C}$ and $\mathcal{C}'$ be chain complexes that vanish below a certain dimension, such that $\mathcal{C}$ is free and $H_i(\mathcal{C})$ is finitely generated in each dimension. Then there is a natural isomorphism*

$$\oplus_{p+q=m}(H^p(\mathcal{C};F)\otimes_F H^q(\mathcal{C}';F)) \xrightarrow{\Theta} H^m(\mathcal{C}\otimes\mathcal{C}';F).$$

Proof. By the usual argument, we may assume that $\mathcal{C}$ itself is finitely generated in each dimension. The chain complexes $\mathcal{E}$ and $\mathcal{E}'$ defined by

$$E_{-p} = \operatorname{Hom}(C_p,F) \qquad \text{and} \qquad E'_{-q} = \operatorname{Hom}(C'_q,F)$$

are vector spaces over F. By the "vector space version" of the Künneth theorem (Theorem 58.4), there exists a natural vector space isomorphism

$$\oplus_{p+q=m}H_{-p}(\mathcal{E})\otimes_F H_{-q}(\mathcal{E}') \rightarrow H_{-m}(\mathcal{E}\otimes_F\mathcal{E}').$$

Since $\mathcal{C}$ and $\mathcal{C}'$ vanish below a certain dimension and $\mathcal{C}$ is free and finitely generated in each dimension, the proof of Lemma 60.2 goes through essentially unchanged to show that the map θ defines a vector space isomorphism of $\mathcal{E}\otimes_F\mathcal{E}'$ with $\operatorname{Hom}(\mathcal{C}\otimes\mathcal{C}',F)$ that is a cochain map. Our theorem follows. □

Corollary 60.7. *Let X and Y be topological spaces; suppose $H_i(X)$ is finitely generated for each i. If F is a field, there is a natural vector space isomorphism*

$$\oplus_{p+q=m}H^p(X;F)\otimes_F H^q(Y;F) \xrightarrow{\times} H^m(X\times Y;F). \quad \square$$

EXERCISES

1. State and prove a Künneth theorem in relative singular cohomology. (See Exercise 2 of §59.)

2. Compute the cohomology groups of the following spaces. You may use either the Künneth theorem, or the universal coefficient theorem applied to the results of Exercise 5 of §59.
   (a) $S^2\times P^5$
   (b) $P^3\times P^5$
   (c) $S^3\times L(n,k)$
   (d) $L(n,k)\times L(m,j)$
   (e) $S^1\times S^1\times S^3$
   (f) $CP^2\times CP^3$

3. Let M be a compact, connected n-manifold such that $H_n(M) \simeq \mathbf{Z}$. There are relations between the homology and cohomology groups of M. Compare the computations of the preceding exercise with those you made in Exercise 5 of §59 to formulate a conjecture concerning these relations.

*§61. APPLICATION: THE COHOMOLOGY RING OF A PRODUCT SPACE

In the preceding section, we defined the cohomology cross product and used it in obtaining a Künneth theorem in cohomology. In this section, we explore some of its properties, and use it to obtain information about the ring structure of cohomology.

Throughout, we shall state our results first for the cross product with coefficients in R, and then explain to what extent they hold for the cross product with $(\mathbf{Z},G)$ coefficients.

Definition. The cross product is defined as the composite of the homomorphisms Θ and ν^* of cohomology. Because these homomorphisms are induced by the cochain maps θ and $\tilde{\nu}$, respectively, the cross product is induced by the cochain map $\tilde{\nu} \circ \theta$. Accordingly, we define

$$c^p \times c^q = \tilde{\nu}\theta(c^p \otimes c^q),$$

and call it the **cochain cross product.**

This operation induces the cohomology cross product. We compute it as follows: Let $m = p + q$. If $T : \Delta_m \to X \times Y$, then

$$\begin{aligned}\langle c^p \times c^q, T\rangle &= \langle \theta(c^p \otimes c^q), \nu(T)\rangle \\ &= \sum_{i=0}^{m} \langle c^p, \pi_1 \circ T \circ l(\epsilon_0, \ldots, \epsilon_i)\rangle \cdot \langle c^q, \pi_2 \circ T \circ l(\epsilon_i, \ldots, \epsilon_m)\rangle.\end{aligned}$$

Now only one of these terms is non-zero, the term for which $i = p$. We conclude the following:

Lemma 61.1. *The cochain cross product is given by the formula*

$$\langle c^p \times c^q, T\rangle = \langle c^p, \pi_1 \circ T \circ l(\epsilon_0, \ldots, \epsilon_p)\rangle \cdot \langle c^q, \pi_2 \circ T \circ l(\epsilon_p, \ldots, \epsilon_m)\rangle. \quad \square$$

This result holds for $(\mathbf{Z},G)$ coefficients as well. In words, it says that the value of $c^p \times c^q$ on T equals the value of c^p on the front face of the first component of T, times the value of c^q on the back face of the second component of T!

We now derive some properties of cross products.

Theorem 61.2. (a) *If* $\lambda: X \times Y \to Y \times X$ *is the map that reverses coordinates, then*

$$\lambda^*(\beta^q \times \alpha^p) = (-1)^{pq}\alpha^p \times \beta^q.$$

(b) *In* $H^*(X \times Y \times Z; R)$, *we have the equation* $(\alpha \times \beta) \times \gamma = \alpha \times (\beta \times \gamma)$.

(c) *Let* $\pi_1: X \times Y \to X$ *be the projection map. Let* 1_Y *be the unity element of the ring* $H^*(Y; R)$. *Then*

$$\pi_1^*(\alpha^p) = \alpha^p \times 1_Y.$$

Similarly,

$$\pi_2^*(\beta^q) = 1_X \times \beta^q.$$

Proof. (a) Consider the chain maps

$$\begin{array}{ccc} (\mathcal{S}(X) \otimes \mathcal{S}(Y))_m & \xleftarrow{\nu} & S_m(X \times Y) \\ \big\downarrow \omega & & \big\downarrow \lambda_\# \\ (\mathcal{S}(Y) \otimes \mathcal{S}(X))_m & \xleftarrow{\nu} & S_m(Y \times X), \end{array}$$

where we define ω by the equation

$$\omega(c_p \otimes c_q) = (-1)^{pq}(c_q \otimes c_p).$$

Then ω is a chain map, as you can check. (Without the sign, it would not be.) This diagram commutes up to chain homotopy, as we now show.

The chain maps $\lambda_\#$ and ν are known to preserve augmentation, and they are natural with respect to homomorphisms induced by continuous maps. One checks readily that ω has the same properties. Consider the functors

$$G(X,Y) = \mathcal{S}(X \times Y), \qquad G'(X,Y) = \mathcal{S}(Y) \otimes \mathcal{S}(X),$$
$$G(f,g) = (f \times g)_\#, \qquad G'(f,g) = g_\# \otimes f_\#.$$

We have already shown that these functors are free and acyclic relative to the collection of model objects $\mathcal{M} = \{(\Delta_p, \Delta_q)\}$. Therefore, any two natural transformations of G to G' are naturally chain homotopic. The chain maps $\omega \circ \nu$ and $\nu \circ \lambda_\#$ are two such natural transformations, as we noted; therefore they are chain homotopic.

Consider the dual diagram

$$\begin{array}{ccccc} \oplus S^p(X;R) \otimes S^q(Y;R) & \xrightarrow{\theta} & \operatorname{Hom}((\mathcal{S}(X) \otimes \mathcal{S}(Y))_m, R) & \xrightarrow{\tilde{\nu}} & S^m(X \times Y; R) \\ \big\uparrow \eta & & \big\uparrow \tilde{\omega} & & \big\uparrow \lambda^\# \\ \oplus S^q(Y;R) \otimes S^p(X;R) & \xrightarrow{\theta} & \operatorname{Hom}((\mathcal{S}(Y) \otimes \mathcal{S}(X))_m, R) & \xrightarrow{\tilde{\nu}} & S^m(Y \times X; R), \end{array}$$

where we define η by the equation

$$\eta(c^q \otimes c^p) = (-1)^{pq} c^p \otimes c^q.$$

The first square of this diagram commutes, as you can check. The second square commutes up to cochain homotopy. Therefore, the induced cohomology diagram actually commutes. Then (a) follows.

The proofs of (b) and (c) may be carried out, as was the proof of (a), using acyclic model arguments. It is easier, however, to use the cochain cross product formula. In particular, the proof of (b) is trivial if one uses this formula. To check (c), let z^p be a cocycle of X that represents the element α^p of $H^p(X;R)$. Then $\alpha^p \times 1_Y$ is represented by the cochain $z^p \times z^0$, where z^0 is the cocyle of Y whose value is 1 on each singular 0-simplex of Y. If $T:\Delta_p \to X \times Y$ is a singular p-simplex, we compute

$$\begin{aligned}\langle z^p \times z^0,T\rangle &= \langle z^p,\pi_1 \circ T \circ l(\epsilon_0,\ldots,\epsilon_p)\rangle \cdot \langle z^0,\pi_2 \circ T \circ l(\epsilon_p)\rangle \\ &= \langle z^p,\pi_1 \circ T\rangle \cdot 1 = \langle \pi_1^\#(z^p),T\rangle.\end{aligned}$$

Therefore, $\alpha^p \times 1_Y = \pi_1^*(\alpha^p)$, as claimed. The proof for $1_X \times \beta^q$ is similar. □

We note that all three of these properties, suitably interpreted, hold for the other version of cross product. Anticommutativity (a) holds if one uses $(\mathbf{Z},G)$ coefficients on the left side of the equation and $(G,\mathbf{Z})$ coefficients on the right. Associativity (b) holds if two of the coefficient groups are $\mathbf{Z}$ and one is G. The first statement in (c) holds if $1_Y \in H^0(Y)$ and $\alpha^p \in H^p(X;G)$, for then $\alpha^p \times 1_Y \in H^p(X \times Y;G)$ and the statement makes sense. A similar remark applies to $1_X \times \beta^q$.

The similarity of the formula for the cross product to the formula for the cup product may make you suspect that there is some relation between the two. It is the following:

Theorem 61.3. *Let $d: X \to X \times X$ be the diagonal map, given by $d(x) = (x,x)$. Then*

$$d^*(\alpha^p \times \beta^q) = \alpha^p \cup \beta^q.$$

This formula holds for both versions of the cohomology cross product.

Proof. Let z^p and z^q be representative cocycles for α^p and β^q, respectively. If $T:\Delta_{p+q} \to X$ is a singular simplex, we compute

$$\begin{aligned}\langle d^\#(z^p \times z^q),T\rangle &= \langle z^p \times z^q,d \circ T\rangle \\ &= \langle z^p,\pi_1 \circ (d \circ T) \circ l(\epsilon_0,\ldots,\epsilon_p)\rangle \cdot \langle z^q,\pi_2 \circ (d \circ T) \circ l(\epsilon_p,\ldots,\epsilon_{p+q})\rangle \\ &= \langle z^p \cup z^q,T\rangle.\end{aligned}$$

Here we use the fact that $\pi_1 \circ d = i_X = \pi_2 \circ d$. □

This theorem is due to Lefschetz. It shows, finally, why cohomology has a ring structure but homology does not. Both homology and cohomology have cross products, but only in cohomology can one form the composite of this cross product with the homomorphism induced by the diagonal map. In homology, one has the diagram

$$H_p(X) \otimes H_q(X) \xrightarrow{\times} H_{p+q}(X \times X) \xleftarrow{d_*} H_{p+q}(X).$$

Here d_* goes in the wrong direction!

Corollary 61.4. *Cup products are anticommutative. That is,*

$$\alpha^p \cup \beta^q = (-1)^{pq}\beta^q \cup \alpha^p.$$

This formula holds for both versions of cup product.

Proof. Let $\lambda : X \times X \to X \times X$ be the map that reverses coordinates. Then $\lambda \circ d = d$. We compute

$$\begin{aligned}\alpha^p \cup \beta^q &= d^*(\alpha^p \times \beta^q) = (\lambda \circ d)^*(\alpha^p \times \beta^q)\\ &= d^*((-1)^{pq}\beta^q \times \alpha^p) = (-1)^{pq}\beta^q \cup \alpha^p. \quad \square\end{aligned}$$

Now we compute cup products in the cohomology of a product space.

Theorem 61.5. *In the cohomology ring $H^*(X \times Y; R)$, the following formula holds:*

$$(\alpha \times \beta) \cup (\alpha' \times \beta') = (-1)^{(\dim \beta)(\dim \alpha')}(\alpha \cup \alpha') \times (\beta \cup \beta').$$

Proof. Let $\pi_1 : X \times Y \to X$ and $\pi_2 : X \times Y \to Y$ be the projection maps.

Step 1. We first prove the formula when β and α' are the unity elements of their respective cohomology rings. Let z^p be a cocycle representing α, and z^q a cocycle representing β'. Then $\alpha \times \beta'$ is represented by $z^p \times z^q$. Now

$$\begin{aligned}\langle z^p \times z^q, T\rangle &= \langle z^p, \pi_1 \circ T \circ l(\epsilon_0, \dots, \epsilon_p)\rangle \cdot \langle z^q, \pi_2 \circ T \circ l(\epsilon_p, \dots, \epsilon_{p+q})\rangle\\ &= \langle \pi_1^\#(z^p), T \circ l(\epsilon_0, \dots, \epsilon_p)\rangle \cdot \langle \pi_2^\#(z^q), T \circ l(\epsilon_p, \dots, \epsilon_{p+q})\rangle\\ &= \langle \pi_1^\#(z^p) \cup \pi_2^\#(z^q), T\rangle.\end{aligned}$$

Thus we have the desired formula,

$$\alpha \times \beta' = \pi_1^*(\alpha) \cup \pi_2^*(\beta') = (\alpha \times 1_Y) \cup (1_X \times \beta').$$

Step 2. We now prove the formula when β and β' equal the unity element of their cohomology ring. By naturality of cup products,

$$\pi_1^*(\alpha) \cup \pi_1^*(\alpha') = \pi_1^*(\alpha \cup \alpha').$$

This formula says that

$$(\alpha \times 1_Y) \cup (\alpha' \times 1_Y) = (\alpha \cup \alpha') \times 1_Y.$$

Step 3. Finally, we prove the general case. We compute

$$\begin{aligned}(\alpha \times \beta) \cup (\alpha' \times \beta') &= (\alpha \times 1_Y) \cup (1_X \times \beta) \cup (\alpha' \times 1_Y) \cup (1_X \times \beta') \\ &= (-1)^{(\dim \beta)(\dim \alpha')}(\alpha \times 1_Y) \cup (\alpha' \times 1_Y) \cup \\ &\qquad\qquad (1_X \times \beta) \cup (1_X \times \beta') \\ &= (-1)^{(\dim \beta)(\dim \alpha')}(\alpha \cup \alpha') \times (\beta \cup \beta'). \quad \square\end{aligned}$$

This theorem enables us to compute the cup product of certain elements of the cohomology ring $H^*(X \times Y; R)$ in terms of the cup product operations in $H^*(X)$ and $H^*(Y)$. In the case where the Künneth theorem for cohomology holds, we can state this fact more formally, as we do in the following theorem.

Definition. Given X and Y, we give the tensor product

$$H^*(X; R) \otimes_R H^*(Y; R)$$

the structure of ring by defining

$$(\alpha \otimes \beta) \cup (\alpha' \otimes \beta') = (-1)^{pq}(\alpha \cup \alpha') \otimes (\beta \cup \beta'),$$

if $\beta \in H^p(Y; R)$ and $\alpha' \in H^q(X; R)$.

You can check this operation is well-defined, and satisfies the axioms for a ring. In particular, the signs come out right for associativity.

Theorem 61.6. *Suppose $H_i(X)$ is finitely generated for all i. Then the cross product defines a monomorphism of rings*

$$H^*(X) \otimes H^*(Y) \to H^*(X \times Y).$$

If F is a field, it defines an isomorphism of algebras

$$H^*(X; F) \otimes_F H^*(Y; F) \to H^*(X \times Y; F). \quad \square$$

Example 1. Consider the cohomology ring of $S^n \times S^m$, where $n,m \geq 1$. Let $\alpha^n \in H^n(S^n)$ and $\beta^m \in H^m(S^m)$ be generators. Then $H^*(S^n \times S^m)$ is free of rank 4, with basis 1×1, $\alpha \times 1$, $1 \times \beta$, and $\alpha \times \beta$. The only non-trivial product of positive dimensional elements is $(\alpha \times 1) \cup (1 \times \beta) = \alpha \times \beta$.

In the special case $n = m = 1$, this space is the torus, and this is the cohomology ring we have already computed. Let us picture these cohomology classes. Represent the torus by the usual diagram, as in Figure 61.1. If X denotes the subspace *abca*, then a generator α for its 1-dimensional cohomology is represented by the cocycle $x^1 = [b,c]^*$. Since the projection map π_1 is simplicial, the cocycle $\pi_1^\#(x^1)$ is the cocycle w^1 of T, as pictured. This cocycle thus represents the cohomology class $\alpha \times 1$. Similarly, the cocycle z^1 pictured in Figure 61.2 represents the cohomology class $1 \times \beta$, where β is represented by the cocycle $[d,e]^*$ of the space $Y = adea$. Now $(\alpha \times 1) \cup (1 \times \beta) = \alpha \times \beta$, and this class generates $H^2(T)$. This fact we already knew, of course.

Example 2. Consider the cohomology ring of $P^2 \times P^2$, with $\mathbf{Z}/2$ coefficients. Let $\alpha \in H^1(P^2; \mathbf{Z}/2)$ be the nonzero element; then $\alpha^2 = \alpha \cup \alpha$ is non-zero. (See the

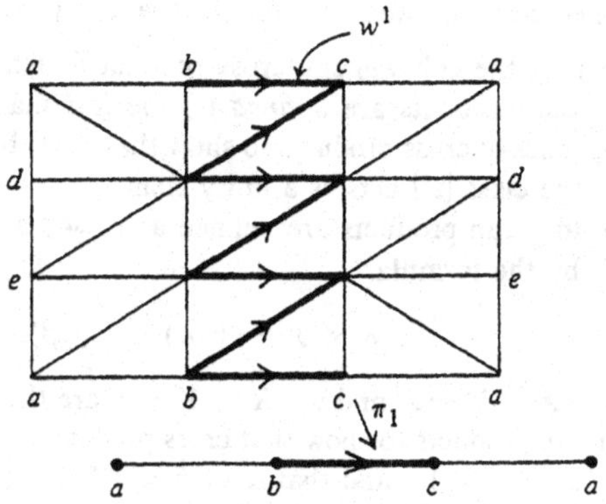

Figure 61.1

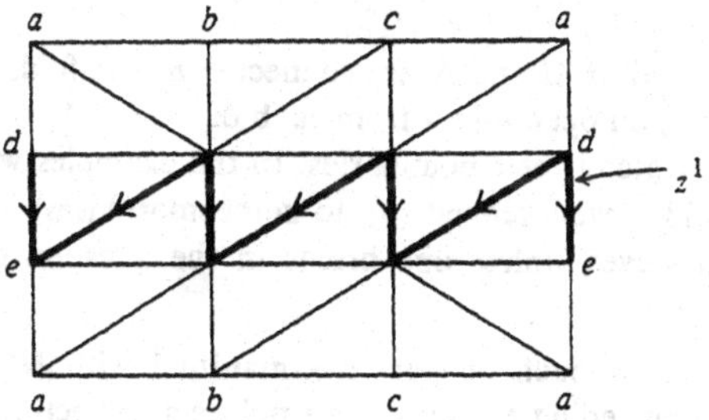

Figure 61.2

exercises of §49.) By the preceding theorem, the vector space $H^*(P^2 \times P^2; \mathbf{Z}/2)$ has dimension 9 with the following basis elements:

dim 0	1×1
dim 1	$\alpha \times 1, 1 \times \alpha$
dim 2	$\alpha^2 \times 1, \alpha \times \alpha, 1 \times \alpha^2$
dim 3	$\alpha^2 \times \alpha, \alpha \times \alpha^2$
dim 4	$\alpha^2 \times \alpha^2$.

The multiplication table is easy to write down. No signs are involved because the coefficient field is $\mathbf{Z}/2$.

EXERCISES

1. (a) Let $A \subset X$ and $B \subset Y$. Show that if $\{X \times B, A \times Y\}$ is excisive, there is a relative cross product

$$H^p(X,A) \otimes H^q(Y,B) \to H^{p+q}(X \times Y, (X \times B) \cup (A \times Y))$$

for our usual sets of coefficients.

(b) Prove Theorems 61.3 and 61.4 for relative cup products.

2. (a) Suppose that the cohomology cross product is defined as in this section, and that cup products are *defined* by the formula $\alpha \cup \beta = d^*(\alpha \times \beta)$. Use properties of cross products to show that $\cup$ is bilinear and associative, and that the class $\{z^0\}$ acts as a unity element.

   (b) Suppose that cup products are defined as in §48 and that we *define* cross products by the formula

$$\alpha \times \beta = \pi_1^*(\alpha) \cup \pi_2^*(\beta),$$

   where $\pi_1 : X \times Y \to X$ and $\pi_2 : X \times Y \to Y$ are the projections. Use properties of cup products to show that cross product is bilinear and associative, that $\pi_1^*(\alpha) = \alpha \times 1_Y$, and that $\alpha \cup \beta = d^*(\alpha \times \beta)$.

3. Compute the following cohomology rings:
   (a) $H^*(S^1 \times S^1 \times S^3)$.
   (b) $H^*((T \# T) \times S^3)$, where T is the torus.
   (c) $H^*((S^1 \times S^5) \# (S^2 \times S^4))$.
   (In general, if M and N are connected n-manifolds, $M \# N$ is obtained by removing an open n-ball from each of M and N and pasting the remnants together along their boundaries. In the examples we have considered, this manifold is well-defined (up to homeomorphism). This is not true in general, however, unless one brings in the question of "orientation," as we shall see.)

4. Let M be a compact, connected n-manifold with $H^n(M) \simeq \mathbf{Z}$. Use the results of the preceding exercise to formulate a conjecture about the cup product operation

$$H^k(M) \otimes H^{n-k}(M) \to H^n(M)$$

5. Let S denote the Klein bottle and let T denote the torus. Compute the following cohomology rings:
   (a) $H^*(T \times S)$.
   (b) $H^*(T \times S; \mathbf{Z}/2)$.

8

Duality in Manifolds

As we have mentioned before, manifolds are among the most familiar and important geometric objects. Therefore, any theorems we can prove about manifolds are likely to be useful, not only in topology, but also in differential geometry and other branches of mathematics. The duality theorems are examples of such theorems.

We already know one relationship that holds between the homology and cohomology groups of an arbitrary topological space; it is expressed by the universal coefficient theorem for cohomology. For manifolds, there is a second such relationship; it is expressed in the famous theorem called the Poincaré duality theorem. If one thinks of the universal coefficient theorem as expressing a kind of algebraic duality between cohomology and homology, then Poincaré duality can be thought of as basically geometric in nature. This duality does not hold for an arbitrary space, but depends specifically on properties possessed by manifolds.

There is a certain amount of interplay between these geometric and algebraic types of duality. This interplay has some interesting applications; we shall apply it to the problem of computing the cohomology ring of a manifold.

There are further duality theorems; they bear the names of Lefschetz, Alexander, and Pontryagin. We shall take them up in due course. All are concerned in one way or another with algebraic-topological properties of manifolds.

Throughout, we assume familiarity with local homology groups (§35).

§62. THE JOIN OF TWO COMPLEXES

In this section, we introduce the notion of the join of two complexes, and study some of its properties.

Definition. Let K and L be nonempty complexes in some euclidean space E^J. Let $s = v_0 \dots v_m$ be the general simplex of K and let $t = w_0 \dots w_n$ be the general simplex of L. Suppose that whenever $s \in K$ and $t \in L$, the points $v_0, \dots, v_m, w_0, \dots, w_n$ are independent. Then we let

$$s * t = v_0 \dots v_m w_0 \dots w_n$$

denote the simplex they span. If the collection consisting of all the simplices $s * t$ and their faces is a simplicial complex, then this complex is called the **join** of K and L, and is denoted by $K * L$.

If K consists of a single vertex v, then $v * L$ is the cone over L, as already defined. If K consists of two points, then $K * L$ is the suspension of L and is denoted $S(L)$. Conditions under which $K * L$ exists in general are given in the following lemma.

Lemma 62.1. *Let K and L be disjoint nonempty complexes in E^J.*

(a) *If $K * L$ exists, then its polytope equals the union of all line segments joining points of $|K|$ to points of $|L|$; two such line segments intersect in at most a common end point.*

(b) *Conversely, if every pair of line segments joining points of $|K|$ to points of $|L|$ intersect in at most a common end point, then $K * L$ exists.*

Proof. Step 1. We prove the following: Let $\sigma = v_0 \dots v_m w_0 \dots w_n$ be a simplex in E^J. Then σ equals the union of all line segments joining points of $s = v_0 \dots v_m$ to points of $t = w_0 \dots w_n$; two such line segments intersect in at most a common end point. Furthermore, Int σ equals the union of all *open* line segments joining points of Int s to points of Int t.

This result was given as an exercise in §1; we give a proof here.

Let us prove the statement about Int σ first. Suppose $p = \lambda x + (1 - \lambda)y$, where $0 < \lambda < 1$ and $x \in \text{Int } s$ and $y \in \text{Int } t$. Then $x = \Sigma\, \alpha_i v_i$ and $y = \Sigma\, \beta_j w_j$, where $\alpha_i > 0$ and $\beta_j > 0$ and $\Sigma\, \alpha_i = \Sigma\, \beta_j = 1$. Hence,

$$p = \Sigma\, (\lambda\alpha_i)v_i + \Sigma\, (1 - \lambda)\beta_j w_j,$$

where the coefficients are positive and their sum is 1. Thus p is in Int σ. Conversely, suppose $p \in \text{Int } \sigma$, so that

$$p = \Sigma\, \gamma_i v_i + \Sigma\, \delta_j w_j,$$

where $\gamma_i > 0$ and $\delta_j > 0$ and $\Sigma\, \gamma_i + \Sigma\, \delta_j = 1$. Set $\lambda = \Sigma\, \gamma_i$; set $\alpha_i = \gamma_i/\lambda$ and $\beta_j = \delta_j/(1 - \lambda)$. Then

$$p = \lambda\, \Sigma\, \alpha_i v_i + (1 - \lambda)\, \Sigma\, \beta_j w_j.$$

Thus p lies interior to a line segment joining a point of Int s and a point of Int t.

Now we prove the statement about σ. Since σ is convex, it contains all line segments joining points of s to points of t. Conversely, suppose $x \in \sigma$ and x is in neither s nor t. Then x lies interior to a face σ' of σ that has vertices in common with both s and t. By the result of the preceding paragraph, x lies on an open line segment joining a point of $\sigma' \cap s$ with a point of $\sigma' \cap t$.

Finally, we show that any two of these line segments intersect in at most a common end point. Suppose p is a point lying on two such line segments—that is, p equals

$$\lambda \Sigma \alpha_i v_i + (1 - \lambda) \Sigma \beta_j w_j \quad = \quad \lambda' \Sigma \alpha_i' v_i + (1 - \lambda') \Sigma \beta_j' w_j,$$

where $0 \leq \lambda, \lambda' \leq 1$, and $\alpha_i \geq 0$, $\beta_j \geq 0$, $\alpha_i' \geq 0$, $\beta_j' \geq 0$, and $\Sigma \alpha_i = \Sigma \beta_j = \Sigma \alpha_i' = \Sigma \beta_j' = 1$. Since the sum of the coefficients on either side of this equation is 1, we conclude from the independence of the vertices of σ that

$$\lambda \alpha_i = \lambda' \alpha_i', \quad \text{and}$$
$$(1 - \lambda)\beta_j = (1 - \lambda')\beta_j'.$$

Summing the first of these, we see that $\lambda = \lambda'$. If $\lambda = 0$ or $\lambda = 1$, then p is an end point of each line segment. Otherwise, these two equations imply that $\alpha_i = \alpha_i'$ and $\beta_j = \beta_j'$, and the two line segments coincide.

Step 2. Now we prove (a). Consider the distinct line segments xy and $x'y'$, where

$$x \in \operatorname{Int} s \quad \text{and} \quad x' \in \operatorname{Int} s', \quad \text{with} \quad s, s' \in K;$$
$$y \in \operatorname{Int} t \quad \text{and} \quad y' \in \operatorname{Int} t', \quad \text{with} \quad t, t' \in L.$$

Since the interior of the line segment xy lies in $\operatorname{Int}(s * t)$, we have the inclusions

$$xy \subset (\operatorname{Int} s) \cup (\operatorname{Int}(s * t)) \cup (\operatorname{Int} t),$$
$$x'y' \subset (\operatorname{Int} s') \cup (\operatorname{Int}(s' * t')) \cup (\operatorname{Int} t').$$

Because $K * L$ is a complex, these six open simplices are disjoint if $s \neq s'$ and $t \neq t'$, whence xy and $x'y'$ are disjoint. If $s = s'$ and $t \neq t'$, they intersect only in the set Int s, so the line segments intersect in at most the end points x, x'. If $s \neq s'$ and $t = t'$, then the line segments intersect in at most the end points y, y'. Finally, if $s = s'$ and $t = t'$, it follows from Step 1 that the line segments intersect in at most a common end point.

Step 3. We prove (b). First, we show that if $v_0 \ldots v_m \in K$ and $w_0 \ldots w_n \in L$, then the points $v_0, \ldots, v_m, w_0, \ldots, w_n$ are linearly independent. Consider the linear map

$$l = l(v_0, \ldots, v_m, w_0, \ldots, w_n),$$

mapping Δ_{m+n+1} into $\mathbf{E}^J$. This map carries each line segment joining a point of $\epsilon_0 \ldots \epsilon_m$ to a point of $\epsilon_{m+1} \ldots \epsilon_{m+n+1}$ linearly onto a line segment joining a point of $v_0 \ldots v_m$ to a point of $w_0 \ldots w_n$. Then by hypothesis, l is injective. As a result, the plane spanned by these points must have dimension $m + n + 1$; otherwise, the linear map l could not be injective. Hence $v_0, \ldots, v_m, w_0, \ldots, w_n$ are independent.

It follows that $K * L$ exists. For if two simplices of $K * L$ have interiors that intersect, one must be of the form $s * t$. Each point of Int $s * t$ lies on an open line segment joining a point of $|K|$ to a point of $|L|$. If Int $s * t$ and Int $s' * t'$ intersect, two such open line segments must intersect. If Int $s * t$ and Int s' intersect in a point y, then y lies interior to one such line segment and is the end point of another. Similarly, if Int $s * t$ and Int t' intersect. Each of these situations is contrary to hypothesis. □

Suppose K and L are complexes, but they are not situated in euclidean space in such a way that $K * L$ is defined. Then one can find complexes K_0 and L_0 isomorphic to K and L, respectively, such that $K_0 * L_0$ *is* defined. For instance, one can take K_0 and L_0 to be disjoint subcomplexes of the standard simplex Δ^J, for some index set J.

Lemma 62.2. *Suppose $K * L$ exists, and suppose K is locally finite. Then the map*

$$\pi : |K| \times |L| \times I \rightarrow |K * L|$$

defined by

$$\pi(x,y,t) = (1 - t)x + ty$$

is a quotient map. For each $x \in |K|$ and $y \in |L|$, it collapses $x \times |L| \times 0$ to a point and $|K| \times y \times 1$ to a point. Otherwise, it is one-to-one.

Proof. The topology of $|K| \times |L| \times I$ is coherent with the subspaces $\sigma \times \tau \times I$, for $\sigma \in K$ and $\tau \in L$. To prove this fact, we apply the results of §20. The topology of $|L| \times I$ is coherent with the subspaces $\tau \times I$, for $\tau \in L$. Because $|K|$ is locally compact Hausdorff, the topology of $|K| \times |L| \times I$ is thus coherent with the subspaces $|K| \times \tau \times I$. Because $\tau \times I$ is locally compact Hausdorff (in fact, compact), the topology of $|K| \times \tau \times I$ is in turn coherent with the subspaces $\sigma \times \tau \times I$, for $\sigma \in K$ and $\tau \in L$. Thus A is closed in $|K| \times |L| \times I$ if and only if its intersection with each space $\sigma \times \tau \times I$ is closed in that space.

Now π is continuous as a map of $\sigma \times \tau \times I$ onto $\sigma * \tau$, by definition. (Both are subspaces of euclidean space.) And inclusion of $\sigma * \tau$ in $|K * L|$ is continuous. Thus $\pi | \sigma \times \tau \times I$ is continuous; because the topology of $|K| \times |L| \times I$ is coherent with these subspaces, π is continuous.

To show π is a quotient map, suppose $\pi^{-1}(C)$ is closed in $|K| \times |L| \times I$. Then its intersection with $\sigma \times \tau \times I$ is closed and hence compact. We conclude that $C \cap (\sigma * \tau)$ is compact and therefore closed. Hence C is closed in $K * L$.

The fact that π carries out precisely the indicated identifications follows from Lemma 62.1. □

Corollary 62.3. *Suppose $K * L$ and $M * N$ are defined, where K is locally finite. If $|K| \approx |M|$ and $|L| \approx |N|$, then $|K * L| \approx |M * N|$.*

Proof. Since $|K|$ is locally compact, so is $|M|$; therefore, M is locally finite. If $h : |K| \to |M|$ and $k : |L| \to |N|$ are homeomorphisms, then $h \times k \times i_I$ induces, via the quotient maps indicated in the preceding lemma, a homeomorphism of $|K * L|$ with $|M * N|$ that equals h on $|K|$ and k on $|L|$. □

Lemma 62.4. *Let J, K, L be complexes. Assume $J * K$ and $(J * K) * L$ exist. Then $J * K = K * J$ and $(J * K) * L = J * (K * L)$.*

Proof. The symmetry of the definition shows that $J * K = K * J$. Similarly, if $(J * K) * L$ is defined, then for $v_0 \dots v_m \in J$ and $w_0 \dots w_n \in K$ and $x_0 \dots x_p \in L$, the points

$$v_0, \dots, v_m, w_0, \dots, w_n, x_0, \dots, x_p$$

are independent and the collection of simplices they span, along with faces of such simplices, forms a complex. But this is just the complex $J * (K * L)$. □

Theorem 62.5. *Assume $K * L$ exists. If $|L| \approx S^{n-1}$, then for all i,*

$$\tilde{H}_{i+n}(K * L) \simeq \tilde{H}_i(K).$$

Proof. In the case $n = 1$, $|L|$ consists of two points and the complex $K * L$ is just the suspension of K. The existence of the isomorphism in question is a consequence of the Mayer-Vietoris sequence. (See Theorem 25.4.)

In general, we suppose the theorem true in dimension n, and prove it true in dimension $n + 1$.

In view of Corollary 62.3, it suffices to prove the theorem for any particular complex L whose space is homeomorphic to S^n. Choose a complex J such that $|J| \approx S^{n-1}$ and let $L = J * \{w_0, w_1\}$ be a suspension of J. Then it is easy to see that $|L| \approx S^n$. Now

$$\begin{aligned} \tilde{H}_i(K) &\simeq \tilde{H}_{i+n}(K * J) && \text{by the induction hypothesis,} \\ &\simeq \tilde{H}_{i+n+1}((K * J) * \{w_0, w_1\}) && \text{as noted earlier,} \\ &\simeq \tilde{H}_{i+n+1}(K * L) && \text{by Lemma 62.4.} \end{aligned}$$ □

Now we apply these results to study local properties of arbitrary simplicial complexes.

Definition. Let s be a simplex of the complex K. The **star** of s in K, denoted St s, is the union of the interiors of all simplices of K having s as a face. The closure of St s is denoted $\overline{\text{St}}\, s$; it is the union of all simplices of K having s as a face and is called the **closed star** of s in K. The **link** of s in K, denoted Lk s, is the union of all simplices of K lying in $\overline{\text{St}}\, s$ that are disjoint from s.

We have already given these definitions when s is a vertex. See §2.

In general, if $s = v_0 \dots v_n$, then St s is the open subset of $|K|$ consisting of those points x for which the barycentric coordinate of x on each of $v_0, \ldots, v_n$ is positive. That is,

$$\operatorname{St} s = \operatorname{St} v_0 \cap \cdots \cap \operatorname{St} v_n.$$

Note that $\overline{\operatorname{St}}\, s$ consists of all simplices of K of the form $s * t$, and the link of s equals the union of all the faces t of such simplices. The sets $\overline{\operatorname{St}}\, s$ and $\operatorname{Lk} s$ are thus polytopes of subcomplexes of K; we often use the notation $\overline{\operatorname{St}}\, s$ and $\operatorname{Lk} s$ to denote these complexes as well as their polytopes.

Example 1. In the 2-dimensional complex pictured in Figure 62.1, the link of the vertex g consists of the hexagon $abcdefa$ and the vertex h. The link of the 1-simplex ag consists of the two vertices b and f. The link of the vertex h is the vertex g.

In the 3-dimensional complex pictured in Figure 62.2, the link of the 1-simplex fg is the pentagon $P = abcdea$, and the link of the vertex f is the cone $P * g$. The link of the 1-simplex ab is the 1-simplex fg, and the link of the vertex a is the union of the 2-simplices bfg and efg.

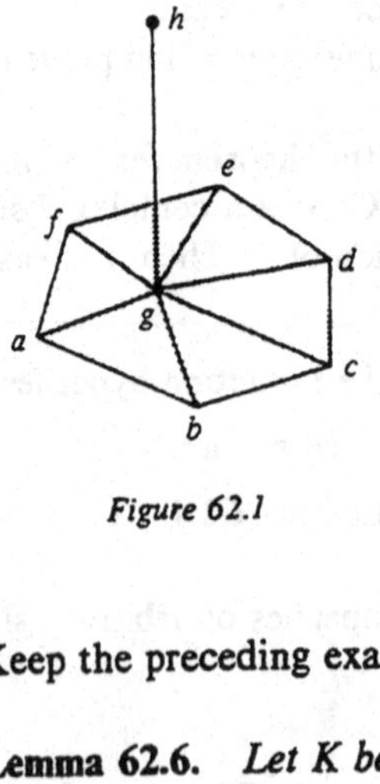

Figure 62.1

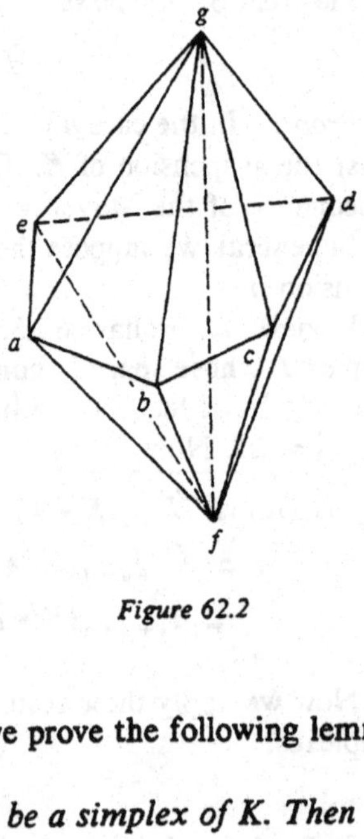

Figure 62.2

Keep the preceding example in mind as we prove the following lemma.

Lemma 62.6. *Let K be a complex. Let s be a simplex of K. Then*

$$\overline{\operatorname{St}}\, s = s * \operatorname{Lk} s,$$

$$\overline{\operatorname{St}}\, s - \operatorname{St} s = \operatorname{Bd} s * \operatorname{Lk} s.$$

Note that here we are considering s and $\operatorname{Bd} s$ and $\operatorname{Lk} s$ not only as subspaces of $|K|$ but also as subcomplexes of K. In order that these equations should hold even if $\operatorname{Lk} s$ or $\operatorname{Bd} s$ is empty, we make the convention that $K * \varnothing = \varnothing * K = K$ for all K.

Proof. The first equation is immediate from the definitions. $\overline{\text{St}}\, s$ is the union of all simplices of K of the form $s * t$, and $\text{Lk}\, s$ is the union of all the faces t of such simplices.

On the other hand, if a simplex of K lies in $\overline{\text{St}}\, s$ but not in $\text{St}\, s$, it must be a face of $s * t$ of the form $s' * t'$, where s' is a *proper* face of s and t' is a face of t. Since t lies in $\text{Lk}\, s$, so does t'; hence $s' * t'$ lies in $\text{Bd}\, s * \text{Lk}\, s$. Conversely, every simplex of this form lies in $\overline{\text{St}}\, s$ but not in $\text{St}\, s$. □

EXERCISES

1. Let s denote the simplex fg in Figure 62.2. Describe $\hat{s} * \text{Lk}\, s$ and $\hat{s} * \text{Bd}\, s$ and $\text{Bd}\, s * \text{Lk}\, s$.

2. If X and Y are topological spaces, let us define $X * Y$ to be the quotient space of $X \times Y \times I$ obtained by identifying each set $x \times Y \times 0$ to a point and each set $X \times y \times 1$ to a point.
   (a) Show that the maps
   $$i(x) = x \times Y \times 0 \qquad \text{and} \qquad j(y) = X \times y \times 1$$
   define imbeddings of X and Y, respectively, into $X * Y$.
   (b) Show that if $X \approx X'$ and $Y \approx Y'$, then $X * Y \approx X' * Y'$.
   (c) Show that $X * Y \approx Y * X$.

*3. Show there is a split short exact sequence
$$0 \to \tilde{H}_{p+1}(X * Y) \to \tilde{H}_p(X \times Y) \to \tilde{H}_p(X) \oplus \tilde{H}_p(Y) \to 0.$$

*4. Show that $(X * Y) * Z \approx X * (Y * Z)$. [*Hint:* The "obvious" proof does not work. You can check that the quotient maps
$$(X \times Y \times I_1) \times Z \times I_2 \to (X * Y) \times Z \times I_2 \to (X * Y) * Z,$$
$$X \times (Y \times Z \times I_2) \times I_1 \to X \times (Y * Z) \times I_1 \to X * (Y * Z)$$
carry out different identifications. Instead, in the space $X \times Y \times Z \times \Delta_2$, introduce the following relations:
$$(x,y,z,w) \sim (x',y,z,w) \quad \text{if} \quad w \in \epsilon_1\epsilon_2,$$
$$(x,y,z,w) \sim (x,y',z,w) \quad \text{if} \quad w \in \epsilon_0\epsilon_2,$$
$$(x,y,z,w) \sim (x,y,z',w) \quad \text{if} \quad w \in \epsilon_0\epsilon_1.$$
Let W be the quotient space; let π be the quotient map. Let $f : I_1 \times I_2 \to \Delta_2$ map $t \times I_2$ linearly onto the line segment from $t \in \epsilon_0\epsilon_1 = [0,1]$ to ϵ_2. Show that
$$X \times Y \times Z \times I_1 \times I_2 \xrightarrow{\text{id} \times f} X \times Y \times Z \times \Delta_2 \xrightarrow{\pi} W$$
carries out the same identifications as the first of the preceding quotient maps. Conclude that W is homeomorphic to $(X * Y) * Z$. Show that W is also homeomorphic to $X * (Y * Z)$.]

§63. HOMOLOGY MANIFOLDS

In this section we define homology manifolds and derive some of their local properties. The class of homology manifolds includes, among other things, all topological manifolds, so it is a broad and important class of spaces. The triangulable homology manifolds will be the basic objects involved in the various duality theorems.

For convenience, we shall deal with the case of a relative homology manifold in this section, even though we shall not prove the duality theorems in the relative case until §70.

Convention. We shall often be dealing in this chapter with triangulable spaces. If X is a triangulable space with a specific triangulation, we shall frequently abuse notation and make no notational distinction between the space and the complex triangulating it. For instance, we may refer to a point of (the space) X, or to a simplex of (the complex) X. The context will in each case make the meaning clear.

Definition. A topological pair (X,A) is called a **relative homology n-manifold** if for each point x of X not in A, the local homology group $H_i(X,X-x)$ vanishes if $i \neq n$ and is infinite cyclic if $i = n$. In the case where A is empty, we refer to X simply as a **homology n-manifold.**

If M is an n-manifold with boundary, then the pair $(M, \text{Bd } M)$ is a relative homology n-manifold. (See §35.) More generally, if (X,A) is any pair such that $X - A$ is an n-manifold, then (X,A) is a relative homology n-manifold.

Now if (X,A) is a relative homology n-manifold that is triangulated, the local homology properties of X are reflected in the homology groups of the links of simplices of X. The connection is given by the following lemma.

Lemma 63.1. *Let s be a k-simplex of the complex K. Let $\hat{s}$ be its barycenter. Then*

$$H_i(|K|,|K| - \hat{s}) \simeq \begin{cases} \tilde{H}_{i-k-1}(\text{Lk } s) & \text{if } \text{Lk } s \neq \varnothing, \\ H_i(s, \text{Bd } s) & \text{if } \text{Lk } s = \varnothing. \end{cases}$$

Proof. If $\text{Lk } s = \varnothing$, then s is a face of no other simplex of K. Hence $\text{Int } s$ is open in $|K|$. It follows that

$$H_i(|K|,|K| - \hat{s}) \simeq H_i(s, s - \hat{s}) \simeq H_i(s, \text{Bd } s).$$

The first isomorphism holds by excision; and the second holds because $\text{Bd } s$ is a deformation retract of $s - \hat{s}$.

Now suppose $\text{Lk } s \neq \varnothing$. Since the barycenter of s lies in the open set $\text{St } s$, we may excise the complement of $\overline{\text{St}}\, s$ to obtain an isomorphism

$$H_i(|K|,|K| - \hat{s}) \simeq H_i(\overline{\text{St}}\, s, \overline{\text{St}}\, s - \hat{s}).$$

From basic results about convex sets, we know that $|s| = |\hat{s} * \text{Bd}\, s|$; from Lemma 62.6 it follows that

$$|\overline{\text{St}}\, s| = |s * \text{Lk}\, s| = |\hat{s} * \text{Bd}\, s * \text{Lk}\, s|.$$

(By our convention, these equations hold even if Bd s is empty.)

It follows that $|\overline{\text{St}}\, s|$ is acyclic, being a cone with vertex $\hat{s}$. Now if we delete the vertex of a cone, what is left can be collapsed onto the base of the cone by a deformation retraction. (See Lemma 35.5.) In particular, $|\text{Bd}\, s * \text{Lk}\, s|$ is a deformation retract of $\overline{\text{St}}\, s - \hat{s}$. We conclude that

$$\begin{aligned} H_i(|K|,|K| - \hat{s}) &\simeq H_i(\overline{\text{St}}\, s, \overline{\text{St}}\, s - \hat{s}) \\ &\simeq \tilde{H}_{i-1}(\overline{\text{St}}\, s - \hat{s}) \\ &\simeq \tilde{H}_{i-1}(\text{Bd}\, s * \text{Lk}\, s) \\ &\simeq \tilde{H}_{i-k-1}(\text{Lk}\, s). \end{aligned}$$

The last isomorphism is an equality if Bd $s = \varnothing$, in which case $k = 0$; otherwise, it holds by Theorem 62.5, since Bd s is topologically a $k - 1$ sphere. □

Theorem 63.2. *Let (X,A) be a triangulated relative homology n-manifold. Let s be a k-simplex of X not in A. If* Lk s *is empty, then $k = n$. If* Lk s *is non-empty, it has the homology of an $n - k - 1$ sphere, where $n - k - 1 \geq 0$. In either case, $k \leq n$.*

Proof. Since s is not in A, its barycenter $\hat{s}$ lies in $X - A$. Therefore, the local homology of X at $\hat{s}$ is infinite cyclic in dimension n, and vanishes otherwise.

If Lk s is empty, then by the preceding lemma, $H_n(s, \text{Bd}\, s) \simeq H_n(X, X - \hat{s})$, which is non-trivial. Hence dim $s = n$. If Lk s is non-empty, then its reduced homology is infinite cyclic in dimension $n - k - 1$ and vanishes otherwise. This implies in particular that $n - k - 1 \geq 0$, or that $k < n$. □

Corollary 63.3. *Let (X,A) be a triangulated relative homology n-manifold.*

(a) *The closure of $X - A$ equals a union of n-simplices.*

(b) *Every $n - 1$ simplex s of X not in A is a face of precisely two n-simplices of X.*

Proof. (a) Let s be a k-simplex of X not in A. The preceding theorem shows that $k \leq n$. If $k < n$, then Lk s is a homology $n - k - 1$ sphere, so it contains an $n - k - 1$ simplex t. Then s is a face of the n-simplex $s * t$.

(b) If s is an $n - 1$ simplex of X not in A, then Lk s is a homology 0-sphere. Since $\overline{\text{St}}\, s$ has dimension n and $\overline{\text{St}}\, s = s * \text{Lk}\, s$, the complex Lk s must have dimension 0. This means that Lk s consists of precisely two points; hence (b) holds. □

This theorem shows that a triangulated relative homology n-manifold (X,A) satisfies the first two conditions in the definition of a relative pseudo manifold. (See the exercises of §43 for the definition.) If the space $X - A$ is connected, then (X,A) satisfies the third condition as well. A proof of this fact is outlined in the exercises; it also be derived from the duality theorems, as we shall see.

Example 1. Let X be the disjoint union of a torus T and a two-sphere S^2; let $S(X)$ be the suspension of X; let $A = \{w_0,w_1\}$ consist of the suspension points. See Figure 63.1. Then $(S(X),A)$ is a relative homology 3-manifold, since $S(X) - A$ is homeomorphic to $X \times (-1,1)$, which is a 3-manifold. The subspaces $(S(T),A)$ and $(S(S^2),A)$ are also relative homology 3-manifolds, and they are relative pseudo 3-manifolds as well. The space $S(S^2)$ is a homology 3-manifold, but the space $S(T)$ is not, for the link of w_0 in $S(T)$ is not a homology 2-sphere but a torus.

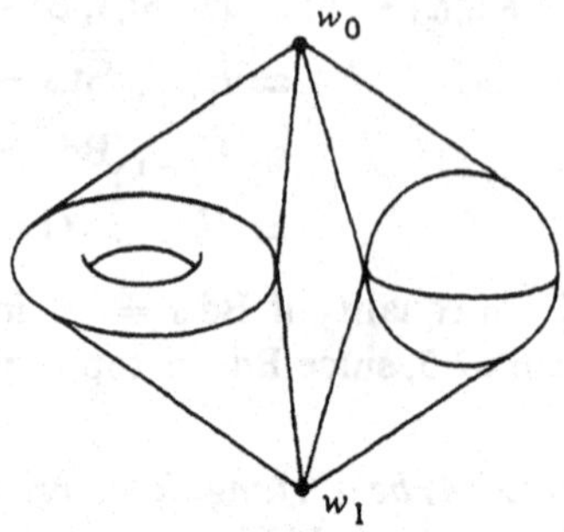

Figure 63.1

Example 2. The pictures we usually draw of manifolds may lead you to conjecture the following: If M is a triangulated *topological* n-manifold, then the link of each k-simplex of M is a *topological* $n - k - 1$ sphere.

Certainly this result is true for our usual triangulations of 2-manifolds, for instance; the links of 1-simplices are 0-spheres and the links of vertices are 1-spheres. This conjecture is one of long standing, and has only recently been answered, in the negative. R. D. Edwards has found a counterexample in dimension 5. The details are complicated.

Example 3. There do exist homology n-manifolds that are not topological manifolds. We outline the construction of one such, but the proof uses tools we have not studied.

The basic fact one needs is that there is a compact triangulated 3-manifold M that is a *homology* 3-sphere, but whose fundamental group $\pi_1(M)$ does not vanish. The construction can be found in [S-T]; the fundamental group of M is the icosahedral group.

Now we form the suspension $S(M) = M * \{w_0,w_1\}$. This space is a homology 4-manifold: Since $S(M) - \{w_0,w_1\}$ is homeomorphic to the 4-manifold $M \times (-1,1)$, the local homology conditions are satisfied at all points except possibly at the suspension points. Since $\text{Lk}\, w_0 = \text{Lk}\, w_1 = M$, and M is a homology 3-sphere, the local homology conditions are also satisfied at w_0 and w_1.

Then one needs a separate argument to show that in a triangulated topological n-manifold, the link of each vertex, while it may not be *topologically* an $n - 1$ sphere, it must have the *homotopy type* of an $n - 1$ sphere. Since $\pi_1(M) \neq 0$, M cannot have the homotopy type of a 3-sphere, so $S(M)$ cannot be a topological 4-manifold.

EXERCISES

Each of the following exercises depends on the preceding ones.

1. Let (X,A) be a triangulated pair. Suppose that for each k-simplex s of X not in A we have: (i) $k \leq n$, (ii) $H_i(\text{Lk}\, s) \simeq H_i(S^{n-k-1})$ if $k < n$. Show that (X,A) is a relative homology n-manifold.

2. Let (X,A) be a triangulated relative homology n-manifold. If s is a k-simplex of X not in A, with $k < n$, show that $\text{Lk}\, s$ is a homology $n-k-1$ manifold. [*Hint:* Consider $\text{Lk}\, s$ as a subcomplex of X. Then if $t \in \text{Lk}\, s$, show that $\text{Lk}(t, \text{Lk}(s,X)) = \text{Lk}(t * s, X)$.]

3. *Theorem. Let (X,A) be a triangulated relative homology n-manifold, with $X - A$ connected. Then (X,A) is a relative pseudo n-manifold.*

   [*Hint:* The result is trivial if $n = 1$; suppose it true for dimensions less than n. Define $\sigma \sim \sigma'$ if there is a sequence
   $$\sigma = \sigma_0, \sigma_1, \ldots, \sigma_k = \sigma'$$
   of n-simplices of X not in A such that $\sigma_i \cap \sigma_{i+1}$ is an $n-1$ simplex not in A for each n. Let X_1 be the union of the elements of one equivalence class; let X_2 be the union of the elements of the remaining equivalence classes. Let s be a simplex of the complex X such that $s \notin A$ and $s \subset X_1 \cap X_2$. If $k = \dim s$, then $k < n - 1$; show $\underline{\text{Lk}}\, s$ is a pseudo $n-k-1$ manifold. Conclude that every two n-simplices of $\overline{\text{St}}\, s$ are equivalent.]

4. *Corollary. Let (X,A) be a triangulated relative homology n-manifold. If s is a simplex of X not in A, then* $\text{Lk}\, s$ *is a finite complex.* [*Hint:* If X is a pseudo m-manifold and $H_m(X) \neq 0$, then X must be finite.]

§64. THE DUAL BLOCK COMPLEX

For simplicity, we are going to restrict ourselves in the next few sections to the case of a homology n-manifold X. The techniques we use will reappear later when we consider the case of a relative homology manifold.

There is associated with the triangulated homology manifold X, a certain partition of X into subsets that are open cells (almost), such that X becomes a regular cell complex (almost). The subsets are not actual topological cells, but only homological cells. Since we cannot call them "cells," we shall call them "blocks," for want of a better term. And we shall call the collection of these blocks the *dual block decomposition* of X. Just as there was with cell complexes, there is a chain complex associated to this block complex that can be used to compute the homology and cohomology of X. We will call it the "dual chain complex"; it will be a crucial tool in proving our duality theorems.

Definition. Let X be a locally finite simplicial complex, and let $\text{sd}\, X$ denote its first barycentric subdivision. The simplices of $\text{sd}\, X$ are of the form

$$\hat{\sigma}_{i_1}\hat{\sigma}_{i_2}\ldots\hat{\sigma}_{i_k}$$

where $\sigma_{i_1} > \sigma_{i_2} > \cdots > \sigma_{i_k}$. We shall partially order the vertices of sd X by decreasing dimension of the simplices of X of which they are the barycenters; this ordering induces a linear ordering on the vertices of each simplex of sd X. Given a simplex σ of X, the union of all open simplices of sd X of which $\hat{\sigma}$ is the *initial* vertex is just Int σ. We define $D(\sigma)$ to be the union of all open simplices of sd X of which $\hat{\sigma}$ is the *final* vertex; this set is called the **block dual to** σ.

The blocks $D(\sigma)$ will play a role similar to that of the open cells of a CW complex. We call the closure $\overline{D}(\sigma)$ of $D(\sigma)$ the **closed block dual to** σ. It equals the union of all simplices of sd X of which $\hat{\sigma}$ is the final vertex; it is the polytope of a subcomplex of sd X. We let $\dot{D}(\sigma) = \overline{D}(\sigma) - D(\sigma)$.

Example 1. Let X be the 2-dimensional complex pictured in Figure 64.1. The complex sd X is indicated by dotted lines. The block dual to any 2-simplex σ consists of its barycenter $\hat{\sigma}$ alone. The block $D(ab)$ dual to the 1-simplex $s = ab$ consists of its barycenter $\hat{s}$ and the two open line segments joining $\hat{s}$ to the barycenters of the two triangles having ab as a face. The block dual to the vertex e is the shaded region indicated; the corresponding closed block consists of this region plus its boundary. Note the interesting fact that $\dot{D}(e)$ is the union of the lower-dimensional blocks $D(s_i)$ as s_i ranges over all 1-simplices and 2-simplices having e as a face. This situation will hold in general, as we shall see.

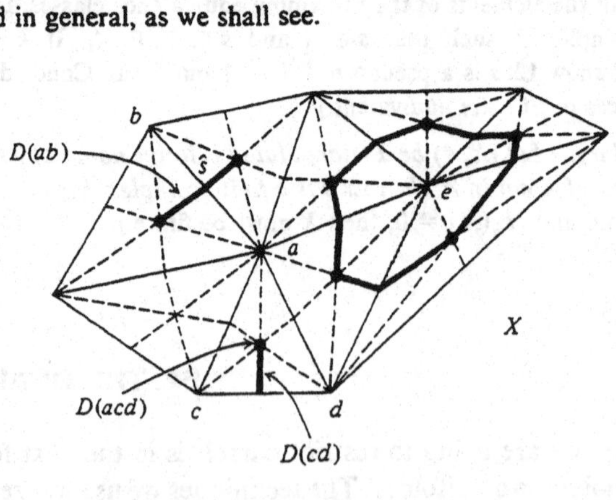

Figure 64.1

Example 2. Let X be the complex pictured in Figure 64.2. It is the join of the polygon *cdefc* with the line segment *ab*. You can picture mentally what the complex sd X looks like. The block dual to each 3-simplex of X is its barycenter. The closed block dual to the 2-simplex *abc* consists of the two line segments joining the barycenter of *abc* with the barycenters of the two 3-simplices having *abc* as a face. The closed block $\overline{D}(ab)$ dual to the 1-simplex *ab* is the closed octagonal region pictured in Figure 64.3. The set $\dot{D}(ab)$ is its boundary; it is the union of the blocks dual to the 2-simplices and 3-simplices of X having *ab* as a face.

Keep these examples in mind as we prove the following theorem:

Theorem 64.1. *Let X be a locally finite simplicial complex that consists entirely of n-simplices and their faces. Let σ be a k-simplex of X. Then:*

(a) *The dual blocks are disjoint and their union is $|X|$.*

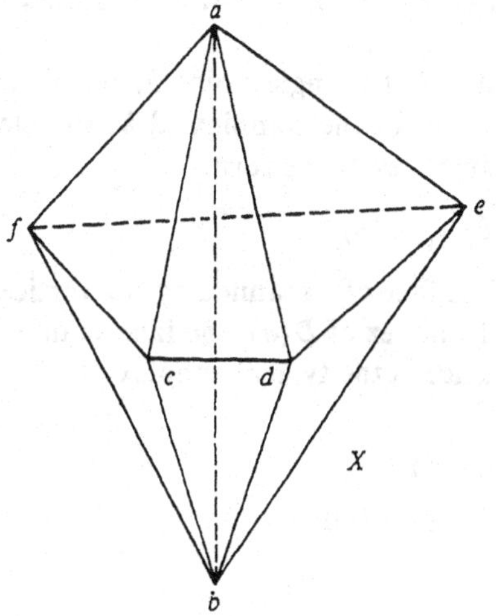

Figure 64.2

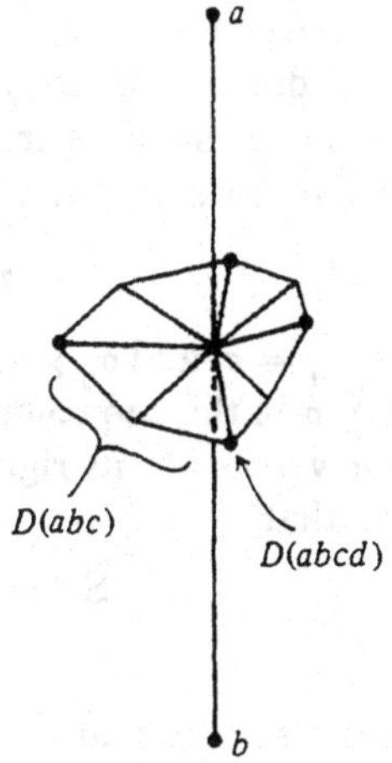

Figure 64.3

(b) *$\overline{D}(\sigma)$ is the polytope of a subcomplex of* sd X *of dimension $n - k$.*

(c) *$\dot{D}(\sigma)$ is the union of all blocks $D(\tau)$ for which τ has σ as a proper face; these blocks have dimensions less than $n - k$.*

(d) *$\overline{D}(\sigma)$ equals the cone $|\dot{D}(\sigma) * \hat{\sigma}|$.*

(e) *If $H_i(X, X - \hat{\sigma}) \simeq \mathbf{Z}$ for $i = n$ and vanishes otherwise, then $(\overline{D}(\sigma), \dot{D}(\sigma))$ has the homology of an $n - k$ cell modulo its boundary.*

Proof. (a) The open simplices of sd X are disjoint, and each one lies in precisely one block $D(\sigma)$—namely, the one such that $\hat{\sigma}$ is its final vertex.

(b) Given σ of dimension k, let σ' be an n-simplex of X having σ as a face. There is a sequence of simplices of X

$$\sigma' = \sigma_n \succ \sigma_{n-1} \succ \cdots \succ \sigma_k = \sigma$$

such that each simplex in the sequence has dimension one less than the preceding simplex. Then the simplex $\hat{\sigma}_n \hat{\sigma}_{n-1} \ldots \hat{\sigma}_k$ of sd X has dimension $n - k$ and lies in $\overline{D}(\sigma)$. Clearly, no simplex of sd X of larger dimension can have σ_k as its final vertex, since X has no simplices of dimension greater than n.

(c)–(d) Now $\overline{D}(\sigma)$ is the union of all simplices of sd X whose final vertex is $\hat{\sigma}$; they are of the form

$$\hat{\sigma}_{i_1} \ldots \hat{\sigma}_{i_p} \hat{\sigma}.$$

The intersection of such a simplex with $\dot{D}(\sigma)$ consists of the face of this simplex obtained by deleting $\hat{\sigma}$. Thus $\dot{D}(\sigma)$ consists of all simplices of the form

$$\hat{\sigma}_{i_1} \ldots \hat{\sigma}_{i_p}$$

for which $\sigma_{i_p} \succ \sigma$. Now the interior of such a simplex is contained in $D(\sigma_{i_p})$; and conversely, any open simplex of sd X lying in $D(\sigma_{i_p})$ is of this form. It follows that $\dot{D}(\sigma)$ is the union of the blocks $D(\sigma_{i_p})$ for $\sigma_{i_p} \succ \sigma$. It also follows that $|\overline{D}(\sigma)| = |\dot{D}(\sigma) * \hat{\sigma}|$.

(e) If $\dim \sigma = n$, then $D(\sigma)$ consists of the single point $\hat{\sigma}$, which is of course a 0-cell. Suppose that $\dim \sigma = k < n$. In the complex $\operatorname{sd} X$, the closed star $\overline{\operatorname{St}}(\hat{\sigma}, \operatorname{sd} X)$ equals the union of all simplices of the form

$$\tau = \hat{\sigma}_{i_1} \dots \hat{\sigma}_{i_p} \hat{\sigma}_{i_{p+1}} \dots \hat{\sigma}_{i_q}$$

where $\sigma_{i_{p+1}} = \sigma$, and $\sigma_{i_j} \succ \sigma_{i_{j+1}}$ for all j. The face of τ spanned by the vertices to the left of $\hat{\sigma}$ in this sequence is the typical simplex of $\dot{D}(\sigma)$; the face spanned by $\hat{\sigma}$ and the vertices to its right in this sequence is the typical simplex of $\operatorname{sd} \sigma$. We conclude that

$$\begin{aligned} \overline{\operatorname{St}}(\hat{\sigma}, \operatorname{sd} X) &= \dot{D}(\sigma) * \operatorname{sd} \sigma \\ &= \dot{D}(\sigma) * \hat{\sigma} * \operatorname{sd}(\operatorname{Bd} \sigma). \end{aligned}$$

Now it is also true that

$$\overline{\operatorname{St}}(\hat{\sigma}, \operatorname{sd} X) = \hat{\sigma} * \operatorname{Lk}(\hat{\sigma}, \operatorname{sd} X).$$

We conclude that

$$\operatorname{Lk}(\hat{\sigma}, \operatorname{sd} X) = \dot{D}(\sigma) * \operatorname{sd}(\operatorname{Bd} \sigma).$$

Since $\operatorname{sd}(\operatorname{Bd} \sigma)$ is topologically a $k - 1$ sphere (or is empty if $k = 0$),

$$\tilde{H}_{i+k}(\operatorname{Lk}(\hat{\sigma}, \operatorname{sd} X)) \simeq \tilde{H}_i(\dot{D}(\sigma)).$$

Because the local homology of X at $\hat{\sigma}$ is infinite cyclic in dimension n and vanishes otherwise, $\operatorname{Lk}(\hat{\sigma}, \operatorname{sd} X)$ is a homology $n - 1$ sphere, by Lemma 63.1. Therefore, $\dot{D}(\sigma)$ is a homology $n - k - 1$ sphere.

Since $\overline{D}(\sigma)$ is acyclic (being a cone), the long exact homology sequence gives us an isomorphism

$$H_i(\overline{D}(\sigma), \dot{D}(\sigma)) \simeq \tilde{H}_{i-1}(\dot{D}(\sigma)).$$

We conclude that $(\overline{D}(\sigma), \dot{D}(\sigma))$ has the homology of an $n - k$ cell modulo its boundary. □

Definition. Let X be a locally finite complex that is a homology n-manifold.[†] Then the preceding theorem applies to each simplex σ of X. The collection of dual blocks $D(\sigma)$ will be called the **dual block decomposition** of X. The union of the blocks of dimension at most p will be denoted by X_p, and called the **dual p-skeleton** of X. The **dual chain complex** $\mathcal{D}(X)$ of X is defined by letting its chain group in dimension p be the group

$$D_p(X) = H_p(X_p, X_{p-1}).$$

Its boundary operator is the homomorphism ∂_* in the exact sequence of the triple (X_p, X_{p-1}, X_{p-2}).

[†]In fact, every complex that is a homology n-manifold is locally finite. See the exercises of §63.

Being the polytope of a subcomplex of sd X, the dual p-skeleton X_p will sometimes be treated as a complex rather than as a space. In particular, we shall for convenience use simplicial homology in the definition of $D_p(X)$.

For all practical purposes, the dual chain complex will play the same role in computing homology and cohomology as the cellular chain complex did for a CW complex. Repeating the pattern of §39, we prove the following:

Theorem 64.2. *Let X be a locally finite complex that is a homology n-manifold. Let X_p be the dual p-skeleton of X. Let $\mathcal{D}(X)$ be the dual chain complex of X.*

(a) *The group $H_i(X_p,X_{p-1})$ vanishes for $i \neq p$ and is a free abelian group for $i = p$. A basis when $i = p$ is obtained by choosing generators for the groups $H_p(\overline{D}(\sigma),\dot{D}(\sigma))$, as $D(\sigma)$ ranges over all p-blocks of X, and taking their images, under the homomorphisms induced by inclusion, in $H_p(X_p,X_{p-1})$.*

(b) *The dual chain complex $\mathcal{D}(X)$ can be used to compute the homology of X. Indeed, $D_p(X)$ equals the subgroup of $C_p(\text{sd } X)$ consisting of those chains carried by X_p whose boundaries are carried by X_{p-1}. And the inclusion map $D_p(X) \to C_p(\text{sd } X)$ induces a homology isomorphism; therefore, it also induces homology and cohomology isomorphisms with arbitrary coefficients.*

Proof. The proof of (a) follows the pattern of Lemma 39.2, but is easier. Because $\overline{D}(\sigma)$ is a cone with vertex $\hat{\sigma}$, its base $\dot{D}(\sigma)$ is a deformation retract of $\overline{D}(\sigma) - \hat{\sigma}$, for each σ. These deformation retractions induce a deformation retraction of $X_p - \bigcup_p \hat{\sigma}$ onto X_{p-1}, where $\bigcup_p \hat{\sigma}$ denotes the union of the barycenters of all simplices σ of X of dimension $n - p$.

If $\Sigma_p(\overline{D}(\sigma),\dot{D}(\sigma))$ denotes the topological sum of the p-blocks of X, we have the commutative diagram

$$\begin{array}{ccccc}
\Sigma_p(\overline{D}(\sigma),\dot{D}(\sigma)) & \longrightarrow & \Sigma_p(\overline{D}(\sigma),\overline{D}(\sigma) - \hat{\sigma}) & \longleftarrow & \Sigma_p(D(\sigma),D(\sigma) - \hat{\sigma}) \\
\downarrow & & \downarrow & & \downarrow = \\
(X_p,X_{p-1}) & \longrightarrow & (X_p,X_p - \bigcup_p \hat{\sigma}) & \longleftarrow & \bigcup (D(\sigma),D(\sigma) - \hat{\sigma})
\end{array}$$

where the horizontal maps induce homology isomorphisms in singular homology and the vertical maps are inclusions. Statement (a) follows for singular theory, and hence for simplicial theory.

Statement (b) is an immediate consequence of Theorem 39.5. □

Now we have the basic tools we need to prove the Poincaré duality theorem.

EXERCISES

1. Consider the usual triangulation of the torus T.
   (a) Sketch the blocks of the dual block decomposition of T; note that they actually make T into a regular cell complex.
   (b) Let $\mathcal{C}(T)$ denote the simplicial chain complex of T, and let $\mathcal{D}(T)$ denote the

dual chain complex. Orient the 2-simplices σ of T counterclockwise; orient the 1-simplices e arbitrarily. Define an isomorphism

$$\phi : C^2(T) \to D_0(T)$$

by setting

$$\phi(\sigma^*) = D(\sigma) = \hat{\sigma}$$

for each oriented 2-simplex σ. Define an isomorphism

$$\phi : C^1(T) \to D_1(T)$$

by letting $\phi(e^*)$ equal a fundamental cycle for $(\overline{D}(e),\dot{D}(e))$ for each oriented 1-simplex e. Show that ϕ may be so chosen that

$$\partial\phi(e^*) = \phi(\delta e^*).$$

(c) Define

$$\phi : C^0(T) \to D_2(T)$$

by letting $\phi(v^*)$ equal a fundamental cycle for $(\overline{D}(v),\dot{D}(v))$. Show that if the 2-cell $\overline{D}(v)$ is oriented counterclockwise, then

$$\partial\phi(v^*) = \phi(\delta v^*).$$

(d) Conclude that $H^p(T) \simeq H_{2-p}(T)$ for all p (which we knew already). This exercise indicates how Poincaré duality is proved.

2. Repeat Exercise 1 for the Klein bottle, using $\mathbf{Z}/2$ coefficients for both homology and cohomology. Note that the proof is easier, since no signs are involved.

§65. POINCARÉ DUALITY

The Poincaré duality theorem is one of the most striking results of topology. In its original form, it stated that for a *compact orientable triangulated n-manifold X*, the betti numbers in dimensions k and $n - k$ were equal and the torsion numbers in dimensions k and $n - k - 1$ were equal. Nowadays, it is stated in a different but equivalent formulation, as a theorem stating that the homology and cohomology groups of X in dimensions k and $n - k$ are isomorphic.

There are also versions of Poincaré duality that hold when X is not orientable, or when X is not compact, which we shall discuss.

We shall give two proofs of Poincaré duality. The first, which we give in this section, is more straightforward and intuitive. But the second, which we give in §67, will provide valuable information on how the duality isomorphism behaves with respect to continuous maps, as well as information about cup products in manifolds.

Definition. Let X be a compact triangulated homology n-manifold. We say that X is **orientable** if it is possible to orient all the n-simplices σ_i of X

so their sum $\gamma = \Sigma\, \sigma_i$ is a cycle. Such a cycle γ will be called an **orientation cycle** for X.

If X is not connected, then each component of X is itself a homology n-manifold. In order that X be orientable, it is necessary and sufficient that each component of X be orientable; a sum of orientation cycles for each component of X is an orientation cycle for X. We will show shortly that orientability of X does not depend on the particular triangulation of X. (See Corollary 65.4.)

Theorem 65.1 (Poincaré duality—first version). *Let X be a compact triangulated homology n-manifold. If X is orientable, then for all p, there is an isomorphism*

$$H^p(X; G) \simeq H_{n-p}(X; G),$$

where G is an arbitrary coefficient group. If X is non-orientable, there is for all p an isomorphism

$$H^p(X; \mathbf{Z}/2) \simeq H_{n-p}(X; \mathbf{Z}/2).$$

Proof. We shall use the simplicial chain complex $\mathcal{C}(X)$ of X to compute the cohomology of X, and the dual chain complex $\mathcal{D}(X)$ of X to compute the homology of X. There is a 1-1 correspondence between the p-simplices of X and the dual $n - p$ blocks of X that maps each simplex to its dual block. Hence the free abelian groups $C^p(X)$ and $D_{n-p}(X)$ are isomorphic, by an isomorphism ϕ that carries the basis element σ^* for $C^p(X)$ (where σ is an oriented p-simplex) to a generator of $H_{n-p}(\overline{D}(\sigma), \dot{D}(\sigma))$. If X is orientable, we shall show that the sign of $\phi(\sigma^*)$ may be so chosen that the following diagram commutes:

$$\begin{array}{ccc} C^{p-1}(X) & \xrightarrow{\ \phi\ } & D_{n-p+1}(X) \\ \delta \downarrow & & \downarrow \partial \\ C^p(X) & \xrightarrow{\ \phi\ } & D_{n-p}(X). \end{array}$$

This will prove the existence of the Poincaré duality isomorphism in the case of integer coefficients.

To define ϕ, we proceed as follows: First, orient the n-simplices of X so their sum γ is a cycle. Orient the other simplices of X arbitrarily.

We begin by defining ϕ in dimension n. The oriented n-simplices σ of X form a basis for $C_n(X)$; their duals σ^* form a basis for $C^n(X)$. The dual block $\overline{D}(\sigma)$ of the n-simplex σ is the 0-cell $\hat{\sigma}$. We define $\phi(\sigma^*) = \hat{\sigma}$, noting that $\hat{\sigma}$ is a generator of the group $H_0(\hat{\sigma})$. Thus we have defined an isomorphism

$$\phi : C^n(X) \to D_0(X).$$

We next define ϕ in dimension $n - 1$. Let s be an oriented $n - 1$ simplex of X. We wish to define $\phi(s^*)$ so that

$$\partial\phi(s^*) = \phi(\delta s^*).$$

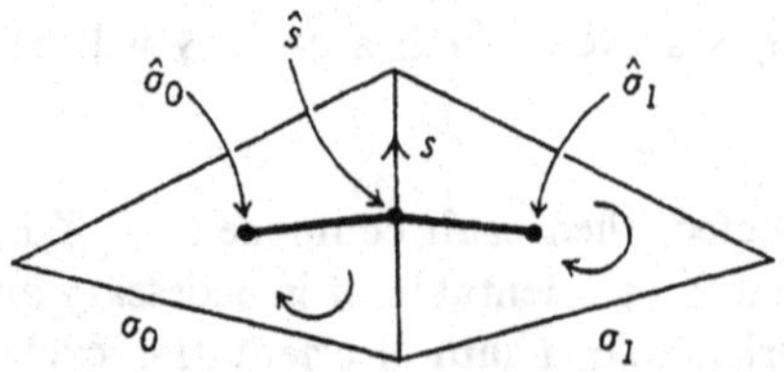

Figure 65.1

Now s is a face of exactly two n-simplices σ_0 and σ_1 of X; since γ is a cycle, they are oriented so that $\partial\sigma_0 + \partial\sigma_1$ has coefficient zero on s. Suppose the indexing chosen so that $\partial\sigma_0$ has coefficient -1 on s and $\partial\sigma_1$ has coefficient 1 on s. Then $\delta s^* = \sigma_1^* - \sigma_0^*$, so that

$$\phi(\delta s^*) = \hat{\sigma}_1 - \hat{\sigma}_0,$$

by definition. The 1-block $\overline{D}(s)$ is the union of the line segments from $\hat{s}$ to $\hat{\sigma}_0$ and $\hat{\sigma}_1$; we define

$$\phi(s^*) = [\hat{\sigma}_0,\hat{s}] + [\hat{s},\hat{\sigma}_1].$$

See Figure 65.1. Then $\phi(s^*)$ is a fundamental cycle for the 1-cell $(\overline{D}(s),\dot{D}(s))$, as desired; and

$$\partial\phi(s^*) = \hat{\sigma}_1 - \hat{\sigma}_0 = \phi(\delta s^*).$$

In general, suppose ϕ defined in dimension $p + 1 < n$. We wish to define it in dimension p so that for each oriented p-simplex s,

$$\partial\phi(s^*) = \phi(\delta s^*).$$

Given the oriented p-simplex s, we compute

$$\delta s^* = \Sigma\, \epsilon_i \sigma_i^*,$$

where the sum extends over all $p + 1$ simplices σ_i of which s is a face, and $\epsilon_i = \pm 1$ is the coefficient of s in the expression for $\partial\sigma_i$. Then

$$\phi(\delta s^*) = \Sigma\, \epsilon_i \phi(\sigma_i^*).$$

We shall prove that $\phi(\delta s^*)$ is a fundamental cycle for $\dot{D}(s)$.

The chain $\phi(\sigma_i^*)$ is by hypothesis a fundamental cycle for the block $\overline{D}(\sigma_i)$ dual to σ_i. This block lies in $\dot{D}(s)$ since $\sigma_i \succ s$. Thus $\phi(\sigma_i^*)$ is carried by $\dot{D}(s)$ for each i. Then the chain $\phi(\delta s^*)$ is carried by $\dot{D}(s)$ also. Furthermore, $\phi(\delta s^*)$ is a *cycle*, since $\partial\phi(\delta s^*) = \phi\delta\delta s^*$ by the induction hypothesis. Now $\dot{D}(s)$ is a homology $n - p - 1$ sphere, where $n - p - 1 > 0$. The cycle $\phi(\delta s^*)$ is non-trivial, and it is not a multiple of another cycle, because its restriction to the block $\overline{D}(\sigma_i)$ equals $\phi(\sigma_i^*)$, which is a fundamental cycle for this block modulo $\dot{D}(\sigma_i)$. Hence $\phi(\delta s^*)$ represents a generator of $H_{n-p-1}(\dot{D}(s))$, as asserted.

Consider the exact sequence

$$0 \longrightarrow H_{n-p}(\overline{D}(s),\dot{D}(s)) \xrightarrow{\partial_*} H_{n-p-1}(\dot{D}(s)) \longrightarrow 0.$$

For dimensional reasons, homology groups equal cycle groups. Define $\phi(s^*)$ to be the fundamental cycle for $(\overline{D}(s),\dot{D}(s))$ whose boundary equals the generator $\phi(\delta s^*)$. Then ϕ is defined for dimension p, and $\partial\phi(s^*) = \phi(\delta s^*)$, as desired.

The theorem is thus proved for the case of integer coefficients. To prove it for arbitrary coefficients, one notes that the isomorphisms

$$\operatorname{Hom}(C_p(X),G) \simeq \operatorname{Hom}(C_p(X),\mathbf{Z}) \otimes G \xrightarrow{\phi \otimes i_G} D_{n-p}(X) \otimes G$$

commute with δ and ∂. (The first of these isomorphisms was proved in Step 2 of Theorem 56.1.)

Finally, we consider the non-orientable case. Roughly speaking, the same proof goes through. One defines an isomorphism

$$\operatorname{Hom}(C_p(X),\mathbf{Z}/2) \xrightarrow{\phi} D_{n-p}(X) \otimes \mathbf{Z}/2,$$

by induction. If σ is an n-simplex, we let σ^* denote the cochain whose value is $[1] \in \mathbf{Z}/2$ on σ, and 0 on all other simplices. We then define $\phi(\sigma^*) = \hat{\sigma} \otimes [1]$. If s is an $n-1$ simplex that is a face of the n-simplices σ_0 and σ_1, we define $\phi(s^*) = ([\hat{\sigma}_0,\hat{s}] + [\hat{s},\hat{\sigma}_1]) \otimes [1]$; signs do not matter because the coefficient group is $\mathbf{Z}/2$. The rest of the proof goes through without change, with the phrase "fundamental cycle" replaced by "unique non-trivial cycle with $\mathbf{Z}/2$ coefficients." □

Let us note that there was nothing special, in this proof, about using the simplicial chain complex to compute the *cohomology* of X and the dual chain complex to compute the *homology* of X. One could just as well have done it the other way. For since ϕ is an isomorphism, the dual of ϕ is an isomorphism

$$\operatorname{Hom}(\operatorname{Hom}(C_p(X),\mathbf{Z}),\mathbf{Z}) \xleftarrow{\tilde{\phi}} \operatorname{Hom}(D_{n-p}(X),\mathbf{Z})$$

that carries the operator δ in $\mathcal{D}(X)$ to the dual of δ in $\mathcal{C}(X)$. Since there is a natural isomorphism

$$C_p(X) \simeq \operatorname{Hom}(\operatorname{Hom}(C_p(X),\mathbf{Z}),\mathbf{Z})$$

that carries ∂ to the dual of δ (see Step 4 of Theorem 56.1), the result is an isomorphism

$$C_p(X) \leftarrow \operatorname{Hom}(D_{n-p}(X),\mathbf{Z})$$

that commutes with δ and ∂.

Corollary 65.2. *Let X be a compact triangulated homology n-manifold. If X is connected, then for any two n-simplices σ, σ' of X, there is a sequence*

$$\sigma = \sigma_0, \sigma_1, \ldots, \sigma_k = \sigma'$$

of n-simplices of X such that $\sigma_i \cap \sigma_{i+1}$ is an $n-1$ simplex of X, for each i.

Proof. Let us define $\sigma \sim \sigma'$ if there is such a sequence connecting them. This relation is clearly an equivalence relation. The sum $\Sigma\, \sigma_i$ of the n-simplices in any one equivalence class is a cycle if its coefficient group is taken to be $\mathbf{Z}/2$. For, given an $n-1$ simplex s of X, the two n-simplices of which it is a face are equivalent, so either both of them appear in this sum, or neither does. In any case $\partial(\Sigma\,\sigma_i)$ has coefficient $[0] \in \mathbf{Z}/2$ on s.

We conclude that there is only one equivalence class, since otherwise X would have more than one non-trivial n-cycle over $\mathbf{Z}/2$, contradicting the fact that

$$H_n(X; \mathbf{Z}/2) \simeq H^0(X; \mathbf{Z}/2) \simeq \mathbf{Z}/2. \quad \square$$

Note that this corollary shows that a compact connected triangulated homology manifold is a pseudo manifold. (See Corollary 63.3 and the comment following.)

Corollary 65.3. *Let X be a compact triangulated homology n-manifold. If X is connected, then $H_n(X) \simeq \mathbf{Z}$ for X orientable and $H_n(X) = 0$ for X non-orientable.*

Proof. If X is orientable, then $H_n(X) \simeq H^0(X) \simeq \mathbf{Z}$. Conversely, we show that if $H_n(X) \neq 0$, then X is orientable. Orient the n-simplices of X arbitrarily. Suppose z is a non-trivial n-cycle of X, with integer coefficients. Now if σ_i and σ_{i+1} are n-simplices of X with an $n-1$ face in common, and if z has coefficient m on σ_i, it must have coefficient $\pm m$ on σ_{i+1}. In view of the preceding corollary, z must have the same coefficient, up to sign, on each n-simplex of X. Suppose this coefficient has absolute value m. Then dividing z by m gives us an orientation cycle for X. $\square$

Corollary 65.4. *Let X be a compact triangulated homology n-manifold. Then X is orientable if and only if for each component X_i of X, we have $H_n(X_i) \simeq \mathbf{Z}$.* $\square$

It follows that orientability of X does not depend on the triangulation of X.

The following corollary depends on the universal coefficient theorems proved in the last chapter.

Corollary 65.5. *Let X be a compact triangulated homology n-manifold; assume X is connected.*

(a) *If X is orientable, then $H_{n-1}(X)$ has no torsion, and for all G,*

$$H_n(X; G) \simeq G \simeq H^n(X; G).$$

(b) *If X is non-orientable, then the torsion subgroup of $H_{n-1}(X)$ has order 2, and for all G,*

$$H_n(X; G) \simeq \ker(G \xrightarrow{2} G) \qquad \text{and} \qquad H^n(X; G) \simeq G/2G.$$

In particular, $H_n(X) = 0$ and $H^n(X) \simeq \mathbf{Z}/2$.

Proof. (a) The computation when X is orientable is immediate, since

$$H^n(X; G) \simeq H_0(X; G) \qquad \text{and} \qquad H_n(X; G) \simeq H^0(X; G)$$

by Poincaré duality; each of the 0-dimensional groups is isomorphic to G because X is connected. It follows that $H_{n-1}(X)$ has no torsion. For

$$H^n(X) \simeq \operatorname{Hom}(H_n(X),\mathbf{Z}) \oplus \operatorname{Ext}(H_{n-1}(X),\mathbf{Z}).$$

Since $\operatorname{Ext}(H_{n-1}(X),\mathbf{Z})$ is isomorphic to the torsion subgroup of $H_{n-1}(X)$ (by Theorem 52.3), this torsion subgroup must vanish.

(b) Now suppose X is non-orientable. We compute $H_n(X; G)$. Let $\Sigma\,\sigma_i$ be the sum of all the n-simplices of X, oriented arbitrarily. If $g \in G$ has order 2, then $\Sigma\, g\sigma_i$ is clearly a cycle of X, since each $n - 1$ simplex is a face of exactly two n-simplices. Conversely, let z be a non-trivial n-cycle of X with G coefficients. If z has coefficient g on the oriented n-simplex σ of X, then z must have coefficient $\pm g$ on each n-simplex having an $n - 1$ face in common with σ; otherwise z would not be a cycle. Application of Corollary 65.2 shows that z has coefficient $\pm g$ on every n-simplex of X. Thus $z = g(\Sigma\,\epsilon_i\sigma_i)$, where $\epsilon_i = \pm 1$ and the sum extends over all n-simplices of X. Now the chain $\Sigma\,\epsilon_i\sigma_i$ cannot be a cycle, since that would imply that X is orientable. Hence $\partial(\Sigma\,\epsilon_i\sigma_i)$ has a non-zero coefficient on at least one $n - 1$ simplex s of X. This coefficient is ± 2, since s is a face of exactly two n-simplices of X. We conclude that ∂z has coefficient $\pm 2g$ on s, which means (since z is a cycle) that g has order 2. In particular, $g = -g$, so that $z = g(\Sigma\,\epsilon_i\sigma_i) = \Sigma\, g\sigma_i$.

Thus the n-cycles of X consist of all chains of the form $\Sigma\, g\sigma_i$, where g is zero or has order 2. Since the cycle group equals the homology group in dimension n, it follows that

$$H_n(X; G) \simeq \ker(G \xrightarrow{2} G).$$

We now show that the torsion subgroup $T_{n-1}(X)$ of $H_{n-1}(X)$ is isomorphic to $\mathbf{Z}/2$. Since $H_n(X) = 0$, it follows from the universal coefficient theorem that

$$H_{n-1}(X) * G \simeq H_n(X; G) \simeq \ker(G \xrightarrow{2} G).$$

Hence

$$T_{n-1}(X) * G \simeq \ker(G \xrightarrow{2} G) \simeq \mathbf{Z}/2 * G.$$

In particular,

$$T_{n-1}(X) * \mathbf{Z}/m \simeq \begin{cases} 0 & \text{if } m \text{ is odd,} \\ \mathbf{Z}/2 & \text{if } m \text{ is even.} \end{cases}$$

It follows that T_{n-1} has no odd torsion coefficient $2p + 1$ (set $m = 2p + 1$ in this expression to obtain a contradiction), that it has only one even torsion coefficient (set $m = 2$), and that this torsion coefficient is not of the form $2k$ for $k > 1$ (set $m = 2k$). Hence $T_{n-1}(X) \simeq \mathbf{Z}/2$.

Finally, it follows from the universal coefficient theorem for cohomology that

$$H^n(X;G) \simeq \text{Ext}(H_{n-1}(X),G) \simeq G/2G. \quad \square$$

EXERCISES

1. Let M and N be compact, connected, triangulable n-manifolds. Let $f: M \to N$. Show that if M is orientable and N is non-orientable, then
$$f_*: H_n(M;\mathbf{Z}/2) \to H_n(N;\mathbf{Z}/2)$$
is trivial.
2. Let X be a compact triangulable orientable manifold, with ith betti number $\beta_i(X)$. Recall that
$$\chi(X) = \Sigma(-1)^i\beta_i(X)$$
is called the *Euler number* of X. (See §22.)
(a) Show that if $\dim X$ is odd, then $\chi(X) = 0$.
(b) If $\dim X$ equals $2k$, express $\chi(X)$ in terms of β_i for $i \leq k$.
(c) Give applications to tangent vector fields. (See Exercise 4 of §22.)
3. Let X be a homology n-manifold (not necessarily compact) that is triangulated by a locally finite complex. Derive the *Poincaré duality isomorphisms for non-compact homology manifolds*, as follows:
(a) Show that
$$H_c^p(X;\mathbf{Z}/2) \simeq H_{n-p}(X;\mathbf{Z}/2),$$
where H_c^p denotes cohomology with compact support. Show that
$$H^p(X;\mathbf{Z}/2) \simeq H_{n-p}^\infty(X;\mathbf{Z}/2),$$
where H_{n-p}^∞ denotes homology based on infinite chains.
(b) Show that if X is connected, then X is a pseudo n-manifold.
(c) We say X is **orientable** if it is possible to orient the n-simplexes σ_i of X so that the (possibly infinite) chain $\Sigma\,\sigma_i$ is a cycle. Show that in this case,
$$H_c^p(X;G) \simeq H_{n-p}(X;G) \qquad \text{and} \qquad H^p(X;G) \simeq H_{n-p}^\infty(X;G)$$
for all G. Conclude that orientability of X depends only on the underlying topological space, not on the particular triangulation.

§66. CAP PRODUCTS

To give our second proof of Poincaré duality, we must first treat a topic that is purely algebraic in nature; it properly belongs in the last chapter, or even in Chapter 5. It is a certain operation called the *cap product* of a cohomology class and a homology class. Its definition is formally similar to the definition of cup product we gave in §48. Its usefulness, which lies in its connection with the duality theorems, will appear shortly.

Throughout this section, let R be a commutative ring with unity element.

Definition. Let X be a topological space. We define a map

$$S^p(X; R) \otimes S_{p+q}(X; R) \xrightarrow{\cap} S_q(X; R)$$

by the following equation: If $T : \Delta_{p+q} \to X$ is a singular simplex on X, and if $\alpha \in R$, then

$$c^p \cap (T \otimes \alpha) = T \circ l(\epsilon_0, \ldots, \epsilon_q) \otimes \alpha \cdot \langle c^p, T \circ l(\epsilon_q, \ldots, \epsilon_{p+q}) \rangle.$$

This chain is called the **cap product** of the cochain c^p and the chain $T \otimes \alpha$.

Roughly speaking, $c^p \cap (T \otimes \alpha)$ equals the restriction of T to the front q-face of Δ_{p+q}, with coefficient the product of α and the value of c^p on the back p-face of T.

As with cup product, there are alternate versions of the cap product. Among these are maps

$$S^p(X; G) \otimes S_{p+q}(X; \mathbf{Z}) \xrightarrow{\cap} S_q(X; G),$$

$$S^p(X; F) \otimes_F S_{p+q}(X; F) \xrightarrow{\cap} S_q(X; F),$$

where G is an abelian group and F is a field. They are all given by the preceding cap product formula. These additional versions of cap product will also be of use to us.

Theorem 66.1. *Cap product is bilinear, and is natural with respect to continuous maps. It satisfies the boundary formula*

$$\partial(d^p \cap c_{p+q}) = (-1)^q(\delta d^p \cap c_{p+q}) + d^p \cap \partial c_{p+q},$$

and is related to cup product by the formula

$$c^p \cap (d^q \cap e_{p+q+r}) = (c^p \cup d^q) \cap e_{p+q+r}.$$

Proof. Bilinearity is easy to check. Naturality is harder to formulate than to prove. Given a continuous map $f : X \to Y$, consider the following diagram:

$$\begin{array}{ccc} S^p(X) \otimes S_{p+q}(X) & \xrightarrow{\cap} & S_q(X) \\ \uparrow f^\# \quad \downarrow f_\# & & \downarrow f_\# \\ S^p(Y) \otimes S_{p+q}(Y) & \xrightarrow{\cap} & S_q(Y). \end{array}$$

(We omit the coefficients for convenience.) Naturality states that if $c^p \in S^p(Y)$ and $d_{p+q} \in S_{p+q}(X)$, then

$$f_\#(f^\#(c^p) \cap d_{p+q}) = c^p \cap f_\#(d_{p+q}).$$

One checks this formula readily by direct computation. The proofs of the coboundary and cup product formulas are straightforward but reasonably messy. We leave the details to you. □

These formulas hold in fact for all three versions of cap product. In the case of the coefficient pairing $G \otimes \mathbf{Z} \to G$, one must add the proviso that the cup product formula makes sense. This will occur if two of c^p, d^q, and e_{p+q+r} have coefficients in $\mathbf{Z}$ and the other has coefficients in G.

Theorem 66.2. *Cap product induces a homomorphism*

$$H^p(X;R) \otimes H_{p+q}(X;R) \xrightarrow{\frown} H_q(X;R).$$

It is natural, and satisfies the formula

$$\alpha^p \frown (\beta^q \frown \gamma_{p+q+r}) = (\alpha^p \smile \beta^q) \frown \gamma_{p+q+r}.$$

The same result holds for the other versions of cap product.

Proof. First, we note that if d^p is a cocycle and c_{p+q} is a cycle, then $d^p \frown c_{p+q}$ is a cycle:

$$\partial(d^p \frown c_{p+q}) = (-1)^q \delta d^p \frown c_{p+q} + d^p \frown \partial c_{p+q} = 0.$$

Second, we show that $\frown$ carries the kernel of the natural projection

$$Z^p(X;R) \otimes Z_{p+q}(X;R) \to H^p(X;R) \otimes H_{p+q}(X;R)$$

into $B_q(X;R)$. This kernel is generated by the elements of the two groups $B^p \otimes Z_{p+q}$ and $Z^p \otimes B_{p+q}$. The boundary formula shows that

$$\text{(cobdy)} \frown \text{(cycle)} = (-1)^{q+1}\delta d^{p-1} \frown z_{p+q} = \partial(d^{p-1} \frown z_{p+q}),$$
$$\text{(cocycle)} \frown \text{(bdy)} = z^p \frown \partial c_{p+q+1} = \partial(z^p \frown c_{p+q+1}),$$

both of which are boundaries, as desired. Naturality, and the cup product formula, are immediate. □

Relation to the Kronecker index

Naturality of the cap product on the homology-cohomology level is expressed by a diagram of the following form (where we omit the coefficients for convenience):

$$\begin{array}{ccc}
H^p(X) \otimes H_{p+q}(X) & \longrightarrow & H_q(X) \\
\uparrow f^* \qquad \downarrow f_* & & \downarrow f_* \\
H^p(Y) \otimes H_{p+q}(Y) & \longrightarrow & H_q(Y).
\end{array}$$

The similarity of this diagram to the corresponding diagram for the Kronecker index makes one suspect there is some relation between the two. There is; we discuss it now.

Recall that the Kronecker index was defined in §45 as the homomorphism

$$\langle\ ,\ \rangle : H^p(X;G) \otimes H_p(X) \to G$$

induced by evaluation of a cochain on a chain.

Theorem 66.3. *Let $\epsilon_* : H_0(X;G) \to G$ be the homomorphism induced by the augmentation map ϵ, which is an isomorphism if X is path connected. Then the Kronecker index equals the composite*

$$H^p(X;G) \otimes H_p(X) \xrightarrow{\frown} H_0(X;G) \xrightarrow{\epsilon_*} G.$$

Proof. The augmentation map $\epsilon : C_0 \to \mathbf{Z}$ gives rise to a homomorphism $C_0 \otimes G \to \mathbf{Z} \otimes G \simeq G$ that carries boundaries to zero. Hence it induces a homology homomorphism ϵ_*. We compute directly as follows: If $T : \Delta_p \to X$, then

$$\begin{aligned}\epsilon(c^p \frown T) &= \epsilon[T \circ l(\epsilon_0) \otimes \langle c^p, T \circ l(\epsilon_0, \ldots, \epsilon_p)\rangle] \\ &= \langle c^p, T\rangle.\end{aligned}$$

The theorem follows. □

It was convenient to define the Kronecker *map*

$$\kappa : H^p(X;G) \to \operatorname{Hom}(H_p(X), G)$$

as the map sending α to the homomorphism $\langle \alpha, \ \rangle$. Its naturality was easier to formulate than that of the Kronecker index. There is a similar version of the cap product. Its naturality is expressed by the commutative diagram

$$\begin{array}{ccc} H_{p+q}(X;R) & \xrightarrow{\frown} & \operatorname{Hom}(H^p(X;R), H_q(X;R)), \\ \downarrow f_* & & \downarrow \operatorname{Hom}(f^*, f_*) \\ H_{p+q}(Y;R) & \xrightarrow{\frown} & \operatorname{Hom}(H^p(Y;R), H_q(Y;R)) \end{array}$$

where the horizontal maps carry the homology class γ_{p+q} to the homomorphism "$\frown \gamma_{p+q}$." This is the formulation of naturality we often use in practice.

Simplicial cap product

Just as with cup products, one needs a formula that holds for simplicial theory if one wants to compute readily. By now it should be clear how to obtain such a formula.

Definition. Given a complex K, choose a partial ordering of the vertices of K that linearly orders the vertices of each simplex. We define a homomorphism

$$C^p(K;R) \otimes C_{p+q}(K;R) \xrightarrow{\frown} C_q(K;R),$$

called **simplicial cap product**, by the formula

$$c^p \frown ([v_0, \ldots, v_{p+q}] \otimes \alpha) = [v_0, \ldots, v_q] \otimes \alpha \cdot \langle c^p, [v_q, \ldots, v_{p+q}]\rangle.$$

Theorem 66.4. *The simplicial cap product satisfies the boundary formula and cup product formula of Theorem 66.1. It induces a homomorphism in simplicial theory*

$$H^p(K;R) \otimes H_{p+q}(K;R) \to H_q(K;R)$$

that corresponds to the singular cap product under the standard isomorphism between simplicial and singular theory.

Proof. Let $\eta : C_p(K) \to S_p(|K|)$ be the chain map defined in §34 that induces an isomorphism of simplicial with singular theory. Since η is a monomorphism onto a direct summand, its dual $\tilde{\eta}$ is surjective.

We first show that η commutes with $\frown$ on the chain level. That is, given $c^p \in C^p(K;R)$, we choose $d^p \in S^p(|K|;R)$ with $\tilde{\eta}(d^p) = c^p$, and verify that for $c_{p+q} \in C_{p+q}(K;R)$, we have

$$\eta(c^p \frown c_{p+q}) = d^p \frown \eta(c_{p+q}).$$

To prove this formula, we assume $v_0 < \cdots < v_{p+q}$, and compute

$$\begin{aligned}\eta(c^p \frown [v_0,\dots,v_{p+q}] \otimes \alpha) &= l(v_0,\dots,v_q) \otimes \alpha \cdot \langle c^p,[v_q,\dots,v_{p+q}]\rangle \\ &= l(v_0,\dots,v_q) \otimes \alpha \cdot \langle d^p, l(v_q,\dots,v_{p+q})\rangle \\ &= d^p \frown \eta([v_0,\dots,v_{p+q}] \otimes \alpha).\end{aligned}$$

This last equation uses the fact that

$$l(v_0,\dots,v_{p+q}) \circ l(\epsilon_0,\dots,\epsilon_q) = l(v_0,\dots,v_q).$$

The boundary formula and cup product formula may be proved by direct computation, as in Theorem 66.1. Alternatively, they can be derived from the similar formulas for singular theory. For example, given c^p, let $\tilde{\eta}(d^p) = c^p$, as before. Then if we apply η to both sides of the desired equation

$$(*) \qquad \partial(c^p \frown c_{p+q}) = (-1)^q \delta c^p \frown c_{p+q} + c^p \frown \partial c_{p+q},$$

we obtain the valid equation

$$\partial(d^p \frown \eta(c_{p+q})) = (-1)^q \delta d^p \frown \eta(c_{p+q}) + d^p \frown \partial\eta(c_{p+q}).$$

Since η is a monomorphism, (*) must hold. A similar argument applies to the cup product formula, using the fact that $\tilde{\eta}$ commutes with cup products. □

Relative cap products

Just as with the cup product, there is a relative version of the cap product. It will be useful in what follows. The most general relative cap product is a bilinear map

$$H^p(X,A) \otimes H_{p+q}(X,A \cup B) \to H_q(X,B),$$

which is defined when $\{A,B\}$ is an excisive couple. (Here we omit the coefficients for convenience.) We shall in fact be interested only in the cases where either A or B is empty, or $A = B$. That is, we are concerned with the following operations:

$$H^p(X) \otimes H_{p+q}(X,B) \xrightarrow{\frown} H_q(X,B),$$

$$H^p(X,A) \otimes H_{p+q}(X,A) \xrightarrow{\frown} H_q(X),$$

$$H^p(X,A) \otimes H_{p+q}(X,A) \xrightarrow{\frown} H_q(X,A).$$

The existence of these cap products is easy to demonstrate. Let $c^p \in S^p(X)$ and $d_{p+q} \in S_{p+q}(X)$. To obtain the first operation, we note simply that if d_{p+q} is carried by B, then so is $c^p \frown d_{p+q}$, by its definition. To check the second, we note that if c^p vanishes on all chains carried by A, and if d_{p+q} is carried by A, then $c^p \frown d_{p+q} = 0$. The third one equals the second one followed by projection $H_q(X) \to H_q(X,A)$.

The boundary formula and the cup product formula hold, as before. Therefore, one has a natural bilinear map on the level of relative homology and cohomology.

One has similar relative cap products in simplicial theory. One simply repeats the arguments already given for the absolute cap product.

EXERCISES

1. Verify the naturality formula of Theorem 66.1, by setting $d_{p+q} = T \otimes \alpha$ and computing directly.
2. Verify the boundary formula of Theorem 66.1 by setting $c_{p+q} = T \otimes \alpha$ and adding the expressions for $(-1)^q \delta d^p \frown c_{p+q}$ and $d^p \frown \partial c_{p+q}$. Most of the terms cancel; one has remaining the expression

$$\sum_{i \leq q} (-1)^i T \circ l(\epsilon_0, \ldots, \hat{\epsilon}_i, \ldots, \epsilon_q) \otimes \alpha \cdot \langle d^p, T \circ l(\epsilon_q, \ldots, \epsilon_{p+q}) \rangle,$$

   where the qth term is left over from $(-1)^q \delta d^p \frown c_{p+q}$ and the terms for which $i < q$ are left over from $d^p \frown \partial c_{p+q}$.
3. Verify the cup product formula of Theorem 66.1 by setting $e_{p+q+r} = T \otimes \alpha$ and computing directly.
4. Let T denote the torus; let α and β be the usual generators for $H^1(T)$; let Γ and Λ generate $H_2(T)$ and $H^2(T)$, respectively. (See Example 1 of §49.)
   (a) Compute $\alpha \frown \Gamma$ and $\beta \frown \Gamma$; draw pictures.
   (b) Compute $\Lambda \frown \Gamma$.
   (c) Verify that $\alpha \frown (\beta \frown \Gamma) = (\alpha \smile \beta) \frown \Gamma$.
5. Let P^2 be the projective plane; let $\Gamma_{(2)}$ denote the non-zero element of $H_2(P^2; \mathbf{Z}/2)$. Compute the values of the homomorphism $\frown \Gamma_{(2)}$ on the groups $H^i(P^2; \mathbf{Z}/2)$ for $i = 1, 2$.

6. Let (X,A) be a triangulated pair; let $\alpha \in H_n(X,A)$. Then $\partial_*\alpha$ is an element of $H_{n-1}(A)$. Consider the following diagram:

$$\begin{array}{ccccccc} H^p(X,A) & \longrightarrow & H^p(X) & \longrightarrow & H^p(A) & \longrightarrow & H^{p+1}(X,A) \\ \downarrow{\scriptstyle \cap\alpha} & & \downarrow{\scriptstyle \cap\alpha} & & \downarrow{\scriptstyle \cap\partial_*\alpha} & & \downarrow{\scriptstyle \cap\alpha} \\ H_{n-p}(X) & \longrightarrow & H_{n-p}(X,A) & \longrightarrow & H_{n-p-1}(A) & \longrightarrow & H_{n-p-1}(X). \end{array}$$

Show that each of the squares commutes up to sign. What are the signs involved?

§67. A SECOND PROOF OF POINCARÉ DUALITY

We now give a second proof of the Poincaré duality theorem. In this proof, the duality isomorphism will be constructed using cap products. There are several interesting consequences, including a naturality formula for the duality isomorphism and a proof of its independence of the triangulation involved.

Definition. Let X be a compact triangulable homology n-manifold. (Note that no specific triangulation of X is assumed.) Let X_i be a component of X. If X is orientable, then $H_n(X_i)$ is infinite cyclic; and a generator $\Gamma^{(i)}$ of $H_n(X_i)$ is called an **orientation class** for X_i. The image of the classes $\Gamma^{(i)}$ under the isomorphism induced by inclusions,

$$\bigoplus H_n(X_i) \simeq H_n(X),$$

is called an **orientation class** for X, and is denoted by Γ. Similarly, if X is not necessarily orientable, and if $\Gamma^{(i)}_{(2)}$ is the unique non-trivial element of $H_n(X_i; \mathbf{Z}/2)$, then the image of these elements in $H_n(X; \mathbf{Z}/2)$ is denoted $\Gamma_{(2)}$ and is called an **orientation class for X over $\mathbf{Z}/2$**.

If X is given a specific triangulation, then $\Gamma^{(i)}$ is represented by the sum of all the n-simplices of the component X_i, suitably oriented, and Γ is represented by the sum of γ of all the n-simplices of X, suitably oriented. Similarly, $\Gamma_{(2)}$ is represented by the sum of the n-simplices of X, each with coefficient $[1] \in \mathbf{Z}/2$.

Theorem 67.1 (Poincaré duality—second version). *Let X be a compact triangulable homology n-manifold. If X is orientable, and if $\Gamma \in H_n(X)$ is an orientation class for X, then*

$$H^p(X; G) \xrightarrow{\cap\,\Gamma} H_{n-p}(X; G)$$

is an isomorphism for arbitrary G. Whether X is orientable or not,

$$H^p(X; \mathbf{Z}/2) \xrightarrow{\cap\,\Gamma_{(2)}} H_{n-p}(X; \mathbf{Z}/2)$$

is an isomorphism.

Proof. The idea of the proof is the same as before. We choose a triangulation of X, partition X into its dual blocks, and construct an isomorphism

$$\phi : C^p(X) \to D_{n-p}(X)$$

that commutes with δ and ∂. The difference is that we are going to construct ϕ by use of the cap product formula.

Step 1. We first prove the following elementary fact: Let K be a complex; let $g : \text{sd } K \to K$ be a simplicial approximation to the identity. Then given $\sigma \in K$, there is exactly one simplex t of sd σ that g maps onto σ; the others are mapped onto proper faces of σ.

Recall that if $\sigma \in K$, then since $\hat{\sigma}$ lies interior to σ, the map g must carry $\hat{\sigma}$ to one of the vertices of σ.

We proceed by induction on the dimension of σ. If σ is a vertex v, then $g(v) = v$, and our result holds. If $\sigma = v_0 v_1$, then $g(v_0) = v_0$ and $g(v_1) = v_1$, while $g(\hat{\sigma})$ equals either v_0 or v_1. In either case, g collapses one of the simplices $v_0\hat{\sigma}$ and $v_1\hat{\sigma}$ to a point, and maps the other onto $v_0 v_1$.

In general, let s be a k-simplex $v_0 \dots v_k$. By the induction hypothesis, each $k-1$ face $s_i = v_0 \dots \hat{v}_i \dots v_k$ of σ contains exactly one $k-1$ simplex t_i of sd K that is mapped by g onto s_i. Thus every k-simplex of sd σ is collapsed by g except possibly for k-simplices of the form $\hat{\sigma} * t_i$, for $i = 0, \dots, k$. Now g carries $\hat{\sigma}$ to one of the vertices of σ. Suppose $g(\hat{\sigma}) = v_j$. Since

$$g(t_i) = s_i = v_0 \dots \hat{v}_i \dots v_k,$$

the map g carries $\hat{\sigma} * t_j$ onto σ, and for $i \neq j$ collapses each simplex $\hat{\sigma} * t_i$ onto s_i.

Step 2. Here is another elementary fact: Let K be a complex; let $\text{sd} : C_p(K) \to C_p(\text{sd } K)$ be the barycentric subdivision chain map defined in §17. If σ is an oriented p-simplex of K, then sd σ is the sum of all the p-simplices of sd K lying in σ, suitably oriented.

We proceed by induction. The formula $\text{sd}(v) = v$ shows the result true for dimension 0. In general, let σ have dimension p. Now the chain $\partial\sigma$ is the sum of the $p-1$ faces of σ, suitably oriented. Then $\text{sd}(\partial\sigma)$ is by the induction hypothesis the sum of all the $p-1$ simplices of sd K lying in Bd σ, suitably oriented. It follows that

$$\text{sd } \sigma = [\hat{\sigma}, \text{sd}(\partial\sigma)]$$

is the sum of all the p-simplices of sd K lying in σ, suitably oriented.

Step 3. Now let X be the manifold of our theorem; choose a specific triangulation of X. Let γ be the cycle representing Γ relative to this triangulation. In the complex sd X, we shall use the standard partial ordering of the vertices, and we orient the simplices of sd X by using this ordering. Orient the simplices of X arbitrarily.

By Step 2, the chain sd γ is the sum of all the oriented n-simplices of sd X, with signs ± 1. Because it is a cycle, sd γ is an orientation cycle for sd X.

Choose $g: \text{sd}\, X \to X$ to be a simplicial approximation to the identity. Consider the diagram

$$\begin{array}{ccc} C^p(X) & \overset{\psi}{\dashrightarrow} & D_{n-p}(X) \\ \downarrow g^\# & & \downarrow j \\ C^p(\text{sd}\, X) & \xrightarrow{\cap\, \text{sd}\,\gamma} & C_{n-p}(\text{sd}\, X) \end{array}$$

where j is inclusion of the dual chain complex of X into the simplicial chain complex of $\text{sd}\, X$. We shall prove that the composite of $g^\#$ and $\cap\ \text{sd}\,\gamma$ carries $C^p(X)$ isomorphically onto the subgroup $D_{n-p}(X)$ of $C_{n-p}(\text{sd}\, X)$. It follows that there is a unique isomorphism ψ making this diagram commute.

First, we prove the preliminary result that if σ is an oriented p-simplex of X, then $g^\#(\sigma^*) \cap \text{sd}\,\gamma$ equals the sum of all $n-p$ simplices of $\text{sd}\, X$ lying in the block $\overline{D}(\sigma)$, with coefficients ± 1.

We prove this fact by direct computation. Given σ, there is by Step 1 exactly one p-simplex t of $\text{sd}\, X$ that is mapped onto σ by g. Then

$$g^\#(\sigma^*) = \pm t^*,$$

the sign depending on the chosen orientations. Let us compute the cap product $t^* \cap \text{sd}\,\gamma$.

Recall the formula for simplicial cap product. Given an ordering of the vertices of a complex K, and given a cochain c^p of K and a $p+q$ simplex τ of K, the chain $c^p \cap \tau$ equals the "front q-face" of τ, multiplied by the value of c^p on the "back p-face" of τ.

Now t is a p-simplex of $\text{sd}\, X$ of the form

$$t = [\hat{s}_p, \dots, \hat{s}_0]$$

where $s_p = \sigma$ and s_i has dimension i for each i. Furthermore, $\text{sd}\,\gamma$ is the sum of all the n-simplices of $\text{sd}\, X$; each is of the form

$$\tau = [\hat{\sigma}_n, \dots, \hat{\sigma}_0]$$

where σ_i has dimension i for each i. The chain $t^* \cap \tau$ vanishes unless t equals the back face of τ; that is, unless

$$t = [\hat{s}_p, \dots, \hat{s}_0] = [\hat{\sigma}_p, \dots, \hat{\sigma}_0].$$

In this case, $t^* \cap \tau$ equals $[\hat{\sigma}_n, \dots, \hat{\sigma}_p]$. It follows that $t^* \cap \text{sd}\,\gamma$ is the sum, with signs ± 1, of all simplices of $\text{sd}\, X$ of the form

$$[\hat{\sigma}_n, \dots, \hat{\sigma}_p]$$

for which $\sigma_p = s_p = \sigma$. These are precisely the $n-p$ simplices of $\overline{D}(\sigma)$.

Now we prove our result. We show that $g^\#(\sigma^*) \cap \text{sd}\,\gamma$ is in fact a *fundamental cycle* for the block $(\overline{D}(\sigma), \dot{D}(\sigma))$.

We know that $g^\#(\sigma^*) \cap \text{sd}\,\gamma$ is a non-trivial $n-p$ chain carried by $\overline{D}(\sigma)$, and that it is not a multiple of any other chain (since its coefficients are all

± 1). To show it is a fundamental cycle, it suffices to show that its boundary is carried by $\dot{D}(\sigma)$. For this, we compute

$$(*)\qquad \begin{aligned}\partial(g^{\#}(\sigma^*) \cap \operatorname{sd}\gamma) &= (-1)^{n-p}\delta g^{\#}(\sigma^*) \cap \operatorname{sd}\gamma + 0\\ &= (-1)^{n-p} g^{\#}(\delta\sigma^*) \cap \operatorname{sd}\gamma.\end{aligned}$$

Now $\delta\sigma^*$ is a sum of terms of the form $\pm\sigma^*_{p+1}$, where σ_{p+1} is a $p+1$ simplex of X having σ as a face. By the result just proved, the chain $g^{\#}(\sigma^*_{p+1}) \cap \operatorname{sd}\gamma$ is carried by the block $\overline{D}(\sigma_{p+1})$ dual to σ_{p+1}. Since $\sigma_{p+1} \succ \sigma$, this block is contained in $\dot{D}(\sigma)$. Thus $g^{\#}(\delta\sigma^*) \cap \operatorname{sd}\gamma$ is carried by $\dot{D}(\sigma)$, as desired.

Step 4. We have thus proved that the isomorphism

$$\psi : C^p(X) \to D_{n-p}(X)$$

exists, and that it carries the basis element σ^* to a fundamental cycle for the dual block to σ. Therefore, ψ resembles the isomorphism ϕ of Theorem 65.1. The only question remaining is whether it commutes with δ and ∂. This it does, but only up to sign. Formula (*) preceding gives us the equation

$$\partial\psi(c^p) = (-1)^{n-p}\psi(\delta c^p).$$

Now the sign $(-1)^{n-p}$ does not affect the groups of cocycles and boundaries. It follows that ψ induces an isomorphism of $H^p(X)$ with $H_{n-p}(\mathcal{D}(X))$, just as ϕ did. (In fact, ψ differs from ϕ only by a sign. See Exercise 1.)

Now since the inclusion $\mathcal{D}(X) \to \mathcal{C}(\operatorname{sd} X)$ induces a homology isomorphism, it follows that the composite map

$$H^p(X) \xrightarrow{g^*} H^p(\operatorname{sd} X) \xrightarrow{\cap\{\operatorname{sd}\gamma\}} H_{n-p}(\operatorname{sd} X)$$

is an isomorphism. Then if we identify the homology of sd X with that of X, by passing to the homology of the triangulable space X, for instance, or to singular homology and cohomology, we see that since g^* is the identity map, this isomorphism can be expressed in the form

$$H^p(X) \xrightarrow{\cap\Gamma} H_{n-p}(X).$$

Step 5. To treat the case of arbitrary coefficients, one checks that the composite

$$C^p(X) \otimes G \simeq C^p(X; G) \xrightarrow{g^{\#}} C^p(\operatorname{sd} X; G) \xrightarrow{\cap\operatorname{sd}\gamma} D_{n-p}(X) \otimes G$$

equals the isomorphism $C^p(X) \to D_{n-p}(X)$ of Step 4 tensored with the identity map of G. Hence it is an isomorphism.

Step 6. In the case where X is not necessarily orientable, the preceding proof works if one reduces all the coefficients mod 2. Steps 1 and 2 proceed unchanged. The proof in Step 3 shows that $g^{\#}(\sigma^*) \cap \operatorname{sd}\gamma_{(2)}$ is the non-trivial $n-p$ cycle (with $\mathbf{Z}/2$ coefficients) of the block $(\overline{D}(\sigma), \dot{D}(\sigma))$. Step 4 is easier, since no signs are involved. □

Our proof of this theorem used simplicial homology throughout. But the fact that simplicial and singular theory are isomorphic, by an isomorphism that preserves cap products, means that this theorem holds for singular theory as well. In particular, the isomorphism given by $\cap\,\Gamma$ depends only on the homology class Γ, not on any particular triangulation involved.

The preceding theorem gives us information about how the Poincaré duality isomorphism behaves with respect to continuous maps. Specifically, one has the following theorem, whose proof is an immediate consequence of the naturality of cap product:

Theorem 67.2. *Let X and Y be compact, connected, triangulable, orientable homology n-manifolds; let Γ_X and Γ_Y be orientation classes for X and Y, respectively. Given $f: X \to Y$, let d be the integer such that $f_*(\Gamma_X) = d \cdot \Gamma_Y$. Then the following diagram commutes:*

$$\begin{array}{ccc} H^k(X;G) & \xrightarrow[\cong]{\cap\,\Gamma_X} & H_{n-k}(X;G) \\ \uparrow f^* & & \downarrow f_* \\ H^k(Y;G) & \xrightarrow{\cap\, d\Gamma_Y} & H_{n-k}(Y;G). \end{array}$$

In particular, if $d = \pm 1$, then f^* is injective and f_* is surjective. The same result holds for $G = \mathbf{Z}/2$ without requiring X and Y to be orientable. □

Definition. The integer d appearing in the statement of this theorem is called the **degree** of f, relative to the orientation cycles Γ_X and Γ_Y. Since Γ_X and Γ_Y are uniquely determined up to sign, so is d. If $X = Y$ and if $\Gamma_X = \Gamma_Y$, then d is uniquely determined.

The second version of Poincaré duality, in which the isomorphism is obtained by use of cap products, has consequences that go far beyond the naturality diagram just proved. Note that everything in the statement of Theorem 67.1 makes sense even if X is not triangulable, provided one knows what one means by an orientation class for X. This fact suggests that by using cap products in singular theory, one might obtain a proof of Poincaré duality for an arbitrary compact topological manifold without assuming triangulability. This is in fact the case; the interested reader may consult [Do] (or alternatively, [G-H] or [V] or [S]).

EXERCISES

1. Given the orientation cycle γ for X, let ψ be constructed as in the proof of Theorem 67.1.

   (a) Define ϕ by the rule

$$\phi(c^p) = (-1)^{\alpha(p)}\psi(c^p),$$

where

$$\alpha(4k + n) = 0, \qquad \alpha(4k + n + 2) = 1,$$
$$\alpha(4k + n + 1) = 0, \qquad \alpha(4k + n + 3) = 1,$$

for all k. Show that ϕ satisfies the condition $\phi\delta = \partial\phi$.

(b) Show that if σ is an n-simplex of X, oriented as it appears in the cycle γ, then $\psi(\sigma^*) = \hat{\sigma}$. [*Hint:* $\epsilon(\psi(\sigma^*)) = \epsilon(g_\#(\psi(\sigma^*))) = \epsilon(\sigma^* \cap \gamma)$.]

(c) Conclude that the homomorphism ϕ of (a) is identical with that in the proof of Theorem 65.1.

2. Let X and Y be compact connected triangulable homology n-manifolds. Let $f : X \to Y$.
   (a) Show that if X and Y are orientable and if f has non-zero degree, then $\beta_i(X) \geq \beta_i(Y)$, where β_i denotes the ith betti number.
   (b) If $f_* : H_n(X; \mathbf{Z}/2) \to H_n(Y; \mathbf{Z}/2)$ is non-trivial, what can you say about $H_i(X; \mathbf{Z}/2)$ and $H_i(Y; \mathbf{Z}/2)$?

3. Let X_n denote the n-fold connected sum of tori; let $X_0 = S^2$. Show that there is a map $f : X_n \to X_m$ of non-zero degree if and only if $n \geq m$.

4. Let X_n be as in Exercise 3; let Y_n be the n-fold connected sum of projective planes, with $n \geq 1$. Given f, consider the condition:

   (*) *f_* is non-trivial as a map of H_2 with $\mathbf{Z}/2$ coefficients.*

   (a) Show that there is a map $f : Y_n \to Y_m$ satisfying (*) if and only if $n \geq m$.
   (b) Show that there is no map $f : X_n \to Y_m$ satisfying (*).

   The case of maps $f : Y_m \to X_n$ will be considered in the exercises of the next section.

*§68. APPLICATION: COHOMOLOGY RINGS OF MANIFOLDS†

We now apply the Poincaré duality theorem to the problem of computing the cohomology ring of a manifold. While we cannot compute the cohomology ring of *every* compact manifold, the theorem we shall prove is strong enough to enable us to compute the cohomology rings of the real and complex projective spaces. This computation leads, in turn, to proofs of such classical theorems of topology as the Borsuk-Ulam theorem and the so-called "Ham Sandwich Theorem."

Definition. Let A and B be free abelian groups of the same rank. Let C be an infinite cyclic group. We say a homomorphism

$$f : A \otimes B \to C$$

†In this section, we assume familiarity with the projective spaces (§40). We also use the universal coefficient theorem for cohomology (§53).

is a **dual pairing** if there are bases $a_1, \ldots, a_m$ for A, and $b_1, \ldots, b_m$ for B, such that

$$f(a_i \otimes b_j) = \delta_{ij}\gamma$$

for all $i, j = 1, \ldots, m$, where γ is a generator of C.

Theorem 68.1. *Let X be a compact, connected, triangulable, orientable homology n-manifold. Let $T^k(X)$ denote the torsion subgroup of $H^k(X)$. The cup product operation induces a homomorphism*

$$\frac{H^k(X)}{T^k(X)} \otimes \frac{H^{n-k}(X)}{T^{n-k}(X)} \to H^n(X)$$

that is a dual pairing.

Proof. Let $\alpha \in H^k(X)$ and $\beta \in H^{n-k}(X)$. If α is an element of finite order, so is $\alpha \cup \beta$; since $H^n(X)$ is infinite cyclic, this means that $\alpha \cup \beta = 0$. A similar remark applies if β has finite order. Therefore (by Lemma 50.3), cup product induces a homomorphism

$$\frac{H^k(X)}{T^k(X)} \otimes \frac{H^{n-k}(X)}{T^{n-k}(X)} \to H^n(X).$$

Choose a generator Λ of $H^n(X)$. Then choose a generator Γ of $H_n(X)$ by choosing its sign so that the isomorphism

$$(*) \qquad H^n(X) \xrightarrow{\cap \Gamma} H_0(X) \xrightarrow{\epsilon_*} \mathbf{Z},$$

maps Λ to 1. Then $\epsilon_*(\Lambda \cap \Gamma) = 1$.

To prove the theorem, we shall find elements $\alpha_1, \ldots, \alpha_m$ of $H^k(X)$ and $\beta_1, \ldots, \beta_m$ of $H^{n-k}(X)$ such that

$$\alpha_i \cup \beta_j = \delta_{ij}\Lambda$$

for $i, j = 1, \ldots, m$, and such that their cosets modulo torsion, denoted $\{\alpha_i\}$ and $\{\beta_j\}$, respectively, form bases for $H^k(X)/T^k(X)$ and $H^{n-k}(X)/T^{n-k}(X)$, respectively.

The idea of the proof is this. First, there is a *geometric* duality isomorphism

$$\cap \Gamma : H^{n-k}(X) \to H_k(X),$$

obtained from cap product with Γ. This isomorphism induces an isomorphism of $H^{n-k}(X)/T^{n-k}(X)$ with $H_k(X)/T_k(X)$.

Second, there is an *algebraic* duality isomorphism κ^* obtained from the Kronecker map κ. It is obtained from the exact sequence

$$0 \leftarrow \operatorname{Hom}(H_k(X),\mathbf{Z}) \xleftarrow{\kappa} H^k(X) \leftarrow \operatorname{Ext}(H_{k-1}(X),\mathbf{Z}) \leftarrow 0,$$

as follows: The left-hand group is isomorphic to $\mathrm{Hom}(H_k(X)/T_k(X),\mathbf{Z})$ and is thus free, while the right-hand group is a torsion group. Therefore, there exists an induced isomorphism

$$\mathrm{Hom}\left(\frac{H_k(X)}{T_k(X)},\mathbf{Z}\right) \xleftarrow{\kappa^*} \frac{H^k(X)}{T^k(X)}.$$

It is related to the cap product by the formula

$$[\kappa^*\{\alpha\}]\{\beta\} = [\kappa(\alpha)](\beta) = \langle\alpha,\beta\rangle = \epsilon_*(\alpha \cap \beta).$$

Combining these two types of duality will give us our theorem.

Consider the sequence of isomorphisms

$$\frac{H^{n-k}}{T^{n-k}} \xrightarrow[\simeq]{\cap \Gamma} \frac{H_k}{T_k} \simeq \mathrm{Hom}\left(\frac{H_k}{T_k},\mathbf{Z}\right) \xleftarrow[\simeq]{\kappa^*} \frac{H^k}{T^k}.$$

(Here we delete X from the notation, for convenience.) Choose, arbitrarily, a basis $\{\beta_1\},\dots,\{\beta_m\}$ for H^{n-k}/T^{n-k}. Let $\{\gamma_1\},\dots,\{\gamma_m\}$ be the corresponding basis for H_k/T_k, defined by the equation $\beta_j \cap \Gamma = \gamma_j$ for all j. Let $\gamma_1^*,\dots,\gamma_m^*$ be the corresponding dual basis for $\mathrm{Hom}(H_k/T_k,\mathbf{Z})$, defined by the equation $\gamma_i^*(\{\gamma_j\}) = \delta_{ij}$. Finally, let $\{\alpha_1\},\dots,\{\alpha_m\}$ be the corresponding basis for H^k/T^k, defined by the equation $\kappa^*(\{\alpha_i\}) = \gamma_i^*$ for all i.

To show that $\alpha_i \cup \beta_j = \delta_{ij}\Lambda$, it suffices (applying the isomorphism of $H^n(X)$ with $\mathbf{Z}$ given by (*)) to show that

$$\epsilon_*((\alpha_i \cup \beta_j) \cap \Gamma) = \delta_{ij}\epsilon_*(\Lambda \cap \Gamma) = \delta_{ij}.$$

We compute as follows:

$$\begin{aligned}\delta_{ij} &= \gamma_i^*(\{\gamma_j\}) = [\kappa^*\{\alpha_i\}](\{\beta_j\} \cap \Gamma)\\ &= \langle\alpha_i, \beta_j \cap \Gamma\rangle\\ &= \epsilon_*(\alpha_i \cap (\beta_j \cap \Gamma)) = \epsilon_*((\alpha_i \cup \beta_j) \cap \Gamma). \quad \square\end{aligned}$$

A similar theorem holds if the coefficients form a field:

Theorem 68.2. *Let X be a compact, connected, triangulable homology n-manifold. Let F be a field; assume F equals $\mathbf{Z}/2$ if X is non-orientable. Let Λ generate $H^n(X;F)$. There are (vector space) bases $\alpha_1,\dots,\alpha_m$ for $H^k(X;F)$, and $\beta_1,\dots,\beta_m$ for $H^{n-k}(X;F)$, such that for $i, j = 1,\dots,m$,*

$$\alpha_i \cup \beta_j = \delta_{ij}\Lambda.$$

Proof. In the orientable case, choose Γ so that the isomorphism

$$H^n(X;F) \xrightarrow{\cap \Gamma} H_0(X;F) \xrightarrow{\epsilon_*} F,$$

carries Λ to $1 \in F$. In the non-orientable case, $\Gamma_{(2)}$ is unique. Then proceed as before, using the diagram

$$H^{n-k}(X; F) \xrightarrow[\simeq]{\cap \Gamma'} H_k(X; F) \simeq \text{Hom}_F(H_k(X; F), F) \xleftarrow[\simeq]{\kappa^*} H^k(X; F).$$

Here Γ' denotes either Γ or $\Gamma_{(2)}$, depending on the case. $\square$

Example 1. Let us consider the torus once again. By Theorem 68.1, there is a basis α_1, α_2 for $H^1(T)$, and a basis β_1, β_2 for $H^1(T)$ such that $\alpha_i \cup \beta_j = \delta_{ij}\Lambda$, where Λ is a generator of $H^2(T)$. The two "picket fence" cocycles w^1 and z^1 pictured in Figure 68.1 give us a basis for $H^1(T)$, and the indicated cocycle σ^* generates $H^2(T)$. Recall that $z^1 \cup w^1 = \sigma^*$. The conclusion of the theorem holds if we let α_1 and α_2 be the cohomology classes of z^1 and w^1, respectively, and we let β_1 and β_2 be the cohomology classes of w^1 and $-z^1$, respectively, and we let Λ be the cohomology class of σ^*. These computations were carried out in Example 1 of §49.

The proof of the theorem tells us that this is the way it ought to work. If one begins with the cocycle w^1, the geometric (Poincaré) duality map ψ carries it to the *cycle* c_1 pictured in Figure 68.2. (We know that the cycle "dual" to w^1 is carried by the blocks dual to the simplices appearing in the carrier of w^1. To check that ψ carries w^1 to c_1 rather than to $-c_1$ requires some care.) The cycle c_1 is homologous to one represented by the cell B along the side edge of the rectangle. Under the *algebraic* duality κ^*, this cycle B in turn corresponds to a cocycle whose value is 1 on B and 0 on the cycle A going around the torus the other way. The cocycle z^1 is such a cocycle. Thus w^1 is carried to z^1 under our combined geometric-algebraic duality isomorphism. This is just as expected, since $\{z^1 \cup w^1\} = \Lambda$, while $z^1 \cup z^1 = w^1 \cup w^1 = 0$.

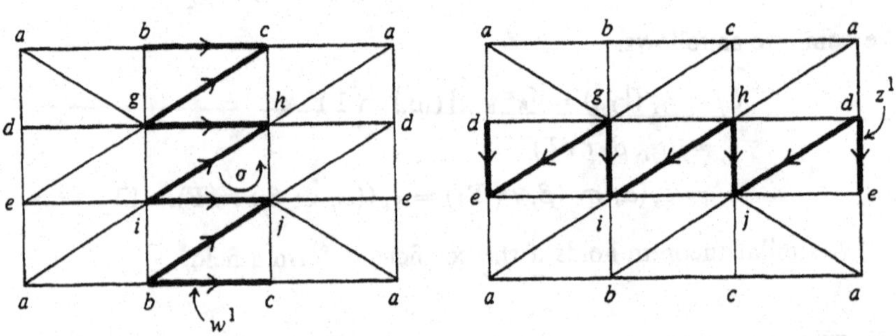

Figure 68.1

Now we apply these results to real projective space P^n. We know P^n is triangulable (see Lemma 40.7), so Poincaré duality applies. The cellular chain complex of P^n is infinite cyclic in each dimension $k = 0, \ldots, n$; and the boundary operator is either 0 or multiplication by 2. Therefore, $H^k(P^n; \mathbf{Z}/2)$ is a vector space of dimension 1, for $k = 0, \ldots, n$. It remains to calculate cup products in the ring $H^*(P^n; \mathbf{Z}/2)$.

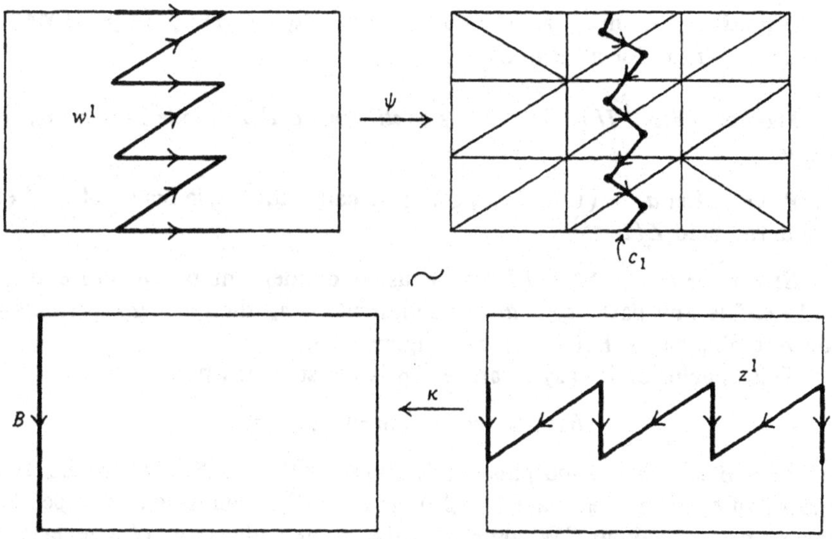

Figure 68.2

Theorem 68.3. *If u is the non-zero element of $H^1(P^n; \mathbf{Z}/2)$, then u^k is the non-zero element of $H^k(P^n; \mathbf{Z}/2)$, for $k = 2, \ldots, n$. Thus $H^*(P^n; \mathbf{Z}/2)$ is a truncated polynomial algebra over $\mathbf{Z}/2$ with one generator u in dimension 1, truncated by setting $u^{n+1} = 0$.*

Proof. We proceed by induction, beginning with $n = 2$. The vector space $H^1(P^2; \mathbf{Z}/2)$ has dimension 1. If u is its non-zero element, then by Theorem 68.2, $u^2 = u \cup u$ must be non-zero.

Suppose the theorem is true for dimension $n - 1$. The inclusion map $j : P^{n-1} \to P^n$ induces homology and cohomology isomorphisms (over $\mathbf{Z}/2$) in dimensions less than n. Therefore, by the induction hypothesis, if $u \in H^1(P^n; \mathbf{Z}/2)$ is non-zero, so are $u^2, \ldots, u^{n-1}$. (Recall that j^* is a ring homomorphism.) It remains to show $u^n \neq 0$.

By Theorem 68.2, there is a basis α_1 for $H^1(P^n; \mathbf{Z}/2)$ and a basis β_1 for $H^{n-1}(P^n; \mathbf{Z}/2)$ such that $\alpha_1 \cup \beta_1$ generates $H^n(P^n; \mathbf{Z}/2)$. Since u and u^{n-1} are the unique non-zero elements of these groups, respectively, we must have $\alpha_1 = u$ and $\beta_1 = u^{n-1}$. Thus $u^n = u \cup u^{n-1} \neq 0$, as desired. □

Corollary 68.4. *$H^*(P^\infty; \mathbf{Z}/2)$ is a polynomial algebra over $\mathbf{Z}/2$ with a single one-dimensional generator.*

Proof. This corollary follows from the fact that inclusion $j : P^n \to P^\infty$ induces a cohomology isomorphism (over $\mathbf{Z}/2$) in dimensions less than or equal to n, and the fact that j^* preserves cup products. □

We now apply this result to prove several classical theorems of topology.

Definition. A map $f:S^n \to S^m$ is said to be **antipode-preserving** if $f(-x) = -f(x)$ for all x in S^n.

Theorem 68.5. *If $f:S^n \to S^m$ is continuous and antipode-preserving, then $n \leq m$.*

Proof. Let $a_n = (1,0,\ldots,0) \in S^n$; we call it the "base point" of S^n. Let F denote the field $\mathbf{Z}/2$.

Step 1. Let $\pi_n : S^n \to P^n$ be the usual quotient map. We prove that if $\alpha : I \to S^n$ is any path from a_n to its antipode $-a_n$, then $\pi_n \circ \alpha$ represents the non-zero element of $H_1(P^n; F)$. See Figure 68.3.

This conclusion is easy to prove if α is the standard path

$$\beta(t) = (\cos \pi t, \sin \pi t, 0, \ldots, 0).$$

For then β is a homeomorphism of $(I,\text{Bd}\, I)$ with (E^1_+, S^0), where E^1_+ is the upper half of S^1. And π_1 maps (E^1_+,S^0) onto (P^1,P^0), collapsing S^0 to a point. By our basic results about CW complexes, the image under $(\pi_1)_*$ of a generator of $H_1(E^1_+,S^0)$ is a fundamental cycle for the 1-cell of P^n. The identity map, considered as a singular simplex $i:\Delta_1 \to I$, generates $H_1(I,\text{Bd}\, I; F)$; therefore, the singular simplex $\pi_1 \circ \beta \circ i = \pi_1 \circ \beta$ generates $H_1(P^n; F)$.

To show the same result for a general path α, we proceed as follows: Consider the singular 1-chain $\alpha - \beta$, where β is the standard path as before. It is a singular *cycle* of S^n since its boundary vanishes. In the case $n > 1$, $H_1(S^n) = 0$, so that $\alpha - \beta$ must bound some 2-chain d. Then

$$\pi_n \circ \alpha - \pi_n \circ \beta = \partial(\pi_n \circ d),$$

so $\pi_n \circ \alpha$ and $\pi_n \circ \beta$ are homologous. Hence they are also homologous as cycles with F coefficients.

If $n = 1$, then we use the fact that the map $\pi_1 : S^1 \to P^1$ has degree 2. Since $\alpha - \beta$ is a singular cycle of S^1, the cycle $\pi_1 \circ \alpha - \pi_1 \circ \beta$ represents an even multiple of the generator of $H_1(P^1)$. In particular, it represents the zero element of $H_1(P^1; F)$. Then $\pi_1 \circ \alpha$ and $\pi_1 \circ \beta$ are homologous as cycles with F coefficients.

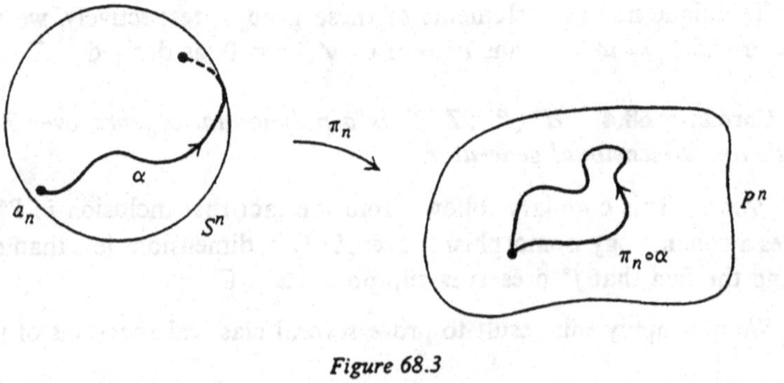

Figure 68.3

Step 2. Let $f : S^n \to S^m$ be an antipode-preserving continuous map. Choose a rotation of S^m, say $\rho : S^m \to S^m$, that carries the point $f(a_n)$ to the base point a_m of S^m. Then the map $g = \rho \circ f$ is continuous and antipode-preserving, and carries a_n to a_m. Via the quotient maps π_n and π_m, the map g induces a continuous map $h : P^n \to P^m$:

$$\begin{array}{ccc} S^n & \xrightarrow{g} & S^m \\ \pi_n \downarrow & & \downarrow \pi_m \\ P^n & \xrightarrow{h} & P^m. \end{array}$$

We show that $h_* : H_1(P^n; F) \to H_1(P^m; F)$ is non-trivial.

Let α be any path in S^n from a_n to $-a_n$. Since g preserves antipodes, $g(-a_n) = -a_m$, so $g \circ \alpha$ is a path in S^m from a_m to $-a_m$. Now $h_\#$ carries the cycle $\pi_n \circ \alpha$ to the cycle $\pi_m \circ g \circ \alpha$; therefore, by Step 1, the homomorphism h_* carries the non-zero element of $H_1(P^n; F)$ to the non-zero element of $H_1(P^m; F)$.

Step 3. Now we prove the theorem. Using naturality of the isomorphism

$$\text{Hom}_F(H_1(P^k; F), F) \xleftarrow{\kappa^*} H^1(P^k; F)$$

for $k = m,n$, it follows from Step 2 that the homomorphism h^* of cohomology

$$H^1(P^n; F) \xleftarrow{h^*} H^1(P^m; F)$$

is non-trivial. Let $u \in H^1(P^m; F)$ be non-trivial; then $h^*(u) \in H^1(P^n; F)$ is non-trivial. Because h^* is a ring homomorphism, $h^*(u^n) = (h^*(u))^n$; by the preceding theorem, the latter element is non-trivial. Therefore, u^n must be non-trivial. It follows that $m \geq n$. □

Theorem 68.6 (The Borsuk-Ulam theorem). *If $h : S^n \to \mathbf{R}^n$ is a continuous map, then $h(x) = h(-x)$ for at least one $x \in S^n$.*

Proof. If $h(x) \neq h(-x)$ for all $x \in S^n$, then the function $f : S^n \to S^{n-1}$ defined by

$$f(x) = \frac{h(x) - h(-x)}{\|h(x) - h(-x)\|}$$

is continuous and antipode-preserving. Such a function does not exist. □

Theorem 68.7 (The baby ham sandwich theorem). *Let A_1 and A_2 be two bounded measurable subsets of $\mathbf{R}^2$. There is a line in $\mathbf{R}^2$ that bisects both A_1 and A_2.*

This result is *not* elementary. Try finding such a line even in the simple case where A_1 and A_2 are triangular regions!

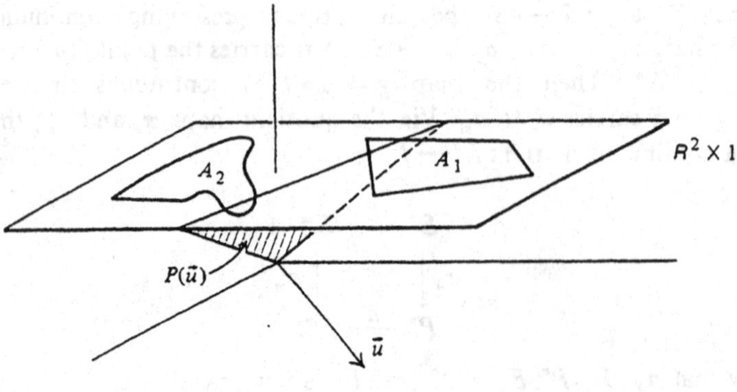

Figure 68.4

Proof. Suppose A_1 and A_2 lie in the plane $\mathbf{R}^2 \times 1$ in $\mathbf{R}^3$. For each unit vector $\vec{u}$ in $\mathbf{R}^3$, consider the plane $P(\vec{u})$ through the origin perpendicular to $\vec{u}$. Let $f_i(\vec{u})$ equal the measure of the part of A_i that lies on the same side of $P(\vec{u})$ as does the vector $\vec{u}$. See Figure 68.4. Note that

$$f_i(\vec{u}) + f_i(-\vec{u}) = \text{measure } A_i.$$

Now the function

$$\vec{u} \to (f_1(\vec{u}), f_2(\vec{u}))$$

is a continuous map of S^2 into $\mathbf{R}^2$. By the Borsuk-Ulam theorem,

$$(f_1(\vec{a}), f_2(\vec{a})) = (f_1(-\vec{a}), f_2(-\vec{a}))$$

for some $\vec{a} \in S^2$. Then

$$f_i(\vec{a}) = \tfrac{1}{2}\,(\text{measure } A_i).$$

The line in which the plane $P(\vec{a})$ intersects the plane $\mathbf{R}^2 \times 1$ is thus the desired line that bisects each of A_1 and A_2. □

The proof of this theorem generalizes to show that for n bounded measurable sets in $\mathbf{R}^n$, there is an $n - 1$ plane in $\mathbf{R}^n$ that bisects them all. If one thinks of a ham sandwich as consisting of two slices of bread and a slice of ham, this theorem in the case $n = 3$ says that one can bisect each slice of bread and the slice of ham with a single whack of the knife!

EXERCISES

1. Let T be as in Example 1. Let $g : \text{sd } T \to T$ be a simplicial approximation to the identity. Check that if sd γ is the sum of all the n-simplices of sd T, oriented counterclockwise, then $g^{\#}(w^1) \cap \text{sd } \gamma$ is the cycle c_1 pictured in Figure 68.2.

2. (a) If $\phi : R \to R'$ is a homomorphism of commutative rings with unity, show that the induced homomorphism

$$H^*(X; R) \to H^*(X; R')$$

is a ring homomorphism.

(b) Use the coefficient homomorphism $\mathbf{Z} \to \mathbf{Z}/2$ to compute the cohomology rings of P^{2k} and P^{2k+1} and P^∞.

3. Show that at any given point in time, there are antipodal points on the surface of the earth at which the temperature and barometric pressure are equal.

4. Compute the cohomology rings of the following:
(a) $S^3 \times P^5$ with $\mathbf{Z}$ and $\mathbf{Z}/2$ coefficients.
(b) $P^2 \times P^3$ with $\mathbf{Z}/2$ coefficients.

5. Assume that $\mathbf{C}P^n$ can be triangulated. (It can.)
(a) Compute the cohomology rings $H^*(\mathbf{C}P^n)$ and $H^*(\mathbf{C}P^\infty)$.
(b) Show that if $f : \mathbf{C}P^n \to \mathbf{C}P^n$, then the degree of f equals a^n for some integer a.
(c) Show that every map $f : \mathbf{C}P^{2n} \to \mathbf{C}P^{2n}$ has a fixed point.
(d) Show that there exists no map $f : \mathbf{C}P^{2n} \to \mathbf{C}P^{2n}$ of degree -1.
(e) If $f : \mathbf{C}P^{2n+1} \to \mathbf{C}P^{2n+1}$ has no fixed point, what is the degree of f?

*6. Suppose we form a manifold M by taking two copies of $\mathbf{C}P^2$, deleting a small open 4-ball from each, and pasting the two remaining spaces together along their boundary 3-spheres. Discuss the cohomology ring of M; compare with the cohomology ring of $S^2 \times S^2$.

7. Let X be a connected, triangulable homology 7-manifold. Suppose that

$$H_7(X) \simeq \mathbf{Z}$$
$$H_6(X) \simeq \mathbf{Z}$$
$$H_5(X) \simeq \mathbf{Z}/2$$
$$H_4(X) \simeq \mathbf{Z} \oplus \mathbf{Z}/3$$

Give all the information you can about the cohomology groups of X, and of the cohomology ring $H^*(X)$.

8. Let X be a compact, orientable, triangulable homology manifold of dimension $4n + 2$. Show that it cannot be true that $H_{2n+1}(X) \simeq \mathbf{Z}$.

9. Let $X_0 = S^2$; let X_n be the n-fold connected sum of T with itself; let Y_n be the n-fold connected sum of P^2 with itself. Given f, consider the condition:

(*) *f_* is a non-trivial homomorphism of H_2 with $\mathbf{Z}/2$ coefficients.*

(a) Show there is no map $f : Y_{2n} \to X_n$ satisfying (*). [*Hint:* If $\phi : A \to B$ is a homomorphism, where A and B are free and rank $A >$ rank B, then ker ϕ has positive rank and is a direct summand in A. Choose a basis $\beta_1, \ldots, \beta_{2n}$ for $H^1(X_n)$ such that $f^*(\beta_1) = 0$. Pass to $\mathbf{Z}/2$ coefficients. Apply the proof of Theorem 68.2.]
(b) Show there is a map $f : Y_{2n+1} \to X_n$ satisfying (*). [*Hint:* First show $Y_3 \approx T \# P^2$.]
(c) Show there is a map $f : Y_m \to X_n$ satisfying (*) if and only if $m \geq 2n + 1$.

*§69. APPLICATION: HOMOTOPY CLASSIFICATION OF LENS SPACES[†]

Throughout this section, let X denote a compact, connected, triangulable n-manifold. If X is orientable, let Γ denote an orientation class for X, and let R denote an arbitrary commutative ring with unity. If X is not orientable, let Γ denote the class $\Gamma_{(2)}$, and let $R = \mathbf{Z}/2$.

Given the ring structure in cohomology, defined by cup product, the Poincaré duality isomorphism

$$H^*(X; R) \xrightarrow{\cap \Gamma} H_*(X; R)$$

gives us an induced ring structure in homology, called the *homology intersection ring*. This ring was known well before cohomology was discovered; the intersection product $\alpha \cdot \beta$ of two homology classes was defined, roughly speaking, as the geometric intersection of two cycles that represent α and β. Defining the homology intersection ring in this way, and deriving its properties, involved a good deal of labor. Furthermore, although one proved that the multiplication operation was a topological invariant, one found that in general it was not preserved by continuous maps, even between manifolds of the same dimension. As a consequence, it seems hardly surprising that all efforts to extend the intersection product from manifolds to arbitrary polyhedra were unsuccessful.

Nowadays we view the matter differently. We consider the cohomology ring as the natural object, and view the existence of the homology intersection ring for manifolds as a happy accident resulting from the Poincaré duality isomorphism.

We are not going to study the homology intersection ring in detail. We shall, however, prove one lemma interpreting the ring operation as a geometric intersection of cycles. This lemma will enable us to make computations in the cohomology rings of the lens spaces.

Definition. We define the **intersection product** of two homology classes of X as follows: Given $\alpha_p \in H_p(X; R)$ and $\beta_q \in H_q(X; R)$, choose α^{n-p} and β^{n-q} so that

$$\alpha^{n-p} \cap \Gamma = \alpha_p \quad \text{and} \quad \beta^{n-q} \cap \Gamma = \beta_q.$$

Then define

$$\alpha_p \cdot \beta_q = (\alpha^{n-p} \cup \beta^{n-q}) \cap \Gamma.$$

Note that when X is orientable, the sign of the result depends on the orientation class Γ. Note also that the product of α_p and β_q has dimension $p + q - n$. This fact makes the connection with geometric intersection plausible, for this is just the dimension of the intersection in $\mathbf{R}^n$ of a general plane of dimension p and a general plane of dimension q.

[†]In this section, we assume familiarity with lens spaces (§40).

Lemma 69.1. *Let α_p and β_q be homology classes of X; let $p + q = n$. Let X be triangulated; let α_p be represented by a cycle c_p of* sd X *that is carried by the dual p-skeleton of X; let β_q be represented by a cycle d_q of X.*

Suppose that the carriers of c_p and d_q meet in a single point, the barycenter of the q-simplex σ of X. If d_q has coefficient $\pm a$ on the simplex σ, and c_p has coefficient $\pm b$ on the dual cell $D(\sigma)$, then $\alpha_p \cdot \beta_q$ is represented (up to sign) by the 0-chain $ab\,\hat{\sigma}$ of sd X.

Proof. Let α^q and β^p be cohomology classes such that

$$\alpha^q \cap \Gamma = \alpha_p \qquad \text{and} \qquad \beta^p \cap \Gamma = \beta_q.$$

Then by definition of the homology intersection ring,

$$\begin{aligned}\alpha_p \cdot \beta_q &= (\alpha^q \cup \beta^p) \cap \Gamma = \alpha^q \cap (\beta^p \cap \Gamma)\\ &= \alpha^q \cap \beta_q.\end{aligned}$$

Because the homomorphism $\epsilon_* : H_0(X; R) \to R$ induced by the augmentation map ϵ is an isomorphism, to prove our result it suffices to show that

$$\pm ab = \epsilon_*(\alpha_p \cdot \beta_q) = \epsilon_*(\alpha^q \cap \beta_q) = \langle \alpha^q, \beta_q \rangle.$$

We evaluate this Kronecker index by finding a representative cocycle for α^q. Orient the simplices of X. Let γ be an n-cycle of X representing Γ. Let $\psi : C^q(X) \to D_p(X)$ be the isomorphism defined by the diagram

$$\begin{array}{ccc} C^q(X) & \xrightarrow{\ \psi\ } & D_p(X) \\ \big\downarrow g^{\#} & & \big\downarrow j \\ C^q(\text{sd } X) & \xrightarrow{\ \cap\, \text{sd}\,\gamma\ } & C_p(\text{sd } X), \end{array}$$

where g is a simplicial approximation to the identity. If σ_i is an oriented q-simplex of X, and σ_i^* is the corresponding elementary cochain, then $\psi(\sigma_i^*)$ is a fundamental cycle for the p-block $\overline{D}(\sigma_i)$ dual to σ_i, which we denote by z_i.

Now the cycle representing α_p has the form

$$c_p = \Sigma\, b_j z_j = \Sigma\, b_j \psi(\sigma_j^*).$$

We assert that α^q is represented by the cocycle defined by the equation

$$c^q = \Sigma\, b_j \sigma_j^*.$$

The fact that $\partial\psi = \pm\psi\delta$ implies that c^q is a cocycle; and the fact that ψ induces the isomorphism $\cap\, \Gamma$ means that c^q represents α^q.

Now the cycle d_q has the form

$$d_q = \Sigma\, a_i \sigma_i.$$

The dual cell $\overline{D}(\sigma_j)$ carrying the chain z_j intersects the simplex σ_i only if $i = j$, in which case they intersect in the barycenter $\hat{\sigma}_i$ of σ_i. The carrier of c_p consists of those dual cells $\overline{D}(\sigma_j)$ for which $b_j \neq 0$, while the carrier of d_q consists of those simplices σ_i for which $a_i \neq 0$. Since these carriers intersect in only one point $\hat{\sigma}$, it follows that the only index i for which both the coefficients b_i and a_i are non-zero is the index for which $\sigma_i = \sigma$. Our lemma follows:

$$\langle \alpha^q, \beta_q \rangle = \langle c^q, d_q \rangle = \Sigma\, b_j a_i \langle \sigma_j^*, \sigma_i \rangle = \Sigma\, b_i a_i = \pm ba. \quad \square$$

To apply this result to the cohomology of lens spaces, we need a certain operation called the *Bockstein*, which we introduced for homology in the exercises of §24.

Definition. Let $\mathcal{C}$ be a free chain complex. Given an exact sequence of abelian groups

$$0 \to G \to G' \to G'' \to 0,$$

consider the associated short exact sequences

$$0 \to C_p \otimes G \to C_p \otimes G' \to C_p \otimes G'' \to 0,$$
$$0 \to \operatorname{Hom}(C_p, G) \to \operatorname{Hom}(C_p, G') \to \operatorname{Hom}(C_p, G'') \to 0.$$

From the zig-zag lemma, we obtain homomorphisms

$$\beta_* : H_p(\mathcal{C}; G'') \to H_{p-1}(\mathcal{C}; G),$$
$$\beta^* : H^p(\mathcal{C}; G'') \to H^{p+1}(\mathcal{C}; G)$$

that are natural with respect to homomorphisms induced by continuous maps. They are called the **Bockstein homomorphisms** associated with the coefficient sequence in question.

Lemma 69.2. *Let X be orientable. Then the Bockstein homomorphisms commute with the Poincaré duality isomorphism, up to sign.*

Proof. Let $\phi : C^k(X) \to D_{n-k}(X)$ be the Poincaré duality isomorphism of Theorem 65.1. Consider the commutative diagram

$$\begin{array}{ccccccc}
0 \to & C^k(X) \otimes G & \to & C^k(X) \otimes G' & \to & C^k(X) \otimes G'' & \to 0 \\
 & \downarrow & & \downarrow & & \downarrow & \\
0 \to & D_{n-k}(X) \otimes G & \to & D_{n-k}(X) \otimes G' & \to & D_{n-k}(X) \otimes G'' & \to 0.
\end{array}$$

where the vertical homomorphisms are induced by ϕ. Since $\partial\phi = \phi\delta$, one has an induced homomorphism of the corresponding exact homology sequences. In particular, the diagram

$$\begin{array}{ccc}
H^k(X; G'') & \xrightarrow{\beta^*} & H^{k+1}(X; G) \\
\simeq \downarrow & & \downarrow \simeq \\
H_{n-k}(X; G'') & \xrightarrow{\beta_*} & H_{n-k-1}(X; G)
\end{array}$$

commutes. Here we use the fact that one has a natural isomorphism

$$C^i(X) \otimes G \simeq \operatorname{Hom}(C_i(X), G)$$

of cochain complexes, so that we may use either to compute the cohomology of

X. We also use the naturality of the zig-zag construction to conclude that the cochain complex $\{C^i(X) \otimes G\}$ can be used to compute β^*, and $\{D_i(X) \otimes G\}$ can be used to compute β_*.

The isomorphism induced by ϕ differs from that induced by $\cap \Gamma$ only by a sign, so our lemma follows. □

Now we apply these results to lens spaces. Recall that if n and k are relatively prime positive integers, then the lens space $X = L(n,k)$ is a compact, connected, orientable, triangulable 3-manifold. Furthermore, X has the structure of CW complex. Its cellular chain complex $\mathcal{D}(X)$ is infinite cyclic in each dimension $i = 0,1,2,3$, and the boundary operator is either 0 or multiplication by n. It follows that

$$H_i(X; \mathbf{Z}/n) \simeq \mathbf{Z}/n \simeq H^i(X; \mathbf{Z}/n)$$

for $i = 1,2,3$.

Theorem 69.3. *Let $X = L(n,k)$. Let Λ generate the infinite cyclic group $H^3(X)$; let $\Lambda_{(n)}$ denote its image in $H^3(X; \mathbf{Z}/n)$ under the coefficient homomorphism induced by $\mathbf{Z} \to \mathbf{Z}/n$. Let*

$$\beta^* : H^1(X; \mathbf{Z}/n) \to H^2(X; \mathbf{Z}/n)$$

be the Bockstein homomorphism associated with the coefficient sequence

$$0 \to \mathbf{Z}/n \xrightarrow{n} \mathbf{Z}/n^2 \to \mathbf{Z}/n \to 0.$$

Consider the ring $H^*(X; \mathbf{Z}/n)$. There is a generator u for $H^1(X; \mathbf{Z}/n)$ such that

$$u \cup \beta^*(u) = \pm \frac{1}{[k]} \Lambda_{(n)},$$

where $1/[k]$ is the unique inverse to $[k]$ in the ring $\mathbf{Z}/n$.

Proof. Let

$$\beta_* : H_2(X; \mathbf{Z}/n) \to H_1(X; \mathbf{Z}/n)$$

be the homology Bockstein associated with the given coefficient sequence. Since the Bockstein homomorphisms commute with Poincaré duality up to sign, it suffices to show that in homology there exists a generator w for $H_2(X; \mathbf{Z}/n)$ such that

$$w \cdot \beta_*(w) = \pm(1/[k])\, \Lambda_{(n)} \cap \Gamma.$$

We know that $\Lambda_{(n)} \cap \Gamma$ is a generator of $H_0(X; \mathbf{Z}/n)$; it equals the homology class of $v_0 \otimes [1]$, where v_0 is a vertex of X and $[1] \in \mathbf{Z}/n$.

First, we define w. Let v, e, and c denote fundamental cycles for the cells of X in dimensions 0,1,2, respectively, as indicated in Figure 69.1. As we showed in the proof of Theorem 40.9, $\partial c = ne$. When c is considered as a chain with $\mathbf{Z}/n$ coefficients, then c is a cycle. Let w be the homology class of c; it generates $H_2(X; \mathbf{Z}/n)$.

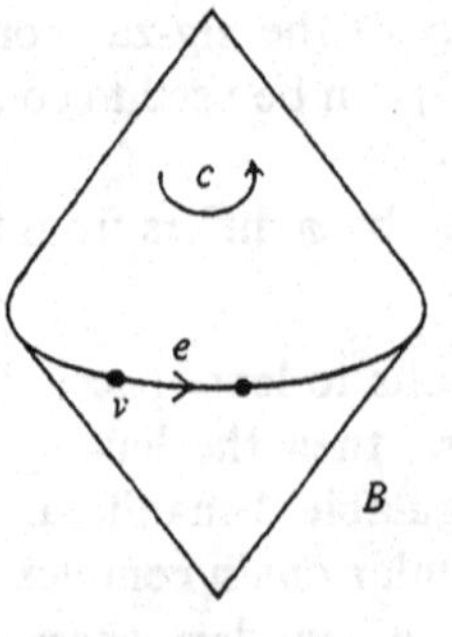

Figure 69.1

We now compute $\beta_*(w)$. We may use the cellular chain complex of X to compute β_*. Applying the zig-zag lemma, we begin with c, which is a cycle with $\mathbf{Z}/n$ coefficients, then we "consider" it as a chain with $\mathbf{Z}/n^2$ coefficients. Next we take its boundary (which is still ne), and finally we divide by n. Thus $\beta_*(w) = \{e\}$, where e is considered as a cycle with $\mathbf{Z}/n$ coefficients.

Finally, we compute the intersection product $w \cdot \beta_*(w) = \{c\} \cdot \{e\}$. Unfortunately, we are not in a position to compute this product directly, for the carrier of e intersects the carrier of c in much more than a single point. We shall replace c and e by homologous cycles in order to carry out the computation.

Note first that c is homologous to the cycle d pictured in Figure 69.2. Indeed, the boundary of the solid upper half of the figure, suitably oriented, equals the chain $c - d$.

Second, let z be the cycle that runs from the top of the polyhedron B to the bottom, as pictured in Figure 69.3. We assert that when considered in X, it is homologous to the cycle $-ke$. For there are k copies of the 1-cell e lying between the vertex v and its image $r_\theta(v)$ under rotation through $\theta = 2\pi k/n$. When computed in B, the boundary of the shaded 2-chain in the figure equals the chain $ke + z + s_1 + s_2$, where s_1 and s_2 are the two slanted 1-chains pictured. These slanted 1-chains cancel each other after the identifications made in forming X. Thus the boundary of the shaded 2-chain equals $ke + z$, when computed in $L(n,k)$.

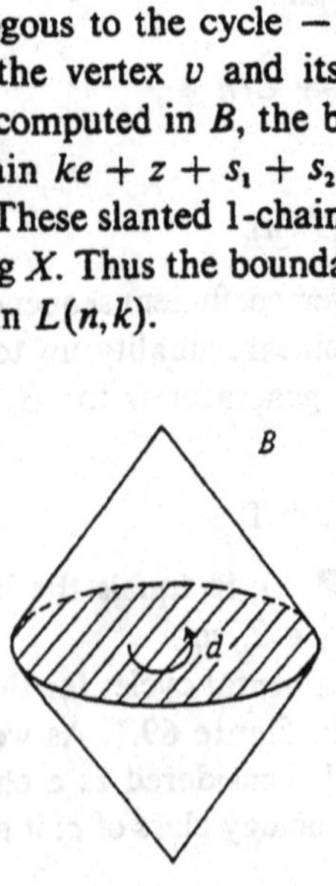

Figure 69.2

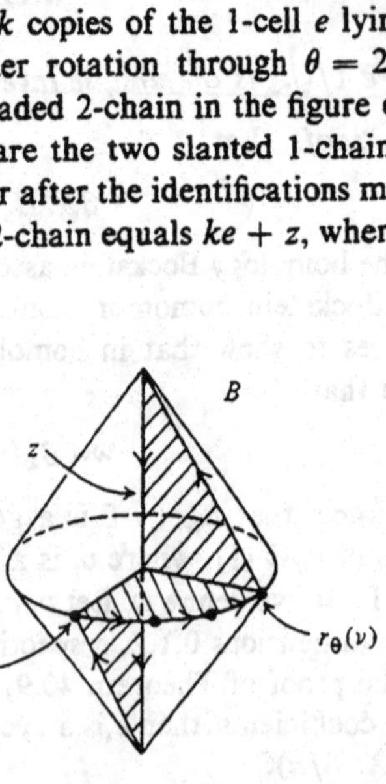

Figure 69.3

Now the intersection product of $\{d\}$ and $\{z\}$ is easy to compute, because their carriers intersect in a single point. Actually, we must triangulate X and "push z off" onto the dual 1-skeleton in order for the preceding lemma to apply. But this is not hard to do in such a way that the resulting carrier still intersects d in only one point. It then follows from Lemma 69.1 that $\{d\} \cdot \{z\}$ equals ± 1 times our specified generator of H_0. Since $\{d\} = w$ and $\{z\} = \{-ke\} = -k\beta^*(w)$, our result is proved. □

Theorem 69.4. *For $X = L(n,k)$ and $Y = L(n,l)$ to have the same homotopy type, it is necessary that*

$$k \equiv \pm a^2 l \pmod n$$

for some a.

Proof. Let $f: X \to Y$ be a homotopy equivalence. Let Λ generate $H^3(X)$ and let Λ' generate $H^3(Y)$. Let $u \in H^1(X; \mathbb{Z}/n)$ and $v \in H^1(Y; \mathbb{Z}/n)$ be chosen as in the preceding lemma, so that

$$u \cup \beta^*(u) = \pm \frac{1}{[k]} \Lambda_{(n)} \quad \text{and} \quad v \cup \beta^*(v) = \pm \frac{1}{[l]} \Lambda'_{(n)}.$$

Now $f^*(v) = au$ for some integer a. Then

$$\begin{aligned} \pm \frac{1}{[l]} f^*(\Lambda'_{(n)}) &= f^*(v \cup \beta^*(v)) \\ &= f^*(v) \cup \beta^*(f^*(v)) \\ &= (au) \cup \beta^*(au) \\ &= a^2(u \cup \beta^*(u)) = \pm \frac{a^2}{[k]} \Lambda_{(n)}. \end{aligned}$$

Now f is a homotopy equivalence, whence f^* maps Λ' to $\pm\Lambda$, so it maps $\Lambda'_{(n)}$ to $\pm\Lambda_{(n)}$. We conclude that

$$1/[l] = \pm a^2/[k],$$

and the theorem is proved. □

The condition stated in this theorem is in fact both necessary *and* sufficient for $L(n,k)$ and $L(n,l)$ to have the same homotopy type; the sufficiency is a theorem of J. H. C. Whitehead. Thus the lens spaces are a rare class of spaces for which both the homeomorphism and the homotopy type classification problems have been solved. (See the exercises of §40.)

At one time, there was a classical conjecture, due to Hurewicz, to the effect that two compact manifolds having the same homotopy type should be homeomorphic as well. This conjecture is certainly true for 2-manifolds. But the lens spaces $L(7,k)$ provide a counterexample in dimension 3. The spaces $L(7,1)$ and $L(7,2)$ have the same homotopy type, since $2 \equiv (3)^2 \cdot 1 \pmod 7$. But they are not homeomorphic, by the results discussed in §40, since

$$2 \not\equiv \pm 1 \pmod 7 \quad \text{and} \quad 2 \cdot 1 \not\equiv \pm 1 \pmod 7.$$

EXERCISES

1. Show that if $f : L(n,k) \to L(n,l)$ is a map of degree d, then
$$kd \equiv \pm a^2 l \pmod{n}$$
for some integer a.
2. Using the results stated here and in the exercises of §40, classify up to homeomorphism and up to homotopy type the manifolds $L(7,k)$ for $k = 1,\ldots,6$, and the manifolds $L(10,k)$ for k relatively prime to 10.
3. Let $X = L(n,k)$.
   (a) Show that if u_i generates $H^i(X;\mathbb{Z}/n)$ for $i = 1,2$, then $u_1 \cup u_2$ generates $H^3(X;\mathbb{Z}/n)$. [*Hint:* First show there *exist* generators u_1, u_2 such that $u_1 \cup u_2$ generates H^3, by showing that $\beta^* : H^1 \to H^2$ is an isomorphism.]
   (b) Show that $u_1 \cup u_1$ is either zero or of order 2.
   This exercise shows that the cohomology ring $H^*(X;\mathbb{Z}/n)$ is not by itself adequate to classify the lens spaces up to homotopy type. One needs also the Bockstein operation β^*.

   The Bockstein is a particular example of what is called a **cohomology operation.** In general, a cohomology operation θ of type $(p,q;G,G')$ is a set map
$$\theta : H^p(X,A;G) \to H^q(X,A;G'),$$
   defined for all topological pairs (X,A), that is natural with respect to homomorphisms induced by continuous maps. The existence of cohomology operations provides the cohomology of a space with a much richer structure than that of a ring. Frequently, as was the case with lens spaces, these operations give one valuable information about the spaces in question.

§70. LEFSCHETZ DUALITY

As one might expect, the Poincaré duality theorem generalizes to relative homology manifolds. The generalization is due to Lefschetz. It takes a particularly nice form in the case of manifolds with boundary, as we shall see.

First, we need a lemma. Recall that A is said to be a *full* subcomplex of the complex X if every simplex of X whose vertices are in A is itself in A.

In general, a subcomplex A of X need not be full. But if we pass to the first barycentric subdivision, we can show easily that sd A is a full subcomplex of sd X: Let $s = \hat{\sigma}_1 \ldots \hat{\sigma}_k$ be a simplex of sd X, where $\sigma_1 \succ \sigma_2 \succ \cdots \succ \sigma_k$; then $s \subset \sigma_1$. If each vertex of s belongs to sd A, then in particular $\hat{\sigma}_1 \in$ sd A, so that Int σ_1 intersects $|A|$. It follows that $\sigma_1 \in A$, so that $s \subset \sigma_1 \subset |A|$, as claimed.

Lemma 70.1. *Let A be a full subcomplex of the finite simplicial complex X. Let C consist of all simplices of X that are disjoint from $|A|$. Then $|A|$ is a deformation retract of $|X| - |C|$, and $|C|$ is a deformation retract of $|X| - |A|$.*

Proof. First, we note that each vertex of X belongs to either A or C. Second, we note that C is a full subcomplex of X: If the vertices of σ are in C, then the simplex σ cannot intersect $|A|$, so it is in C. Third, we note that A consists of all simplices X disjoint from $|C|$: Every simplex of A is disjoint from $|C|$; and conversely, if σ is disjoint from $|C|$, then each vertex of σ is in A, whence σ belongs to A.

By symmetry, it suffices to show $|A|$ is a deformation retract of $|X| - |C|$. Let σ be a simplex of X that belongs to neither A nor C. Then $\sigma = s * t$, where s is spanned by the vertices of σ that belong to A, and t is spanned by the vertices that belong to C. Then $s \in A$ and $t \in C$. Each point x of $\sigma - s - t$ lies on a unique line segment joining a point of s with a point of t; let us denote the end point of this line segment that lies in $|A|$ by $f_\sigma(x)$. We extend f_σ to the simplex s by letting it equal the identity on s; then f_σ is continuous on $\sigma - t$. (Indeed, if $s = v_0 \dots v_p$ and $t = v_{p+1} \dots v_n$, and if x lies in $\sigma - t$, then

$$x = \sum_{i=0}^{n} \alpha_i v_i \quad \text{implies} \quad f_\sigma(x) = \sum_{i=0}^{p} \frac{\alpha_i}{\lambda} v_i,$$

where $\lambda = \Sigma_{i=0}^{p} \alpha_i$. See the proof of Lemma 62.1.)

It is immediate that any two of the functions f_σ agree on the common part of their domains. Therefore, we can define a continuous function

$$f: |X| - |C| \to |A|$$

that retracts $|X| - |C|$ onto $|A|$, by the equations:

$$f(x) = \begin{cases} x & \text{if } x \in |A|, \\ f_\sigma(x) & \text{if } x \in \sigma - |C| \text{ and } \sigma \notin A. \end{cases}$$

The function $F(x,t) = (1 - t)x + tf(x)$ is then the required deformation retraction. □

The preceding lemma holds without assuming X is finite, but in this case it requires some care to show that F is continuous. The problem does not arise in the finite case.

Definition. Let (X,A) be a compact triangulated relative homology n-manifold. We say that (X,A) is **orientable** if it is possible to orient all the n-simplices σ_i of X not in A so that their sum $\gamma = \Sigma\sigma_i$ is a cycle of (X,A). Such a cycle γ will be called an **orientation cycle** for (X,A).

Theorem 70.2 (Lefschetz duality). *Let (X,A) be a compact triangulated relative homology n-manifold. If (X,A) is orientable, there are isomorphisms*

$$H^k(X,A;G) \simeq H_{n-k}(|X| - |A|;G),$$
$$H_k(X,A;G) \simeq H^{n-k}(|X| - |A|;G),$$

for all G. If (X,A) is not orientable, these isomorphisms exist for $G = Z/2$.

In the statement of this theorem, the homology and cohomology groups of $|X| - |A|$ should be interpreted as singular groups, since $|X| - |A|$ is not the polytope of a subcomplex of X. (It *is* triangulable, but that we have not proved.)

Proof. Let X^* denote the subcomplex of the first barycentric subdivision of X consisting of all simplices of sd X that are disjoint from $|A|$. Now $|A|$ is the polytope of a full subcomplex of sd X. By the preceding lemma, $|X^*|$ is a deformation retract of $|X| - |A|$. Therefore, we may replace $|X| - |A|$ by X^* in the statement of the theorem.

We consider, as in the proof of Poincaré duality, the collection of blocks $D(\sigma)$ dual to the simplices of X. We shall prove the following:

The space $|X^*|$ equals the union of all those blocks $D(\sigma)$ dual to simplices σ of X that are not in A.

To prove this fact, let $s = \hat{\sigma}_1 \ldots \hat{\sigma}_k$ be a simplex of sd X, where $\sigma_1 \succ \cdots \succ \sigma_k$. Then s lies in the dual block $\overline{D}(\sigma_k)$. If s is disjoint from $|A|$, then the vertex $\hat{\sigma}_k$ is in particular not in $|A|$; hence the simplex σ_k does not belong to A. Conversely, if σ_k does not belong to A, then neither do the simplices $\sigma_1, \ldots, \sigma_{k-1}$, since they have σ_k as a face. Hence s does not intersect $|A|$. See Figure 70.1.

Now for each block $D(\sigma)$ for which σ is not in A, the point $\hat{\sigma}$ is in $|X| - |A|$, so that $H_i(|X|, |X| - \hat{\sigma})$ is infinite cyclic for $i = n$ and vanishes otherwise. It follows that $(\overline{D}(\sigma), \dot{D}(\sigma))$ is a homological cell of dimension $n - \dim \sigma$. Let $\mathcal{D}(X^*)$ denote the dual chain complex of X^*; the group $D_p(X^*)$ is generated by fundamental cycles for the homological cells $(\overline{D}(\sigma), \dot{D}(\sigma))$ of dimension p. Just as before, inclusion $\mathcal{D}(X^*) \to \mathcal{C}(X^*)$ induces an isomorphism in cohomology and in homology. (See Theorem 64.2.)

Now we consider the case where (X, A) is orientable. Recall that the relative cohomology group $C^k(X, A)$ can be naturally considered as the subgroup of $C^k(X)$ consisting of all cochains of X that vanish on simplices of A. It is therefore free abelian with a basis consisting of the cochains σ^*, as σ ranges over all k-simplices of X not in A. The argument used in the first proof of Poincaré duality goes through with no change, to give us an isomorphism

$$C^k(X, A) \xrightarrow{\phi} D_{n-k}(X^*),$$

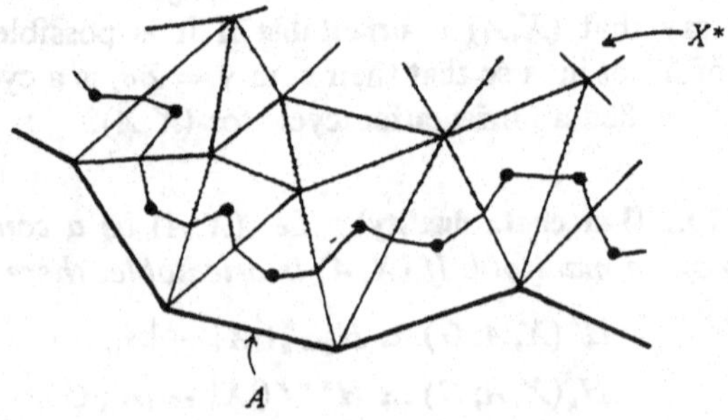

Figure 70.1

having the property that $\phi\delta = \partial\phi$. The existence of an isomorphism

$$H^k(X,A) \simeq H_{n-k}(X^*) \simeq H_{n-k}(|X| - |A|)$$

follows at once. This is the first of our duality isomorphisms.

To obtain the second duality isomorphism, we recall that the group $\text{Hom}(C^k(X,A),\mathbf{Z})$ is naturally isomorphic to the group $C_k(X,A)$. (See Step 4 of Theorem 56.1.) Since ϕ is an isomorphism, so is its dual

$$\text{Hom}(C^k(X,A),\mathbf{Z}) \xleftarrow{\tilde{\phi}} \text{Hom}(D_{n-k}(X^*),\mathbf{Z}).$$

Therefore, one has an isomorphism

$$H_k(X,A) \underset{\simeq}{\overset{\phi^*}{\leftarrow}} H^{n-k}(X^*) \simeq H^{n-k}(|X| - |A|).$$

The proof for arbitrary coefficients if (X,A) is orientable, or for $\mathbf{Z}/2$ coefficients if (X,A) is not orientable, goes through without difficulty, just as in the proof of Poincaré duality. □

Corollary 70.3. *Let (X,A) be a compact triangulated relative homology n-manifold, with $|X| - |A|$ connected. If σ and σ' are two n-simplices of X not in A, there is a sequence*

$$\sigma = \sigma_0, \sigma_1, \dots, \sigma_k = \sigma'$$

of n-simplices not in A, such that $\sigma_i \cap \sigma_{i+1}$ is an $n-1$ simplex not in A, for each i.

Proof. We define two n-simplices of X not in A to be equivalent if there is such a sequence joining them. The sum of the members of any one equivalence class is a relative cycle of (X,A) with $\mathbf{Z}/2$ coefficients. We conclude that there is only one equivalence class, since

$$H_n(X,A;\mathbf{Z}/2) \simeq H^0(|X| - |A|;\mathbf{Z}/2) \simeq \mathbf{Z}/2. \quad \square$$

Corollary 70.4. *Let (X,A) be a compact triangulated relative homology n-manifold. Assume $|X| - |A|$ is connected. Then $H_n(X,A) \simeq \mathbf{Z}$ if (X,A) is orientable and $H_n(X,A) = 0$ if (X,A) is non-orientable.*

Proof. The proof follows the pattern of Corollary 65.3. □

Corollary 70.5. *The compact triangulated relative homology n-manifold (X,A) is orientable if and only if for each component X_i of $|X| - |A|$, one has $H_n(\bar{X}_i, \bar{X}_i \cap A) \simeq \mathbf{Z}$.* □

It follows that orientability of (X,A) does not depend on the triangulation of (X,A).

Definition. Let (X,A) be a compact triangulable relative homology n-manifold. Let $X_1,\dots,X_k$ be the components of $X - A$; let $A_i = \bar{X}_i \cap A$. If

(X,A) is orientable, then $H_n(\overline{X}_i,A_i)$ is infinite cyclic; a generator $\Gamma^{(i)}$ of this group is called an **orientation class** for $(\overline{X}_i,A_i)$. The image of the classes $\Gamma^{(i)}$ under the isomorphism

$$\oplus H_n(\overline{X}_i,A_i) \simeq H_n(X,A)$$

induced by inclusion is called an **orientation class** for (X,A) and is denoted by Γ. Similarly, if (X,A) is not necessarily orientable, and if $\Gamma^{(i)}_{(2)}$ is the unique non-trivial element of $H_n(\overline{X}_i,A_i;\mathbf{Z}/2)$, the image of these classes in $H_n(X,A;\mathbf{Z}/2)$ is denoted $\Gamma_{(2)}$ and is called an **orientation class** for (X,A) **over** $\mathbf{Z}/2$.

If (X,A) is given a specific triangulation, then Γ is represented by the sum γ of all n-simplices of X not in A, suitably oriented. And $\Gamma_{(2)}$ is represented by the sum of all these n-simplices, each with coefficient $[1] \in \mathbf{Z}/2$.

A natural question to ask at this point is whether the second version of the proof of Poincaré duality, which involves cap products, generalizes to the relative case. The answer is that it does not. Let us examine where the difficulty lies.

We begin as in the proof of Theorem 67.1. Choose a map

$$g: (\text{sd } X, \text{sd } A) \to (X,A)$$

which is a simplicial approximation to the identity. Let γ be an orientation cycle for (X,A). Consider the diagram

$$\begin{array}{ccc} C^k(X,A) & \overset{\psi}{\dashrightarrow} & D_{n-k}(X^*) \\ \downarrow{\scriptstyle g^\#} & & \downarrow{\scriptstyle j} \\ C^k(\text{sd } X, \text{sd } A) & \xrightarrow{\cap \text{ sd } \gamma} & C_{n-k}(\text{sd } X), \end{array}$$

where j is inclusion, as before. The proof of Theorem 67.1 goes through to show that the composite of $g^\#$ and $\cap$ sd γ carries $C^k(X,A)$ isomorphically onto the subgroup of $C_{n-k}(\text{sd } X)$ which is the image of $D_{n-k}(X^*)$ under j. Up to sign, the resulting isomorphism ψ is a suitable candidate for our isomorphism ϕ, just as before. (Here we are of course using the relative cap product

$$C^k(\text{sd } X,\text{sd } A) \otimes C_n(\text{sd } X,\text{sd } A) \to C_{n-k}(\text{sd } X)$$

in defining the map $\cap$ sd γ.)

So far, no difficulties have appeared. The problem arises when one passes to the cohomology-homology level. On this level, we have a homomorphism

$$H^k(X,A) \xrightarrow{\cap \Gamma} H_{n-k}(X).$$

The range of this map is *not* the group $H_{n-k}(|X| - |A|)$ that we wish it to be!

Thus Lefschetz duality cannot in general be expressed by cap product with an orientation class for (X,A). All one can say is the following.

Theorem 70.6. *Let (X,A) be a compact triangulable relative homology n-manifold. If (X,A) is orientable, let Γ denote an orientation class for (X,A);*

then the following diagram commutes, up to a sign depending on k and n:

$$\begin{array}{ccc} H^k(X,A;G) & \xrightarrow[\cong]{\phi_*} & H_{n-k}(X-A;G) \\ & {\scriptstyle \cap\Gamma} \searrow & \downarrow {\scriptstyle j_*} \\ & & H_{n-k}(X;G). \end{array}$$

(Here ϕ_* is the Lefschetz duality isomorphism and j is inclusion.) The same result holds if (X,A) is not orientable if one replaces Γ by $\Gamma_{(2)}$, and G by $\mathbf{Z}/2$.

Proof. Choose a triangulation of (X,A). Then consider the diagram

$$\begin{array}{ccccc} C^k(X,A) & \xrightarrow[\cong]{\phi} & D_{n-k}(X^*) & \xrightarrow{\eta} & S_{n-k}(|X^*|) \\ \downarrow {\scriptstyle g^\#} & & \downarrow & & \downarrow \\ & & & & S_{n-k}(|X|-|A|) \\ & & & & \downarrow \\ C^k(\mathrm{sd}\,X,\mathrm{sd}\,A) & \xrightarrow{\cap\,\mathrm{sd}\,\gamma} & C_{n-k}(\mathrm{sd}\,X) & \xrightarrow{\eta} & S_{n-k}(|X|). \end{array}$$

Here η is the chain map carrying simplicial chains to singular chains; the unlabelled maps are induced by inclusion. The proof of Theorem 67.1 applies to show the first square commutes up to sign; the rest of the diagram commutes. □

One situation in which the Lefschetz duality isomorphism is actually given by cap product is the case of a manifold with boundary. We consider that situation now.

Definition. Let M be an n-manifold with boundary. We say $\mathrm{Bd}\,M$ has a **product neighborhood** in M if there is a homeomorphism

$$h : \mathrm{Bd}\,M \times [0,1) \to U$$

whose image is an open set in M, such that $h(x,0) = x$ for each $x \in \mathrm{Bd}\,M$.

It is in fact true that such a product neighborhood always exists, but the proof is decidedly non-trivial. (See [B_2].)

Theorem 70.7 (Poincaré-Lefschetz duality). *Let M be a compact triangulable n-manifold with boundary, such that* $\mathrm{Bd}\,M$ *has a product neighborhood in M.*

If $(M,\mathrm{Bd}\,M)$ is orientable, let $\Gamma \in H_n(M,\mathrm{Bd}\,M)$ be an orientation class for $(M,\mathrm{Bd}\,M)$. Then there are isomorphisms

$$H^k(M,\mathrm{Bd}\,M;G) \xrightarrow{\cap\Gamma} H_{n-k}(M;G),$$

$$H^k(M;G) \xrightarrow{\cap\Gamma} H_{n-k}(M,\mathrm{Bd}\,M;G),$$

for arbitrary G. If $(M,\mathrm{Bd}\,M)$ is not orientable, the same result holds if G is replaced by $\mathbf{Z}/2$ and Γ is replaced by $\Gamma_{(2)}$.

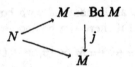

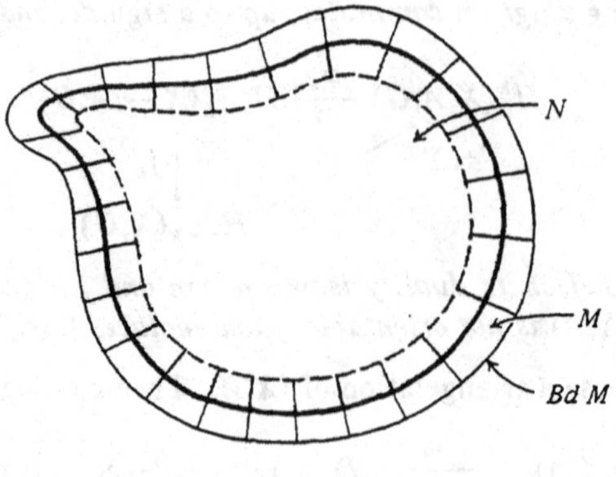

Figure 70.2

Proof. We first prove that inclusion induces an isomorphism

$$j_* : H_{n-k}(M - \text{Bd}\, M; G) \to H_{n-k}(M; G).$$

Let $h : \text{Bd}\, M \times [0,1) \to M$ be a product neighborhood of the boundary. Let

$$N = M - h(\text{Bd}\, M \times [0,\tfrac{1}{2})).$$

See Figure 70.2. Now $\text{Bd}\, M \times [\tfrac{1}{2},1)$ is a deformation retract of both $\text{Bd}\, M \times [0,1)$ and of $\text{Bd}\, M \times (0,1)$. Hence N is a deformation retract of both M and $M - \text{Bd}\, M$. Thus the left-hand inclusions in the diagram

$$\begin{array}{ccc} & & M - \text{Bd}\, M \\ & \nearrow & \downarrow j \\ N & \searrow & M \end{array}$$

are homotopy equivalences. Then so is j.

The first of the two isomorphisms stated in our theorem now follows from the naturality diagram of Theorem 70.6, setting $A = \text{Bd}\, M$.

We now derive the other isomorphism. Let us triangulate $(M, \text{Bd}\, M)$. The space $\text{Bd}\, M$ is a union of simplices of M (by Theorem 35.3); being an $n-1$ manifold, it is the union of $n-1$ simplices. Now we claim that each $n-1$ simplex s of $\text{Bd}\, M$ is a face of exactly one n-simplex σ of M. It must be a face of at least one, because M is a union of n-simplices. And it cannot be a face of more than one, since by Lemma 63.1,

$$H_n(M, M - \hat{s}) \simeq \tilde{H}_0(\text{Lk}(s, M)),$$

and the former group vanishes because $\hat{s}$ is a point of $\text{Bd}\, M$.

Suppose γ is an orientation cycle for $(M, \text{Bd}\, M)$. Then γ is a sum of *all* the n-simplices of M. Since each $n-1$ simplex of $\text{Bd}\, M$ is the face of exactly one n-simplex of M, the chain $\partial\gamma$ has coefficient ± 1 on each $n-1$ simplex of $\text{Bd}\, M$. Since $\partial\gamma$ is a cycle, it must be an orientation cycle for $\text{Bd}\, M$.

Let $\Gamma = \{\gamma\}$. Then the class $\partial_*\Gamma = \{\partial\gamma\}$ is an orientation class for Bd M. Consider the following diagram:

$$\begin{array}{ccccccccc}
H^{k-1}(\text{Bd } M) & \to & H^k(M,\text{Bd } M) & \to & H^k(M) & \to & H^k(\text{Bd } M) & \to & H^{k+1}(M,\text{Bd } M) \\
\downarrow \cap \partial_*\Gamma & & \downarrow \cap \Gamma & & \downarrow \cap \Gamma & & \downarrow \cap \partial_*\Gamma & & \downarrow \cap \Gamma \\
H_{n-k}(\text{Bd } M) & \to & H_{n-k}(M) & \to & H_{n-k}(M,\text{Bd } M) & \to & H_{n-k-1}(\text{Bd } M) & \to & H_{n-k-1}(M)
\end{array}$$

where all the groups are assumed to have G coefficients. It is easy to check that this diagram commutes up to sign. The first and fourth vertical maps are isomorphisms by Poincaré duality, and the second and fifth are isomorphisms by Poincaré-Lefschetz duality. Therefore, the third map is an isomorphism as well.

The same argument applies if M is non-orientable provided one replaces G by $\mathbf{Z}/2$, and Γ by $\Gamma_{(2)}$. □

As an application of Poincaré-Lefschetz duality, we consider the following question: Given a compact n-manifold, under what conditions is it the boundary of a compact $n + 1$ manifold M? There is an entire theory that deals with this question; it is called *cobordism* theory. Here we prove just one elementary result.

Theorem 70.8. *Let M be a compact triangulable manifold with boundary; suppose* $\dim M = 2m + 1$. *Suppose* Bd M *is nonempty and has a product neighborhood in M. Then the vector space $H_m(\text{Bd } M; \mathbf{Z}/2)$ has even dimension.*

Proof. Step 1. Consider an exact sequence of vector spaces and linear transformations

$$\cdots \longrightarrow A_{k-1} \xrightarrow{\phi_{k-1}} A_k \xrightarrow{\phi_k} A_{k+1} \longrightarrow \cdots,$$

where A_k and ϕ_k are defined for all integers k, and $\dim A_k = 0$ for $|k|$ sufficiently large. We assert that

$$\text{rank } \phi_k = \dim A_k - \dim A_{k-1} + \dim A_{k-2} - \cdots \quad \text{and}$$
$$\text{rank } \phi_{k-1} = \dim A_k - \dim A_{k+1} + \dim A_{k+2} - \cdots.$$

The proof is straightforward. Exactness at A_i tells us that

$$\dim A_i = \text{rank } \phi_{i-1} + \text{rank } \phi_i.$$

Summation gives us the desired equations.

Step 2. Using $\mathbf{Z}/2$ coefficients (which we suppress from the notation), we have an exact sequence of vector spaces

$$\cdots \longrightarrow H^k(M) \xrightarrow{\mu_k} H^k(\text{Bd } M) \xrightarrow{\nu_k} H^{k+1}(M,\text{Bd } M) \longrightarrow \cdots.$$

Let

$$\beta_k = \dim H^k(M),$$
$$\gamma_k = \dim H^k(\text{Bd } M),$$
$$\alpha_{k+1} = \dim H^{k+1}(M, \text{Bd } M).$$

We apply Step 1 to this sequence. Beginning at the term $H^k(\text{Bd } M)$ and summing first to the left and then to the right, we have

$$(*) \qquad \text{rank } \nu_k = \sum_{i=0}^{\infty} (-1)^i (\gamma_{k-i} - \beta_{k-i} + \alpha_{k-i}),$$

$$(**) \qquad \text{rank } \mu_k = \sum_{i=0}^{\infty} (-1)^i (\gamma_{k+i} - \alpha_{k+i+1} + \beta_{k+i+1}).$$

Now Lefschetz duality and algebraic duality give us isomorphisms

$$H^j(M, \text{Bd } M) \simeq H_{n-j}(M) \simeq H^{n-j}(M),$$
$$H^j(\text{Bd } M) \simeq H_{n-j-1}(\text{Bd } M) \simeq H^{n-j-1}(\text{Bd } M).$$

Therefore, $\alpha_j = \beta_{n-j}$ and $\gamma_j = \gamma_{n-j-1}$. Substituting these results in (**) and comparing with (*), we see that

$$\text{rank } \mu_k = \text{rank } \nu_{n-k-1}.$$

In particular, since $n = 2m + 1$, $\text{rank } \mu_m = \text{rank } \nu_m$. Then

$$\dim H_m(\text{Bd } M) = \dim H^m(\text{Bd } M) = \gamma_m$$
$$= \text{rank } \mu_m + \text{rank } \nu_m = 2(\text{rank } \nu_m),$$

which is even. □

Corollary 70.9. *The manifold P^{2m} is not the boundary of a compact triangulable $2m + 1$ manifold.* □

EXERCISES

1. Let (X,A) be a compact triangulable relative homology n-manifold. Assume $X - A$ is connected. Show that if X is orientable,

$$H_n(X,A; G) \simeq G \simeq H^n(X,A; G),$$

while if X is non-orientable,

$$H_n(X,A; G) \simeq \ker(G \xrightarrow{2} G)$$
$$H^n(X,A; G) \simeq G/2G.$$

2. Let M be a compact triangulable orientable n-manifold with boundary. Suppose Bd M is the disjoint union of two $n - 1$ manifolds V_0 and V_1, and Bd M has a product neighborhood in M. Let Γ be an orientation cycle for $(M, \text{Bd } M)$. Show that

$$H^k(M, V_0) \xrightarrow{\cap \Gamma} H_{n-k}(M, V_1)$$

is an isomorphism. [*Hint:* Consider the exact sequence of the triple $(M, \text{Bd } M, V_1)$.]

3. Using the fact that CP^n can be triangulated, show that the manifold CP^{2m} does not bound.

4. Which compact 2-manifolds bound?

5. Let M be a compact triangulable orientable n-manifold with boundary. Show that cup product defines a dual pairing

$$\frac{H^k(M,\text{Bd } M)}{T^k(M,\text{Bd } M)} \otimes \frac{H^{n-k}(M)}{T^{n-k}(M)} \to H^n(M,\text{Bd } M),$$

where T^i denotes the torsion subgroup of H^i.

6. Let M be the torus with two open discs removed. See Figure 70.3.
   (a) Compute $H^1(M)$ and $H^1(M,\text{Bd } M)$; draw generating cocycles.
   (b) Verify the existence of the dual pairing of Exercise 5.
   (c) Show that cup product does *not* define a dual pairing

$$\frac{H^1(M,\text{Bd } M)}{T^1(M,\text{Bd } M)} \otimes \frac{H^1(M,\text{Bd } M)}{T^1(M,\text{Bd } M)} \to H^2(M,\text{Bd } M).$$

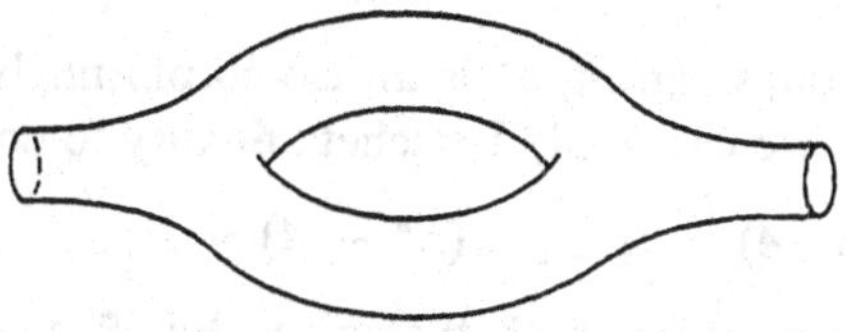

Figure 70.3

7. Let X be a compact triangulable homology n-manifold. Let A and B be polytopes of subcomplexes of a triangulation of X, with $B \subset A$. Let G be arbitrary if X is orientable; let $G = \mathbf{Z}/2$ otherwise. Show there are isomorphisms

$$H^k(A,B;G) \simeq H_{n-k}(X-B, X-A; G)$$
$$H_k(A,B;G) \simeq H^{n-k}(X-B, X-A; G).$$

[*Hint:* Triangulate X so A and B are polytopes of full subcomplexes. If C is a full subcomplex of X, let X_C denote the collection of all simplices of sd X disjoint from C. Consider the following diagram and its dual:

$$\begin{array}{ccccccccc} 0 \leftarrow & C^k(A,B) & \leftarrow & C^k(X,B) & \leftarrow & C^k(X,A) & \leftarrow 0 \\ & & & \downarrow \phi & & \downarrow \phi & \\ 0 \leftarrow & D_{n-k}(X_B,X_A) & \leftarrow & D_{n-k}(X_B) & \leftarrow & D_{n-k}(X_A) & \leftarrow 0. \end{array}$$]

*8. Generalize Lefschetz duality to the non-compact case.

9. Let (X,A) be a triangulated relative homology n-manifold, not compact. Show that if $|X| - |A|$ is connected, then (X,A) is a relative pseudo n-manifold.

§71. ALEXANDER DUALITY

The Alexander duality theorem is almost as old as the Poincaré duality theorem. It predates Lefschetz duality by some years. In its original form, it dealt with the relation between the betti numbers and torsion coefficients of a subcomplex A of the n-sphere S^n, and the betti numbers and torsion coefficients of the complement $S^n - A$. Nowadays, the Alexander duality theorem is formulated in terms of cohomology, and it is proved by using Lefschetz duality.

Theorem 71.1 (Alexander duality). *Let A be a proper, nonempty subset of S^n. Suppose (S^n,A) is triangulable. Then there is an isomorphism*

$$\tilde{H}^k(A) \simeq \tilde{H}_{n-k-1}(S^n - A).$$

Proof. Assume $n > 0$ to avoid triviality. Triangulate the pair (S^n,A).

Step 1. We prove the theorem first in the case $k \neq n, n-1$. Consider the exact sequence

$$\tilde{H}^{k+1}(S^n) \leftarrow H^{k+1}(S^n,A) \xleftarrow{\delta^*} \tilde{H}^k(A) \leftarrow \tilde{H}^k(S^n).$$

Because the end groups vanish, δ^* is an isomorphism. Now since (S^n,A) is a relative n-manifold, we can apply Lefschetz duality to conclude that

$$H^{k+1}(S^n,A) \simeq H_{n-k-1}(S^n - A) = \tilde{H}_{n-k-1}(S^n - A).$$

Combining this isomorphism with that given by δ^* gives us our desired isomorphism.

Step 2. We show the theorem holds in the case $k = n - 1$. The preceding argument needs some modification. As before, one has the exact sequence

$$H^n(S^n) \xleftarrow{j^*} H^n(S^n,A) \xleftarrow{\delta^*} \tilde{H}^{n-1}(A) \longleftarrow 0,$$

where $j : S^n \to (S^n,A)$ is inclusion. We conclude that

$$\tilde{H}^{n-1}(A) \simeq \ker j^*.$$

Let Γ be an orientation class for S^n and let $k : (S^n - A) \to S^n$ be inclusion. We apply Poincaré and Lefschetz duality to obtain the isomorphisms in the following diagram:

$$\begin{array}{ccc}
H^n(S^n,A) & \xrightarrow{\;j^*\;} & H^n(S^n) \\
\simeq \downarrow \phi_* & \searrow {\scriptstyle \cap j_*\Gamma} & \simeq \downarrow \cap \Gamma \\
H_0(S^n - A) & \xrightarrow[\;k_*\;]{} & H_0(S^n).
\end{array}$$

Since this diagram commutes up to sign, it follows that

$$\ker j^* \simeq \ker k_*.$$

It now follows from exactness of the sequence

$$0 \longrightarrow \tilde{H}_0(S^n - A) \longrightarrow H_0(S^n - A) \xrightarrow{k_*} H_0(S^n) \longrightarrow 0,$$

whose proof we leave as an exercise, that

$$\ker k_* \simeq \tilde{H}_0(S^n - A).$$

Combining these isomorphisms gives us our desired isomorphism.

Step 3. It remains to show that the theorem holds for $k = n$. We show first that inclusion $i : A \to S^n$ induces the zero homomorphism

$$H^n(A) \xleftarrow{i^*} H^n(S^n).$$

The group $H^n(S^n)$ is infinite cyclic, and is generated by the cohomology class of σ^*, where σ is any oriented simplex of S^n. Since A is a *proper* subcomplex of S^n, we can choose σ to be outside A; then $i^\#(\sigma^*)$ is the zero cochain.

Now consider the exact sequence

$$0 = H^{n+1}(S^n,A) \xleftarrow{\delta^*} H^n(A) \xleftarrow{i^*} H^n(S^n).$$

Since i^* is the zero homomorphism, it follows that $H^n(A) = 0$. Therefore, the groups

$$\tilde{H}^n(A) = H^n(A) \qquad \text{and} \qquad \tilde{H}_{-1}(S^{n-1} - A)$$

are isomorphic, because both vanish. □

Corollary 71.2 (The polyhedral Jordan curve theorem). *Let $n > 0$. If A is a subset of S^n that is homeomorphic to S^{n-1}, and if (S^n,A) is triangulable, then $S^n - A$ has precisely two path components, of which A is the common boundary.*

Proof. A is a proper subset of S^n, since S^{n-1} is not homeomorphic to S^n. The fact that $S^n - A$ has exactly two path components follows from the fact that $\tilde{H}^{n-1}(A)$ is infinite cyclic because $A \approx S^{n-1}$, and the fact that

$$\tilde{H}^{n-1}(A) \simeq \tilde{H}_0(S^n - A).$$

The fact that A is the common boundary of the two path components C_1 and C_2 follows, as in the proof of Theorem 36.3, from the fact that if s is an $n-1$ simplex in a (very fine) subdivision of A, then letting $B = A - \operatorname{Int} s$, we have

$$0 = \tilde{H}^{n-1}(B) \simeq \tilde{H}_0(S^n - B).$$

Thus one can connect a point of C_1 to a point of C_2 by a path that intersects A in points of Int s. □

EXERCISES

1. Let A be a nonempty subset of X. Show that if X is path connected, then
$$0 \to \tilde{H}_0(A) \to H_0(A) \to H_0(X) \to 0$$
is exact. This fact completes the proof of Alexander duality.
2. Suppose A is a proper nonempty subset of S^n and (S^n,A) is triangulable. Show that if A is acyclic, then $S^n - A$ is acyclic.
3. Let A be a homology $n - 1$ manifold. Show that if A is homeomorphic to a subset B of S^n and (S^n,B) is triangulable, then A is orientable.
4. Prove the following version of Alexander duality: Let A be a proper nonempty subset of S^n. If (S^n,A) is triangulable, then
$$\tilde{H}_k(A) \simeq \tilde{H}^{n-k-1}(S^n - A).$$
[*Hint:* Find a complex C that is a deformation retract of $S^n - A$.]
5. Prove Alexander duality for the n-ball: Let A be a proper nonempty subset of B^n. Let $\partial A = A \cap \operatorname{Bd} B^n$. If (B^n,A) is triangulable, show that
$$H^k(A,\partial A) \simeq \tilde{H}_{n-k-1}(B^n - A).$$
[*Hint:* Consider the subset $(A \times 1) \cup (\partial A \times I) \cup (B^n \times 0)$ of $\operatorname{Bd}(B^n \times I)$.]

§72. "NATURAL" VERSIONS OF LEFSCHETZ AND ALEXANDER DUALITY

The naturality property of the Lefschetz duality isomorphism
$$H^k(X,A) \xrightarrow{\phi_*} H_{n-k}(X - A)$$
stated in Theorem 70.6 is rather unsatisfactory in general. It is not clear as it stands, for instance, that the isomorphism ϕ_* is independent of the triangulation of (X,A), or even that it is invariant under subdivision of X!

This lack of naturality carries over to Alexander duality; it is reflected in the fact that we were able in the last section to prove only the polyhedral version of the Jordan curve theorem.

In this section, we obtain more natural versions of the Lefschetz and Alexander isomorphisms. These will enable us, after we have introduced another version of cohomology due to Čech, to construct the Alexander duality isomorphism in the situation where A is an arbitrary closed subset of S^n.

First, a definition and a lemma:

Definition. Let A be the polytope of a subcomplex of the finite complex X. Define $\operatorname{St}(A,X)$ to be the union of all the sets $\operatorname{St}(\sigma,X)$, as σ ranges over all simplices of X lying in A. It is called the **star of A in X**.

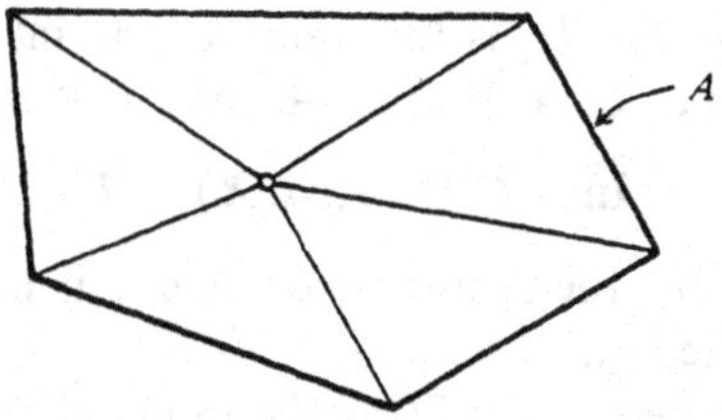

Figure 72.1

If C is the collection of all simplices of X disjoint from A, then it is immediate that

$$\mathrm{St}(A,X) = |X| - |C|.$$

If A is the polytope of a *full* subcomplex of X, it follows from Lemma 70.1 that A is a deformation retract of $\mathrm{St}(A,X)$. Note however that A need not be a deformation retract of the *closed* star $\overline{\mathrm{St}}(A,X)$. See Figure 72.1.

Lemma 72.1. *Let X be a finite complex. Let A be the polytope of a full subcomplex of X; let C be the union of all simplices of X disjoint from A. Let* $\mathrm{sd}\,X$ *be the first barycentric subdivision of X.*

(a) *The space $|X|$ is the disjoint union of the three sets*

$$U_A = \mathrm{St}(A, \mathrm{sd}\,X),$$
$$U_C = \mathrm{St}(C, \mathrm{sd}\,X),$$
$$B = \overline{U}_A - U_A = \overline{U}_C - U_C.$$

(b) *The following inclusions are homotopy equivalences:*

$$A \to U_A \to \overline{U}_A \to \mathrm{St}(A,X),$$
$$C \to U_C \to \overline{U}_C \to \mathrm{St}(C,X).$$

Proof. The sets U_A and U_C are pictured in Figures 72.2 and 72.3. Let

$$s = \hat{\sigma}_1 \ldots \hat{\sigma}_k$$

denote the general simplex of $\mathrm{sd}\,X$, where $\sigma_1 \succ \cdots \succ \sigma_k$.

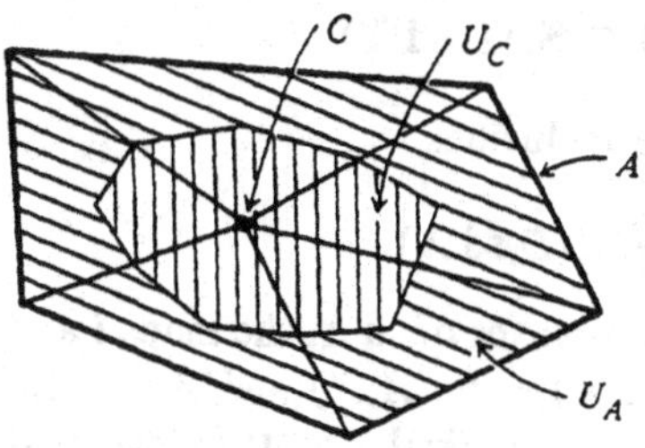

Figure 72.2

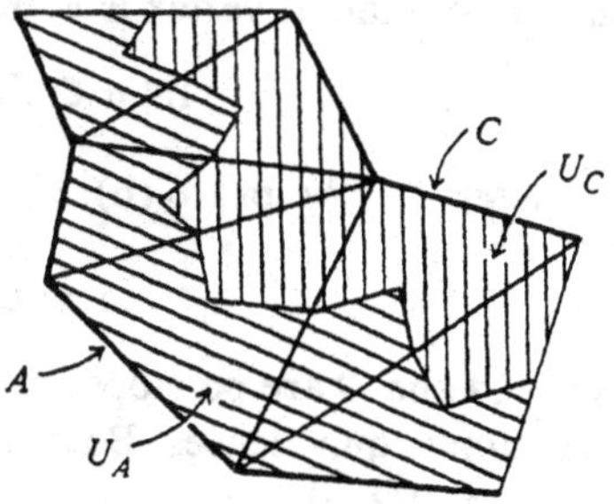

Figure 72.3

Step 1. We show that U_A is the union of the interiors of those simplices $s = \hat{\sigma}_1 \dots \hat{\sigma}_k$ for which $\sigma_k \subset A$. If $\sigma_k \subset A$, then $\hat{\sigma}_k \in A$, so that by definition

$$\operatorname{Int} s \subset \operatorname{St}(\hat{\sigma}_k, \operatorname{sd} X) \subset U_A.$$

Conversely, if $\operatorname{Int} s \subset U_A$, then some vertex $\hat{\sigma}_j$ of s is in A; it follows that σ_j lies in A and so does its face σ_k.

It follows, by symmetry, that U_C is the union of the interiors of those simplices s for which $\sigma_k \subset C$. Therefore, U_A and U_C are disjoint open sets in $|X|$.

Step 2. We show that if σ_k lies in neither A nor C, then $s = \hat{\sigma}_1 \dots \hat{\sigma}_k$ lies in $\bar{U}_A - U_A$ and in $\bar{U}_C - U_C$. Then (a) is proved.

The simplex σ_k must have a vertex v in A and a vertex w in C. Then

$$\hat{\sigma}_1 \dots \hat{\sigma}_k \hat{v} \qquad \text{and} \qquad \hat{\sigma}_1 \dots \hat{\sigma}_k \hat{w}$$

are simplices of $\operatorname{sd} X$ whose interiors lie in U_A and U_C, respectively. Their common face s thus lies in $\bar{U}_A - U_A$ and in $\bar{U}_C - U_C$.

Step 3. Let $D = \bar{U}_A$. By (a), $D = X - U_C$. We show that D is the polytope of a full subcomplex of $\operatorname{sd} X$, and that

$$\operatorname{St}(D, \operatorname{sd} X) = \operatorname{St}(A, X).$$

First, we show that D is full. If the interior of $s = \hat{\sigma}_1 \dots \hat{\sigma}_k$ lies in U_C, then $\sigma_k \subset C$, so that $\hat{\sigma}_k \notin D$. It follows that if all the vertices of s lie in D, then s must lie in D.

We show that the vertex $\hat{\sigma}$ of $\operatorname{sd} X$ lies in D if and only if σ intersects A. Now a vertex $\hat{\sigma}$ of $\operatorname{sd} X$ lies in U_C if and only if it lies in C, and this occurs if and only if $\sigma \subset C$. Therefore, $\hat{\sigma}$ lies in D if and only if σ does not lie in C, that is, if and only if σ intersects A.

Now we prove that $\operatorname{St}(D, \operatorname{sd} X) \subset \operatorname{St}(A, X)$. Suppose $s = \hat{\sigma}_1 \dots \hat{\sigma}_k$ has a vertex $\hat{\sigma}_i$ in D. Then as just proved, σ_i must intersect A, so that σ_i has a vertex v in A. Since $\sigma_1 \succ \sigma_i \succ v$, we must have

$$\operatorname{Int} s \subset \operatorname{Int} \sigma_1 \subset \operatorname{St}(v, X) \subset \operatorname{St}(A, X).$$

To prove the reverse inclusion, let σ be a simplex of X having a vertex v in A. Then $\hat{\sigma}$ is in D. Now $\operatorname{Int} \sigma$ is the union of the interiors of those simplices of $\operatorname{sd} X$ whose initial vertex is $\hat{\sigma}$. Thus

$$\operatorname{Int} \sigma \subset \operatorname{St}(\hat{\sigma}, \operatorname{sd} X) \subset \operatorname{St}(D, \operatorname{sd} X).$$

Step 4. We prove (b). Consider the inclusions

$$A \xrightarrow{i} U_A \xrightarrow{j} D \xrightarrow{k} \operatorname{St}(D, \operatorname{sd} X).$$

Since A and D are polytopes of full subcomplexes of $\operatorname{sd} X$, the maps i and k are homotopy equivalences. Because $\operatorname{St}(D, \operatorname{sd} X) = \operatorname{St}(A, X)$ and A is a full subcomplex of X, the map $k \circ j \circ i$ is a homotopy equivalence. It follows that j is a homotopy equivalence.

By symmetry, the inclusions

$$C \to U_C \to \overline{U}_C \to \operatorname{St}(C,X)$$

are also homotopy equivalences. □

Definition. Let (X,A) be a triangulable pair. If $D \subset X$, we say that D is a **polyhedron in** (X,A) if there is some triangulation of the pair (X,A) relative to which D is the polytope of a subcomplex. If $A = \varnothing$, we say simply that D is a polyhedron in X.

Note that if C and D are polyhedra in X, it does not follow that there is a single triangulation of X relative to which both C and D are polytopes of subcomplexes. For example, both the x-axis C and the set

$$D = (0,0) \cup \{(x, x\sin(1/x)) \mid x \neq 0\}$$

are polyhedra in $\mathbf{R}^2$, but there is no triangulation of $\mathbf{R}^2$ relative to which both are polytopes of subcomplexes.

Lemma 72.2. *Let D be a polyhedron in the compact triangulable space X. Then there are arbitrarily small neighborhoods U of D such that:*

(1) *$\overline{U}$ and $X - U$ are polyhedra in X.*

(2) *The following inclusions are homotopy equivalences:*

$$D \to U \to \overline{U} \quad \textit{and} \quad (X - \overline{U}) \to (X - U) \to (X - D).$$

In fact, if X is triangulated so that D is the polytope of a full subcomplex of X, then $U = \operatorname{St}(D, \operatorname{sd}^N X)$ satisfies these conditions for all $N \geq 1$.

Proof. It suffices to show that if D is the polytope of a full subcomplex of X, then $U = \operatorname{St}(D, \operatorname{sd} X)$ satisfies the requirements of the lemma.

The preceding lemma shows that the inclusions $D \to U \to \overline{U}$ are homotopy equivalences. To consider the other inclusions, let A be the union of all simplices of X disjoint from D. Then one has the following diagram:

$$\begin{array}{ccccccc} A \to \operatorname{St}(A, \operatorname{sd} X) & \to & \overline{\operatorname{St}}(A, \operatorname{sd} X) & \to & \operatorname{St}(A,X) \\ \| & & \| & & \| \\ (X - \overline{U}) & \to & (X - U) & \to & (X - D) \end{array}$$

where the first two equalities follow from (a) of the preceding lemma, and the third by definition. By the preceding lemma, the inclusion maps of this diagram are homotopy equivalences. □

Theorem 72.3 (Lefschetz duality). *Let (X,A) be a compact triangulable relative homology n-manifold. There is a function assigning to each polyhedron D in (X,A) that contains A, an isomorphism*

$$\lambda_D : H^k(X,D;G) \to H_{n-k}(X - D; G).$$

This assignment is natural with respect to inclusions of polyhedra. The group G is arbitrary if (X,A) is orientable, and $G = \mathbf{Z}/2$ otherwise.

Proof. Let Γ be an orientation class for (X,A) in the orientable case; let it denote an orientation class over $\mathbf{Z}/2$ otherwise.

Step 1. If U is any neighborhood of A such that $\bar{U}$ is a polyhedron in (X,A), let $X_U = X - U$; then X_U is also a polyhedron in (X,A). Let Γ_U denote the image of Γ under the homomorphism

$$H_n(X,A) \xrightarrow{m_*} H_n(X,\bar{U}) \underset{\simeq}{\xrightarrow{k_*^{-1}}} H_n(X_U, \text{Bd } U).$$

(Here m and k are inclusion maps, and $\text{Bd } U = \bar{U} - U$.) We show that Γ_U is an orientation class for the relative homology manifold $(X_U, \text{Bd } U)$ in the orientable case, and an orientation class over $\mathbf{Z}/2$ in the non-orientable case.

The proof is easy. Triangulate (X,A) so that $\bar{U}$ is the polytope of a subcomplex. Then Γ is represented by the sum of all n-simplices of X not in A, suitably oriented. Its image under $k_*^{-1} \circ m_*$ is represented by the sum of those simplices not in $\bar{U}$; it is automatically a cycle of $(X_U, \text{Bd } U)$.

We think of Γ_U as the "restriction" of Γ to X_U.

Step 2. Let D be a polyhedron in (X,A) containing A. Let U be any neighborhood of D satisfying the conditions of the preceding lemma, such that A, D, and $\bar{U}$ are all polytopes of subcomplexes of a triangulation of X. See Figure 72.4.

The following diagram commutes up to sign:

$$\begin{array}{ccc} H^k(X_U, \text{Bd } U) & \underset{\simeq}{\xrightarrow{\phi_*}} & H_{n-k}(X_U - \text{Bd } U) \\ & \searrow^{\cap \Gamma_U} & \downarrow j_* \\ & & H_{n-k}(X_U). \end{array}$$

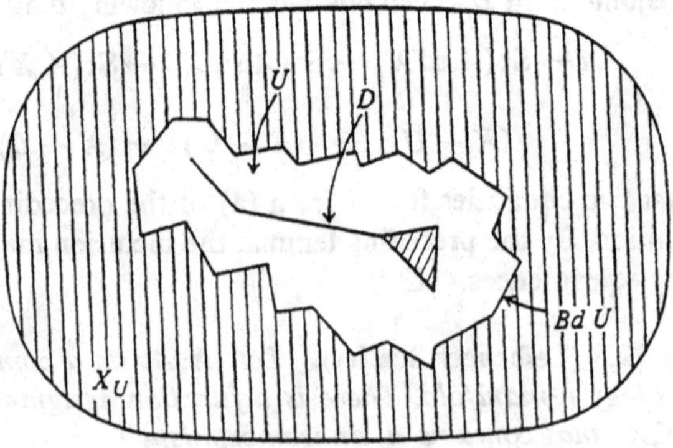

Figure 72.4

Here ϕ_* is the Lefschetz duality isomorphism for the triangulated relative homology manifold $(X_U, \text{Bd } U)$; we delete G from the notation for simplicity. Now $X_U = X - U$ and $X_U - \text{Bd } U = X - \bar{U}$; therefore, the inclusion j is a homotopy equivalence. It follows that $\cap\, \Gamma_U$ is an isomorphism in simplicial theory; then it is also an isomorphism in singular theory.

Consider the following diagram in singular theory:

$$H^k(X,D) \xleftarrow{i^*} H^k(X,\bar{U}) \xrightarrow{k^*} H^k(X_U,\text{Bd } U) \xrightarrow{\cap\, \Gamma_U} H_{n-k}(X_U) \xrightarrow{l_*} H_{n-k}(X - D)$$

where i, k, and l are inclusions. The homomorphism k^* is an isomorphism because k is an excision map. The homomorphisms i^* and l_* are isomorphisms because $i : D \to \bar{U}$ and $l : (X - U) \to (X - D)$ are homotopy equivalences.

We denote the composite isomorphism for the present by

$$\lambda_{D,U} : H^k(X,D) \to H_{n-k}(X - D).$$

Note that it does not depend on the specific triangulation involved, as did the Lefschetz isomorphism ϕ_*.

Step 3. We check a version of naturality for the isomorphism $\lambda_{D,U}$. Suppose that E is a polyhedron in (X,A) such that $A \subset E \subset D$, and suppose that V is a neighborhood of E chosen as in Step 2, such that $\bar{V} \subset U$. (Such a neighborhood of E always exists, by the preceding lemma.) See Figure 72.5. We show that $\lambda_{D,U}$ and $\lambda_{E,V}$ commute with the homomorphisms induced by inclusion.

To prove this fact, we consider the following formidable diagram:

$$\begin{array}{ccccccccc}
H^k(X,D) & \leftarrow & H^k(X,\bar{U}) & \rightarrow & H^k(X_U,\text{Bd } U) & \xrightarrow{\cap\, \Gamma_U} & H_{n-k}(X_U) & \rightarrow & H_{n-k}(X-D) \\
\downarrow & & \downarrow & \searrow & \uparrow & & \downarrow & & \downarrow \\
 & & & & H^k(X_V,\bar{U}-V) & \searrow{\scriptstyle \cap\, \Gamma'} & & & \\
 & & & & \downarrow & & & & \\
H^k(X,E) & \leftarrow & H^k(X,\bar{V}) & \rightarrow & H^k(X_V,\text{Bd } V) & \xrightarrow{\cap\, \Gamma_V} & H_{n-k}(X_V) & \rightarrow & H_{n-k}(X-E).
\end{array}$$

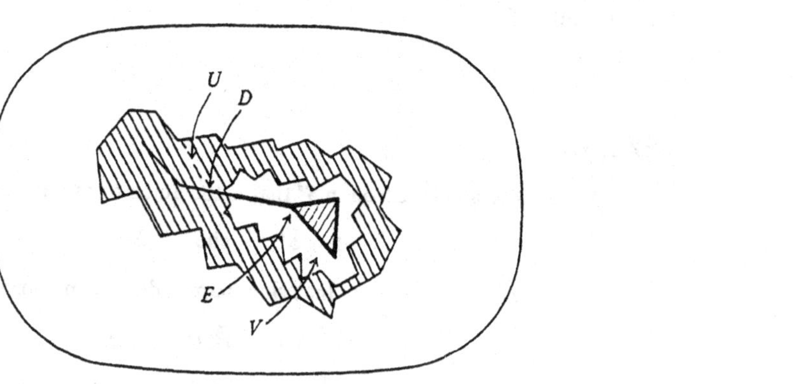

Figure 72.5

Here we *must* interpret these groups as singular groups, because the pair $(X_V, \overline{U} - V)$ is not necessarily triangulable. This arises from the fact that $\overline{U}$ and $\overline{V}$ may not be polytopes of subcomplexes of the *same* triangulation of X. The top row of maps in this diagram defines $\lambda_{D,U}$, and the bottom row defines $\lambda_{E,V}$. All unlabelled maps are induced by inclusion. The class Γ' is the appropriate "restriction" of Γ—that is, the image of Γ under

$$H_n(X,A) \longrightarrow H_n(X,\overline{U}) \xleftarrow{\approx} H_n(X_V, \overline{U} - V).$$

All squares and triangles in the diagram commute, by naturality of cap product. Our result follows.

Step 4. We now show that $\lambda_{D,U}$ is independent of the choice of U. We apply the preceding step to the case $D = E$. If U and U' are two neighborhoods of D satisfying the conditions of Lemma 72.2, we can choose V (also satisfying these conditions) so that $D \subset \overline{V} \subset U \cap U'$. Then by Step 3,

$$\lambda_{D,U} = \lambda_{D,V} = \lambda_{D,U'}.$$

We define $\lambda_D = \lambda_{D,U}$. Naturality of λ_D with respect to inclusions $E \to D$ now follows from Step 3. □

Corollary 72.4 (Alexander duality). *Let n be fixed. There is a function assigning to each proper nonempty polyhedron A in S^n, an isomorphism*

$$\alpha_A : \tilde{H}^k(A) \to \tilde{H}_{n-k-1}(S^n - A).$$

This assignment is natural with respect to inclusions.

Proof. The case $k < n - 1$ follows by noting that α_A may be defined as the composite of the isomorphisms

$$\tilde{H}^k(A) \xrightarrow{\delta^*} H^{k+1}(S^n, A) \xrightarrow{\lambda_A} H_{n-k-1}(S^n - A),$$

both of which are natural. The case $k = n - 1$ follows similarly from naturality of the diagrams used in defining the Alexander isomorphism. The case $k = n$ is trivial. □

EXERCISE

1. Let D be a polyhedron in S^n that is the union of two disjoint sets A and B, with

$$A \approx S^k \quad \text{and} \quad B \approx S^{n-k-1},$$

where $0 \leq k \leq n - 1$. Inclusion maps induce homomorphisms

$$\phi : \tilde{H}_k(A) \to \tilde{H}_k(S^n - B),$$
$$\psi : \tilde{H}_{n-k-1}(B) \to \tilde{H}_{n-k-1}(S^n - A).$$

Alexander duality implies that these groups are all infinite cyclic. Show that up to sign, ϕ and ψ both equal multiplication by the same integer m. (See Exercise 3 of §36.) This integer measures how many times A "links" B. [*Hint:* Triangulate S^n so A and B are polytopes of full subcomplexes, and no simplex intersects both A and B. Let C be the union of all simplices that are disjoint from A. Begin with the following diagram:

$$\begin{array}{ccc} \tilde{H}^{n-k-1}(C) & \longrightarrow & \tilde{H}^{n-k-1}(B) \\ \downarrow \alpha_C & & \downarrow \alpha_B \\ \tilde{H}_k(S^n - C) & \longrightarrow & \tilde{H}_k(S^n - B). \end{array}$$]

§73. ČECH COHOMOLOGY

Until now we have studied two homology and cohomology theories—namely, the simplicial theory and the singular theory. There are a number of others, most notably the Čech theory. This theory turns out to be particularly satisfactory in the case of cohomology, which is what we consider now.

We shall construct the Čech cohomology groups of a topological space and show that they agree with the simplicial groups when both are defined. We also construct a topological space for which the Čech and singular theories disagree. We shall use Čech cohomology in the next section to prove a generalized version of Alexander duality.

We begin with the concept of a *directed set.*

Definition. A **directed set** J is a set with a relation $<$ such that:

(1) $\alpha < \alpha$ for all $\alpha \in J$.

(2) $\alpha < \beta$ and $\beta < \gamma$ implies $\alpha < \gamma$.

(3) Given α and β, there exists δ such that $\alpha < \delta$ and $\beta < \delta$.

The element δ is called an **upper bound** for α and β.

Example 1. Any simply-ordered set is a directed set under the relation $\leq$.

Example 2. Consider the family whose elements are open coverings $\mathcal{A}$ of a topological space X. We make this family into a directed set by declaring $\mathcal{A} < \mathcal{B}$ if $\mathcal{B}$ is a **refinement** of $\mathcal{A}$. This means that for each element B of $\mathcal{B}$, there is at least one element A of $\mathcal{A}$ containing it. (1) and (2) are immediate; to check (3), we note that given open coverings $\mathcal{A}$ and $\mathcal{B}$ of X, the collection

$$\mathcal{D} = \{A \cap B \mid A \in \mathcal{A} \quad \text{and} \quad B \in \mathcal{B}\}$$

is an open covering of X that refines both $\mathcal{A}$ and $\mathcal{B}$.

Definition. A **direct system** of abelian groups and homomorphisms, corresponding to the directed set J, is an indexed family $\{G_\alpha\}_{\alpha \in J}$ of abelian groups, along with a family of homomorphisms

$$f_{\alpha\beta} : G_\alpha \to G_\beta,$$

defined for every pair of indices such that $\alpha < \beta$, such that:

(1) $f_{\alpha\alpha} : G_\alpha \to G_\alpha$ is the identity.

(2) If $\alpha < \beta < \gamma$, then $f_{\beta\gamma} \circ f_{\alpha\beta} = f_{\alpha\gamma}$; that is, the following diagram commutes:

$$\begin{array}{ccc} G_\alpha & \xrightarrow{f_{\alpha\gamma}} & G_\gamma \\ {\scriptstyle f_{\alpha\beta}}\searrow & & \nearrow{\scriptstyle f_{\beta\gamma}} \\ & G_\beta & \end{array}$$

Definition. Given a direct system of abelian groups and homomorphisms, we define a group called the **direct limit** of this system as follows: Take the disjoint union of the groups G_α, and introduce an equivalence relation by declaring $g_\alpha \sim g_\beta$ (for $g_\alpha \in G_\alpha$ and $g_\beta \in G_\beta$) if, for some upper bound δ of α and β, we have

$$f_{\alpha\delta}(g_\alpha) = f_{\beta\delta}(g_\beta).$$

The direct limit is the set of equivalence classes; it is denoted

$$\varinjlim_{\alpha \in J} G_\alpha.$$

We make it into an abelian group by defining

$$\{g_\alpha\} + \{g_\beta\} = \{f_{\alpha\delta}(g_\alpha) + f_{\beta\delta}(g_\beta)\},$$

where δ is some upper bound for α and β.

It is easy to see that this operation is well-defined and makes the direct limit into an abelian group. We note the following elementary facts:

(1) If all the maps $f_{\alpha\beta}$ are isomorphisms, then $\varinjlim G_\alpha$ is isomorphic to any one of the groups G_α.

(2) If all the maps $f_{\alpha\beta}$ are zero-homomorphisms, then $\varinjlim G_\alpha$ is the trivial group. More generally, if for each α there is a β such that $\alpha < \beta$ and $f_{\alpha\beta}$ is the zero-homomorphism, then $\varinjlim G_\alpha$ is the trivial group.

Definition. Let J and K be two directed sets. Let $\{G_\alpha, f_{\alpha\beta}\}$ and $\{H_\gamma, g_{\gamma\delta}\}$ be associated direct systems of abelian groups and homomorphisms. A **map Φ of direct systems** is first, a set map $\phi : J \to K$ that preserves the order relation, and second, for each $\alpha \in J$, a homomorphism

$$\phi_\alpha : G_\alpha \to H_{\phi(\alpha)},$$

such that commutativity holds in the following diagram, where $\gamma = \phi(\alpha)$ and $\delta = \phi(\beta)$ and $\alpha < \beta$:

$$\begin{array}{ccc} G_\alpha & \xrightarrow{f_{\alpha\beta}} & G_\beta \\ \phi_\alpha \downarrow & & \downarrow \phi_\beta \\ H_\gamma & \xrightarrow{g_{\gamma\delta}} & H_\delta. \end{array}$$

Such a map Φ induces a homomorphism, called the **direct limit** of the homomorphisms ϕ_α:

$$\underset{\rightarrow}{\Phi} : \varinjlim_{\alpha \in J} G_\alpha \longrightarrow \varinjlim_{\gamma \in K} H_\gamma.$$

It maps the equivalence class of $g_\alpha \in G_\alpha$ to the equivalence class of $\phi_\alpha(g_\alpha)$.

Example 3. Let $\{G_\alpha, f_{\alpha\beta}\}$ be a direct system of abelian groups and homomorphisms; let H be an abelian group. If for each α, one has a homomorphism $\phi_\alpha : G_\alpha \to H$, and if $\phi_\beta \circ f_{\alpha\beta} = \phi_\alpha$ whenever $\alpha < \beta$, one has an induced homomorphism

$$\underset{\rightarrow}{\Phi} : \varinjlim G_\alpha \to H.$$

It is a special case of the preceding construction, in which the second direct system consists of the single group H.

Example 4. Suppose one has a sequence of abelian groups and homomorphisms

$$G_1 \xrightarrow{f_1} G_2 \xrightarrow{f_2} G_3 \longrightarrow \cdots.$$

It becomes a direct system with index set $J = \mathbf{Z}_+$ if we define

$$f_{mn} = f_{n-1} \circ f_{n-2} \circ \cdots \circ f_m$$

whenever $m < n$.

For example, if each group G_i equals the integers, and if each map ϕ_i is multiplication by 2, one has the direct system

$$\mathbf{Z} \xrightarrow{2} \mathbf{Z} \xrightarrow{2} \mathbf{Z} \to \cdots.$$

Its direct limit is readily seen to be isomorphic to the group H of dyadic rationals (which is the additive group of all rationals of the form $m/2^n$, for m and n integers): Define

$$\phi_n : G_n \to H$$

by the equation $\phi_n(m) = m/2^n$. Then one checks that $\phi_n \circ f_{n-1} = \phi_{n-1}$. It is easy to check that $\underset{\rightarrow}{\Phi}$ is both injective and surjective.

Definition. If J is a directed set, a subset J_0 of J is said to be **cofinal** in J if for each $\alpha \in J$, there exists $\delta \in J_0$ with $\alpha < \delta$.

Lemma 73.1. *Suppose one is given a direct system $\{G_\alpha, f_{\alpha\beta}\}$ of abelian groups and homomorphisms, indexed by the directed set J. If J_0 is cofinal in J, then J_0 is a directed set, and inclusion induces an isomorphism*

$$\varinjlim_{\alpha \in J_0} G_\alpha \simeq \varinjlim_{\alpha \in J} G_\alpha.$$

Proof. The axioms for a directed set are easy to check. Given α, β in J_0, they have an upper bound in J, and hence an upper bound in J_0. Let $\phi : J_0 \to J$ be the inclusion map, and for each $\alpha \in J_0$, let $\phi_\alpha : G_\alpha \to G_{\phi(\alpha)}$ be the identity. Then one has an induced homomorphism

$$\Phi : \varinjlim_{\alpha \in J_0} G_\alpha \longrightarrow \varinjlim_{\alpha \in J} G_\alpha.$$

It is surjective, since given g_α, it is equivalent to some g_δ for $\delta \in J_0$. To show its kernel vanishes, suppose $\alpha \in J_0$ and $g_\alpha \sim 0_\beta$ for some $\beta \in J$. Then there is some element δ of J such that

$$f_{\alpha\delta}(g_\alpha) = f_{\beta\delta}(0_\beta) = 0_\delta.$$

Choose $\epsilon \in J_0$ with $\delta < \epsilon$. Then

$$f_{\alpha\epsilon}(g_\alpha) = f_{\delta\epsilon}(0_\delta) = 0_\epsilon,$$

so $g_\alpha \sim 0_\epsilon$, where $\epsilon \in J_0$. $\square$

Now we define the Čech cohomology groups.

Definition. Let $\mathcal{A}$ be a collection of subsets of the space X. We define an abstract simplicial complex called the **nerve** of $\mathcal{A}$, denoted by $N(\mathcal{A})$. Its vertices are the elements of $\mathcal{A}$ and its simplices are the finite subcollections $\{A_1, \ldots, A_n\}$ of $\mathcal{A}$ such that

$$A_1 \cap A_2 \cap \cdots \cap A_n \neq \varnothing.$$

Now if $\mathcal{B}$ is a collection of sets refining $\mathcal{A}$, we can define a map $g : \mathcal{B} \to \mathcal{A}$ by choosing $g(B)$ to be an element of $\mathcal{A}$ that contains B. If $\{B_1, \ldots, B_n\}$ is a simplex of $N(\mathcal{B})$, then $\{g(B_1), \ldots, g(B_n)\}$ is a simplex of $N(\mathcal{A})$, because $\bigcap B_i$ is nonempty and contained in $\bigcap g(B_i)$. Thus the vertex map g induces a simplicial map

$$g : N(\mathcal{B}) \to N(\mathcal{A}).$$

There is some arbitrariness in the choice of g, but any other choice g' for g is contiguous to g, since

$$\bigcap B_i \subset \bigcap (g(B_i) \cap g'(B_i)).$$

Thus we can make the following definition:

Definition. If $\mathcal{B}$ is a refinement of $\mathcal{A}$, we have uniquely defined homomorphisms

$$g_* : H_k(N(\mathcal{B}); G) \to H_k(N(\mathcal{A}); G)$$
$$g^* : H^k(N(\mathcal{A}); G) \to H^k(N(\mathcal{B}); G).$$

induced by the simplicial map g satisfying the condition $g(B) \supset B$ for all $B \in \mathcal{B}$. We call them the **homomorphisms induced by refinement.**

Definition. Let J be the directed set consisting of all open coverings of the space X, directed by letting $\mathcal{A} < \mathcal{B}$ if $\mathcal{B}$ is a refinement of $\mathcal{A}$. Construct a direct system by assigning to the element $\mathcal{A}$ of J, the group

$$H^k(N(\mathcal{A}); G),$$

and by assigning to the pair $\mathcal{A} < \mathcal{B}$, the homomorphism

$$f_{\mathcal{A}\mathcal{B}} : H^k(N(\mathcal{A}); G) \rightarrow H^k(N(\mathcal{B}); G),$$

induced by refinement. We define the **Čech cohomology group** of X in dimension k, with coefficients in G, by the equation

$$\check{H}^k(X; G) = \varinjlim_{\mathcal{A} \in J} H^k(N(\mathcal{A}); G).$$

We define the reduced Čech cohomology of X in a similar manner, as the direct limit of the reduced groups $\tilde{H}^k(N(\mathcal{A}); G)$.

Since these groups depend only on the collection of open coverings of X, they are obviously topological invariants of X.

In spite of the abstractness of this definition, it gives us nothing new in the case of a triangulable space, as we now show.

Theorem 73.2. *Let K be a simplicial complex. Then*

$$\check{H}^k(|K|; G) \simeq H^k(K; G).$$

The same is true for reduced cohomology.

Proof. For any complex K, let $\mathcal{A}(K)$ be the covering of $|K|$ by the open stars of its vertices. The vertex correspondence f_K that assigns to the vertex v of K, the vertex $\operatorname{St} v$ of $N(\mathcal{A}(K))$, defines an isomorphism between the complex K and the abstract complex $N(\mathcal{A}(K))$. For $v_0 \ldots v_n$ is a simplex of K if and only if

$$\operatorname{St} v_0 \cap \cdots \cap \operatorname{St} v_n \neq \varnothing;$$

this is equivalent to the statement that $\{\operatorname{St} v_0, \ldots, \operatorname{St} v_n\}$ is a simplex of $N(\mathcal{A}(K))$.

If K' is a subdivision of K, then there is a simplicial approximation to the identity $h : K' \rightarrow K$; it is induced by a vertex map h specified by the condition

$$\operatorname{St}(w,K') \subset \operatorname{St}(h(w),K).$$

The same vertex correspondence can be used to define a simplicial map $g : N(\mathcal{A}(K')) \rightarrow N(\mathcal{A}(K))$; it assigns to the vertex $\operatorname{St}(w,K')$ of $N(\mathcal{A}(K'))$, the vertex $\mathrm{St}(h(w),K)$ of $N(\mathcal{A}(K))$. Then the following diagram commutes:

$$\begin{array}{ccc} K' & \xrightarrow[\simeq]{f_{K'}} & N(\mathcal{A}(K')) \\ \downarrow h & & \downarrow g \\ K & \xrightarrow[\simeq]{f_K} & N(\mathcal{A}(K)). \end{array}$$

All four of the induced maps in cohomology are isomorphisms.

Now consider the family J_0 of open coverings of $|K|$ of the form $\mathcal{A}(K')$, as K' ranges over all subdivisions of K. The general simplicial approximation theorem implies that this family is cofinal in the family J of all open coverings of $|K|$. (If K is finite, the finite simplicial approximation theorem will suffice.) Therefore,

$$\check{H}^k(|K|;G) \simeq \varinjlim H^k(K';G),$$

where the direct system on the right is indexed by the family of all subdivisions of K. Each of the maps h^* in the latter direct system is an isomorphism, so the direct limit is isomorphic to $H^k(K;G)$, as desired. □

This theorem shows that Čech cohomology agrees with simplicial cohomology when both are defined. But what does Čech cohomology look like for more general spaces? In order to make the required computations, we shall prove a basic "continuity property" of Čech cohomology. First, we need a lemma.

Lemma 73.3. *Let Y be a compact subspace of the normal space X. Consider the directed set L consisting of coverings $\mathcal{A}$ of Y by sets that are open in X (rather than in Y). Then*

$$\check{H}^k(Y;G) \simeq \varinjlim_{\mathcal{A}\in L} H^k(N(\mathcal{A});G).$$

The same is true in reduced cohomology.

Proof. Step 1. Let $U_1,\dots,U_n$ be an open covering of X. We show there exists an open covering $V_1,\dots,V_n$ of X such that $\bar{V}_i \subset U_i$ for each i.

The set $X-(U_2\cup\cdots\cup U_n)$ is a closed subset of U_1; by normality, we can choose an open set V_1 containing it so that $\bar{V}_1 \subset U_1$. Then $V_1,U_2,\dots,U_n$ covers X. Apply the same construction to choose V_2 containing $X-(V_1\cup U_3\cup\cdots\cup U_n)$ such that $\bar{V}_2\subset U_2$. Then $V_1,V_2,U_3,\dots,U_n$ covers X. Similarly continue.

Step 2. Let $\mathcal{C}=\{C_1,\dots,C_n\}$ and $\mathcal{D}=\{D_1,\dots,D_n\}$ be two indexed families having the same index set. We say that their nerves are **naturally isomorphic** provided

$$C_{i_1}\cap\cdots\cap C_{i_p}\neq\varnothing \quad\Leftrightarrow\quad D_{i_1}\cap\cdots\cap D_{i_p}\neq\varnothing.$$

This means that the vertex map $C_i \to D_i$ induces a simplicial isomorphism $N(\mathcal{C})\to N(\mathcal{D})$.

We prove the following: Let $\{C_1,\ldots C_n\}$ be a collection of closed sets in X, and let $\{U_1,\ldots,U_n\}$ be a collection of open sets in X, with $C_i \subset U_i$ for all i. Then there exists a collection $\{W_1,\ldots,W_n\}$ of open sets of X such that $C_i \subset W_i$ and $\overline{W}_i \subset U_i$ for all i, and the nerves of $\{C_1,\ldots,C_n\}$ and $\{\overline{W}_1,\ldots,\overline{W}_n\}$ are naturally isomorphic.

Consider the collection $\mathscr{E}$ of all subsets of X of the form

$$E_{i_1,\ldots,i_p} = C_{i_1} \cap \cdots \cap C_{i_p}.$$

Let E be the union of all such sets that do not intersect C_1. Then E is closed in X and disjoint from C_1. Choose W_1 to be an open set of X containing C_1 whose closure is disjoint from E and lies in U_1. We show that the nerves of $\{\overline{W}_1,C_2,\ldots,C_n\}$ and $\{C_1,C_2,\ldots,C_n\}$ are naturally isomorphic. For this purpose, it suffices to show that

$$\overline{W}_1 \cap C_{i_1} \cap \cdots \cap C_{i_p} \neq \varnothing \quad \Leftrightarrow \quad C_1 \cap C_{i_1} \cap \cdots \cap C_{i_p} \neq \varnothing.$$

The implication $\Leftarrow$ is trivial, since $C_1 \subset \overline{W}_1$. We prove the other implication. If the right-hand set is empty, then the set $C_{i_1} \cap \cdots \cap C_{i_p}$ is disjoint from C_1. Thus it is contained in E, by definition. Then it is disjoint from $\overline{W}_1$, by definition of W_1. Hence the left-hand set is empty as well.

Apply this construction a second time to choose W_2 so that $C_2 \subset W_2$ and $\overline{W}_2 \subset U_2$, and the nerves of

$$\{\overline{W}_1,\overline{W}_2,C_3,\ldots,C_n\} \quad \text{and} \quad \{\overline{W}_1,C_2,C_3,\ldots,C_n\}$$

are naturally isomorphic. Similarly continue.

Step 3. A covering $\mathcal{W} = \{W_1,\ldots,W_n\}$ of Y by sets open in X is said to be *adapted to* Y if the nerves of $\{W_1,\ldots,W_n\}$ and $\{W_1 \cap Y,\ldots,W_n \cap Y\}$ are naturally isomorphic. Let L be the family of all coverings of Y by sets open in X; let L_0 be the subfamily consisting of those coverings that are adapted to Y. We show L_0 is cofinal in L.

Let $\mathcal{A}$ be a covering of Y by sets open in X. Pass to a finite subcollection $\{U_1,\ldots,U_n\}$ covering Y. Consider the covering

$$\{U_1 \cap Y,\ldots,U_n \cap Y\}$$

of Y by sets open in Y. Applying Step 1 to the normal space Y, choose a collection $\{V_1,\ldots,V_n\}$ of sets open in Y that cover Y, such that $\overline{V}_i \subset (U_i \cap Y)$ for each i. Let $C_i = \overline{V}_i$, and apply Step 2 to find a collection of open sets $\mathcal{W} = \{W_1,\ldots,W_n\}$ of X such that $C_i \subset W_i$ and $\overline{W}_i \subset U_i$, and such that the nerves of $\{C_1,\ldots,C_n\}$ and $\{\overline{W}_1,\ldots,\overline{W}_n\}$ are naturally isomorphic.

Now $\mathcal{W}$ refines $\mathcal{A}$, and $\mathcal{W}$ covers Y. The fact that $\mathcal{W}$ is adapted to Y follows from the implications

$$\begin{array}{ccc} C_{i_1} \cap \cdots \cap C_{i_p} \neq \varnothing & \Leftrightarrow & \overline{W}_{i_1} \cap \cdots \cap \overline{W}_{i_p} \neq \varnothing \\ \Downarrow & & \Uparrow \\ W_{i_1} \cap \cdots \cap W_{i_p} \cap Y \neq \varnothing & \Rightarrow & W_{i_1} \cap \cdots \cap W_{i_p} \neq \varnothing. \end{array}$$

Step 4. In general, if $\mathcal{A} = \{A_1, \ldots, A_n\}$ is a covering of Y by sets open in X, let $\mathcal{A} \cap Y$ denote the covering $\{A_1 \cap Y, \ldots, A_n \cap Y\}$ of Y by sets open in Y. Consider the following four directed sets:

$$\begin{aligned} L &= \{\mathcal{A} \mid \mathcal{A} \text{ is a covering of } Y \text{ by sets open in } X\}, \\ L_0 &= \{\mathcal{W} \mid \mathcal{W} \in L \text{ and } \mathcal{W} \text{ is adapted to } Y\}, \\ J &= \{\mathcal{B} \mid \mathcal{B} \text{ is a covering of } Y \text{ by sets open in } Y\}, \\ J_0 &= \{\mathcal{W} \cap Y \mid \mathcal{W} \in L_0\}. \end{aligned}$$

Note that every element $\mathcal{B}$ of J equals $\mathcal{A} \cap Y$ for some $\mathcal{A}$ in L.

We have shown that L_0 is cofinal in L. It follows at once that J_0 is cofinal in J: Given $\mathcal{B} \in J$, it equals $\mathcal{A} \cap Y$ for some $\mathcal{A} \in L$. Then if $\mathcal{W}$ is an element of L_0 that refines $\mathcal{A}$, it follows that $\mathcal{W} \cap Y$ refines $\mathcal{A} \cap Y = \mathcal{B}$.

The lemma follows. For

$$\begin{aligned} \check{H}^k(Y; G) &= \underrightarrow{\lim}_J\, H^k(N(\mathcal{B}); G) && \text{by definition,} \\ &\simeq \underrightarrow{\lim}_{J_0}\, H^k(N(\mathcal{W} \cap Y); G) && \text{since } J_0 \text{ is cofinal in } J, \\ &\simeq \underrightarrow{\lim}_{L_0}\, H^k(N(\mathcal{W}); G) && \text{since } N(\mathcal{W}) \simeq N(\mathcal{W} \cap Y), \\ &\simeq \underrightarrow{\lim}_L\, H^k(N(\mathcal{A}); G) && \text{since } L_0 \text{ is cofinal in } L. \quad \square \end{aligned}$$

Theorem 73.4. *Let X be a compact triangulable space. Let $D_1 \supset D_2 \supset \ldots$ be a sequence of polyhedra in X whose intersection is Y. Then*

$$\check{H}^k(Y; G) \simeq \underrightarrow{\lim}\, H^k(D_n; G).$$

The same result holds in reduced cohomology.

Proof. We shall delete G from the notation, for simplicity. The direct limit in the conclusion of the theorem does not depend on the particular triangulations of the spaces D_n. Let X_n be a triangulation of X chosen so that D_n is the polytope of a full subcomplex of X_n, so that the maximum diameter of a simplex of X_n is less than $1/n$, and so that the inclusion map $|X_{n+1}| \to |X_n|$ satisfies the star condition. Let us denote the subcomplex of X_n whose polytope is D_n by K_n. Then the isomorphism

$$\underrightarrow{\lim}\, H^k(D_n) \simeq \underrightarrow{\lim}\, H^k(K_n)$$

holds trivially.

Step 1. Given n, consider the covering of D_n by sets open in D_n defined by

$$\mathcal{A}'_n = \{\operatorname{St}(v, K_n) \mid v \in K_n\}.$$

Then, as we noted in the proof of Theorem 73.2, one has a simplicial isomorphism $f_n : K_n \to N(\mathcal{A}'_n)$, defined by $f_n(v) = \operatorname{St}(v, K_n)$. The same argument as was used there can be applied to pass to an isomorphism of the cohomology direct limits, as we now show.

Let $h: K_{n+1} \to K_n$ be a simplicial approximation to the inclusion map $|K_{n+1}| \to |K_n|$; we then define $g: \mathcal{A}'_{n+1} \to \mathcal{A}'_n$ by the equation

$$g(\text{St}(w, K_{n+1})) = \text{St}(h(w), K_n).$$

The map g induces a simplicial map $g: N(\mathcal{A}'_{n+1}) \to N(\mathcal{A}'_n)$ such that the following diagram commutes:

$$\begin{array}{ccc} K_{n+1} & \xrightarrow[\cong]{f_{n+1}} & N(\mathcal{A}'_{n+1}) \\ \downarrow h & & \downarrow g \\ K_n & \xrightarrow[\cong]{f_n} & N(\mathcal{A}'_n). \end{array}$$

We conclude that

$$\varinjlim H^k(K_n) \simeq \varinjlim H^k(N(\mathcal{A}'_n)),$$

just as before. *However*, in this case we cannot identify the left-hand group with a particular simplicial cohomology group, since h^* need not be an isomorphism.

Step 2. Given n, consider the covering of D_n defined by

$$\mathcal{A}_n = \{\text{St}(v, X_n) \mid v \in K_n\}.$$

This is a covering of D_n by sets open in X, whereas the collection $\mathcal{A}'_n$ was a covering of D_n by sets open in D_n. The collection $\mathcal{A}_{n+1}$ is a refinement of $\mathcal{A}_n$, because inclusion $|X_{n+1}| \to |X_n|$ satisfies the star condition.

We show that the nerves of $\mathcal{A}_n$ and $\mathcal{A}'_n$ are naturally isomorphic, the vertex map in question carrying $\text{St}(v, K_n)$ to $\text{St}(v, X_n)$ for each $v \in K_n$. Since $\text{St}(v, K_n) \subset \text{St}(v, X_n)$, it is immediate that

$$\bigcap \text{St}(v_i, K_n) \neq \varnothing \quad \Rightarrow \quad \bigcap \text{St}(v_i, X_n) \neq \varnothing.$$

We prove the reverse implication. If $\bigcap \text{St}(v_i, X_n) \neq \varnothing$, then the vertices v_i span a simplex σ of X_n. Because the v_i are vertices of K_n and K_n is full, σ must belong to K_n. Then $\bigcap \text{St}(v_i, K_n) \neq \varnothing$, as desired.

It follows at once that

$$\varinjlim H^k(N(\mathcal{A}'_n)) \simeq \varinjlim H^k(N(\mathcal{A}_n)).$$

Step 3. We now show that the family of coverings $\{\mathcal{A}_1, \mathcal{A}_2, \ldots\}$ of Y is cofinal in the family L of all coverings $\mathcal{B}$ of Y by sets open in X.

First, we note the following: If D is a compact subset of the metric space X, and if $\mathcal{B}$ is a covering of D by sets open in X, then there is a $\delta > 0$ such that any set of diameter less than δ that intersects D necessarily lies in an element of $\mathcal{B}$. This is an extension of the "Lebesgue number lemma" of general topology; its proof is left as an exercise.

Now let $\mathcal{B}$ be an arbitrary covering of Y by sets open in X. First, we show that $\mathcal{B}$ covers $D_n = |K_n|$ for some n, say $n = N$. For if not, we can choose for each n, a point $x_n \in D_n$ such that x_n is not in any element of $\mathcal{B}$. Some subsequence x_{n_i} converges, necessarily to a point of Y (since $Y = \bigcap D_n$). But $\mathcal{B}$ covers a neighborhood of Y, so it must cover the point x_{n_i} for i sufficiently large.

We now apply the extended Lebesgue number lemma. Choose δ so that any set of diameter less than δ that intersects D_N necessarily lies in an element of $\mathcal{B}$. Then choose $m > N$ large enough so that $1/m < \delta/2$. It follows that $\mathcal{A}_m$ refines $\mathcal{B}$. For, given a vertex v of K_m, we have

$$\operatorname{diam} \operatorname{St}(v, X_m) \leq 2/m < \delta.$$

Since $v \in |K_m| \subset |K_N|$, the set $\operatorname{St}(v, K_m)$ intersects $|K_N|$, so that it lies in an element of $\mathcal{B}$.

Step 4. We prove the theorem. We have

$$\begin{aligned} \varinjlim H^k(D_n) &\simeq \varinjlim H^k(K_n) \\ &\simeq \varinjlim H^k(N(\mathcal{A}'_n)) && \text{by Step 1,} \\ &\simeq \varinjlim H^k(N(\mathcal{A}_n)) && \text{by Step 2,} \\ &\simeq \varinjlim{}_{\mathcal{B} \in L} H^k(N(\mathcal{B})) && \text{by Step 3.} \end{aligned}$$

The latter group is isomorphic to $\check{H}^k(Y)$, by the preceding lemma. □

We use this theorem to compute the Čech cohomology of the following non-triangulable space.

Corollary 73.5. *Let X be the closed topologist's sine curve. Then*

$$H^1(X) = 0 \qquad \textit{(singular cohomology)},$$
$$\check{H}^1(X) \simeq \mathbf{Z} \qquad \textit{(Čech cohomology)}.$$

Proof. For convenience, we express X as a union of line segments in the plane, as pictured in Figure 73.1.

Step 1. Let us first compute the singular homology and cohomology of X.

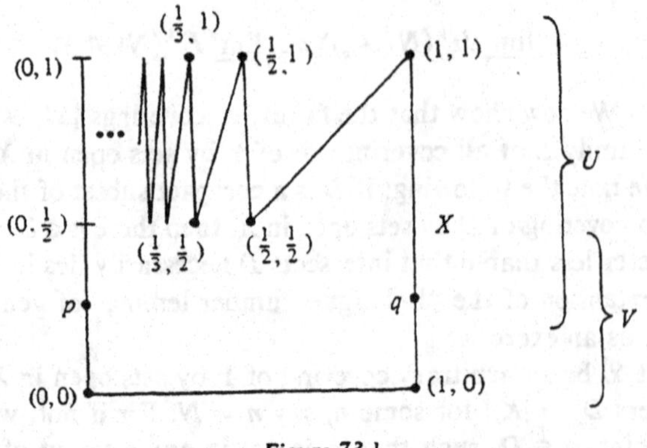

Figure 73.1

Let U be the intersection of X with the set of all points (x,y) of $\mathbf{R}^2$ for which $y > 1/8$, and let V be the intersection of X with those (x,y) for which $y < 3/8$. Because U and V are open, we have a Mayer-Vietoris sequence in singular homology:

$$H_1(U) \oplus H_1(V) \to H_1(X) \to \tilde{H}_0(U \cap V) \xrightarrow{i_*} \tilde{H}_0(U) \oplus \tilde{H}_0(V).$$

Now V is homeomorphic to an open interval, so $H_1(V) = \tilde{H}_0(V) = 0$. The space U has two path components U_1 and U_2, one of which is homeomorphic to an open interval, and the other to a half-open interval. Therefore,

$$H_1(U) \simeq H_1(U_1) \oplus H_1(U_2) = 0.$$

The group $\tilde{H}_0(U)$ is infinite cylic and is generated by the 0-chain $q - p$, where $p = (0,1/4) \in U_1$ and $q = (1,1/4) \in U_2$. The space $U \cap V$ consists of two disjoint open line segments, so $\tilde{H}_0(U \cap V)$ is also infinite cyclic and is generated by $q - p$. It follows that i_* is an isomorphism, so that $H_1(X) = 0$. The universal coefficient theorem for cohomology now implies that $H^1(X) = 0$. (Alternatively, one can use a Mayer-Vietoris sequence in cohomology to prove this fact.)

Step 2. Now let us compute the Čech cohomology of X. We express X as an intersection of polyhedra, as follows: For each n, let C_n be the closed rectangular region pictured in Figure 73.2. Let $D_n = X \cup C_n$; then D_n is a polyhedron. By the preceding theorem,

$$\check{H}^1(X) \simeq \varinjlim H^1(D_n).$$

Now D_n is a space having the homotopy type of a circle, so that $H^1(D_n) \simeq \mathbf{Z}$. Furthermore, D_{n+1} is readily seen to be a deformation retract of D_n, so that inclusion i induces an isomorphism

$$H^1(D_{n+1}) \xleftarrow{i^*} H^1(D_n).$$

It follows that $\check{H}^1(X) \simeq \mathbf{Z}$. □

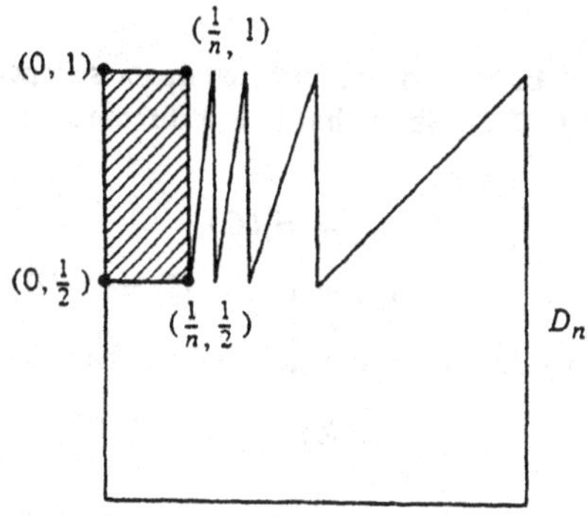

Figure 73.2

We have only touched on the subject of Čech cohomology. For a full-scale development, one needs to define the homomorphism induced by a continuous map (which is easy), extend the definition to the cohomology of a pair (X,A), and verify the Eilenberg-Steenrod axioms (which requires some work). The interested reader is referred to Chapters 9 and 10 of [E-S].

EXERCISES

1. Check the details of Example 4.
2. The direct limit of the system
$$\mathbf{Z} \xrightarrow{2} \mathbf{Z} \xrightarrow{3} \mathbf{Z} \xrightarrow{4} \mathbf{Z} \to \cdots$$
is a familiar group. What is it?
3. Check that the system of groups and homomorphisms used in defining Čech cohomology is a direct system.
4. Verify the extended Lebesgue number lemma quoted in the proof of Theorem 73.4.
5. Let X denote the solid torus $B^2 \times S^1$, and let $f: X \to X$ be the imbedding pictured in Figure 73.3. Let $X_0 = X$, let $X_1 = f(X_0)$, and in general, let $X_{n+1} = f(X_n)$. Let $S = \bigcap X_n$; the space S is often called a *solenoid*.
(a) Describe the space X_2.
(b) Compute $\check{H}^1(S)$.

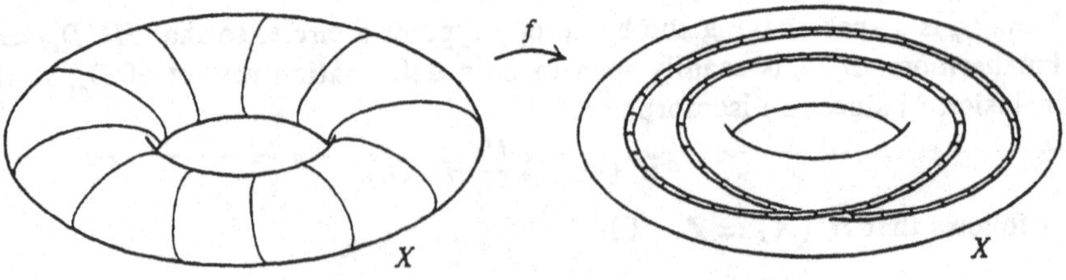

Figure 73.3

6. Given X, let $\mathcal{C}$ be the collection of compact subspaces of X, directed by letting $C < D$ if $C \subset D$. Show that $\mathcal{C}$ is a directed set. Define a direct system by the rules
$$G_C = H_i(C),$$
$$f_{C,D} = i_* : H_i(C) \to H_i(D),$$
where i is inclusion and H_i denotes singular homology. Show that
$$H_i(X) \simeq \varinjlim H_i(C).$$
(This is just another way of expressing the compact support property of singular homology.)

§74. ALEXANDER-PONTRYAGIN DUALITY

Now we prove a generalized version of the Alexander duality theorem, which applies to an arbitrary closed subspace of S^n, not just to a polyhedron in S^n. As an application, we derive a version of the Jordan curve theorem more general than any we have seen up to now.

Theorem 74.1 (Alexander-Pontryagin duality). *Let A be a proper, nonempty closed subset of S^n. Then*

$$\check{\tilde{H}}^k(A) \simeq \tilde{H}_{n-k-1}(S^n - A).$$
$$(\text{Čech cohomology}) \qquad (\text{singular homology})$$

Proof. Choose a triangulation X of S^n fine enough that not every simplex of X intersects A. Consider the sequence $\operatorname{sd} X, \operatorname{sd}^2 X, \ldots$ of successively finer subdivisions of X.

Let A_m denote the subcomplex of $\operatorname{sd}^m X$ consisting of those simplices that intersect A, along with their faces. Then $|A_m|$ is a proper nonempty polyhedron in X, and $|A_{m+1}| \subset |A_m|$ for all m. Furthermore, A is the intersection of the sets $|A_m|$.

The Alexander duality theorem gives us an isomorphism

$$\alpha_m : \tilde{H}^k(A_m; G) \to \tilde{H}_{n-k-1}(X - A_m; G).$$

Since this isomorphism is natural with respect to inclusions, it induces an isomorphism of the direct limits

$$\alpha : \varinjlim \tilde{H}^k(A_m; G) \longrightarrow \varinjlim \tilde{H}_{n-k-1}(X - A_m; G).$$

The left-hand group is isomorphic to the reduced Čech cohomology group $\check{\tilde{H}}^k(A; G)$, by Theorem 73.4. We show that the right-hand group is isomorphic to the singular homology group $\tilde{H}_{n-k-1}(X - A; G)$; then the theorem is proved.

Consider the homomorphisms induced by inclusion,

$$\tilde{H}_{n-k-1}(X - A_m) \to \tilde{H}_{n-k-1}(X - A).$$

They induce a homomorphism of the direct limit

$$\varinjlim \tilde{H}_{n-k-1}(X - A_m) \longrightarrow \tilde{H}_{n-k-1}(X - A).$$

The fact that this homomorphism is an isomorphism follows from the compact support property of singular homology: Every singular simplex $T : \Delta_p \to X - A$ lies in a compact set and hence lies in one of the open sets $X - A_m$. Thus every singular cycle of $X - A$ is carried by $X - A_m$ for some m. Furthermore, any homology between two such singular cycles is carried by $X - A_{m+p}$ for some p. $\square$

Corollary 74.2. *Let $n > 1$. Let M be a compact connected triangulable $n-1$ manifold; suppose $h : M \to S^n$ is an imbedding. Then M is orientable, and $S^n - h(M)$ has precisely two path components, of which $h(M)$ is the common boundary.*

Proof. By Alexander duality, we know that $\check{H}^{n-1}(M) \simeq \tilde{H}_0(S^n - h(M))$. Since M is triangulable, its Čech and simplicial cohomology groups are isomorphic. If M were non-orientable, we would have $H^{n-1}(M) \simeq \mathbf{Z}/2$, although the group $\tilde{H}_0(S^n - h(M))$ is free abelian. We conclude that M is orientable. Hence $H^{n-1}(M) \simeq \mathbf{Z}$, and $S^n - h(M)$ has precisely two path components U and V.

The proof that $h(M)$ is the common boundary of U and V proceeds just as in the proof of the Jordan curve theorem (§36). All one needs is the fact that if we delete the interior of a small open $n-1$ simplex s from M, then $h(M - \text{Int } s)$ does not separate S^n. And that follows from the fact that (letting $M_0 = h(M - \text{Int } s)$) we have

$$\tilde{H}_0(S^n - M_0) \simeq \check{H}^{n-1}(M_0) \simeq H^{n-1}(M_0) = 0. \quad \square$$

EXERCISES

1. Let X be the closed topologist's sine curve. If $f : X \to S^n$ is an imbedding, calculate $\tilde{H}_i(S^n - f(X))$.

2. Show that Alexander-Pontryagin duality does not hold if Čech cohomology is replaced by singular cohomology.

3. Show that neither the Klein bottle nor the projective plane can be imbedded in $\mathbf{R}^3$, but they can be imbedded in $\mathbf{R}^4$. [*Hint:* If one removes an open disc from P^2, what remains is a Möbius band.]

4. Let A and B be closed sets in S^n. Show that if $A \approx B$, then

$$H_i(S^n - A) \simeq H_i(S^n - B)$$

for all i. [*Note:* It does not follow that $S^n - A$ and $S^n - B$ are themselves homeomorphic. Figure 74.1 pictures an imbedding of S^1 in S^3 called an "overhand knot." Its complement C is not homeomorphic to the complement D of the standard imbedding of S^1 in S^3; for although the homology groups of C and D are isomorphic, their fundamental groups are different. See [C-F], Chapter 6.]

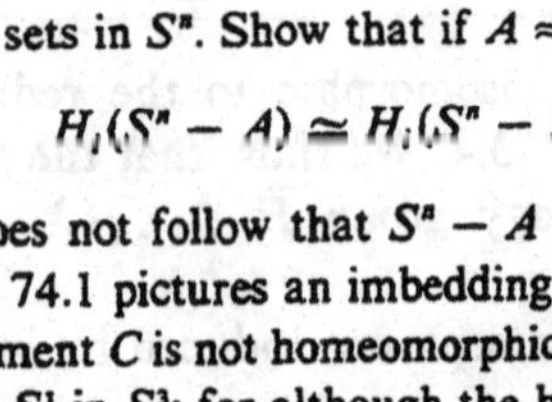

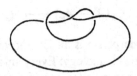

Figure 74.1

Bibliography

[B] Brown, M., "A proof of the generalized Schoenflies theorem," *Bull. Am. Math. Soc.*, **66**, pp. 74–76 (1960).

[B_2] Brown, M., "Locally flat imbeddings of topological manifolds," *Ann. Math.*, **75**, pp. 331–341 (1962).

[C-F] Crowell, R. H., and Fox, R. H., *Introduction to Knot Theory*, Ginn, 1963; Springer-Verlag, 1977.

[Do] Dold, A., *Lectures on Algebraic Topology*, Springer-Verlag, 1972.

[D] Dugundji, J., *Topology*, Allyn & Bacon, 1966.

[E-S] Eilenberg, S., and Steenrod, N., *Foundations of Algebraic Topology*, Princeton Univ. Press, 1952.

[F] Fuchs, M., "A note on mapping cylinders," *Mich. Math. J.*, **18**, pp. 289–290 (1971).

[G-H] Greenberg, M. J., and Harper, J. R., *Algebraic Topology: A First Course*, Benjamin/Cummings, 1981.

[H-W] Hilton, P. J., and Wylie, S., *Homology Theory*, Cambridge Univ. Press, 1960.

[H-Y] Hocking, J. G., and Young, G. S., *Topology*, Addison-Wesley, 1961.

[K] Kelley, J. L., *General Topology*, Van Nostrand Reinhold, 1955; Springer-Verlag, 1975.

[MacL] MacLane, S., *Homology*, Springer-Verlag, 1963.

[Ma] Massey, W. S., *Algebraic Topology: An Introduction,* Harcourt Brace Jovanovich, 1967; Springer-Verlag, 1977.

[Mu] Munkres, J. R., *Topology, A First Course,* Prentice-Hall, 1975.

[N] Newman, M. H. A., *Topology of Plane Sets of Points,* Cambridge Univ. Press, 1951.

[S-T] Seifert, H., and Threlfall, W., *Lehrbuch der Topology,* Teubner, 1934; Chelsea, 1947; translated as *A Textbook of Topology,* Academic Press, 1980.

[S] Spanier, E. *Algebraic Topology,* McGraw-Hill, 1966; Springer-Verlag, 1982.

[V] Vick, J., *Homology Theory,* Academic Press, 1973.

[Wh] Whitehead, G. W., *Elements of Homotopy Theory,* Springer-Verlag, 1978.

[W] Willard, S., *General Topology,* Addison-Wesley, 1970.

Index

Printed in the United States
by Baker & Taylor Publisher Services